Chemistry of Learning
Invertebrate Research

Chemistry
of Learning
Invertebrate Research

EDITED BY

W. C. CORNING
Department of Psychology
Fordham University
New York, New York

AND

S. C. RATNER
Department of Psychology
Michigan State University
East Lansing, Michigan

Proceedings of a Symposium, Sponsored by the American Institute of Biological Sciences, Held at Michigan State University, September 7-10, 1966.

℗ Springer Science+Business Media, LLC 1967

ISBN 978-1-4899-6261-4   ISBN 978-1-4899-6565-3 (eBook)
DOI 10.1007/978-1-4899-6565-3

Library of Congress Catalog Card Number 67-25103

© *1967 Springer Science+Business Media New York*
Originally published by Plenum Press in 1967.
Softcover reprint of the hardcover 1st edition 1967

Preface

To some extent the rapid developments in brain research have been made possible by the ingenious exploitation of certain lower-animal preparations. For example, early research on nervous system functions was aided considerably by the particular characteristics of the giant axon in the squid and the lateral eye of the horse-shoe crab. More recently, we have been witnessing an even wider use of less advanced organisms to study the complex operations of nervous systems. The present volume is concerned with research on the chemical bases of learning and memory and the invertebrate preparations which are currently being studied or which have potential experimental value for this problem. The contents of the volume are derived from the presentations and discussions from a three-day symposium held at Michigan State University.

A major portion of the book is devoted to planarian research, partly because the planarian was one of the first organisms to be used to investigate the molecular bases of memory and partly because there has been considerable discussion and controversy over the findings obtained with this animal. In fact, the impetus for organizing the symposium began with a number of letters, some of which were unfairly critical of planarian research and others of which contained requests for "memory molecules" usually to help some senile relative. The various findings seemed to be subject to frequent misinterpretation, especially concerning their implications for memory mechanisms. It also appeared that the flatworm was being held responsible for much of the frustration and difficulty encountered in this area of investigation. A clarification of findings was necessary along with a consideration of the relationship of certain studies to the more general body of data and theory relevant to learning and its physiological mechanisms. Our purpose was not to pass judgment on a research area, but to permit as much exposure of relevant information as possible, including an exploration of other potentially useful invertebrate preparations.

To place the detailed treatments of the various contributors in their proper perspective, the initial portion of the book is concerned with an elucidation of the problems, general concepts, techniques, and data of the behavioral and biological sciences which are pertinent in attempts to relate behavioral changes to physiological processes. The first section of Part II contains a specific treatment of planarian ecology, anatomy, physiology, biochemistry, and general behavior. The information in this section should be of value to anyone working with Platyhelminthes regardless of whether they are interested in learning mechanisms or in more traditional biological problems. The second and third sections of Part II deal with attempts to demonstrate behavioral modifications in planarians and with studies relevant to molecular bases of learning and memory. The material of these sections should clarify the areas of disagreement and permit an objective assessment of what actually has been accomplished to date. The third part of the book examines the research possibilities in a number of other invertebrates and reviews recent biological, biochemical, and behavioral data related to learning and memory. The preparations discussed range from the "micrometazoa" studied by P. Applewhite to the very interesting and important cockroach preparation of M. Cohen.

In organizing the book, we decided to do away with the "conversational" type format characteristic of so many symposia volumes. Each of the contributors was asked to provide an extensive and detailed paper dealing with his respective topic. In this way we hope to make the volume a useful source of basic information for a number of years. So as not to lose some of the more important comments and exchanges which occurred during the sessions we have included at the end of each section a condensed version of these discussions.

We are indebted to Dr. Leroy Augenstein for his invaluable advice concerning the intricacies of obtaining financial support for the symposium and for suggestions on its organization. Considerable help in the early organizing stages was also provided by Dr. James McConnell, especially in the selection of speakers and topics.

The American Institute of Biological Sciences deserves special thanks for its generous financial support of the proceedings. Mr. R. Beem, Miss Mary-Frances Thompson, and Miss Donna Clarke were among those at AIBS who assisted us in the planning and execution of the symposium.

Lastly, we wish to thank Mr. Bruce Alderman and the Continuing Education Service at Michigan State University for making the necessary arrangements for the participants, providing comfortable conference space, and for the tedious job of transcribing thirty hours of tape.

W. C. Corning
New York   S. C. Ratner
July 1967

Contributors

Dr. Philip Applewhite, *Department of Molecular Biophysics, Yale University*
Dr. Leroy Augenstein, *Department of Biophysics, Michigan State University*
Dr. E. L. Bennett, *Lawrence Radiation Laboratory, University of California (Berkeley)*
Dr. J. B. Best, *Department of Physiology, Colorado State University*
Dr. H. M. Brown, *Department of Neurology, University of Utah*
Dr. M. Clay, *Mental Health Research Institute, University of Michigan*
Dr. M. Cohen, *Department of Biology, University of Oregon*
Dr. W. C. Corning, *Department of Psychology, Fordham University*
Dr. P. Cornwell, *Animal Behavior Laboratory, Pennsylvania State University*
Dr. S. Coward, *Department of Zoology, University of Georgia*
Dr. F. T. Crawford, *Department of Psychology, Florida State University*
Dr. M. Ray Denny, *Department of Psychology, Michigan State University*
Dr. P. M. Driver, *Mental Health Research Institute, University of Michigan*
Dr. J. Gaito, *Department of Psychology, York University*
Dr. E. Halas, *Department of Psychology, University of North Dakota*
Dr. A. Jacobson, *Department of Psychology, University of California (Los Angeles)*
Dr. M. Jenkins, *Department of Biology, Madison College*
Dr. D. Jensen, *Department of Psychology, Indiana University*
Dr. R. Kenk, *Division of Worms, Smithsonian Institute*
Dr. H. Lenhoff, *Department of Biology, University of Miami*
Dr. J. V. McConnell, *Mental Health Research Institute, University of Michigan*
Dr. S. C. Ratner, *Department of Psychology, Michigan State University*
Dr. N. Rushforth, *Department of Biology, Western Reserve University*
Dr. P. Sengel, *Zoology Laboratory, Domaine University, France*
Dr. D. M. Vowles, *Institute of Experimental Psychology, Oxford University, England*
Dr. P. Wells, *Department of Biology, Occidental College*

Contents

ix

PART IIB. Planarian Research
Demonstrations of Learning and Problems of Interpretation

Introduction

Chapter 1

The Search for Learning and Memory Mechanisms

Stanley C. Ratner[1]

Department of Psychology
Michigan State University
East Lansing, Michigan

The understanding of behavior usually moves along several different but, hopefully, related tracks. One track involves investigations of the behaviors themselves. Another track involves formulations, theories, about the processes involved in these behaviors. At different times in the history of this search for understanding, these two tracks have assumed different magnitudes and slopes, so the train of understanding has moved at different speeds and shifted from track to track. At the present time, investigations and formulations are both moving rapidly. However, effective engineering requires some appreciation of the route previously traveled. Many eminent philosophers, psychologists, and biologists have dealt with the problems of learning and memory. In the context of the present volume, we are primarily concerned with the types of formulations that these individuals have suggested as mechanisms associated with the understanding of learning and memory.

Mechanisms have been postulated that range from blatant analogy based on mechanical principles to energetically defended explanations derived from observation of chemical or electrical changes during learning. In the present chapter, we shall do three things: (1) review some of the formulations that have been proposed, (2) explore these formulations for problems that have slowed the progress of understanding, and (3) relate the problems of understanding to the material in the present book.

A SURVEY OF MECHANISMS

As suggested above, investigators have proposed a great range of mechanisms to account for the processes of learning and memory. However, a problem exists in the classification of these efforts. A classification by the individual associated with the formulation provides a system that is simple, structural, and based on a distinct unitary characteristic. The same is true for a chronological classification. But such classifications glance off the problem of the analysis of the kinds of mechanisms that have been

[1]Present address: Beloit College, Beloit, Wisconsin.

proposed to account for learning and memory. Thus, the survey of mechanisms is organized by the relations between the proposed mechanisms and the behavioral data. This organization is derived from one used by Kantor in *Problems of Physiological Psychology* (1947, p. 138).

Four types of postulations about mechanisms can be identified. These are mechanisms based on *analogy, correlation, symptom,* and *participation.* Each type of formulation involves a range of specific mechanisms, and an individual theorist frequently uses formulations of different types at different points in his career. But our goal is to indicate the range and types of mechanisms rather than to exhaust the universe of mechanisms and theorists.

Mechanisms by Analogy

The oldest and certainly the most free swinging formulations regarding mechanisms of learning and memory are those based on analogy. In these cases observations of behavior are likened to other processes such as telephone switchboards, computers, or signet-on-wax imprinting. Both Plato and Aristotle dealt at length with learning and memory, and both proposed mechanisms by analogy. Their mechanism of long-term memory was likened to a signet imprint on wax. However, both analysts commented on the difference between short-term and long-term memory and suggested different mechanisms for the two types.

Popular writings are full of suggestions for mechanisms of learning and memory that use analogy. The computer is probably the present leader in this type of postulation, followed closely by diagrams of substitution or association. But more detailed and analytic analogies have been proposed. As Kantor notes (1947, p. 138), "Excellent examples of such analogies are found in the attempts to relate summations and inhibitions of behavior with synaptic mechanisms worked out with physiological preparations. One instance is the use of Lorento de No's finding regarding the recurrent nervous circuit to describe complexities of total organismic psychological behavior." Clark L. Hull (1943) and more recently David D. Smith (1965) developed elaborate mechanisms regarding learning and memory that are based on analogies with electrical circuits and reflex physiology.

Mechanisms by Correlation

The postulation of mechanisms for learning and memory by correlation obviously has more restrictions than postulations by analogy. In the present class, the investigator must find some biological state that covaries with occurrence of learning and/or memory. Ordinarily, the formal requirements of correlation as a statistical index are not met; rather, a limited number of observations are made and mechanisms are postulated. In addition, causal relations between the two variables are sometimes assumed. This also violates the formal properties of correlation.

Examples of mechanisms by correlation are drug and fatigue postulations in which the changes in performance that covary with these states are overinterpreted and suggested as general mechanisms. Lashley's mass action hypothesis is another example. Clearly, any observations of correlations are relevant and important, but they are strengthened by systematic experimental study and attention to the restrictions on interpretations that are appropriate.

Esper (1964, p. 172) reports an example of mechanisms by correlation that Aristotle proposed. In translation, at least, it illustrates a number of the characteristics of such formulations. The statement is as follows: ". . . bodily actions are all accompanied by a change in temperature, some in a particular member, others in the body generally. So memory and anticipations, using, as it were, the reflected images of these pleasures and pains, are now more and now less causes of the same changes of temperature."

Mechanisms by Symptom

The principal characteristic of this class of mechanisms is that ". . . neural or muscular incidents are taken as symptomatic of the larger total event and are presumed to give a descriptive or predictive insight into the whole psychological happening" (Kantor, 1947, p. 139). Thus, mechanisms by symptom are formulations in which one aspect of an event is taken as an index of a larger and more general state. This class of mechanisms requires more theoretical sophistication than the previous classes. In this case, total processes are assumed and valid indexes of the processes are required. However, deficiencies in either of these aspects of the formulation lead to inappropriate formulation.

Investigators have proposed a variety of mechanisms by symptom. The formulations of the past several decades have emphasized *electrical indexes* and *chemical indexes*, both of which are taken from measurement of CNS (central nervous system) activity. Hebb (1949) formulates mechanisms of learning in relation to electrical activity. He uses the EEG (electroencephalogram) as an index. Deutsch and Deutsch (1966, Chapter 8) summarize a number of other formulations that follow the same general idea. Considerable importance is attached to the form and changes in form of EEG records. Hebb (1949, p. 122), for example, suggests that ". . . there appears to be two extremes in the organization of cortical activity that correspond exactly with the two extremes of learning ability. . . one is established early (infancy) . . . (is) manifested by large potentials in the EEG . . . the other is slowly acquired (adult) . . . and produces a flattened, irregular EEG . . . ."

Mechanisms for learning and memory based on chemical symptoms are assuming increasing importance in current theory and research. However, they are not a new form of postulation. Analyses of behavior in terms of humors and other chemical constructs held an important place more than a hundred years ago. The recent formulations have become more data oriented and subject to experimental investigation, though. In this way, some mechanisms have been set aside for lack of experimental support. For example, glutamic acid was a prominent contender as an index of cerebral chemistry, and both animal and human research followed from the original formulation.

Presently, several other chemical symptoms associated with learning and memory are under consideration. Overton (1959) makes an analysis of cerebral chemistry and focuses on the relevance of the amount of calcium as an index related to learning and memory. Cholinesterase has been implicated as an index of learning in a number of studies. Rosenzweig, Krech, and Bennett (1960) summarize some of the data and thinking on these mechanisms. Finally, the ribonucleic acid (RNA) molecule has come into great prominence and can be viewed as a mechanism by symptom. Hydén (1959) and others have elaborated hypotheses of learning and memory in terms of the RNA molecule, and the present volume will deal in detail with this mechanism. The RNA mechanism can be used in a number of ways by the investigators, that is, as analogy, correlation, or symptom. In some cases, an individual investigator has shifted the form of his formulation. But it seems appropriate to classify some of the recent analyses as mechanisms by symptom because RNA is frequently viewed as one aspect in a biological complex that is involved in learning and memory.

Mechanism of Participation

The mechanisms in this class represent a rather different sort of postulation from those above. The idea in this case is that biological, chemical, or electrical processes participate in the behavior process but are not basic to the behavior process nor isomorphic with it. Kantor (1947), for example, argues strongly for this view in his treatment of physiological psychology. Ratner and Denny (1964) treat biological processes as factors affecting behavior and assign to them the same theoretical status as other factors such as prior learning, deprivation schedules, and temporal factors.

In these cases, mechanisms for learning and memory are not sought in biological processes, rather they are sought in the identification of *necessary conditions for learning and memory*. So, for example, Denny and Weisman (1964) and Weisman *et al.* (1966) analyze avoidance learning and retention of a learned avoidance response in terms of the actions and durations of the actions that follow the performance of an escape or avoidance response. The necessary condition, or mechanism, for such learning is labeled "relaxation," and it clearly implicates internal factors as participating factors.

SOME PROBLEMS IN THE SEARCH

Formulations regarding mechanisms for learning and memory span several thousand years and range from mechanisms by analogy to mechanisms of participation. While in principle the search must be drawing closer to the goal of understanding, we can identify a number of problems that have led to digressions and uncertainty. These problems are relevant regardless of the type of mechanism that is eventually used to explicate learning and memory. The problems associated with the digressions were identified by examining the early search for their predominant themes and styles, some of which seem to have interfered with progress towards general understanding.

Emphasis on Higher Animal Forms

Until very recently analyses of learning and memory concentrated on humans and a few other mammals. This concentration, probably arising from theological and philosophical biases, has limited the kinds of investigations of learning processes. It has also severely limited the control and manipulation of environmental and genetic factors that may be necessary in order to go beyond the present state of understanding.

Paucity of Research Preparations

A second theme and its problems are closely related to the emphasis on higher animal forms. Early analyses were primarily introspective and armchair. While these were good starting places, their deficiencies for detailed analysis are presently well accepted. Thus, experimental methods and analyses were used. But the experimental analyses have focused on a few learning devices and a few species, primarily the human and the rat. While these preparations have revealed some variables that are associated with learning, rigorous specification of all relevant variables and detailed study of the reliability of the methods have been weak.

Examination of successful areas in biology, for example, suggest that a variety of preparations are necessary for the understanding of processes. Some preparations lend themselves to the study of one process; others lend themselves to the study of others. In addition, powerful preparations often lack initial face validity in the processes they permit the investigator to understand. So, e.g., the sweet pea and fruit fly may not at first have looked like obvious preparations for understanding genetics.

In summary, other preparations are presently required by investigators of learning and memory. These new preparations may provide the understanding that is obscured by the study of the present preparations with the present methods.

Insulation Between Research Areas

A third characteristic of early searches for learning and memory mechanisms is the dedication of investigators to their individual fields of inquiry with consequent disregard for other related fields. Thus, with few exceptions, the vast amount of background information available to some investigators was unknown or unused by other investigators. This was particularly true, if the background information and the investigator were from different research areas. So, e.g., facts about a species, its structures, and habitat were frequently unknown to students of learning who were using these very species. Conversely, methods of behavior measurement and techniques for studying memory in unconfounded designs were often unknown to investigators who were poorly acquainted with psychology.

Such insulation can be abundantly documented, but this would serve a negative purpose. The main point of this section is to identify the problem of insulation and point out the fact that theories and research on behavior are facilitated when all of the appropriate background information is available and used.

CONCLUSIONS

The main body of this volume and the focus of the discussions reported in it are directed toward enhancing the train of understanding about learning and memory mechanisms. The search and associated postulations that are presently available span a range of forms and include many insights as well as false starts. For example, Aristotle hypothesized short-term and long-term memory but proposed unproductive mechanisms for the analyses of these processes.

Several interlocking characteristics of the search for learning and memory mechanisms appear to have blocked effective understanding. These characteristics include emphases on human and higher mammals, paucity of research preparations, and insulation between research areas. Each section in the present volume deals in detail with moving these blocks. One section involves critical examination of theories and mechanisms; one involves detailed reporting of background information and studies of learning about a relatively new and provocative preparation, the planarian; and the final section provides reports on a number of other invertebrates that are not yet commonly studied but may permit new understanding of learning and memory.

REFERENCES

Denny, M. R., and R. G. Weisman, Avoidance behavior as a function of length of nonshock confinement. *J. Comp. Physiol. Psychol.*, 1964, **58**, 252.

Deutsch, J. A., and D. Deutsch, *Physiological Psychology*. Homewood, Ill.: Dorsey Press, 1966.

Esper, E. A., *A History of Psychology*. Philadelphia: W. B. Saunders Co., 1964.

Hebb, D. O., *The Organization of Behavior*. New York: John Wiley & Sons, Inc., 1949.

Hydén, H., 1959. Biochemical changes in glial cells and nerve cells at varying activity, *in* F. Brucke, ed., *Proc. 4th Intern. Congr. Biochem.*, *III: Biochemistry of the Central Nervous System*, London: Pergamon Press, 1959.

Hull, C. L., *Principles of Behavior*. New York: Appleton-Century, 1943.

Kantor, J. R., *Problems of Physiological Psychology*. Bloomington, Ind.: Principia Press, 1947.

Overton, R. K., *Thought and Action: A Physiological Approach*. New York: Random House, 1959.

Ratner, S. C., and M. R. Denny, *Comparative Psychology*. Homewood, Ill.: Dorsey Press, 1964.

Rosenzweig, M. R., D. Krech, and E. L. Bennett, A search for relations between brain chemistry and behavior. *Psychol. Bull.*, 1960, **57**, 476.

Smith, D. D., *Mammalian Learning and Behavior*. Philadelphia: W. B. Saunders, 1965.

Weisman, R. G., M. R. Denny, S. A. Platt, and D. J. Zerbolio, Facilitation of extinction by a stimulus associated with long nonshock confinement periods. *J. Comp. Physiol. Psychol.*, 1966, **62**, 26.

Chapter 2

Invertebrates and the Study of Learning and Memory Mechanisms

W. C. Corning

Department of Psychology
Fordham University
New York, New York

If it were possible for each of us to fabricate an experimental subject best suited for our particular research interests, the result would be a bizarre array of specimens. For one of my own research projects, for instance, I would stipulate that my specimens have only one sensory modality, which would avoid sensory interaction problems and allow easier control of unwanted influences in the experimental situation. The central nervous system would be anatomically and functionally differentiated, easily accessible, and the entire system encased in a tight-fitting bony shell to allow the chronic implantation of electrodes at any point. The effector side would be quite simple—one rigid leg to stand on so as to limit movement and one arm with enough torque to depress a lever. It would also be necessary that the beast be furless, since I have a particular affection for furred animals, and that it be unable to vocalize or exhibit other signs of discomfort when in pain.

We are, however, a considerable distance from achieving such a state in spite of occasional oracular pronouncements issued by some cyberneticists. To be able to design a system implies that we understand its principles of operation, and the mechanisms by which neural aggregates perform their functions remain enigmatic. Besides that, the "perfect" subject like that I have described would not only be dull but also quite unlike any natural organism, a burlesque of our current state of knowledge concerning biological systems.

One way we can structure our subjects, so to speak, is by selecting those animals which possess characteristics advantageous to certain experimental goals. The general impression that is obtained from an overview of current research in psychobiology is that there is an increasing interest in the invertebrate as an experimental preparation. This trend is gratifying to those who believe that the information obtained from these lower organisms may prove useful in attempts to understand some of the complex properties of higher brains, particularly learning and memory mechanisms.

The neurophysiologists have enjoyed a certain degree of success in applying the information and principles derived from such preparations as the squid axon to the properties of neurons in higher organisms. Can we expect similar success when we turn to the problem of how biological systems process, store, and retrieve experiential data? There are some basic questions, which have been raised by others, that need to be considered when information obtained at one level is applied to levels of greater or lesser complexity. We have arrived at some fairly definite ideas in psychology concerning the laws of learning and retention, laws which have been based for the most part upon research involving the rat and other higher animals. Accordingly, when we study learning in lower animals, can we be certain that the same laws apply? Is there a danger of our overlooking some phenomenon or misclassifying it because of too limited and rigid definitions of learning? For that matter, are the neuronal operations the same? There are certainly marked differences between various groups in their neuronal structure and chemistry. Can we assume that the formal operations of neural aggregates are universal phenomena differing only in amount as organisms become more complex? Or, when a structure such as the cortex emerged, was there actually a new and unique neuronal "language" added to the system? More relevant to the interests of the present symposium is the question of whether the cellular processes responsible for information storage in the planarian, for instance, are the same for the cockroach, the rat, or man.

In the following chapters the concern is with invertebrate research and the chemistry of learning. The planarian has received a great deal of attention in considering this topic. For the purpose of summary, the relevant findings are the following:

1. The demonstration that an acquired response is retained after regeneration (McConnell et al., 1959).

2. The finding that ribonuclease can disrupt retention of the acquired habit in regenerating planarians (Corning and John, 1961).

3. Studies which have indicated a "transformation" effect, i.e., a facilitation of learning when naive animals have ingested trained animals or when they are injected with RNA obtained from trained subjects (McConnell, 1962).

These findings have been highly provocative in both a positive and negative sense; in fact, in some quarters they seem to have elicited inspired malevolence. The most heated exchange appears to be over the issue whether planarians can learn. It would seem logical that, if this preparation is used to study the chemistry of learning, then evidence of learning should be clearly established. There are other issues involved here: one concerning the basic assumptions about cellular and subcellular mechanisms that encode information and another concerning the uniqueness of RNA as a "memory molecule." These problems are considered in the papers that follow, and several experimental suggestions are made for questions where data are lacking or for areas where the controversy is unresolved. It is essential that planarian research be evaluated with respect to current thinking and data concerning the chemistry of learning. For example, there are now data available on "transformation" in rats (Babich *et al.*, 1965; Fjerdingstad *et al.*, 1965; and Gross and Carey, 1965); there is evidence on the effects of RNA disruption on learning and retention (Cohen and Barondes, 1966, and Dingman and Sporn, 1961); and there are studies dealing with the effects of RNA injections on behavior and physiology (Cameron and Solyom, 1961, and Cook *et al.*, 1963). As indicated by some of the citations above, there is negative as well as positive evidence—planarian research is not unique for controversy.

There is also a particular need in this area of research for a maximum exchange of information among the various disciplines. If one is assessing the effects of light and shock on an organism, then it is necessary that the physiological consequences of these inputs be taken into consideration in the experimental program. Such data are available: Jay Best reports some interesting results and speculations concerning the relationships between shock parameters and neuronal structures (in Chapter 17 of this book), while Brown presents data on photoreceptor processes along with comments on the relevance of this information to behavioral studies using light as a stimulus (Chapter 11). Related to this is the frequently heard complaint that many of the data are unavailable or reported in bulletins of limited circulation. The material contained in the following chapters should reduce some of these communication problems.

Another point, which was raised by Dr. E. L. Bennett, relates to the problem of replication; what was worrisome was that some of the critical studies appeared to have been done only once. The extent to which various findings have or have not been replicated is examined by several of the contributors here, and in some cases the specific factors that are essential for replication are clarified.

Finally, we might consider whether certain problems in the study of learning mechanisms could be better investigated in other preparations. The planarian may possess characteristics which preclude its being used in some types of experiments. There are potentially useful preparations to be found at practically all invertebrate levels. In the *Hydra*, Passano and McCulloch (1962) have demonstrated the feasibility of obtaining electrophysiological recordings in the intact animal, which opens the way for correlative investigations of bioelectrical and behavioral patterns in a coelenterate. In the *Aplysia* (phylum Mollusca), we have the possibility of studying the electrophysiological consequences of conditioning in a small number of neurons (Frazier *et al.*, 1965).

Among the arthropods there are several exciting preparations. In the orb weaver, for instance, it is found that the web construction is such a well-programmed event that the web geometry can be used as a sensitive measure of drug effects (Bercel, 1960, and Witt and Reed, 1965). An outstanding example of how the central nervous system of an arthropod lends itself to the study of integrative properties may be found in a series of studies by Rowell (1965). The grooming reflex of the locust was found to be programmed in the prothoracic segment. In the intact animal this reflex was rarely elicited. By selectively removing different segments of the ventral cord, it was possible for Rowell to measure accurately the degree of inhibition each ganglionic segment exerted on the reflex. Techniques for the chronic implantation of electrodes in the locust have also been described (Rowell, 1963). Luco (1964) has reported on what looks to be postural learning in the cockroach. Isolated ganglia preparations have also been receiving considerable attention. These preparations could be very important in the search for cellular changes peculiar to information storage, for they limit the number of relevant cells. The work of Horridge (1962), Hoyle (1965), and Eisenstein and Cohen (1965) has provided strong evidence that learning is possible at the segmental level. To summarize, it would seem that "higher brain" functions may be fruitfully studied in the invertebrate.

One final point on the controversial nature of many topics that are discussed: Our purpose is not to draw any definite conclusions regarding the current status of RNA and memory or whether or not flatworms can learn. The main goal is to provide information (much of which has been heretofore unavailable) and to discuss the potentials in various invertebrate preparations with respect to the problem of the chemistry of learning. In the course of doing this it is hoped that some of the controversy will be resolved. It is unlikely that any basic biases will be altered, but the material will at least force those biases to be structured in the light of available evidence.

REFERENCES

Babich, F. R., A. L. Jacobson, S. Bubash, and A. Jacobson, Transfer of learning to naive rats by injection of ribonucleic acid extracted from trained rats. *Science*, 1965, **149**, 656.

Bercel, N. A., A study of the influence of schizophrenic serum on the behavior of the spider: *Zilla-x-Notata, in* D. Jackson, ed., *The Etiology of Schizophrenia*. New York: Basic Books Inc., 1960.

Cameron, D. E., and L. Solyom, Effects of ribonucleic acid on memory. *Geriatrics*, 1961. **16**, 74.

Cohen, H. D., and S. H. Barondes, Further studies of learning and memory after intracerebral actinomycin-D. *J. Neurochem.*, 1966, **13**, 207.

Cook, L., A. B. Davidson, D. J. Davis, H. Green, and E. J. Fellows, Ribonucleic acid: effect on conditioned behavior in rats. *Science*, 1963, **141**, 268.

Corning, W. C., and E. R. John, Effect of ribonuclease on retention of conditioned response in regenerated planarians. *Science*, 1961, **134**, 1363.

Dingham, W., and M. B. Sporn, The incorporation of 8-azaguanine into rat brain RNA and its effect on maze-learning by the rat: an inquiry into the biochemical basis of memory. *J. Psychiat. Res.*, 1961, **1**, 1.

Eisenstein, E. W., and M. J. Cohen, Learning in an isolated prothoracic ganglion. *Animal Behaviour*, 1965, **13**, 104.

Fjerdingstad, E. J., Th. Nissen, and H. H. Roigaard-Petersen, Effect of ribonucleic acid extracted from the brain of trained animals on learning in rats. *Scand. J. Psychol.*, 1965, **6**, 1.

Frazier, W. T., R. Waziri, and E. R. Kandel, Alterations in the frequency of spontaneous activity in *Aplysia* neurons with contingent and non-contingent nerve stimulation. *Fed. Proc.*, 1965, **24**, 2171.

Gross, C. G., and F. M. Carey, Transfer of learned response by RNA injection: Failure of attempts to replicate. *Science*, 1965, **150**, 1749.

Horridge, G. A., Learning of a leg position by headless insects. *Nature*, 1962, **193**, 697.

Hoyle, G., Neurophysiological studies on "learning" in headless insects, *in* J. E. Treherne and J. W. L. Beament, eds., *The Physiology of the Insect Central Nervous System*. New York: Academic Press Inc., 1965.

Luco, J. V., Plasticity of neural function in learning and retention, *in* M. Brazier, ed., *RNA and Brain Function*. Berkeley: University of California Press, 1964.

McConnell, J. V., Memory transfer through cannibalism in planarians. *J. Neuropsychiatry*, 1962, **3**, S42.

McConnell, J. V., A. L. Jacobson, and D. P. Kimble, The effects of regeneration upon retention of a conditioned response in the planarian. *J. Comp. Physiol. Psychol.*, 1959, **52**, 1.

Passano, L. M., and C. B. McCulloch, The light response and the rhythmic potentials of *Hydra*. *Proc. Nat. Acad. Sci.*, 1962, **48**, 1376.

Rowell, C. H. F., A method for implanting chronic stimulating electrodes in the brains of locusts and some results of stimulation. *J. Exp. Biol.*, 1963, **40**, 271.

Rowell, C. H. F., The control of reflex responsiveness and the integration of behaviour, *in* J. E. Treherne and J. W. L. Beament, eds., *The Physiology of the Insect Central Nervous System*. New York: Academic Press, Inc., 1965.

Witt, P. N., and C. F. Reed, Spider-web building. *Science*, 1965, **149**, 1190.

PART I
Biological and Learning Models

Chapter 3

A Brief Survey of Possible Mechanisms in Information Processing

Leroy Augenstein

Biophysics Department
Michigan State University
East Lansing, Michigan

I shall try to give an outsider's view of what invertebrate studies may contribute to learning research. Although I have not conducted studies on invertebrates, I have been interested in them for a long time. When I was with the Atomic Energy Commission, one of the first proposals which came across my desk was from Jim McConnell on planarian research. Considering the trouble that I had getting that proposal approved, I have had an interest in this field ever since, mainly to see whether half of what he proposed back in 1958 has been feasible or not.

My own particular interest is human engineering, as we have investigated the general questions: How much information can a human transmit, and how does he go about it? This is quite far removed from your general area of consideration, and yet I should like to use, as a framework, some of our findings to indicate places where I think invertebrate research could be quite useful. In particular, I shall briefly describe the elements shown in Fig. 1 which we conclude are part of the human operator.

When we started this research in 1952 at the University of Illinois, our first inquiries were: What is the optimum amount which a human can process? Specifically we had experts perform well-learned tasks, such as typing, piano playing, mental arithmetic, etc. By using random texts so they could not rely simply on memory, we found that most people have a capacity of fifteen to twenty-five bits per second (Quastler and Wulff, 1955; Quastler, 1956; and Augenstein and Quastler, 1967); only when outstanding readers are recognizing random words are performances observed as high as about forty bits per second (Pierce and Karlin, 1957).

We next asked: What determines this overall capacity? At least three operations must be performed to transduce information: inputting, processing, and outputting.

We could immediately rule out the outputting as being rate-limiting simply because, if you allowed the pianists to memorize their random text, they could improve their performance by a factor of 2 or 3.

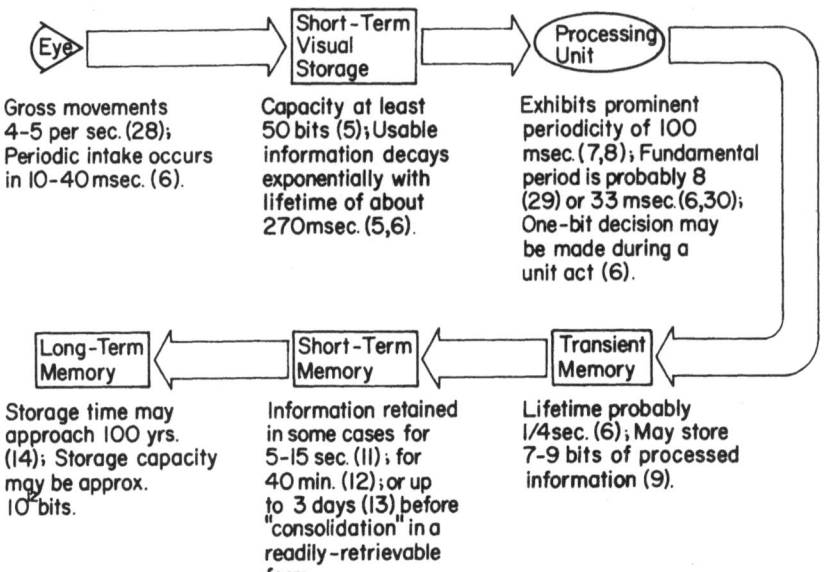

Fig. 1. Schematic representation of elements involved in information processing by humans.

To pursue further these questions, we (and other laboratories) ran a number of experiments to isolate, insofar as possible, the input mechanism and define its capabilities. Let me describe just one experiment carried out at Bell Laboratories which tells us a lot about the input mechanism in humans (Auerbach and Coriell, 1961). They displayed on a television screen a card with 16 randomized letters in two rows of 8 letters each. This was flashed for 40 msec, and then another card with a black marker either below or just above one of the positions was placed in front of a second camera and flashed for 40 msec onto the same monitor. As soon as this latter marker appeared, the subjects were to identify the letter designated.

When the marker actually was shown 40 msec before the letters, the subjects identified the correct letter 70% of the time. When letters and marker came simultaneously, they got 65% correct, this corresponded to an intake of over 50 bits. When the marker was flashed after the letters appeared, the percentage correct decreased as the interval increased; in fact, when the percentage correct was plotted versus the interval between display and marker, the data could be fitted with an exponential having a decay constant of 270 msec.

We did similar experiments with playing cards, but in these the subject found out which item of the display he was to identify only after the display had been terminated for 2 sec. Interestingly enough when we exposed our cards for 40 msec, the value of 15 bits which our subjects could transmit agreed very closely with the value obtained by the Bell

Laboratory group when they used an interval of 2 sec between letters and marker. We also found that the amount of information which can be transmitted about a display illuminated for 100 msec is the same as that for 40 msec. Only for display durations of more than 250 msec is there an increase in performance, and for 10-msec illumination the transmission is only eight bits (Augenstein, 1956).

From these results, we conclude that in 10 to 40 msec the eye can take in and temporarily store away in a usable form at least 50 bits of information, but that this short-term storage decays away with a lifetime of about $\frac{1}{4}$ sec. Further, these results plus earlier information on eye movements during reading imply that the eye can take in such large amounts of information four or five times per second, presumably in conjunction with gross eye movements. These results clearly indicate that the input mechanism is not the rate-limiting process, and thus the limitations must occur in the processing unit.

Let me describe one of a number of experiments we used to investigate the properties of data processing in humans. A card containing a column of randomized letters and numbers was put up in front of the subjects in the dark. Their task was to push a switch (i.e., the task was self-paced) to illuminate the display and then to scan down until they found the first number, release the switch, and tell us the number. When we made a histogram of how often a given response time occurred, the distribution was not at all random. Rather, we found very strong evidence of a 100-msec periodicity, which we initially assumed to be associated with the alpha rhythm (Augenstein, 1955). However, it is now fairly clear that this is the predominant but not the fundamental periodicity; presumably it is a collection of three 33-msec or twelve 8-msec or seventeen 6-msec periods, etc. (Augenstein, 1958). Of great interest is the fact that we found the same kind of periodic behavior, independent of the complexity of the task, i.e., for simple scanning experiments, adding columns of numbers, or typing.

This immediately suggested, although it certainly is not proven, that during the performance of simple tasks, at least, the human processing involves some common type of decision, presumably either on a binary or trinary basis. In fact, the basic hypothesis we continue to test is that during one of these unit acts a human either takes in a large batch of data or makes a one-bit decision.

There is fragmentary evidence that, once decisions are made by the computer unit, the processed information goes into a storage unit, again having a characteristic duration of about $\frac{1}{4}$ sec. Such a unit would be consistent with Miller's seven- to nine-bit chunk hypothesis (Miller, 1956). Also the data for delayed auditory feedback may depend upon such a storage, providing that a feed-back checking mechanism connects this unit and the ear [note that optimum interference occurs when the delay between speaking and hearing is 210 to 300 msec (Fairbanks, 1955)].

Following this temporary storage the information apparently resides in a so-called short-term memory for times varying presumably from 5 sec

(Chorover and Schiller, 1965 and 1966) to 40 min (Jarvik, 1964, and McGaugh, 1966)—or perhaps even 3 days in some cases (Flexner *et al.*, 1965)—before "incorporation" into the "permanent" memory. Recall elicited from implanted electrodes suggests that the duration of storage in the permanent memory may even be as much as 100 years (Penfield, 1959).

This cursory description is not meant to be necessarily all inclusive or even to convince you of the validity of the representation in Fig. 1. Rather, it was designed to call attention to specific elements in the information-processing scheme about which invertebrate research might provide information.

Biochemical studies in planaria have already generated important questions about the chemical basis of memory (Corning, Chapter 18 in this book). In spite of the controversy about these experiments I am confident they will be continued. Also, I have always watched the planaria cannibalism experiments (McConnell, Chapters 14 and 20 in this book) with great interest and hope that this phenomenon can be shown to operate conclusively. If so, it could provide a system for studying how one cell, or at least a small group of cells, may adapt or modify neighboring cells so as to create a processing unit which would ultimately modify behavior. This whole question must be attacked in a very rigorous way in a number of different organisms since clearly neurons can be adapted, i.e., their behavior can be drastically changed. Upon what does this depend? Are there cell-wall changes similar to those induced in paramecia by different agents (Sonneborn, 1963) or similar to those induced by virus or sperm transformation (Smith, 1963, and Rothschild, 1956)? If so, is this what is important in the storage of long-term memory?

In planaria the gut is so primitive that apparently whole cells are ingested (Quastler, 1962). Further, 20 to 25 % of planarian cells are essentially undifferentiated (McConnell, 1965). Thus, ·if learning is transferred by cannibalism, it should be of great interest to investigate whether, when a "learned cell" is introduced into a host, it can act as an organizer to determine the differentiation of these "uncommitted" cells and thus "consolidate" the transferred information. That certain cells can be crucial in adapting others is well known in a number of embryological systems. Once a cell is adapted and information is stored, it still must be retrieved to be of any value. How does this occur? By a change in membrane resistance (Augenstein, 1962, and Augenstein and Van Zytveld, 1964)? By an antigen-antibody type of reaction (Silverstein, 1963)? Whether invertebrate preparations can be used to pursue this question remains to be seen.

Our results on periodicity in human processing call attention to another important problem: What mechanisms account for immediate processing? Since the unit processing time is of the order of a few milliseconds, data processing cannot involve the synthesis of a macromolecule (Augenstein 1962; Augenstein and Quastler, 1967; and Augenstein and Van Zytveld 1964); i.e., to synthesize a protein or nucleic acid requires at least 10 to 30 msec *per monomer unit* incorporated. Thus, the unit operating time for immediate processing must reflect a time constant for a network of cells (McCulloch and Pitts, 1943) or a coupled enzyme system or perhaps conformation changes in proteins which control the flow of current at a synapse (Augenstein, 1962, and Augenstein and Van Zytveld, 1964). Perhaps the Limulus preparation used by Corning, Feinstein, and Haight (1965) may provide some insight into this problem.

Of course the biggest question in all behavioral research is: What is the code by which information is represented internally? Is it binary, trinary, or something more esoteric? Unfortunately, I don't know how to do critical experiments in this area, and, again, this must be attacked at many levels. Thus, if one of you could find some means of determining the code by which information is stored in a single cell, this would be a tremendous step forward. More specifically, is only one bit stored in a single adapted cell or at a single synapse, or is much more than that stored? Once we can say meaningful things about this question, I am sure it will become much easier to determine the chemical basis of memory storage.

REFERENCES

Auerbach, E., and A. S. Coriell, *Short-term Visual Memory*. Film available at Bell Telephone Laboratories, New York, 1961.
Augenstein, L., *Rep. R*-78, Control Systems Laboratory, University of Illinois, 1955.
Augenstein, L., Evidences of Periodicities in Human Task Performance, *in* H. Quastler, ed., *Information Theory in Psychology*. New York: Free Press, 1955, p. 208.
Augenstein, L., *Rep. R*-69, Control Systems Laboratory, University of Illinois, 1956.
Augenstein, L., *Rep. R*-75, Control Systems Laboratory, University of Illinois, 1958.
Augenstein, L., Controlled conformation changes in protein molecules: a possible mechanism of information storage, *in* F. O. Schmitt, ed., *Macromolecular Specificity and Biological Memory*. Cambridge, Mass.: MIT Press, 1962, p. 21.
Augenstein, L., and H. Quastler, Information processing and decision making by man, I: Limitations on transmission rate in sequential actions. *Brain Res.*, 1967.
Augenstein, L., and J. Van Zytveld, Macromolecular conformation changes as possible information processing mechanisms, *in Proceedings of the Symposium on the Role of Macromolecules in Complex Behavior*: Kansas State University Publication, Manhattan, Kansas, 1964.
Buswell, G., *How People Look at Pictures*. Chicago: University of Chicago Press, 1955.
Chorover, S. L., and P. H. Schiller, Short-term retrograde amnesia in rats. *J. Comp. Physiol. Psychol.*, 1965, **58**, 73.
Chorover, S. L., and P. H. Schiller, Re-examination of prolonged retrograde amnesia in one-trial learning. *J. Comp. Physiol. Psychol.*, 1966, **61**, 34.
Corning, W. C., D. A. Feinstein, and J. R. Haight, Anthropod preparation for behavioral, electrophysiological and biochemical investigations. *Science*, 1965, **148**, . 394.
Fairbanks, G., Selective vocal effects of delayed auditory feedback. *J. Speech Hearing*, 1955, **20**, 333.
Flexner, L. B., J. B. Flexner, G. de la Haba, and R. B. Roberts, Loss of memory as related to inhibition of cerebral protein synthesis. *J. Neurochem.*, 1965, **12**, 535.
Freed, S., Endogenous biochemistry of planarians correlated with learning, *Rep. BNL 981*, Upton, L.I., N.Y.: Brookhaven National Laboratory (1964).
Jarvik, M. E., The influence of drugs upon memory, *in* H. Steinberg, ed., *Animal Behavior and Drug Action*. Boston: Little, Brown and Co., 1964.
Leet, D., and L. Augenstein, Further studies of the unit act in information processing by humans (*in progress*).

McConnell, J. V., ed., *A Manual of Psychological Experimentation on Planarians.* Ann Arbor, Mich.: Worm Runner's Digest, 1965.

McCulloch, W. S., and W. Pitts, A logical calculus of the ideas imminent in nervous activity. *Bull. Math. Biophys.*, 1943, **5**, 115.

McGaugh, J. L., Time-dependent processes in memory storage. *Science*, 1966, **153**, 1351.

Miller, G. A., The magical number seven plus or minus two: some limits on our capacity for processing information. *Psychol. Rev.*, 1956, **63**, 81.

Penfield, W., The interpretive cortex. *Science*, 1959, **129**, 1719.

Pierce, J. R., and J. F. Karlin, *Bell Tech. J.*, 1957, **36**.

Quastler, H., Studies on human channel capacity, *Rep. R*-71, Control Systems Laboratory, University of Illinois, 1956.

Quastler, H., *Personal communication* concerning ingestion studies utilizing cells labeled with H^3 cytidine, 1962.

Quastler, H., and V. J. Wulff, Human performance in information transformation, *Rep. R*-62, Control Systems Laboratory, University of Illinois, 1955.

Rothschild, L., *Fertilization.* New York: John Wiley & Sons, Inc., 1956.

Silverstein, A. M., Immunologic and psychic memory. *Neurosci. Res. Progress Bull.*, 1963, **1**, 1.

Smith, W., ed., *Mechanisms of Virus Infection.* New York: Academic Press, Inc., 1963.

Sonneborn, T. M., Does preformed cell structure play an essential role in cell heredity? *in* J. M. Allen, ed., *The Nature of Biological Diversity.* New York: McGraw-Hill Book Company, 1963.

Chapter 4

The Possible Role of RNA in Learning and Memory Events

John Gaito

Department of Psychology
York University
Toronto, Canada

Learning events are of great interest to biologists because of their basic involvement in behavior. Up to several decades ago the major approaches to learning were of a psychological nature. In the last 20 years, however, biological approaches have achieved prominence. The earlier systematic treatments utilized neurological concepts as exemplified by Hebb's system in 1949. Although Halstead (1951) and Katz and Halstead (1950) invoked a systematic set of chemical hypotheses and others spoke of chemical events during learning (e.g., Gerard, 1953), the chemical approach followed the neurological one.

One prominent systematic chemical approach was begun about 1953 at the University of California at Berkeley. Krech and his colleagues concerned themselves with the chemicals involved in intercellular communication, acetylcholine (ACh) and acetylcholinesterase (AChE). Although their interest began in neurochemistry, their most important contributions appear to be of an anatomical nature, e.g., the findings that varied experiences result in a thickening of the cortex in certain parts of the brain (Bennett *et al.*, 1964).

A second chemical approach was precipitated by the rapid and exciting developments in molecular biology, an intracellular concern with the nucleic acids, DNA and RNA, and their involvement in protein synthesis (Gaito, 1966). Even though the intracellular and intercellular approaches developed independently, they are not contradictory. In fact, both approaches probably will converge soon, for the two ultimately must handle both the intra- and intercellular phenomena. The one that I shall be concerned with is the molecular intracellular treatment.

Inasmuch as my discussion is intimately related to protein synthesis, a quick summary of the role of nucleic acids in this event is necessary at this time (Gaito, 1966). With the exception of the viruses which contain only RNA (e.g., polio virus, influenza virus, tobacco mosaic virus), all RNA appears to be synthesized from a DNA template. The three RNA species involved in protein synthesis are messenger RNA, transfer RNA, and ribosomal RNA. The genetic code in DNA is transmitted to messenger RNA in the nucleus. Messenger RNA moves to the cytoplasm where it attaches to ribosomes, which consist of ribosomal RNA and protein. Transfer RNA's gather amino acids and adhere to the ribosome and to their appropriate site on messenger RNA. There are supposed to be transfer RNA's for each amino acid. Thus many transfer RNA's with their associated amino acids "recognize" the appropriate site on messenger RNA by complementary base pairings, and the amino acids become attached to form a specific polypeptide or protein. Notice that in protein synthesis *the three RNA's serve only as intermediaries between DNA and protein.*

The basic work in molecular biology concerned with the nucleic acids had a profound effect on a number of psychologists and other behavioral scientists who were interested in complex behavior, and it has led to the development of an area which might be appropriately titled *molecular psychobiology*, an integration of the ideas and methods of neuropsychology with those from molecular biology.

Based on experiments by Hyden (1961), Corning and John (1961), and others, the idea developed for many individuals that RNA plays a unique role in learning phenomena. These results, however, are subject to multiple interpretations. Thus a basic question in this area of research is: *Does RNA have a unique role in learning and memory events?* I think there are three most likely answers to this question. They are (see Table 1):

1. *RNA does have a unique role beyond its usual protein synthesis function.*[1] The best example of an individual supporting this possibility is Hyden. In his earliest theory (1959), he hypothesized that some bases on RNA are replaced by other bases during neural activity. Thereby one or more new RNA's result and unique protein would be synthesized, the new RNA coding for memory.

Gaito (1961), Corning and John (1961), Landauer (1964), McConnell (1964), Jacobson *et al.* (1966), Pribram (1966), and Albert (1966) also hypothesized a unique role for some RNA species.

2. *RNA has only a role in protein synthesis, but there are specific DNA sites that synthesize unique RNA species which are involved in the production of unique protein species for a given class of behavioral events.* This possibility would mean that there are specific DNA sites, RNA species, and proteins which are functional during vision. Other DNA's, RNA's, and proteins would be functional during audition, etc. Thus for learning there would be unique DNA, RNA, and protein species.

Examples of this possibility are the DNA activation model of Gaito (1966) and the DNA derepression model of Bonner (1966). Bonner suggested that in neurons there is a gene (or a few particular genes) which is repressed,

[1]The following discussion excludes a fourth type of RNA which Bonner (1965) found adhering to histones, apparently providing a specificity for the latter's complexing with DNA. This histone RNA is not pertinent to this paper.

but which is derepressable by certain substances as a result of electrical stimulation of specific dendrites. Once derepressed, the gene makes more RNA and ultimately more enzyme. The enzyme then makes more of the substance which carries the effect of the dendritic input to the repressed DNA site. These genes, once derepressed, remain derepressed permanently. Such derepression accounts for the increased rate of RNA and protein synthesis which is reported in learning experiments.

TABLE 1

Role of RNA in Learning (Acquisition) and Memory (Retention)

Question: Has RNA a Unique Role?

	Possible Answers	
1. RNA has a unique role beyond that in the usual protein synthesis sequence.	2. RNA has a unique role only as an intermediary in protein synthesis.	3. RNA has no unique role.
	Proponents	
Hyden (1959), Gaito (1961), Corning and John (1961), Landauer (1964), McConnell (1964), Pribram (1966), Albert (1966), Jacobson *et al.* (1966)	Hyden and Lange (1965 and 1966), Bonner (1966), Gaito (1966)	Barondes (1965), Dingman and Sporn (1964), Gaito (this work)
	Evidence	
Corning and John (1961), Hyden and Egyhazi (1962)	Actinomycin D experiments which show retarded learning (Appel, 1964, and Meyerson *et al.*, 1965)	Actinomycin experiments which show improved learning or no effect on learning
RNA transfer experiments showing positive transfer effects (Jacobson *et al.*, 1966, and several others)	Hyden and Lange (1965 and 1966)	Appel (1964), Barondes and Jarvik (1964), Barondes and Cohen (1966a), Landauer and Eldridge (1966), Batkin *et al.* (1966)
Tissue transfer experiments (Albert, 1966)		Puromycin experiments which show no effects on learning, Barondes and Cohen (1966b), Agranoff and Klinger (1964), Davis and Agranoff (1966)
		Learning possible even though RNA synthesis is decreased (Gaito *et al.*, 1966) RNA transfer experiments showing no transfer effects (Byrne *et al.*, 1966, and numerous others)

Hyden's latest theory (Hyden and Lange, 1965 and 1966) falls within this category also. He now speaks of DNA sites in glia and in neurons being stimulated by environmental factors such that unique RNA species are synthesized. During the early part of the learning process, the RNA that is synthesized is high in adenine and uracil, being DNA-like in composition. This RNA is formed during the establishment of functional synapses for the new behavior, and this stage represents short-term memory. During the later portion of the learning process an RNA rich in guanine and cytosine (similar to ribosomal RNA) is formed. This stage is supposed to constitute the fixation of long-term memory with a high synthesis of transmitters at the synapse.

3. *RNA has only a role in protein synthesis, and there are no unique DNA's, RNA's, and proteins for each class of behavioral events.* One would expect, however, that there would be DNA sites, and RNA and protein species, which would be involved in the morphology of a nerve cell, i.e., in making a nerve cell different from a liver or kidney cell. Likewise, there would be DNA sites which would synthesize messenger RNA's for the enzymes involved in synaptic events, choline acetylase (for the production of ACh) and AChE. Other DNA's, RNA's, and proteins might be specific to nerve cells but would be involved in behavior in general. The ideas of Barondes (1965) and those of Dingman and Sporn (1964), emphasizing neural circuitry, illustrate this possibility.

Although the first possibility is the most interesting of the three, there is no strong conclusive evidence to support it and it seems to be the one which has the lowest probability of being confirmed. The third alternative seems most likely because it requires no new assumptions beyond those which have strong credibility in molecular biology today. But the second alternative is still a definite possibility.

It is difficult to choose experimentally between the three possibilities because the experimental results are not conclusive. The major evidence cited to support possibility 1 is the Corning and John study with planaria regenerating in ribonuclease (1961), the early Hyden work on base changes as a result of learning (e.g., Hyden and Egyhazi, 1962), and the RNA and tissue transfer experiments from which positive transfer effects have been reported (Jacobson *et al.*, 1966; Fjerdingstad *et al.*, 1965; and Albert, 1966).

The Hyden and Lange experiments (1965 and 1966) analyzing the RNA of rats during various portions of the learning process are offered by those authors to support possibility 2. The actinomycin-D experiments which show a retardation of learning (Appel, 1964, and Meyerson *et al.*, 1965) are consistent with this possibility also.

There is much information which suggests that alternatives 1 and 2 are not appropriate. These two alternatives imply that RNA synthesis is an event which is necessary if learning is to occur. If one were to prevent RNA synthesis, would learning still result? One means of inhibiting RNA synthesis is with the antibiotic, actinomycin D, which binds specific sites on DNA molecules. In an interesting experiment with this chemical, Barondes and Jarvik (1964) found that mice were able to learn a shock-avoidance task even though RNA synthesis was inhibited 83 % throughout the brain. In later work, Barondes and Cohen (1966a) stated that animals with 94 to 96 % inhibition of cerebral RNA synthesis learned Y- and T-mazes as well as control animals did and remembered as well up to 4 hours later. Landauer and Eldridge (1966) replicated the Barondes and Jarvik experiments and obtained the same results. Other research, however, does sometimes show a deleterious effect on learning after actinomycin injections, as has been indicated above (Appel, 1964, and Meyerson, Kruglikov, and Kolomeitseva, 1965). To complicate the picture, Batkin et al. (1966) reported that this antibiotic facilitated the learning of a T-maze for carp. Such inconsistent results may be due to the toxicity of actinomycin and its possible effect on general functions of the cell.

Another research effort which suggests that alternatives 1 and 2 are not likely is another experiment by Barondes and Cohen (1966b). They found that inhibition of protein synthesis by puromycin injections in the temporal lobe area did not affect the acquisition of a shock-avoidance response 5 hours later. However, 45 minutes after acquisition, retention was decreased by more than 50 %. This result and other work by Agranoff and Klinger (1964) and Davis and Agranoff (1966) suggest that RNA and protein synthesis are not required for learning to occur but that protein synthesis is necessary for the maintenance of "memory traces." Research by Flexner et al. (1965) indicated also that protein synthesis appears to be necessary for adequate memory functions. Later studies by Barondes and by Flexner (personal communication) showed that memory was not permanently impaired by another protein-synthesis inhibitor, acetoxy-cycloheximide. However, Agranoff (personal communication) finds that acetoxycycloheximide can produce as great a deficit in retention as does puromycin. These contradictory results need to be clarified by further research.

If protein synthesis is required for the maintenance of memory, there is the implication that some RNA would be necessary during this synthesis. Such RNA need not be unique, however, for any species might suffice. Alternatively, a small number of unique RNA molecules might become more resistant to degradation as a result of the learning events so that they could continue to function for longer periods of time. Some research indicates that hormones may act in this manner (Bendana and Galston, 1965). Thus longer-lived RNA's could handle the protein synthesis requirement if research shows that protein synthesis is necessary.

A number of experiments in our laboratory (Gaito et al., 1966) show that rats are able to learn a one-way active avoidance response even when RNA and protein synthesis are decreased 10 to 25 % throughout the brain because of the shock or the fear developed in the task; such a result is in keeping with the actinomycin-D and puromycin results, although the reduction in synthesis is at a lower level.

The many RNA transfer experiments which show no transfer effect (Byrne *et al.*, 1966; Luttges *et al.*, 1966; Gross and Carey, 1965; and two experiments in our laboratory with labeled RNA) might be evidence for alternative 3.

Other results with visual stimulation are not consistent with the idea that RNA is necessary for behavioral events. Brodsky (cited by Pevzner, 1966) found that during light adaptation RNA synthesis occurred; however, the RNA synthesis followed, rather than preceded, the neural functions in light adaptation. He concluded that the changes in RNA content probably reflected a secondary process and that the physiological activity of the neurons were based directly on other neurochemical reactions. Thus these results appear to be more consistent with the last alternative.

One methodology which is specific to the question concerning the uniqueness of RNA species in learning is the DNA-RNA hybridization procedure (Gillespie and Spiegelman, 1965). If one separates two-stranded DNA into single strands by heating or with alkali, the single-stranded DNA can be trapped on nitrocellulose membranes. If RNA is incubated with the membrane at 66°C, DNA will complex with those portions of RNA which are complementary to it. Thus one might perform a simple learning experiment in which radioisotope-labeled RNA is extracted from critical brain tissue of learning animals and nonlabeled RNA from the same loci in nonlearning animals and also DNA is extracted from the small intestine.[2] The RNA from the nonlearning animal would be hybridized with denatured DNA from the small intestine. The labeled RNA from the learning animal would then be added to the DNA-RNA hybrid on the nitrocellulose membrane. If the learning animal had RNA species not present in the nonlearning animal, such result would be indicated by the presence of radioisotope in the twice-hybridized DNA. Obviously, a number of experiments would be required to provide a definitive answer to the question. At the moment we are doing preliminary work with these procedures and hope to initiate experiments soon.

Before one can undertake these hybridization experiments, it is necessary to determine which brain areas make important contributions during learning behavior. Several years ago we began a research program to uncover these areas. We have conducted 11 learning experiments (one-way active avoidance) with rats using three different methodologies:

1. with normal animals
2. with magnesium pemoline to enhance learning capacities (Plotnikoff, 1966)
3. with hypophysectomized rats, so as to impair avoidance learning (M. H. Appley, personal communication)

[2]Obtaining biologically active DNA from brain tissue is not always successful. Furthermore, the DNA should be labeled for quantitative purposes, and DNA turnover in brain tissue is low. However, DNA from the small intestine is easy to extract and shows high turnover rates. The linear sequence of bases in DNA is supposed to be the same in all tissues of the organism.

In these experiments we have used behavioral-dependent variables (number of avoidances, latency of response, and trial of first avoidance) and neurochemical variables [RNA amounts per gram of tissue; DNA amounts per gram of tissue; protein amounts per gram of tissue; RNA per DNA; protein per DNA; protein per RNA; specific activity of three fractions: RNA, protein, and TCA (low molecular weight constituents of tissues); and the relative specific activity of the RNA and protein fractions]. Protein, ·RNA per DNA, protein per DNA, protein per RNA, and the relative specific activity of the protein fraction have suggested in a number of these experiments that the medial ventral cortex is involved in learning. This portion of the brain has been implicated in memory events in mice by Flexner et al. (1965) and by Barondes and Cohen (1966b) and in cats during acquisition by Adey et al. (1960). We are initiating studies to determine if electrolytic lesions in this area adversely affect avoidance conditioning; preliminary results with a few animals have indicated that a moderate lesion in this area impairs acquisition, whereas the same lesion in white matter just above the medial ventral cortex has no effect on this behavior.

If one accepts possibility 3, the intracellular events of all nerve cells involving DNA, RNA, and proteins would be the same. Therefore, one needs to specify the mechanisms involved in differentiating various behavioral events. To do so, one could suggest intercellular aspects. The ultimate contributions to learning and other behavioral events probably reside in interneuron communications of a synaptic and nonsynaptic nature and in regional localization.

In closing I should like to emphasize that, although this paper has been concerned with learning events, it has focused on a narrow segment of the problem. For example in a specific learning situation a sequence of biological events is involved, say, a, b, c, . . . , x. Event a could be stimulation of receptors; b, the transmission of nerve impulses into the central system, etc. Another event, say g, would involve DNA; h, RNA synthesis; and i, protein synthesis. This paper has been concerned with events g, h, and i of the overall sequence. What occurs before g and after i is not specified. However, an understanding of certain aspects of g, h, and i may help to suggest the antecedent and postcedent conditions prevailing.

ACKNOWLEDGMENTS

This work was supported by funds from the Office of Naval Research (USA) and the National Research Council (Canada).

REFERENCES

Adey, W. R., C. W. Dunlop, and C. E. Hendrix, Hippocampal slow waves. *A.M.A. Arch., Neurol.*, 1960, 3, 74.
Agranoff, B. W., and P. D. Klinger, Puromycin effect on memory fixation in goldfish. *Science*, 1964, 146, 952.

Albert, D. J., Memory in mammals: evidence for a system involving nuclear ribonucleic acid. *Neuropsychologia*, 1966, **4**, 79.

Appel, S., A critical appraisal of the role of RNA in information storage in the nervous system, *in Symposium on the Role of Macromolecules in Complex Behavior*, Kansas State University Publication, 1964.

Barondes, S. H., Relationship of biological regulatory mechanisms to learning and memory. *Nature*, 1965, **205**, 18.

Barondes, S. H., and H. D. Cohen, Further studies of learning and memory after intracerebral actinomycin-D. *J. Neurochem.*, 1966a, **13**, 207.

Barondes, S. H., and H. D. Cohen, Puromycin effect on successive phases of memory storage. *Science*, 1966b, **151**, 594.

Barondes, S. H., and M. E. Jarvik, The influences of actinomycin-D on brain RNA synthesis and on memory. *J. Neurochem.*, 1964, **11**, 187.

Batkin, S., W. T. Woodard, R. E. Cole, and J. B. Hall, RNA and actinomycin-D enhancement of learning in the carp. *Psychon. Sci.*, 1966, **5**, 345.

Bendana, F. E., and A. W. Galston, Hormone-induced stabilization of soluble RNA in pea-stem tissue. *Science*, 1965, **150**, 69.

Bennett, E. L., M. C. Diamond, D. Krech, and M. R. Rosenzweig, Chemical and anatomical plasticity of brain. *Science*, 1964, **146**, 610.

Bonner, J., The next new biology. *Plant Sci. Bull.*, 1965, **11**, 1.

Bonner, J., Molecular biological approaches to the study of memory, *in* J. Gaito, ed., *Macromolecules and Behavior*. New York: Appleton-Century-Crofts, Inc., 1966.

Byrne, W. L., *et al.* (23 authors), RNA and memory transfer. *Science*, 1966, **153**, 658.

Corning, W. C., and E. R. John, Effect of ribonuclease on retention of conditioned response in regenerated planarians. *Science*, 1961, **134**, 1363.

Davis, R. E., and B. W. Agranoff, Stages of memory formation in goldfish: evidence for an environmental trigger. *Proc. Nat. Acad. Sci.*, 1966, **55**, 555.

Dingman, W., and M. B. Sporn, Molecular theories of memory. *Science*, 1964, **144**, 26.

Fjerdingstad, E. J., T. Nissen, and H. H. Roigaard-Petersen, Effect of ribonucleic acid (RNA) extracted from the brain of trained animals on learning in rats. *Scand. J. Psychol.*, 1965, **6**, 1.

Flexner, L. B., J. B. Flexner, G. de la Haba, and R. B. Roberts, Loss of memory as related to inhibition of cerebral protein synthesis. *J. Neurochem.*, 1965, **12**, 535.

Gaito, J., A biochemical approach to learning and memory. *Psychol. Rev.*, 1961, **68**, 288.

Gaito, J., *Molecular Psychobiology: A Chemical Approach to Learning and Other Behavior*. Springfield, Ill.: Charles C. Thomas, Publisher, 1966.

Gaito, J., J. Mottin, E. Schaeffer, J. Davison, and J. Rigler, Effects of avoidance conditioning on brain neurochemistry. *Techn. Rep. MPL* 4, York University, 1966.

Gerard, R. W., What is memory? *Sci. Amer.*, 1953, **189**, 118.

Gillespie, D., and S. Spiegelman, A quantitative assay for DNA-RNA hybrids with DNA immobilized on a membrane. *J. Mol. Biol.*, 1965, **12**, 829.

Gross, C. G., and F. M. Carey, Transfer of learned response by RNA injection: failure of attempts to replicate. *Science*, 1965, **150**, 1749.

Halstead, W. C., Brain and intelligence, *in* L. A. Jeffress, ed., *Cerebral Mechanisms in Behavior*. New York: John Wiley & Sons, Inc., 1951.

Hebb, D. O., *The Organization of Behavior*. New York: John Wiley & Sons, Inc., 1949.

Hyden, H., Biochemical changes in glial cells and nerve cells at varying activity, *in* F. Brucke, ed., *Proc. 4th Intern. Congr. Biochem.*, III: *Biochemistry of the Central Nervous System*. London: Pergamon Press, 1959.

Hyden, H., Satellite cells in the nervous system. *Sci. Amer.*, 1961, **205**, 62.

Hyden, H., and E. Egyhazi, Nuclear RNA changes of nerve cells during a learning experiment in rats. *Proc. Nat. Acad. Sci.*, 1962, **48**, 1366.

Hyden, H., and P. W. Lange, A differentiation in RNA response in neurons early and late during learning. *Proc. Nat. Acad. Sci.*, 1965, **53**, 946.

Hyden, H., and P. W. Lange, A genic stimulation with production of adenine-uracil rich RNA in neurons and glia in learning. The question of transfer of RNA from glia to neurons. *Die Naturwiss.*, 1966, **53**, 64.

Jacobson, A. L., F. R. Babich, S. Bubash, and C. Goren, Maze preference in naive rats produced by injection of ribonucleic acid from trained rats. *Psychon. Sci.*, 1966, 4, 3.

Katz, J. J., and W. C. Halstead, Protein organization and mental function. *Comp. Psychol. Monogr.*, 1950, **20**, 103.

Landauer, T. K., Two hypotheses concerning the biochemical basis of memory. *Psychol. Rev.*, 1964, **71**, 167.

Landauer, T. K., and L. Eldridge, Failure of actinomycin-D to inhibit passive avoidance learning: a confirmation. Unpublished paper, 1966.

Luttges, M., T. Johnson, C. Buck, J. Holland, and J. McGaugh, An examination of "transfer of learning" by nucleic acid. *Science*, 1966, **151**, 834.

McConnell, J. V., RNA and memory, *in Symposium on the Role of Macromolecules in Complex Behavior*, Kansas State University Publication, 1964.

Meyerson, F. Z., R. I. Kruglikov, and I. A. Kolomeitseva, The role of the synthesis of nucleic acids in the mechanism of stabilization of conditioned reflexes and memory. *Bull. Exp. Biol. Med.*, 1965, **12**, 6.

Pevzner, L. Z., Nucleic acid changes during behavioral events, *in* J. Gaito, ed., *Macromolecules and Behavior*. New York: Appleton-Century-Crofts, Inc., 1966.

Plotnikoff, N., Magnesium pemoline: enhancement of learning and memory of a conditioned avoidance response. *Science*, 1966, **151**, 703.

Pribram, K. H., Some dimensions of remembering: steps toward a neuropsychological model of memory, *in* J. Gaito, ed., *Macromolecules and Behavior*. New York: Appleton-Century-Crofts, Inc., 1966.

Chapter 5

A Learning Model

M. Ray Denny

Department of Psychology
Michigan State University
East Lansing, Michigan

I am not going to talk about several different models, although others will be referred to as I discuss the model I am most interested in. This particular model has been called *elicitation theory*, and it originated at Michigan State University in the early 1950's. At that time it was an attempt to integrate the neobehavioristic theories of Hull, Guthrie, Skinner, and Tolman, plus the older ideas of Pavlov (Denny, 1966; Denny and Adelman, 1955).

The position represents a conglomerate of the main points that each of these theorists was promulgating with a few new twists added. Hopefully, it is a true integration in trying to get all the parts together that seem to fit the data. In essence it is a monistic position—a unitary conception of learning, as opposed to what others suggest, namely, that there may be several kinds of learning. In a sense, it will come out that there is probably more than one kind of learning but not in any fundamental sense. Rather, all learning will be viewed in a contiguity framework in which classical conditioning principles are exploited.

The position states that learning depends upon the consistent elicitation of the response that is being learned. In lower animals, at least, this means that learning involves the presentation of unconditioned stimuli which immediately force out, or elicit, a response. Now in operant, or instrumental, learning situations several responses are being learned. As a result, the main unconditioned response may be hidden. Because of this, two-factor approaches to learning are common, i.e., instrumental, or operant, learning is distinguished from classical, or respondent, conditioning. I am going to try to show that something like the classical conditioning notion of an unconditioned response is present in all learning and that such response accounts for all learning, at least all infrahuman learning. When one speaks about learning at the human level, the data nicely fit the position but it is usually not necessary to invoke classical conditioning principles, as such.

Before we get any deeper into the theory, I should like to talk about a definition of learning that most psychologists would probably accept; then I should like to elaborate on this in terms of the concepts of stimulus and response. Learning, as you know, is frequently defined as the modification of behavior through experience, or through responding to a stimulus situation. This is a broad definition, and it is not very clear what kinds of modifications are involved. I would like to explicate these modifications, but first would like to talk a bit about stimulus and response.

In the elicitation framework, stimulus is defined as anything which, via the afferent nervous system, is potentially capable of eliciting a response in a particular class of organisms. In other words, we know that such and such is a stimulus for a particular type of organism when the object or event in question has reliably elicited a particular response some time in the life history of the organism or some representative of this class of organisms. There are at least two points being emphasized here: (1) The afferent nervous system is clearly involved; what I am asserting is that direct innervation to the muscle or motor nerve is not the same thing as the concept of stimulus, at least as the psychologist uses it. When the definition of stimulus always involves the afferent nervous system, the stimulus-response relationships, or laws, can be quite different from the relationships one observes when the concept of stimulus includes elicitors which are applied directly to the muscle or motor nerve, as physiologists and psycho-physiologists are now discovering. (2) Defining stimulus in terms of response is important because psychologists do mean, e.g., every time they describe the procedure of an experiment, that a stimulus *can* elicit a response. This, however, does not mean that the definition is circular. In the stimulus-response law in question, stimulus is not defined in terms of the response of the law, or the dependent variable of the experiment, but in terms of some other response, very possibly some other response that was elicited in other members of the same class of organisms many years before. And, in the case of kinesthetic stimuli, the notion of stimulus is typically inferred from response occurrences which are both antecedent and consequent to the event in question.

In much the same way, response class, or response as an abstract concept, is inferred from some aspect of the stimulus situation; but again, as described in the argument above, no circularity need be involved. The presence of the word *bar* in the response class, *bar-pressing rate*, is a convenient example of the use of stimuluslike concepts to define response class. In the present framework, overt or at least measurable response occurrences are necessary to infer a particular response class. This does not mean, however, that the response class itself must represent overt behavior. Perception, thinking, feelings, etc., are all considered to be legitimate response classes which are inferred from verbal report or whatever observable activity is available. The present definition of response is about as broad as possible.

The present definitions of stimulus and response are recommended because obviously these are the concepts that enter into stimulus-response laws and because laws relate objectively defined abstract concepts to each other. Thus rigorous, but abstract, definitions of stimulus and response are necessary if we are going to have good, empirical laws of behavior.

Given these definitions of stimulus and response and the assertion that learning involves the modification of behavior, I want to emphasize that the term modification implies that learning does not bring about behavior *de novo*. There is no manufacture or creation of new responses. Learning simply involves changing the behaviors which are present initially, or through maturation. Such a view is obvious to many biologists and such a view, I think, cuts through a lot of pseudo problems that have been subsumed under the label of nature versus nurture. The work of the ethologists, the work on imprinting, etc., has highlighted the relevance of this sort of position. Psychologists when they have defined learning as a modification of behavior have implied this all along, but many seem to have forgotten this implication. Strictly speaking, learning can only affect behaviors that are already in the organism at some stage or other; the raw material must be there to begin with.

I have listed in Table 1 four kinds of changes that fall under the rubric of learning. To me, this is an exhaustive list, but the "categories" are by no means mutually exclusive and should not be construed as representing special *kinds* of learning. Item 1 is simply the establishment of a new response tendency; i.e., the organism does not have the tendency to make the response in the presence of said stimulus, but through learning, through conditions of contiguity, this tendency is established. Included here is the possibility that the response tendency may be present initially at some weak level and is then strengthened markedly. In either case, this type of change is best exemplified by classical, or respondent, conditioning.

Item 2 represents a broad look at response differentiation, including trial and error learning and operant conditioning. Even simple bar pressing by a rat is an example in which initially there are many responses occurring to a stimulus situation. Then through learning, a response lower down in the hierarchy is differentiated from the other responses and becomes the main response to occur. This, of course, is a simple restatement of selective learning. I want to stress, however, that training a rat to press a bar in a Skinner box does involve an element of response differentiation. The

TABLE 1
The S-R* Changes Involved in Learning

1. The establishment of a new S-R
2. The specification of a particular R to S
3. The organization of R's into particular sequences (chaining)
4. The specification of R to a particular aspect of S or to a specific stimulus element

* S-R, stimulus-response.

Skinnerians have usually reserved the term response differentiation for training a pigeon, let us say, to turn 360° rather than to turn part way, i.e., to reinforce or specify a nice, discrete response. To me, this is just an extension of the notion that most learning simply involves the specification of a particular response. Sometimes it is necessary to go through all kinds of tricks in order to get out the response that is to be specified. Shaping and fading are examples of such techniques.

Type 3 is also a very common type of behavior change that is typically involved to some extent in even the simplest example of response differentiation. For example, chaining or sequencing of responses is clearly present when a rat learns to press the bar in a Skinner box. The rat learns to approach the bar, address the bar, press down on the bar, leave the bar, and finally go to the food tray or water dipper, as the case may be. All changes in the organization of responses including complex language habits at the human level, may be subsumed under this type of stimulus-response modification.

Type 4 represents the type of stimulus-response change that is present to some extent in any learning situation. The response being learned is always specified for a particular stimulus, even if it is practically the whole stimulus situation. When a rat learns a chain, even such a simple chain as learning to press a bar in the Skinner box, there are stimuli at each point along the chain which specify the next response. Many of these are response-produced stimuli, kinesthetic, auditory, olfactory, visual, or tactual and all of them compose the discriminative stimulus for the next response in the chain. But, most specifically, this category is best exemplified by conventional discrimination learning experiments.

At this juncture, let us take chaining in the Skinner box as an example of the application of the theory. What I am saying about learning is that the response must be made consistently in order for that response to win out over other possible responses. Or, conversely, a good deal of what is involved in learning is the exclusion or elimination of competing responses. Guthrie made the same kind of point, but he did not exploit incentives or reinforcers as the important elicitors in the learning situation, elicitors which produce the response you want over and over again and tend to eliminate competing or alternative responses.

When a rat is learning to press the bar in a Skinner box, the first thing the rat learns, if you watch the rat closely in the early or preliminary stages, is to make a clear-cut approach response to the food tray or dipper. With successive reinforcements, this response tendency gets stronger. What I am pointing out is that the characteristic unconditioned response that is involved in instrumental or operant learning is approach, approach to the place where the food (water) is.

Focusing on the unconditioned response of approach rather than unconditioned eating or any other terminal member of the chain is the critical element in elicitation theory. Much of the kind of "backchaining" that we are going to be talking about in the typical appetitive or positive learning situation was described much earlier by Hull, but Hull focused on the consummatory response. Thus fractional consummatory responses (little eating, little swallowing, and little salivating responses) were being conditioned rather than approach, per se. Now, fractional consummatory responses could very well be occurring and getting conditioned; the point I am making is that this probably is not the main thing occurring (it's just sort of frosting on the cake, really). The data on this point are pretty complicated, but at least that is one acceptable way to look at it. When the main unconditioned response is assumed to be approach, then it is possible to show that approach can mediate the rest of the learning: instrumental or operant learning can be derived from a classical conditioning base.

Back to the Skinner box. The rat (pigeon) first learns to approach the food compartment whether or not the experimenter E has filled the tray with food ahead of time. (In preliminary training, E often fills the tray with food, just to get the critical approach part of the learning going quickly.) In any event, the rat, through exploratory or manipulatory behavior, eventually presses the bar and eventually perceives food in the food tray, which it approaches.

The analysis so far gives some indication of the fundamental role of exploratory behavior in operant or instrumental conditioning. Exploratory behavior is also a class of approach response, though a highly inconsistent one because it is elicited by novel stimuli. What is important is that the occurrence of a wide variety of manipulatory-exploratory response (approach) makes possible the learning of a variety of behavioral or operant routes to the goal (where the goal is an unconditioned stimulus for eliciting *consistent* approach).

With the early strengthening of food approach, the rat learns to become a goal sticker, making strong approaches to the food tray without first pressing the bar. Thus there is no food in the tray when it arrives there, and at this point another characteristic response typically gets elicited. The omission of the food reinforcer is itself an unconditioned stimulus which elicits antagonistic responses, i.e., responses directly away from the tray. These comprise aggressive or tangential responses, which typically include pressing the bar as something to bang away at (the bar is an object of aggression). After bar-pressing aggression occurs, the rat sooner or later goes to the tray and finds a pellet of food. What is going on here is a kind of discrimination learning in which the rat is learning to press the bar *before* visiting the food tray. This, in turn, means the rat is learning to inhibit direct approaches to the food tray. Gradually the animal learns this discrimination, approaching via the indirect route of first going to the bar, pressing it, and then going to the food tray.

Operant learning always involves some degree of response differentiation, which is another way of saying that operant or instrumental learning is a type of discrimination learning in which the animal supplies its own discriminative stimulus. The specific discriminative stimuli associated with getting a pellet of food are those stimuli which occur when the bar is pressed. Here, the experimenter usually helps the rat out by throwing in a "click" so that, when the rat presses the bar, there is the distinctive sound of the food magazine. Thus the click is typically one of the main discriminative stimuli for approaching the food tray. But the click is not necessary. A rat will learn with a quiet food mechanism or with clicks going on indiscriminately. The essential ingredient for building a discriminative chain is a set of stimuli which consistently occurs when the animal looks at the bar, addresses the bar, touches the bar, presses the bar, etc. These response-produced stimuli identify the appropriate behavior route to take and include the discriminative stimuli for when to approach the food tray. In sum, the discriminative stimuli involved in a rat's learning to press a bar are the kinesthetic, tactual, auditory, and visual stimuli which accompany bar pressing.

An unpublished study by Harry Hurwitz nicely points out that the rat first learns to approach the food tray and that this learning is essential to subsequent bar-pressing learning. He used a Skinner box in which there were both a lever L and a panel to be pressed; behind the panel was a food trough T. Early in training all animals showed fast learning of trough approach responses T, so that a rat's typical pattern of responses looked something like this: T-T-T-L-T (three nonreinforced visits to the trough before pressing the lever and before being reinforced with the final trough response). Gradually T-T-T-L-T reduced to T-L-T, and finally a reinforced trough response was followed immediately by an L-T pattern. In other words, the rats finally learned the discrimination emphasized by the model, namely, that the only way to find food in the trough is to press the bar first. The long series of T's which initially preceded the L represented a strong approach tendency to the trough, and in Hurwitz' study the animals never learned to make L-T responses exclusively; the typical discrimination for L-T did not exceed 85 to 90 %. Every once in a while the approach tendency to the trough, which, after all, was consistently being reinforced, would manifest itself, and the rat would go to the trough twice in a row.

But the most important aspect of Hurwitz' experiment concerns one group in which finding food behind the panel was contingent upon making *only* L-T responses. That is, if the L of a new series were preceded by a T, then reinforcement was canceled. This contingency obtained from the very beginning of training, and from the Thorndikian law of effect one would predict learning (the L-T contingency satisfies the usual statement of the law of effect). Hurwitz ran the animals interminably, and no animal ever learned to press the lever. A rat would occasionally make only L-T responses, but as soon as this happened the rat was reinforced for going to the trough. Thus the rat next made direct approaches to the trough and was consistently nonreinforced. Approach to the panel extinguished before it could ever get strengthened. According to the present model, failure to learn with the L-T contingency is the expected result, for the approach response must be well conditioned before it can mediate chaining

a. The steps by means of which T-maze learning proceeds

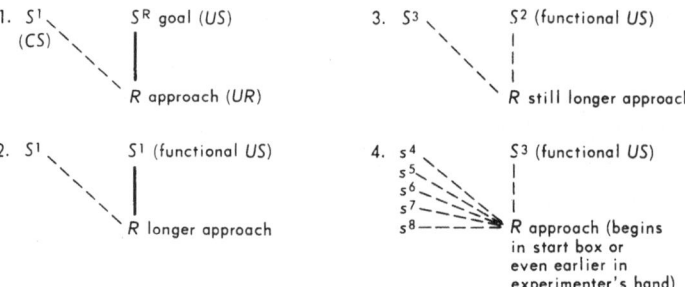

b. Analysis of T-maze

Stimuli, which, to a varying extent, can be conditioned to avoidance. Initially, they are associated with exploratory behavior.

First set of stimuli to be chained or conditioned to approach. Initially associated with exploratory.behavior.

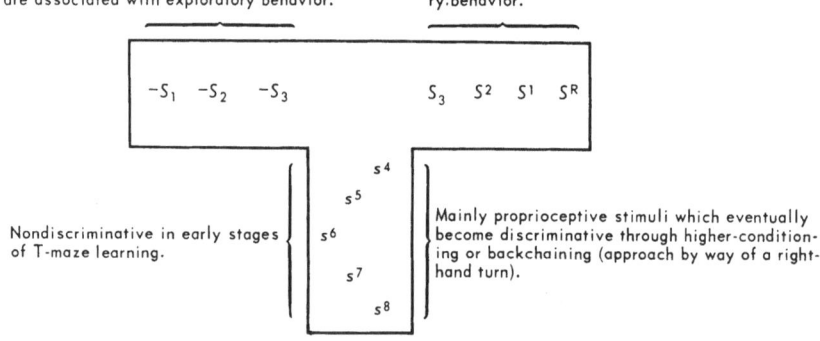

Nondiscriminative in early stages of T-maze learning.

Mainly proprioceptive stimuli which eventually become discriminative through higher-conditioning or backchaining (approach by way of a right-hand turn).

c. The final chain

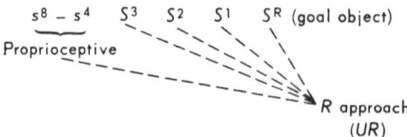

Fig. 1. Chained or conditioned approach analysis of T-maze learning (elicitation theory).

or higher-order conditioning (each stimulus in a chain must become a functional unconditioned stimulus for learning to proceed; see Fig. 1).

All of this, I think, can be more clearly explicated by looking at a rat learning to turn to the right in a T maze. Let us assume a small piece of familiar food is at the end of the right-hand alley, no food is on the left, and the rat is hungry. Figure 1 presents a symbolic representation of the conditioning that presumably goes on in such a T maze. The conditioning begins with step 1 (Fig. 1a) and proceeds as presented. One thing that is immediately apparent is that T-maze learning is not really very simple; it is often called simple, but it is really a complex sequence of learned responses. For any lower animal, T-maze learning is probably more complex than instructing a college student in learning a list of German words and their English equivalents. In the latter case the responses to be learned are directly elicited in close temporal contiguity with the stimuli: the sight of the English word elicits saying the English word in close temporal association with the German word. But in the maze all responses presumably have to be built up gradually in an indirect, backward fashion.

The omnipresent unconditioned response of approach to food— i.e., before the animal can eat the food, it has to approach it—is represented by a solid line in Fig. 1, the conditioned associations by dashed lines. In passing, it should be reiterated that escape and withdrawal are also important classes of unconditioned response in instrumental approach learning. This was implied when it was stressed that the rat learns to inhibit direct approaches to the food tray during bar-press learning. The avoidance aspect of T-maze learning is inadequately covered in Fig. 1.

Another point that should be emphasized is that the unconditioned response UR of the present model involves the whole organism's moving toward or away from the unconditioned stimulus US. The unconditioned response in question is not a segmental response such as an eye blink or knee jerk. It is extremely difficult, if not impossible, to condition single, segmental responses such as the knee jerk or pupillary response. For example, Francis Young (1958) in a series of well-controlled experiments has shown that the simple pupillary response, contraction to a bright light, is not conditionable. When, however, the pupillary response is part and parcel of a large response complex, then it is readily conditionable (the unconditioned stimulus is shock which elicits pupillary dilation plus many other response components of fear; Gerall, Sampson, and Boslov, 1957). This digression is relevant because the conditioning of large response classes such as approach and withdrawal should proceed quickly and effectively, as required by the data of instrumental learning.

Back to Fig. 1. S^1, as a conditioned stimulus, represents that part of the goal box which immediately precedes the region where the food is presented and is close enough in time to be conditioned to the approach response which is unconditionally elicited by S^R, the reinforcing stimulus. Thus the first response learned in the T maze—which you readily notice if you watch the animal—is to approach the food tray as soon as entering the goal box. This is the dashed-line connection between S^1 and R_{approach} in step 1 of Fig. 1a. There are even data (Denny and Martindale, 1956) to indicate that approach is conditioned even if the food is not eaten immediately. It is only critical that the animal make a definite approach response to the food tray; with approach there is an increase in the probability of its selecting the reinforced side on the succeeding trial, regardless of when the food was eaten. Such a finding or interpretation obviously calls drive reduction theory into question. But this does not mean that a broad interpretation of the law of effect has been left out of the model. The incentive is critical for consistently eliciting the approach response. The present model is a reinterpretation of the Thorndikian or Hullian analysis.

As soon as S^1 has acquired good eliciting value, it then serves as a functional US, eliciting approach in close temporal contiguity with S^2. Next, S^2 becomes a functional US. Since S^2 closely follows S^3 in time, S^3 acquires conditioned approach value and becomes a functional US. At this point, the analysis is much the same as described by Tolman years ago: in the early stages of T-maze learning the rat learns to go to a particular place. In the present analysis, the rat learns to approach the cues on the side of the maze where the food is. Thus, in the very early stages, when approach is only conditional to S^1 or S^2, the rat makes a lot of errors. As learning continues, as S^3 becomes a cue, the rat looks first one way and then the other while at the point of choice. This behavior Tolman called *vicarious trial and error* (VTE); with continued trials VTE diminishes.

Up to this point in T-maze learning (and you see this very nicely in its behavior) the rat does not zoom to the right as it leaves the stem of the T. It simply comes up to the choice point (Fig. 1b) and stops and looks around before selecting an alley. But the chain continues to develop, involving response-produced kinesthetic stimuli which are represented in Fig. 1 by s. So, after S^3, s^4 becomes a functional US and then s^5, etc., all the way back to wherever stimuli remain discriminative. Thus a typical well-trained rat starts to turn right as soon as it leaves the start box. At this point in learning, it becomes legitimate to label the response being learned a "right-turn" response. Prior to this stage the rat was learning to approach cues on the right side of the maze; this is different from the response class of *turning* right (in the model appropriate stimulus concepts are used in defining a response class).

In the early stages of learning, none of the stimuli involved in traversing the stem is discriminative; these stimuli are not closely associated in time with consistent approach to food, and the animal goes roughly 50% of the time to the right (reinforcement) and 50% of the time to the left (nonreinforcement). Not until the chain is built back to s^4 do stem stimuli start to become discriminative, specifying a right turn before the rat has even arrived at the choice point (note the centrifugal swing of s^8 to s^4 in Fig. 1). In the Skinner box the rat finally learns to approach the food tray by way of a bar press. In the T maze the rat finally learns to approach the food by way of a right (left) turn.

To my way of thinking this sort of analysis of instrumental learning subsumes the learning of "observing responses" as described and studied by Wyckoff (1952). Such an analysis is quite complicated, but the gist of it is that the external discriminative stimuli produced by an observing response are not that much different from the internal response-produced discriminative stimuli of a partially reinforced, though readily learned, operant. The observing response and the terminal operant form a chain even though the observing response only produces a positive cue 50 % of the time. Such a cue is not really irrelevant or inoperative unless it is also correlated with nonreinforcement 50 % of the time, which is not the case in the observing-response paradigm. Thus conditioning or backchaining takes place with respect to the positive cue whenever it is produced by an observing response (the learning of an observing response is no more paradoxical than instrumental learning via partial reinforcement).

So far we have referred to operant or exploratory behavior as supplementary to the unconditioned responding (food-elicited approach) that typically mediates instrumental learning. But there is not necessarily any basic difference between behaviors elicited by food and those elicited by novel stimuli. If the conditions are appropriately arranged, stimuli to be explored or manipulated can also assume the function of an unconditioned stimulus and mediate instrumental conditioning. I am referring especially to a series of studies performed at Michigan State University, in which we exploited the eliciting value of stimulus satiation or boredom. In these experiments we forced hungry rats to go to the same side of an enclosed T maze, say 10 times in a row, and on each trial they found food at the end of the alley. Forcing meant that the door was closed to the other alley. At the end of the forced trials the animals were given a series of test trials in which they were free to go either way. If the test trial came immediately after the forced trials, virtually 100 % went in the opposite direction. A day later or 16 days later, the rats still went in the opposite direction 75 % of the time. And they continued to go opposite for as many as 10 test trials, so long as the test trials came one per day. According to the model, the rat behaves this way because it is motivated by many considerations (responds to many different stimuli), including exploration of the maze. If the rat is forced to explore one side over and over again, the stimuli on this side undergo satiation (habituation) and the animal attempts to avoid or escape these stimuli and approach those of the other side (those of enhanced approach value).

This satiation-mediated habit is even more pronounced when the forced trials to one side are given one per day rather than in a massed fashion. The preference for the opposite side can be as high as 90 to 95 % for the first two test trials (Denny and Leckart, 1965). Again extinction is slow when the trials are given one per day (all visits to the opposite side are nonreinforced and all visits to the original, forced side continued to be reinforced). The main point I am making is that stimulus satiation is capable of producing almost as good T-maze learning as food reinforcement and that such a finding does not readily fit traditional reinforcement theorizing.

A closely related point in working with lower organisms is that the tendency to explore and manipulate the environment, so that instrumental learning can take place, increases as you go up the phyletic scale. This seems to be important in considering the application of the present model to research with lower organisms. I think however, that the model does apply to many lower organisms. For example the model readily subsumes imprinting as a learned phenomenon. During the critical period certain stimuli elicit or release imprinting responses in association with other stimulus aspects of the situation. Thus the latter (conditioned) stimuli determine the particular object or animal to be imprinted. Thus the model handles a type of learning where traditional reinforcement notions do not readily apply; the consistent elicitation of the imprinting response is all that is critical. There are three or four experiments in the literature that have shown that animals will learn an instrumental act in order to be presented with the imprinting stimulus (approach elicitor). The parallel with the present analysis of food-reinforced instrumental learning is apparent.

When the model describes S^1 as the functional unconditioned stimulus which mediates the conditioning of approach to S^2, this is the same as saying, in traditional terminology, that S^1 is a secondary reinforcing stimulus. And the chain of Fig. 1c could be described as secondary reinforcement chain. The main advantage of the present model is that it seems to cover all possible learning situations in a simple straightforward manner. Contiguity is a basic notion, but so is reinforcement as a consistent elicitor; the part that Guthrie left out has been put back in.

REFERENCES

Denny, M. R., A theoretical analysis and its application to training the mentally retarded, *in* N. R. Ellis, ed., *International Review of Research in Mental Retardation, Vol. 2.* New York: Academic Press, 1966.

Denny, M. R., and H. M. Adelman, Elicitation theory: I. An analysis of two typical learning situations. *Psychol. Rev.*, 1955, **62**, 290.

Denny, M. R., and B. T. Leckart, Alternation behavior: learning and extinction one trial per day. *J. Comp. Physiol. Psychol.*, 1965, **60**, 229.

Denny, M. R., and R. L. Martindale, The effects of the initial reinforcement on response tendency. *J. Exp. Psychol.*, 1956, **52**, 95.

Gerall, A. A., P. B. Sampson, and G. L. Boslov, Classical conditioning of human pupillary dilation. *J. Exp. Psychol.*, 1957, **54**, 467.

Wyckoff, L. B., Jr., The role of observing responses in discrimination learning, Part I. *Psychol. Rev.*, 1952, **59**, 431.

Young, F. A., Studies of pupillary conditioning. *J. Exp. Psychol.*, 1958, **55**, 97.

Chapter 6

Polythetic Operationism and the Phylogeny of Learning[1]

Donald D. Jensen

Department of Psychology
Indiana University
Bloomington, Indiana

During the last two decades, learning has been reported in several simpler invertebrates, and these findings have been criticized and attacked. Because this controversy has been both heated and prolonged, one can suspect that more is involved than differences in the facts available to different investigators. Instead, unexpressed differences in assumptions, definitions, and conceptual modes may be the basis of the controversy and confusion. Because of this possibility, it seems worthwhile to turn from attempting to answer the question, "Do simpler invertebrates learn?" and to concern ourselves with the principles involved in giving precise and scientific meaning to these and related questions.

The principle of *operational definition* is clearly relevant here. Operationism is the point of view associated with the writings of the physicist, P. W. Bridgman (1927); it is the view that the scientific meaning of a concept lies in the operations and observations necessary to determine whether the concept applies, that the scientific meaning of a phenomenon lies in the operations and observations necessary to exemplify it reliably, and that the scientific meaning of a question lies in the operations and observations necessary to answer it.

Operationism has already entered into the discussion of experiments on learning with simpler invertebrates. The failure of a number of workers (i.e., James and Halas, 1964, and Bennett and Calvin, 1964) to replicate the findings of earlier workers is related to the principle (Underwood, 1957, p. 82) that unreliable events or changes are not scientific phenomena in the operational sense. In addition, descriptions of experimental procedures and results which would demonstrate associative learning in planaria (Special Discussion, *Animal Behaviour, Suppl.*, 1965, **1**, pp. 112–114) constitute operational definitions of learning.

[1]Prepared while a NIMH Research Training Fellow at the Anatomical Institute, University of Oslo, Norway, under USPH Special Fellowship 1-F3-MH-7,346-01 (MTLH).

43

Operational analysis of scientific concepts is valuable because it prevents argument about, and research with, scientifically meaningless concepts. It is also desirable because of the refinements in experimental design and conceptualization which may result from it. For example, an earlier analysis of the operations implicit in the conclusion that a behavior has been learned (Jensen, 1961) suggested the use in classical conditioning experiments of a control group receiving training with conditioned and unconditioned stimuli unpaired before the use of this control group had become common. This analysis also indicated, as has Denny at this symposium, that it is inaccurate to speak of behaviors being learned, as is normally done. Considering the operations actually performed, differences in the probability of a behavior may be said to result from learning, or stimulus control of behavior may be said to be learned, but the responses themselves cannot be said to be learned because they typically occur prior to training as unconditioned responses or operant behaviors. It is likely that additional refinements of design and terminology will result from further operational analysis.

In summary, operationism aids the process of giving scientific meaning to our questions by leading us to specify our operations, to demand reliable effects of these operations, and to refine the operations by which we produce effects and the terminology which we use to discuss them. While operationism is of great utility, it has certain limitations. As long as the procedures by which manipulations are made, observations taken, and conclusions drawn are all specified, the requirements of operationism are met. These requirements ensure only that the concepts involved have *some* scientific meaning, but they do not ensure that those concepts have either maximal or even substantial meaning. Operationism provides no criterion for differentiating the important from the unimportant; it provides no protection against triviality. To aid in recognizing the important, to help us avoid the trivial, additional criteria must be sought. Such criteria appear to be available in a more recent methodological development, the principle of *polythetic definition.*

Polytheticism is the point of view associated with recent innovations in biological classification (i.e., in the dividing of the diversity of organisms into species, genera, and higher taxa). This point of view, which has been recently reviewed by Sokal and Sneath in a book entitled *Principles of Numerical Taxonomy* (1963) and by Sokal in *Scientific American* (1966), constitutes an approach which differs markedly from early attempts at biological classification. The early attempts were based upon Aristotelian principles and were a search for the *essence* of a taxonomic group. One or a few *key characters* were sought which expressed these essences, and membership in a group was determined by the presence of these key characters (i.e., a chordate has a notochord, an annelid has setae, a sponge has choanocytes, etc.) In contrast, the polythetic point of view holds that taxonomic concepts are not effectively defined by one or a few key characters even though those key characters may in fact serve to identify members of different groups of animals. According to this newer view, taxonomic classification divides organisms into *natural groups* on the basis of overall similarity, and membership in such natural groups is correlated with many characters, no one of which is *a priori* more important than any other.

Sneath (1962) suggested that the older approach, with its reliance upon one or a few characters, be termed *monothetic*, in contrast with the newer approach which considers many correlated characters and is termed *polythetic*. One example contrasting the monothetic and polythetic approaches is the monothetic definition of man as "the featherless biped which talks." It is possible to criticize this definition by pointing out that it would include a plucked, trained parrot in the ranks of man, but this is not the only criticism that the polythetic taxonomist would make. He would also point out that a one-legged deaf-mute who has feathers instead of hair is still a man, even though he has none of the three characters in the definition. He is a man because he has hundreds of other characters (the shape of the nose, the number and kind of teeth, the human pelvis, ABO blood group, etc.) which are typically found in other men. From the polythetic point of view, no one or no small number of characters are ever either necessary or sufficient to define a class or taxon. The entire set of characters which are typical of members of the taxon and not of members of other equivalent taxa defines the class or taxon. Polythetic and monothetic definitions differ in more than the number of characters which are used in determining classification; they differ as well in the logical relationship between characters and group membership. In monothetic classes more than one character may be relevant but all these characters *must* be present in a member of the class. In polythetic classes, group membership is determined by overall similarity, i.e., by the presence of a sizable number of the many characters typical of the group.

The basic premise of polytheticism is not new. As early as 1763, Adanson "correctly realized that natural taxa are based on the concept of 'affinity' . . . which is measured by taking all characters into consideration . . . and that taxa are separated from each other by means of correlated features" (Sokal and Sneath, 1963, p. 16). Montgomery (1906) wrote, "(Distribution), habits, and structure should always be considered conjointly, and no one of them can be adequately comprehended by itself . . .. An organism must be interpreted from its many aspects." Wittgenstein's principle of "family resemblance" is similar, and Woodger (1952, p. 23) made a similar point when he wrote, "For in natural science, unlike mathematics, we do not deal with sets in the abstract, but only with certain empirically specified sets" and pointed out that, while property words (key characters) are used to specify sets, one must talk about the sets and not the properties used to specify them. Mayr (1965) succinctly expressed the polythetic point of view by the axiom, "One can never trust a single char-.acter."

While polythetic concepts are obviously not new in science, what *is* new is the availability of sophisticated statistical techniques and electronic computational devices which make practicable the use of polythetic concepts, and much of the discussion of numerical taxonomy concerns the practical problems of exploiting the polythetic point of view with multivariate methods (cluster analysis, etc.). Our concern is not with these matters but with the conceptual advantages accruing from polythetic thinking. What then are the advantages of polythetic concepts in scientific endeavors?

Sokal and Sneath (1963, p. 15) write, "the advantages of polythetic groups are that they are 'natural,' have a high content of information, and are useful for many purposes." Because of this, natural classes have the advantage of effective extrapolation or generalization. Anything which is true of one member of such a class is probably, though not necessarily, true of all other members of that natural class. In contrast, membership in a monothetically derived or "arbitrary" class may provide information only about the character defining the class; membership provides assurance only about a few characteristics and thus expresses little information.

It seems likely that the advantages of polythetic classes are also those of all important concepts and variables and theories; the important in science differs from the trivial by containing more information and therefore being of greater general utility. It would appear that choices among alternative concepts, variables, and theories are made upon the criterion of maximal information and that scientists have acted according to the principle of maximal information, even when the principle has not been verbalized. The independent variables (i.e., the single comparisons between experimental and control groups) which have the greatest information value, which have the greatest correlation with dependent variables, are preferred, and the dependent variables which contain the greatest information, which are correlated with the greatest number of possible dependent variables, are preferred. The theories which are preferred are those which express or abbreviate the relations between many independent variables and many dependent variables and therefore abbreviate or express a great deal of information. Just as taxonomists select polythetic groups because they have maximal information value, so other scientists, working with other problems, choose concepts, phenomena, and theories with maximal information value.

It is possible to relate polytheticism and operationism on the basis of their influence upon the information value of scientific concepts. Operationism ensures that scientific concepts have some information value by requiring definition in terms of performable operations and possible observations; polytheticism encourages the development of taxonomic concepts of high information value by allowing definition on the basis of overall mutual similarity.

The use of polythetic concepts and natural classes has not been restricted to taxonomic investigations but has also occurred in the study of morphology. The concept of homology, which antedated evolutionary theory, was a recognition that the "same" structure could be found repeated in one animal (serial homology) or in different animals (special homology) (Boyden, 1947). Homology can be contrasted with analogy; analogous organs have similar functions but not the mass of similar features shared by homologous organs.

Let us take a simple example: A human arm, a chimpanzee's arm, and a monkey's arm all have many characters in common; the arm of an octopus obviously has only a very few features in common with the other arms. If we define *arm* monothetically as any prehensile appendage, we shall classify all four arms together, but shall abbreviate very little information. If we define the concept *arm* polythetically, we shall compare the four arms on a large number of characters, or features, and discover that three are very similar and that one is different from the rest, and we shall therefore include only the arms of man, ape, and monkey in our scientific concept, but shall thereby abbreviate much more information.

Homologous organs form natural or polythetic classes; analogous organs do not. The arms of men, apes, and monkeys are homologous with and share an immense number of attributes with the wings of birds and bats; all of these structures form a single natural group, the vertebrate forelimb. In contrast, the wings of insects and the wings of birds, even though they share certain specific characteristics appertaining chiefly to their function as aerodynamic surfaces, are *not* homologous, but analogous. They do not form a natural group, and the biological concept of wing, if it includes those of birds and those of bees, has little information value. An immense amount of investigation in biology has involved the search for homologies, for natural classes which express considerable information and within which generalization or extrapolation, from one particular organ in one particular organism to other organs and other organisms, is likely to be effective.

The concepts of both homology and affinity involve membership in natural classes and general similarity; the concepts differ in the units which are classified and compared. Homology expresses the general similarity of organs, and affinity expresses the general similarity of organisms. Both concepts appear to be dimensional rather than dichotomous; i.e., they seem to be matters of degree. Man and apes are more closely related and show greater affinity than do man and rodents; all mammals are more closely related and have closer affinities than does any mammal with any bird, but birds, reptiles, and mammals are all more closely related and show greater affinities than do any of these to teleost fishes, etc. Similarly organs may show degrees of general similarity; the arm of man resembles greatly the arm of other higher primates, resembles somewhat the front limb of an early reptile, resembles less the front fin of a coelocanth fish, resembles still less the front fin of an early sharklike fish, and resembles hardly at all one of the arms of an octopus. While all the vertebrate forelimbs are homologous and the octopus arm is analogous to them, the degree of general similarity varies; therefore it may be best in some contexts to discuss, not *homology* with its implied dichotomy with *analogy*, but *homolity*, which refers to the relative general similarity. High homolity would refer to very exact and nearly complete similarity between organs; moderate homolity would refer to moderate similarity, with many but not necessarily even the majority of characters in common; very, very low homolity with similarity of function would be equivalent to *analogy*.

The concepts of homology and analogy have been applied to behaviors as well as to organs. Behaviors are not spatial subdivisions of organisms but are subdivisions in time and space of the larger set of characters which polythetically define an organism. They are groups of characters which occur close together in time and space within the sequence of changing characters of an organism. Medewar's phrase is "natural behavior structures or episodes." Human walking involves a complex spatio-temporal pattern—left arm and right leg forward, etc. This behavior is clearly homologous to the quadripedal walking of cats, which involves similar crossed-extensor movement patterns, similar spinal mechanisms, etc. The running of a lizard and the swimming of a shark are also homologous, though less completely similar and therefore of lower homolity. All of these behaviors involve similar patterns of torso flexion, action of reciprocally innervated muscles, equivalent neurological mechanisms, etc. In contrast, the locomotion of a planarian or paramecium would appear to be analogous rather than homologous to mammalian locomotion.

The concept of homology has been implicit in the study of behavior of laboratory mammals and subsequent extrapolation to the behavior of other animals, including man. The information available upon digestive behaviors of rats can usually be applied to the digestive behaviors of other mammals; the information relevant to the breathing of rabbits can usually be applied to the breathing of other mammals; a specific behavior in one mammal typically resembles that behavior in another mammal in motoric organization, internal mechanisms, and environmental controls. Much of the work of ethologists can be understood as the search for behavioral homologies, as among the various courtship ceremonies of ducks (Lorenz, 1958), the different displays of cichlid fishes (Baerends and Baerends, 1950), and the nest building behaviors of species of lovebirds. (Dilger, 1962). The differentiation between homology and analogy is frequently applied to behaviors by ethologists.

We have so far applied the principles of polythetic definition, of natural classes, to organisms and to parts of organisms, either spatial aspects (organs) or complex temporal-spatial aspects (behaviors). The same principles can also be applied to influences upon behavior; one can group events which influence behavior into natural classes, i.e., stimulation, adaptation, learning, maturation, and inheritance. According to the principles of polythetic definition, these influence processes would be expected to differ in many aspects, not simply in a particular key character or aspect. These natural classes can in turn be subdivided into natural classes. One can ask: what are the polythetic, natural classes of instances of learning? This, to be sure, is not how we have usually dealt with this topic. Typically learning has been treated monothetically, and any modification of behavior has been termed learning; it is treated monothetically when any single aspect or characteristic defines the concept. The reliance of McConnell (1965) upon "a chemical change," the emphasis of James and Halas (1964) upon resistance to extinction, my own emphasis upon the importance of conditioned- and unconditioned-stimulus association (Jensen, 1965), all are examples of reliance upon key characters and therefore of monothetic thinking.

The possibility of a polythetic approach to the study of learning was hinted at by certain comments in the 1964 symposium. Pantin (1965, p. 3) pointed out, "we have no guarantee that the machinery of learning, the adaptive modification of behavior through past experience, always belongs to a single class"; in the special discussion at that symposium, I pointed out that the problem of identifying learning was a problem of distinguishing between homology and analogy in behavioral processes; the dictum given to look at all the data from an experiment implied that no single aspect of the data is to be trusted. It is time that we did not hint about the possibility of polythetic study of behavioral processes but shouted out its necessity. Only by discovering the natural classes of behavioral modification are we likely to be able to guide extrapolation from experimental results on any one example and restrict generalization from particular experiments appropriately.

What does one need to determine the polythetic classes of behavior modification? One needs the same information that is necessary for the determination of homology or organs: several examples and much information (many phenomena and many characters) for each example. We *have* a sizable number of examples of behavioral modification, ranging from trigram learning in the human and operant conditioning of the rat to the work that is the subject of this symposium. We do not have, however, a great deal of reliable information about many aspects of any given example, and, when it is available, it is not typically available in any summary or review form. The reviews of the literature on learning experiments with simpler invertebrates (i.e., McConnell, 1966, and Jacobson, 1963 and 1965) have typically been monothetic in approach and do not provide the kind of information necessary for polythetic analysis. Original reports are frequently no more helpful, since most of the research conducted by comparative and experimental psychologists with simpler invertebrates has involved recording of only one or very few behavioral characteristics and manipulating relatively few independent variables.

It must be recognized that the polythetic method requires a considerable number of characters; unless a case of behavioral influence can be described by much more than the key character of the effect of one independent variable (i.e., training or no training) upon one dependent variable (i.e., probability of response), the polythetic method can not be applied. This fact has enormous consequences for the evaluation of past research on learning and for the planning of future research. It suggests that we must increase the amount and variety of information which we gather about behavioral modification, though we must do this without interfering with the interpretability of results which comes from experimental controls and standard experimental designs.

There are several techniques for increasing the information we obtain through our experimental studies of behavioral modification. One technique, familiar to both psychologists and biologists, is the factorial experiment in which several independent variables are simultaneously manipulated. Another technique is the addition of physiological independent variables (lesions, intracranial stimulation, drug treatment, etc.) and dependent variables (catecholamine levels, eosinophilia, etc.). A third technique, and the one most relevant to the evaluation of learning, is that of simultaneous observation and recording of many behavioral variables. This is the technique of ethological observation and recording which has developed in the context of naturalistic observation of behavior under field conditions. This method of observation has been used in laboratory experiments by a number of workers (e.g., Grant, 1963, and Silverman, 1965) studying social behavior and the effects of drugs; it would appear to be ideally suited to the study of the behavior of animals in learning experiments.

Let me mention an example of such research, an unpublished study done recently by Mrs. Mahmuda Khanum under my direction. By direct observation of the behavior of laboratory rats in simple learning situations (straight alley maze and shock box), 17 separate, commonly occurring behaviors were distinguished. Each behavior was given an alphabetic designation (e, eating; a, air-sniffing; d, defecating; r, resting, etc.). Every 3 seconds during an experimental trial the rat being studied was observed, its behavior classified, and the appropriate alphabetical symbol recorded. Several groups of rats were given acquisition and extinction training with food in a straight alley; other groups were given noncontingent reinforcement training by presentation of pellets, milk, shock, or no reinforcement 30 seconds after the rat was placed in a distinctive compartment. This is not the place to discuss the detailed outcome of this study, but one of the findings can be mentioned as indicative of the potential value of polythetic data. This finding is that similar patterns of behavioral effects appear during extinction following food-rewarded maze acquisition training and during shock training. In both cases there was decreased locomotion, increased defecation and urination, decreased sniffing, increased crouching with hair erect, etc. Here is an example of homology in behavioral effects, a case of very different environmental events producing polythetically similar changes in behavior. What Mowrer (1960) termed "disappointment" and "fear" are polythetically similar and may form a single natural class of events; further, the nature of the behavioral effect suggests very strongly that orthosympathetic nervous-system arousal is prominent in disappointment and fear, since that division of the sympathetic nervous system is responsible for defecation, urination, and hair erection. The polythetic data forced us to this view in a way that conventional data (i.e., choice and latencies) could hardly do. Collection of polythetic data seems desirable in a wide variety of behavioral situations because of the efficiency with which it determines and limits theoretical statements.

Another way of expressing the advantages of polythetic operationism is to say that it makes *special pleading* very difficult. Since a large number of characters or variables are considered necessary to exemplify an effect, the operations implied in the dictum, "Look at all of the data from the experiment," are automatic. It follows from this that polythetic principles can be helpful in examining controversial phenomena. We can, for example, consider the reports of transfer of memory by injection and ask the polythetic question: what is the total pattern of behavioral effects produced by training of various kinds (shock to light signal, food to light signal, and unreinforced light presentation) and by injection of RNA extracted from the animals given various kinds of training? It could be that animals injected with RNA behave in a manner polythetically similar to the behavior of the donor animals, and, if this occurs, the evidence for transfer of training by injection would be very strong. It could also be that the behavior of recipients and donors shows very limited similarities and that there is no homology, but only analogy, between the behavioral effects of training and those of injection of RNA from trained animals. It is possible that RNA injections have many effects, and that by monothetic data, by restricting attention to one or a very few variables, one can make a strong case for the view that learning is transferred by injection, or by choosing other monothetic data, by attending to other variables, one can make a strong case against that proposition. But, since one can never trust a single character, one can effectively interpret the experimental results only when those results are polythetic, when the outcome is a complex of many behavioral characters rather than a single character (e.g., the increased speed of response).

The literature of comparative psychology, especially that concerned with the occurrence of learning in various major animal groups, awaits interpretation according to polythetic precepts. The data available suggests that learning, polythetically defined as the entire pattern of behavioral modifications which typically occurs when mammals are given reinforcement training, is *not* widely and uniformly distributed in the animal kingdom. To be sure, adaptive modification of behavior is shown by a number of groups (mammals, reptiles, teleost fish, sharks, cephalopods, bees, digger wasps, ants, annelids, planaria, and protozoa). But to determine whether these instances fulfill the criteria of homologous events and processes is quite another matter and one which requires polythetic data of a kind not generally available. The modification processes of a rat, an octopus, and an insect may well be analogous, as are the limbs of these three organisms, and may well be units among which effective generalization is severely limited.

The finite distribution of homologous units is illustrated by the phylogeny of locomotor behaviors. There is evidence that the evolutionary sequence from flagellate to fish is, in terms of modern representatives of the groups involved, flagellate, ciliate, flatworm, nemertine, hagfish, lamprey, and shark (Jensen, 1963). Very different locomotor behaviors are found along this sequence. Ciliary locomotion occurs from flagellate to nemertine; muscular contractions shortening the entire body are found in flatworms and nemertines; swimming by muscular flexion of the stiffened body is found in nemertines, hagfish, lampreys, and sharks, but only in the three vertebrate groups is it known to be stimulus directed. One behavior, backward swimming by sinusoidal waves sweeping forward, is reported within this sequence only in the hagfish. Obviously any particular behavior may be unique to one group or found in many groups; only research with many organisms can determine the phylogenetic distribution of behaviors and of the polythetically defined processes of behavioral modification.

There are some empirically correlated sets of characteristics (i.e., some polythetic concepts) which are distributed very widely. A morphological example is the cilium with its ultrastructural pattern of nine outer fibrils; this complex structure is remarkably similar in forms ranging from flagellate to man. A physiological example is the process of neuronal depolarization, which seems very similar in a great variety of metazoan forms. It is, of course, possible that learning is like the ciliary apparatus and the nerve impulse in being widely distributed; it is *possible* that the extrapolations from simpler invertebrates to man are justified, but that seems unlikely.

Examples of neural transmission are generally very similar from crustacean to octopus to man; there has been little recent controversy regarding the generality of the principles of neuronal excitation; there seems to be a single natural class of neuronal excitation, and generalization of findings among animal groups is reasonably effective. In sharp contrast there is great variation in the characteristics of different cases of learning both within specific organisms and among organisms of different taxonomic groups. Approach tendencies and withdrawal tendencies may well form two different natural classes, with very different behavioral organization, functional relationships to environmental variables, and internal mechanisms (Schneirla, 1965). It may be useful to differentiate a larger number of natural classes of mammalian behaviors, such as postural reflexes, protective reflexes, orientation behaviors, breathing, ingestive behaviors, approach locomotion and manipulation, withdrawal and fear, aggressive behaviors, and reproductive behaviors. All of these involve different behavioral patterns and different organ systems for terminal portions of behavior episodes, different sense organs and neural pathways for primary control, and, in the few cases where different behaviors have been directly compared, different results to similar experimental manipulations. Pavlov's dogs, reinforced with food, and Solomon's dogs, reinforced with traumatic shock, showed drastically different acquisition and extinction. Dobrzecka, Szwejkowska, and Konorski (1966) report that dogs show rapid learning of a "go–no go" discrimination to qualitative signals and slow or very little learning of the same task to directional stimuli; in contrast, the dogs learn a right-left discrimination easily if the stimuli are directional, but the discrimination is very difficult if the stimuli differ only qualitatively. Another example is the apparent impossibility of training amphibians to perform locomotor tasks or to strike nonmoving objects and the ease of obtaining learning with nictitating membrane responses (Goldstein, Spies, and Sepinwall, 1964). Not only are different behaviors in the same organisms influenced differently by training, but different animals respond to similar training of the same behavior somewhat differently; relevant here are the observations of Breland and Breland (1961) regarding intermittent reinforcement of manipulation of coins by pigs, raccoons, and primates. Pigs would pick up coins and carry them to the appropriate place and drop them into a hopper, but the tendency to "root" the coin about eventually replaced this tendency. Raccoons would pick up coins, but would persistently rub two coins together and, though they would place coins into the opening of the hopper, could not be trained to release the coins. Only primates could easily be trained to pick up coins, carry them to the hopper, and release them there. When findings regarding food-reinforced learning can be generalized only incompletely within the class of mammals, great caution seems appropriate when generalizing across phyla.

Returning to the experiments on simpler invertebrates and differences between widely separated taxa, the possibility that seems most in keeping with the "apperceptive mass" of experimental data is that there are at least three natural classes of events: vertebrate learning, the modifications observed in planaria and annelids, and the modifications observed in protozoa. In each case the behaviors involved do not appear to be homologous: The motoric organs responsible are not homologous (i.e., somite muscles, nonsomite muscles, and cilia); the central nervous system is either tubular, solid cord, or absent; and association of condition and unconditioned stimuli is either necessary, unimportant (Evans, 1966a and b), or impossible because of the absence of specialized receptors. The behavior and physiology of the three groups are so disparate that I feel that these behavioral phenomena are analogous rather than homologous. Behavioral modification processes observed in annelids and planaria, on the other hand, show resemblances and may be homologous; accordingly what is found about one may reasonably be expected of the other. This classification, with its separation of vertebrate and earthworm processes, is not in keeping with a recent discussion (Bitterman, 1965) but is consistent with the original work involved. The results of Datta (1962) suggest that modifications of behavior in earthworms and maze learning in rats are analogous rather than homologous. In sharp contrast to the standard results with rats, Datta observed in earthworms a low asymptote (70% correct in a single-unit T maze), no carry-over between daily sessions, no or possibly an adverse effect from enriching the maze with discriminative stimuli, and no improvement within days if 25 minutes elapsed between trials. There appears to be little evidence of overall similarity between maze learning of rats and modification processes of earthworms. (This is fully discussed in Chapter 24 of this volume.)

Separation of vertebrate learning from the modifications of behavior found in planaria and annelids and from the modifications found in protozoa challenges extrapolation both from worm to mammal and from mammal to worm. As polythetic principles guide our research, additional groupings will be called for; e.g., the relations of these three classes of phenomena to those of octopuses and insects await analysis. A research strategy appropriate to such problems is now available in polythetic operationism and numerical taxonomy.

In conclusion, my interpretation of specific experiments on learning with simpler invertebrates has changed little since the 1964 symposium. Now, however, I feel that the controversy over the research with planaria has been of some value, because it led to an appreciation of the difficulties engendered by definitions which are not both operational and polythetic. It seems clear that in studying learning we must beware of monothetic concepts. Otherwise we risk the error of calling a fish whatever swims in the sea (including thereby jellyfish and whales) and the error of generalizing from the blood of a beet to the blood of a man because both are red.

REFERENCES

Baerends, G. P., and J. M. Baerends, An introduction to the ethology of cichlid fishes. *Behavior, Suppl.*, 1950, **1**, 1.

Bennett, E. L., and M. Calvin, Failure to train planarians reliably. *Neurosci. Res. Program Bull.*, 1964, 2(4), 3.

Bitterman, M. E., Phyletic differences in learning. *Amer. Psychol.*, 1965, **20**, 396.

Boyden, A., Homology and analogy. *Amer. Midland Natur.*, 1947, **37**, 648.

Breland, K., and Breland, M., The misbehavior of organisms. *Amer. Psychol.*, 1961, **16**, 681.

Bridgman, P. W., *The Logic of Modern Physics.* New York: The Macmillan Company, 1927.

Datta, L. G., Learning in the earthworm, *Lumbricus terrestris. Amer. J. Psychol.*, 1962, **75**, 531.

Dilger, W. C., The behavior of lovebirds. *Sci. Amer.*, 1962, **206**(1), 88.

Dobrzecka, C., G. Szwejkowska, and J. Konorski, Qualitative versus directional cues in two forms of differentiation. *Science*, 1966, **153**, 87.

Evans, S. M., Non-associative avoidance learning in nereid polychaetes. *Anim. Behav.*, 1966a, **14**, 102.

Evans, S. M., Non-associative behavioral modifications in the polychaete, *Nereis diversicolor. Anim. Behav.*, 1966b, **14**, 107.

Goldstein, A. C., G. Spies, and J. Sepinwall, Conditioning of the nictitating membrane in the frog. *J. Comp. Physiol. Psychol.*, 1964, **57**, 456.

Grant, E. C., Social behaviour of the laboratory rat. *Behaviour*, 1963, **22**, 246.

Jacobson, A. L., Learning in flatworms and annelids. *Psychol. Bull.*, 1963, **60**, 74.

Jacobson, A. L., Learning in planarians: current status. *Anim. Behav., Suppl.*, 1965, **1**, 76.

James, R. L., and E. S. Halas, No difference in extinction behavior in planaria following various types and amounts of training. *Psychol. Rec.*, 1964, **14**, 1.

Jensen, D. D., Operationism and the question "Is this behavior learned or innate?" *Behaviour*, 1961, **17**, 1.

Jensen, D. D., Hoplonemertines, Myxinoids, and Vertebrate Origins, *in* E. C. Dougherty, ed., *The Lower Metazoa: Comparative Biology and Phylogeny.* San Francisco, Calif.: University of California Press, 1963, 113–126.

Jensen, D. D., Paramecia, planaria, and pseudo-learning. *Anim. Behav., Suppl.,* 1965, **1**, 9.

Lorenz, K. S., The evolution of behavior. *Sci. Amer.,* 1958, **199**(6), 67.

Mayr, E., Classification and phylogeny, *Am. Zool.,* 1965, **5**, 165.

McConnell, J. V., Cannibals, chemicals, and contiguity. *Anim. Behav., Suppl.,* 1965, **1**, 61–66.

McConnell, J. V., Comparative physiology: learning in invertebrates. *Annu. Rev. Physiol.,* 1966, **28**, 107.

Montgomery, T. H., Jr., *The Analysis of Racial Descent in Animals.* New York: Henry Holt (now Holt, Rinehart & Winston, Inc.), 1906.

Mowrer, O. H., *Learning Theory and Personality.* New York: John Wiley & Sons, Inc., 1960.

Pantin, C. F. A., Learning, world-models, and pre-adaptation. *Anim. Behav., Suppl.,* 1965, **1**, 1.

Schneirla, T. C., Aspects of stimulation and organization in approach/withdrawal processes underlying vertebrate behavioral development. *Advances Stud. Behav.,* 1965, **1**, 1.

Silverman, A. P., An analysis of the actions of some drugs on the social behavior of laboratory rats. Paper presented at 9th International Ethological Congress, Zurich, 1965.

Sneath, P. H. A., The construction of taxonomic groups, *in* G. C. Ainsworth and P. H. A. Sneath, eds., *Microbial Classification, 12th Symposium of the Society for General Microbiology.* Cambridge, Mass.: Cambridge Univ. Press, 1962, 289.

Sokal, R. R., Numerical taxonomy. *Sci. Amer.,* 1966, **215**, 106.

Sokal, R. R., and P. H. A. Sneath, *Principles of Numerical Taxonomy.* San Francisco, Calif.: W. H. Freeman, 1963.

Underwood, B. J., *Psychological Research.* New York: Appleton-Century-Crofts, Inc., 1957.

Woodger, J. H., *Biology and Language.* Cambridge, Mass.: Cambridge Univ. Press, 1952.

Discussion

E. Bennett: From Dr. Gaito's scheme (Table 1) it is apparent that the best thing to do for the present is to cover the waterfront. We really have very little firm evidence to go on for any of the theories that have been discussed. All of us are casting around for the best systems to work with. The area is still quite young—we are talking about a span of only seven years which is quite short for an area as difficult as this one. Some of the experiments are very crucial to the problem and they need confirmation, in particular, the ribonuclease experiments of Corning and John (1961), the transfer experiments of McConnell (1962), and the more recent mammalian transfer studies (Babich *et al.*, 1965). If these experiments could be substantiated *in toto*—if it can be shown that any purely defined relatively simple chemical compound can maintain the sort of learning we have been talking about—then I think all of us would be willing to support a theory of a unique memory molecule.

Our work at Berkeley has used a higher animal—the rat. If you provide the rat with stimulation you actually find that in certain areas there is a measurably thicker cortex. These are not large differences; I am referring to differences of perhaps 5 % between experienced and deprived rats. The primary observation was with wet weight measures performed on littermate pairs of animals. In addition to histological determinations, we have looked at a number of enzymatic changes such as cholinesterase and acetylcholinesterase, and have found small but statistically significant differences between what we call the ECT or trained animal and the isolated-naïve animal (Bennett *et al.*, 1964). More important are the observations made at the anatomical level. We have found, for instance, an increase in the number of glial cells in the experienced rat (Diamond *et al.*, 1966) and have some evidence for increased dendritic branching (Holloway, 1966). These dendritic changes are difficult to substantiate but if they turn out to be true, then we can begin to accept an RNA role in memory as Gaito has in the right-hand column of his table without upsetting any great dogmas. RNA would be required to synthesize protein, new types of cells, new connections, etc.

One experiment that was not mentioned is the recent paper (Zemp *et al.*, 1966) in which they reported an increased uptake of uridine in mouse brain when the animal was conditioned. These were transient increases found approximately 15 to 30 min after the learning task. Within an hour the change disappeared. If these observations are true then we can begin to have some idea of the time scale we have to consider in order to make fruitful studies in this area.

W. Corning: The notion of growth that Dr. Bennett puts forth finds support from a number of different investigations. I believe Weiss (1964) has discussed the concept of perpetual growth in relation to his finding of proximo-distal movement of intracellular constituents in the axon. From these studies it is clear that the neuron is not a static cell. Furthermore, Rose, Malis, and Baker (1961) have demonstrated that dendritic growth is possible. Precise laminar lesions were made in the cortex of adult rabbits by means of deuteron bombardment. The affected layer was soon devoid of cells but after several months they observed the area to be heavily invaded with dendritic sprouts. More relevant to some of Bennett's comments are the recent studies of Altman (1965) which suggest that there is considerable postnatal migration and differentiation of cells in the brains of rats. What I would like to suggest here is that neurones may be in various developmental states—some are pretty well specified at birth whereas others remain "plastic" for some time. The neuroelectrical activity might then bring about a further specification of the more embryonic cells. These changes do not have to be at a very gross level but can be subtle molecular alterations. In other words, it is possible that the neuroelectrical activity can specify connections between cells by inducing them to differentiate in certain ways.

M. Cohen: I would like to support the suggestion of Dr. Corning and not limit it to the immature animal. It has been suggested that synaptic boutons in adult mammals are areas of continued degeneration and regeneration and that the synapse is an anatomically labile area. If it is labile then it could be structurally modified through use and disuse. We should not close our minds to the fact that the synapse may not be a fixed connection.

Caspersson (1950) pointed out several years ago that adult vertebrate neurons are unique; they are special cells in that they produce protein at a rate similar to what they were doing when they were embryonic cells. The question seems to be: What is the neuron doing with all this protein? Is it simply making more of the substances necessary for its nonlearning activity (for instance, transmission through a reflex pathway that is not being conditioned) or is this protein specifically used to modify the cell as a consequence of previous use? Why should the nerve cell retain its embryonic capacity to produce protein? Adult mammalian neurons are rather special in that they lose their mitotic ability quite early. This loss of mitosis is correlated with retention of embryonic protein synthetic ability. I think this peculiarity has some significance in central nervous system functioning.

J. Gaito: A report by Coleman and Riesen (*unpublished*) indicates that visual stimulation can induce a change in dendritic branching—there was a greater degree of dendritic interaction in the visual cortex of stimulated animals than in dark-reared ones. Only certain parts of the striate area were affected (layer IV stellate cells).

E. Bennett: With reference to Altman's work, we would certainly like to agree with his experiments, but we have certain reservations about them in terms of how he matched his animals in size. He may have biased his results by not using a littermate matching. However, we need longitudinal studies of this kind—we need to study cells over time. What makes research difficult with respect to learning is that, in any system, the percentage of elements that are going to be changed at any given time will be small. I think this is one of the reasons that we have gotten the idea that the brain is a static thing. We are not going to observe massive changes with each experience or any one event. They will be small changes, and this makes it difficult to pin down.

J. Best: One of the things that you keep running into is the comment that brains and neurons have a high rate of RNA synthesis. This is difficult to reconcile with the difficulty of trying to label RNA with precursors. If you compare rates with other tissues, neurons are rather slow. We need to distinguish between RNA and protein synthesis. In a well differentiated organ that is involved in a large amount of synthesis of a simple unchanging protein, such as the silk glands of a caterpillar, you find high rates of protein synthesis with little RNA turnover. It may well be that there is a stable template in the case of proteins that are not changing in type even though they are made in large amounts.

M. Cohen: I would agree with Dr. Best's comment. There is additional evidence to support the idea that the state of organization of cytoplasmic RNA may have a bearing on protein synthesis. Ribosomes aggregated on endoplasmic reticulum synthesize protein at one level and when the ribosomes are dispersed, they synthesize at a much higher level or in what Caspersson calls a "superactive state." In this case you do not have to have any new synthesis of RNA, but a redistribution. We do have to be careful about equating RNA synthesis with protein synthesis.

J. McConnell: In reference to Gaito's Table 1, I find that I don't understand the three positions. There are a great many more than three and furthermore I don't see that there is a clear distinction among them. Another thing that bothers me is that we use the word "learning" as if it were a psychological entity. By now we are sophisticated enough to know that there must be a variety of types of behavioral changes, some of which you might call learning and some of which you might choose to call something else. There is not just one type of learning and if there is no one type of learning, it would be naïve of us to expect that there would be just one type of chemical change to mediate all the behavioral changes. RNA might be involved in one and not in another. We are not even sure that the underlying chemical changes will be the same at different positions on the phylogenetic scale. I think that the interesting part of this whole conference is that there are now growing numbers of scientists who will be looking at the chemical changes. And if you look at the transfer studies in particular, there is evidence accumulating that different types of behaviors may be mediated by different agents. Low-molecular-weight molecules may be involved in some types of transfers and larger molecules in others. The evidence here is by no means clear, but I think it is a possibility. Finally, there is some recent interesting work by the Russians, who have been injecting ribonuclease, actinomycin, trypsin, and a variety of other agents into the brains of trained animals and finding that they could knock out conditional responses with ribonuclease, but not affect the unconditioned response. We will probably hear more about this work later.

S. Ratner: The main point I would like to make in reference to learning models and Denny's paper is that we started with a neutral tone—"we are modifying behavior"—and the next thing we have switched to is learning. Take, for instance, the sentence, "The animal has learned a right-turn response." Why can't we say that the animal now makes a right-turn response with a higher frequency or something similar. Now Denny says that there was no learning in a certain situation—I can't imagine any organism being in a box for many trials with no behavioral modification. But this highlights the problem. We suddenly expect a change in RNA when what *we* define as learning occurs. But the animal has modified his behavior to some extent and we certainly would expect that something systematic has gone on. Thus, I am suggesting that behavior modification and associated chemical changes may occur in association with any long series of trials. Insofar as this is true, new or different control procedures become necessary.

The important thing is that Denny has started us toward the important neutral vocabulary of behavioral modification that omits the mention of the word learning. Learning often gets defined from those situations in which school teachers and psychologists have been clever enough to beat the animal into submission. When we don't succeed we call it "not learning." This is not a systematic use of concepts, so I would like to re-emphasize the importance of the neutral vocabulary. Organisms are modifying their behavior and it is our mission to uncover how. We need to avoid a vocabulary that puts a premium on certain kinds of modifications and by implication denies that others have occurred.

D. Jensen: I would like to emphasize that the characteristics outlined by Denny apply to mammalian learning. One wonders if these same stages would apply in describing the learning of a shark, for instance. I am now studying an animal which is a primitive vertebrate, the myxinoid fish, and, as far as we can determine, it demonstrates no exploratory behavior. This might present some difficulties for the model. Whether the animal can learn or not is yet to be determined.

F. Crawford: Dr. Ratner raised a question in his paper about doing demonstrations. In our research we work with animals different from those found in most behavioral laboratories and I believe I *do* demonstrations. When I can get a snake to go down a runway and eat a chicken I have demonstrated something. Isn't it the same when you arrange a burrow for an earthworm to modify its behavior?

A second comment concerns Denny's assumption that our implicitly accepted laws of learning are equally applicable to all organisms. I don't think we have enough data to assume this. Examples of this are our attempts to operantly condition tarantulas. We have found that the longer we work with the tarantula the more persistently they sit and wait for reinforcement. Thus far, we have not been able to shape and operantly train them. In the T maze, furthermore, they do not show approach behavior—to get them to move through the maze we must rely upon escape behavior.

S. Ratner: In many instances, I submit that the demonstrations provide the background material. If we need, for instance, a pH of a certain sort to get the planarian to live happily then that would be a demonstration and fit into this broad pot of material with which the investigator works.

M. Denny: What I was putting forth was a model built on the rat and human which might have relevance or application to many other organisms. I was also pointing out some of the limiting conditions such as the presence of exploratory behavior and the presence of a good *UR*. One thing that I did not elaborate upon was that classical conditioning is really very complicated. The contiguity principle was exploited as it applies in the typical classical conditioning situation, but classical conditioning has on top of it all kinds of stimulus differentiation not only with respect to the *CR* but with respect to *UR* responding between trials, etc. Before conditioning occurs it is necessary to get rid of indiscriminate *UR* elicitations. When this is accomplished you then have the appropriate contiguity conditions for *CS-US* pairing. I think that the identification of the *UR* and its properties would be the main thing that would determine the learnability of a lower organism. I have seen no data that would contradict these basic principles, and they may hold throughout the animal kingdom. You have to focus first on the *UR*, the response capabilities of the animal. Now in some animals you might not have an approach response, but approach behavior is just an example of a response class that fits the general model. (For further considerations of this theory see Denny, 1966, and Denny and Adelman, 1955.)

J. McConnell: Also, I think the model would have difficulty in handling inhibition. Denny has referred to extinction and getting rid of competing responses but didn't mention whether these are *learned* extinctions or *learned* inhibitions.

M. Cohen: We need the neutral vocabulary and by neutral I assume a nonanthropomorphic vocabulary. Yet, the definition of learning as a modification of behavior through experience bothers me because if you look at behavior at various levels you find facilitation at the synapse, sensory adaptation at receptors, post-tetanic potentiation in the spinal cord—in all of these situations you are modifying behavior. Some aspect of output changes as a result of experience. Yet a physiologist does not call facilitation "learning," and I don't think a psychologist would either. I wonder if psychologists could provide us with some way of differentiating these kinds of modifications from what a psychologist would agree is learning. Some of these changes I have referred to even fit what we might call "long-term" memory; they last for days.

M. Denny: Extinction or inhibition is the result of competition of two or more responses. The competing response could be a single response but more likely a whole family of them. In extinction, the original response tendency would lose out—it would not be unhooked as Guthrie would have it but would remain intact and could be reactivated by changing conditions. On occasion, a resting response would be the competing response that is pitted against the original response. Doing nothing is, in a sense, one kind of response class, as is sitting or sleeping. These classes have a neural basis as much as active responses have.

To distinguish learning from some of the phenomena that Cohen mentions is difficult. What we say is that learning is a relatively permanent modification of behavior involving the stimulus-response changes I've talked about. It is not due to fatigue; it is not sensory adaptation. What we do is throw in a bunch of "nots" at the end of the definition.

J. Best: There are some difficulties here when you try to delineate some of the peripheral events such as sensory fatigue from central events. Negative after-images are believed to be essentially due to receptor fatigue and yet you can demonstrate interocular transfer of negative after-images. Even this sort of effect involved central systems. Also, I believe it has been demonstrated that, in the auditory system, central mechanisms could either shut down or facilitate transmission of auditory impulses almost out at the primary sensory level.

P. Driver: What we started with were definitions with which I could agree but as the discussion has gone on, I have become more confused. As Denny described response, I was in complete agreement; as he described learning, I was in complete agreement as an ethologist working with animals in the field. I have worked with animals in their natural habitats and the definition of learning, as I understood Denny, fits in beautifully not only with mammals but with birds as well. If we are to have a definition of learning we are going to have to take notice of evolution and what animals do under natural circumstances. I would say that learning is a modification of the stimulus-response complex which results in a new favorable adaptation of the whole animal to its environment.

N. Rushforth: A question of information; several variables that Dr. Jensen has defined are quite different. Some of them are continuous, such as height and length, and others are discrete variables which could be classified as "0-1" variables, or the presence or absence of something. How will you define a homology statistically, and what procedures are you going to use to combine all of these various variables?

D. Jensen: The combination of variables gets us into the concept of multi-dimensional scaling. What one can say is that there is more of one thing than another and this can handle both continuous and discrete variables. Each bit of information has the effect of constricting where in a spatial representation one may place animals in order to represent the differences you've found. We wind up with a three-dimensional graph representing the relationship of entities to entities. The important thing is that one does not deal with key characters but with correlated characters, and, when these characters go together from instance to instance, we give them a name.

P. Driver: I think we probably all realize that in much of his work Darwin was actually performing the sort of operation that Dr. Jensen is considering. His approach was a polythetic one. Further, when a taxonomist stresses one particular character for purposes of classification, it is because he can't state the sum total of all the characters and the balance of the important and less important characters all at the same time. Jensen also states that ethology is the search for behavioral homology, and this is partly true, but Pringle and others have pointed out that the same behavioral processes may be carried out by different organs. One example is distraction display: Annelids may shed bits of themselves, octopuses squirt out an ink cloud, birds fake injury, etc. All of these have the same function but are subserved by different organ systems and processes.

D. Jensen: These behaviors are analogs; the distraction display is not a natural group; it is a special classification. The distraction displays are not members of the same kind of natural classification that we have when we study something like the sinusoidal wave locomotion, which can be traced from shark to man (see R. N. Shepard, *Psychometrica*, 1962, **27**, 125, for a computer program utilizing nonmetric information to reconstruct metric configurations of entities).

P. Wells: Regardless of whether one is interested in phylogenetic comparisons, Dr. Jensen's important point is that we rarely use polythetic approaches in any studies of behavior. This is the sort of approach which we have used recently in reanalyzing the honey-bee dance. We have had to drastically revise our conceptions of this behavior pattern. I should also add that we now have the statistical tools to apply this approach.

REFERENCES

Altman, J., and G. D. Das, Post-natal origin of microneurones in the rat brain. *Nature*, 1965, **207**, 953.
Babich, F. R., A. L. Jacobson, S. Bubash, and A. Jacobson, Transfer of learning to naïve rats by injection of ribonucleic acid extracted from trained rats. *Science*, 1965, **149**, 656.
Bennett, E. L., M. C. Diamond, D. Krech, and M. Rosenzweig, Chemical and anatomical plasticity of brain. *Science*, 1964, **146**, 610.
Caspersson, E., *Cell Growth and Function*. New York: W. W. Norton and Company, Inc., 1950.
Coleman, P. D., and A. H. Riesen, Environmental effects on cortical dendritic fields: I. Rearing in the dark. *Unpublished paper*.
Corning, W. C., and E. R. John, Effect of ribonuclease on retention of conditioned response in regenerated planarians. *Science*, 1961, **134**, 1363.

Denny, M. R., A theoretical analysis and its application to training the mentally retarded, *in* N. R. Ellis, ed., *International Review of Research in Mental Retardation*, Vol. 2. New York: Academic Press, Inc., 1966.

Denny, M. R., and H. M. Adelman, Elicitation theory, I: An analysis of two typical learning situations. *Psychol. Rev.*, 1955, **87**, 317.

Diamond, M. C., J. Law, H. Rhodes, B. Lidner, M. R. Rosenzweig, D. Krech, and E. L. Bennett, Increase in cortical depth and glia number in rats subjected to enriched environment. *J. Comp. Neurol.*, 1966, **128**, 117.

Holloway, R. L., Dendritic branching: some preliminary results of training and complexity in rat visual cortex. *Brain Res.*, 1966, **2**, 393.

McConnell, J. V., Memory transfer through cannibalism in planarians. *J. Neuropsychiat.*, 1962, **3** (*Suppl.* 1), 542.

Ratner, S. C., Research and theory on conditioning of annelids. *Anim. Behav. Suppl.*, 1965, **1**, 101.

Rose, J. E., L. I. Malis, and C. P. Baker, Neural growth in the cerebral cortex after lesions produced by monoenergetic deuterons, in W. A. Rosenblith, ed., *Sensory Communication.* Cambridge and New York: MIT Press and John Wiley & Sons, Inc., 1961.

Weiss, P., Self-renewal and proximo-distal convection in nerve fibers, in E. Guttman and P. Hnik, eds., *The Effect of Use and Disuse on Neuromuscular Functions.* New York: Elsevier Publishing Co., 1964.

Zemp, J. W., J. E. Wilson, K. Schlesinger, W. O. Boggan, and E. Glassman, Brain function and macromolecules, I: Incorporation of uridine into mouse brain during short-term training experiment. *Proc. Nat. Acad. Sci.*, US, 1966, **55**, 1423.

PART IIA
Planarian Research
Ecological, Structural, and Physiological Factors

Chapter 7

Species Differentiation and Ecological Relations of Planarians

Roman Kenk

Division of Worms
Smithsonian Institution
Washington, D.C.

The planarians, or freshwater triclads, belong, in the zoological classification, to the class Turbellaria (free-living flatworms) of the phylum Platyhelminthes. Thus they are among the lowest, or most primitive, Metazoa with bilateral symmetry and a centralized nervous system. This primitiveness in their structure and, presumably, physiology makes them eminently suitable for the study of many fundamental biological phenomena of a general nature, applicable in many ways to higher animals as well. They have long been favored experimental animals in laboratories. The classical studies of the principles of regeneration (T. H. Morgan, C. M. Child, and many other investigators, see Brøndsted, 1955) were made on planarians. Other investigations in which planarians played a prominent role were studies of the differential action of radiations on various types of cells and tissues (C. R. Bardeen, W. C. Curtis, F. Stéphan-Dubois, etc.), effects of pharmacological and carcinogenic agents (many investigations widely scattered in the literature), the morphogenetic interrelations of various organs during their development (E. Wolff and his collaborators), and relations of cytogenetics to speciation (M. Benazzi and his school). During the last decade we have seen a very intensive upswing in behavioral studies performed on planarians, centering on the mechanisms and possible biochemical correlates of learning and memory, which are the topics of the present symposium.

In the following I should like to point out some general aspects concerning the procurement and handling of planarians for experimental purposes. There are at present about forty species of planarians known to occur in North America. They are found in a great variety of ecological habitats, in running waters (springs, brooks, rivers) and in standing waters (lakes, ponds, ditches), above ground as well as in subterranean habitats (caves and wells). The chief limitations on their occurrence appear to be industrial pollution and extremes of acidity or alkalinity of the water.

67

Most planarians are negatively phototactic or photonegative, i.e., they avoid bright daylight and are found in daytime on the lower, shaded surfaces of stones and other objects, on the undersides of leaves of water plants, or in the angles formed by the leaves and the stem. They are more active at night than during the day, although various stimuli (food) easily induce them to leave their sheltered resting places, even in broad daylight. The various species differ greatly in their ecology, their physiological responses to environmental agents, their sensitivity to fluctuations of ecological factors, and, therefore, in their suitability for laboratory experimentation. For these reasons it is important to identify the species used in order to obtain repeatable experimental results. The modern system of classification of planarians is based on anatomical characteristics of the individual species. Nevertheless, it is possible to identify living specimens from external features if the planarians have been collected in a well-studied geographic region.

Some planarian species are widely distributed over the North American continent—such as *Dugesia tigrina* and probably *D. dorotocephala*. Others are typical of either the eastern half or the western half of the United States. A relatively large number of species are very much restricted in their distribution, having been found in one locality or in a few closely adjacent localities only. This is particularly true of subterranean species inhabiting caves. The primary factor affecting the occurrence of a given species is the temperature of the water and its fluctuation over the annual seasons. Some species, e.g., *D. tigrina*, may thrive in a wide range of temperatures; they are eurythermic species. Planarians living in running waters, particularly in springs and spring brooks, are usually sensitive to sudden temperature variations and generally prefer cold-water habitats; we call them *cold-stenothermic*. Often we observe a regular succession of planarian species as we proceed from the source of a river to its mouth, one species replacing another in the various sections of the river. This is brought about by the progressive changes of the water temperature in the course of the stream. The source and the upper section may be inhabited by a species *A*; at a lower level there is a mixture of species *A* and species *B*; farther down species *B* may be the only planarian present. As we proceed further, species *B* may be joined and finally replaced by a third species, *C*.

To collect planarians in nature, various methods may be employed. They may be obtained from streams, shallow ponds, and lake shores by examining the undersurfaces of flat stones, pieces of wood, leaves of water plants, and other objects, picking them up with a soft brush or with the finger tip, and placing them in a jar. Another method, often rewarding, is that of collecting fallen leaves and other debris in a jar filled with water and letting the jar stand overnight in a cool place. As the oxygen in the water is depleted, the planarians, if present, will collect near the upper surface, mainly on the shady side, away from the window or light source, and may then be picked up individually with a pipette. Planarians living at the bottom of a lake can be caught with a dredge or dragnet with the upper layer of silt or mud and may be separated by a sieve with proper mesh. Baiting with meat, crushed snails, earthworms, etc., may also attract planarians living in running waters, less successfully in ponds or lakes. From a practical standpoint I have found the outflow streams of fish hatcheries to be among the most profitable collecting sites. Hatcheries are always built on unpolluted streams and the organic matter added to the water as fish food serves as a kind of bait and often supports very dense planarian populations.

Transportation of planarians to the laboratory is best done in insulated or thermos flasks to avoid greater fluctuations of the water temperature. This is particularly important in the case of cold-stenothermic species which do not tolerate higher temperatures.

In the laboratory planarians are kept best in shallow glass aquaria (finger bowls) or enameled pans in which the water is changed every two or three days. Chlorinated tap water must be avoided or, for some resistant species, may be used if modified by the addition of chlorine-binding chemicals or by bubbling air through it. Spring water or filtered pond water is to be preferred. Great care should be taken to keep the water temperature within the tolerance levels of the species cultured; if necessary, the culture dishes should be kept in a constant-temperature chamber or refrigerator. The planarians are fed once or twice a week. As most species are scavengers in nature, small pieces of meat, fresh beef liver, cut-up earthworms, or small oligochaetes such as *Tubifex* (sold in aquarium stores as fish food) will be taken readily. Only one common species refuses to accept dead meat; that is *Procotyla fluviatilis*. This species actively attacks living Crustacea, such as *Daphnia*, *Asellus*, and *Gammarus*, or small aquatic insect larvae. Planarians may live without food for several months; in the course of starvation they grow smaller, simplify their internal structure, reduce their reproductive system, and assume the external aspect of young planarians.

The life cycle of freshwater planarians is rather simple, without larval forms or metamorphosis. In species with sexual reproduction, the young hatch from egg capsules or cocoons, which are rather large (about 1 mm), usually spherical or ellipsoidal capsules with a hard, horny shell, attached to the substrate by a thin stalk or by a gelatinous secretion. Each capsule contains several, perhaps up to 20, egg cells and thousands of yolk cells, which furnish food for the developing embryos.

The young hatch from the egg capsules one to six or more weeks after their deposition, the duration of the embryonic development depending chiefly on the temperature. The young have more or less the shape of the adult animals, apart from their size, but completely lack any trace of a differentiated reproductive system. This system develops only after they are almost fully grown, and, in species with asexual propagation, may not develop at all. All sexually mature freshwater planarians are hermaphroditic, having both male and female structures developed in the same individual at the same time. In sexual reproduction, egg capsules are deposited, normally after copulation and cross-fertilization of two individuals.

Autofecundation, by a migration of sperm into the female tract of the same individual, is a rare exception (*Cura foremanii*). In nature sexual reproduction usually occurs in a seasonal cycle correlated apparently with the temperature cycle; this cyclic activity may continue also in the laboratory.

Another mode of multiplication is asexual, by detaching parts, usually the posterior parts of the body or "tails" (fission), or by a breaking-up of the body into small pieces (fragmentation). Each of these pieces then regenerates the missing parts to form a new complete individual. Some species have only sexual reproduction; others are capable of both sexual and asexual propagation, frequently depending on the season of the year or environmental conditions. In a few species there are strains or races which reproduce only asexually generation after generation, as observations of several years' duration have demonstrated.

Related to the asexual multiplication of planarians is their often almost limitless regenerative capacity. Obviously all species multiplying by fission or fragmentation must be capable of regenerating missing parts. However, this capacity is not confined to fissioning planarians alone; it is observed as well in species which reproduce only sexually. There are only a few forms in which the regenerative capacity is limited, among the American species particularly *Procotyla fluviatilis*; this species regenerates a new tail end if this has been removed or injured, but it is not capable of regenerating a new head end unless the level of severance is rather far in front of the pharynx.

Finally, I wish to review a few representative American planarian species which are easily obtainable and appear to be suitable for laboratory work. (For additional species and illustrations see Hyman, 1953.)

Cura foremanii (Girard), formerly known as *Planaria* (or *Curtisia*) *simplissima* Curtis or *simplicissima*, also erroneously confused with the European *Planaria* or *Dugesia lugubris*. This is a uniformly pigmented planarian, brown to black. The head of the quietly gliding animal forms a rather rounded triangle, the two lateral angles of which bear each a lighter oblique dash (auricular sense organ). It is common in cool running waters of the eastern and midwestern states and reproduces by egg capsules only.

Dugesia tigrina (Girard), the *Planaria maculata* Leidy of the older literature. The color and pattern are highly variable but always appear spotted. There is generally a brownish ground color with white patches and usually scattered small blackish spots, frequently with a dark longitudinal stripe on either side of the midline. The head is roughly triangular, with pointed lateral projections or auricles. The two eyes are rather close together (the distance between the black eye cups is less than one-fourth the transversal diameter at eye level). This is a eurytopic species found in standing as well as slowly running waters, commonly in warm habitats with a temperature optimum (for fission) of about 25°C. It occurs in at least two physiological races, apparently genetically different, an asexual race which reproduces by fission and a sexual race which deposits egg capsules mainly in spring and fissions during the warmer part of the year.

Dugesia dorotocephala (Woodworth), a species closely related to *D. tigrina*. Its pigmentation may resemble that of *D. tigrina* but appears to the naked eye more uniform and lacking the black spots scattered over the dorsal surface. The head of the gliding animal is triangular, the front angle somewhat more acute than in *D. tigrina*, the lateral auricles longer and more pointed. It is generally found in rather warm running waters and multiplies by fission and/or sexually. It is easily confused with the preceding species, particularly in the case of young specimens or regenerating posterior fission pieces; also, animals which are not in the best physiological condition may show a head form very closely resembling that of *D. tigrina*. Anatomically the two species may be distinguished by the structure of the copulatory apparatus.

In general, it can be said that the morphological differences between the two species are quantitative rather than qualitative. Some behavioral differences between them have been pointed out recently by R. C. DeBold, W. R. Thompson, and C. Landraitis (1965) concerning their reactions to light stimuli, adherence to the substrate, speed and regularity of gliding locomotion, etc.

Phagocata morgani (Stevens and Boring), the *Planaria truncata* Leidy of older authors, is an unpigmented planarian and, therefore, appears white when its digestive system is empty. After ingesting colored food material, however, the area of the intestinal branches shines through the body wall, leaving only the head, the lateral margins of the body, and the areas of the pharynx and, behind it, the copulatory organs, purely white. The head is truncated, with a slightly bulging frontal margin and rounded lateral edges, without projecting auricles. The pair of eyes is situated close together, removed from the frontal end. This species inhabits cold streams in the eastern half of the United States, is cold-stenothermic and, therefore, should be kept in the laboratory in refrigerated cultures at 10 to 15°C. It reproduces sexually and by fission, depending on season and locality.

Phagocata gracilis (Haldeman), a grayish-brown to black planarian, has a head shape similar to that of *P. morgani*. The distinguishing character of this species, apart from its pigmentation, is the presence of more than one pharynx in the pharyngeal pouch, i.e., up to a dozen or more pharynges (their number increases during the development) which may be faintly seen in the living animal. It lives in cool running waters in the southeastern states. Its propagation is, as far as we know, exclusively sexual.

Polycelis coronata (Girard) is a brown planarian widely distributed in the western parts of North America. It has a rounded head with prominent but not pointed auricles and is at once recognizable by the many eyes which form a band along the head and lateral margins in the anterior third of the prepharyngeal portion of the body. Its reproduction is sexual and asexual.

Procotyla fluviatilis Leidy, a white planarian, inhabits running and standing waters in the eastern half of the United States. It is distinguished by having a well-developed adhesive organ in the center of the frontal margin of the head. The number of eyes is not fixed; there may be one to eight eyes on either side. This is a eurytopic and eurythermic species. Its food consists of living aquatic arthropods (crustaceans and insect larvae). It catches its prey with the aid of the adhesive organ by thrusting its head forward and enveloping the captured animal in a sticky slimy secretion. It does not seem to be attracted by chemical food stimuli, as we observe the great majority of planarians are, but by tactile stimuli produced by the movements of the food animals. Its reproduction is sexual only, and its regenerative powers are limited.

REFERENCES

Bardeen, C. R., and F. H. Baetjer, The inhibitive action of the Roentgen rays on regeneration in planarians. *J. Exp. Zool.*, 1904, **1**, 191.

Benazzi, M., Evoluzione cromosomica e differenziamento razziale e specifico nei tricladi, *in* Evoluzione e Genetica, Colloquio internazionale, Roma, 1959. *Accad. naz. Lincei, Problemi attuali di Scienza e di Cultura, Quad.*, 1960, **47**, 273.

Brøndsted, H. V., Planarian regeneration. *Biol. Rev.*, 1955, **30**, 65.

Curtis, W. C., and J. Hickman, The effects of X-rays upon regeneration in planarians. *Science*, 1926, **63**, 505.

DeBold, R. C., W. R. Thompson, and C. Landraitis, Differences in responses to light between two species of planaria: *Dugesia tigrina* and *D. dorotocephala*. *Psychonomic Sci.*, 1965, **2**, 79.

Dubois (later Stéphan-Dubois), F., and E. Wolff, Sur une méthode d'irradiation localisée permettant de mettre en évidence la migration des cellules de régénération chez les planaires. *C. R. Soc. Biol. (Paris)*, 1947, **141**, 903.

Hyman, Libbie H., Turbellaria (flatworms), *in* R. W. Pennak, ed., *Fresh-Water Invertebrates of the United States*. New York: The Ronald Press Company, 1953, p. 114.

Wolff, E., and T. Lender, Les néoblastes et les phénomènes d'induction et d'inhibition dans la régénération des planaires. *Année biol.*, 1962, (4)**1**, 499.

Chapter 8

Aspects récents de la morphogenèse chez les Planaires

P. Sengel

Laboratoire de Zoologie
Faculté des Sciences de Grenoble
Grenoble, France

Depuis les travaux des pionniers que furent Morgan (1900) et Child (1920), les Planaires n'ont cessé de faire l'objet de très nombreuses et fructueuses recherches. Les Turbellariés Triclades sont en effet des Vers doués d'un pouvoir de régénération peu commun, à l'étude duquel les morphologistes se sont attachés depuis longtemps. Aussi la somme de nos connaissances sur les mécanismes de la régénération est-elle déjà considérable. Les principaux problèmes que les chercheurs des 20 dernières années se sont efforcés de résoudre ont trait d'une part à la nature et aux propriétés des cellules de régénération, d'autre part aux facteurs intervenant dans la morphogenèse régénératrice.

Plusieurs revues récentes et très complètes résument les résultats obtenus dans ce domaine jusqu'en 1963 (Lender, 1956c; Wolff, 1962; Wolff et Lender, 1962; Lender, 1962; Wolff, Lender et Ziller-Sengel, 1964; et Stéphan-Dubois, 1965). Il n'y donc pas lieu de reprendre en détail l'ensemble de la question. Je me contenterai de résumer ici l'essentiel de nos connaissances:

1. Sur la nature et les propriétés des cellules de régénération
2. Sur les phénomènes d'induction et d'inhibition au cours de la régénération de divers organes (tête, cerveau, yeux et pharynx)

Par ailleurs, une part importante de cet exposé sera consacrée aux recherches récentes sur le développement embryonnaire des Planaires et, d'une façon générale, aux résultats nouveaux acquis depuis 1962.

NATURE ET PROPRIETES DES CELLULES DE REGENERATION

Parmi les Turbellariés Triclades, animaux dulcicoles, marins ou terrestres, ce sont surtout les Planaires d'eau douce qui sont douées au plus haut degré du pouvoir de régénération total (p. ex., *Polycelis nigra, Dugesia lugubris, D. gonocephala et Crenobia alpina*). Chez ces Planaires, n'importe quel fragment du corps régénère un individu complet. Chez d'autres, la fréquence de régénération de tête varie en fonction du niveau de l'amputation (*Dugesia dorotocephala et Dendrocoelum lacteum*). D'autres enfin sont incapables de régénérer une tête (*Phagocata*).

Quand on coupe en deux parties, par une section transversale par exemple, une Planaire capable de régénérer sa tête, le fragment postérieur forme vers l'avant, au niveau de la section, un blastème de régénération de tête. De même, le fragment antérieur édifie vers l'arrière un blastème de régénération de queue. L'examen histologique d'un tel blastème révèle qu'il est formé d'une part d'un épithélium cicatriciel bordant très pauvre en cellules, d'autre part de nombreuses petites cellules arrondies, piriformes ou fusiformes, au cytoplasme très basophile, au noyau volumineux pourvu d'un gros nucléole (Fig. 1, 2 et 3). Ce sont les cellules de régénération, qu'on convient d'appeler les "néoblastes." Ils peuvent être caractérisés histochimiquement (Pedersen, 1959 et Lender et Gabriel, 1960) par la coloration au vert de méthyle-pyronine. Le noyau se colore en vert, grâce à son contenu en acide désoxyribonucléique, le cytoplasme et le nucléole en rouge, en raison de leur forte teneur en acide ribonucléique.

Ces cellules de régénération se différencient en cellules épidermiques, musculaires, nerveuses, oculaires, intestinales, etc. et reconstituent finalement la totalité des tissus et des organes amputés.

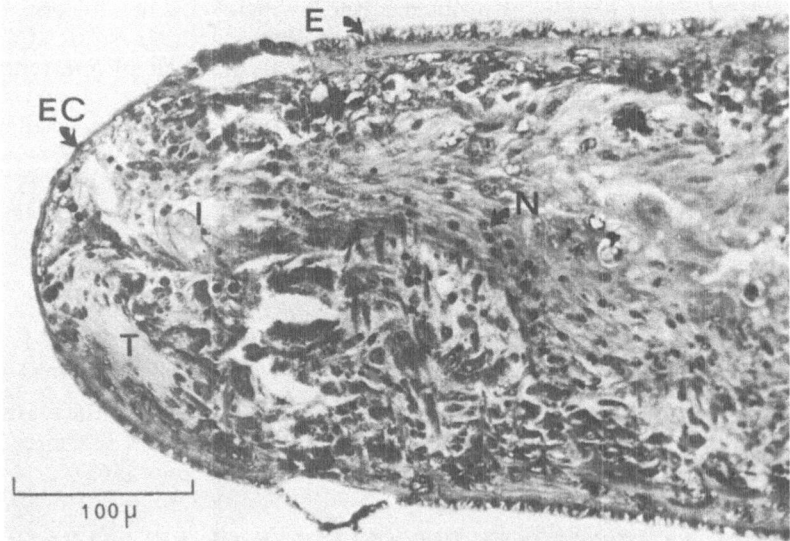

Fig. 1. Coupe parasagittale d'un blastème de régénération antérieur, 48 heures après une section pratiquée au milieu de la région prépharyngienne de *Crenobia alpina*; régénération à 10°C. *E*, épiderme ancien; *EC*, épiderme cicatriciel; *I*, intestin; *N*, néoblastes en cours de migration vers l'avant; *T*, tronc nerveux ventral sectionné. De Beauchamp, hémalun-éosine.

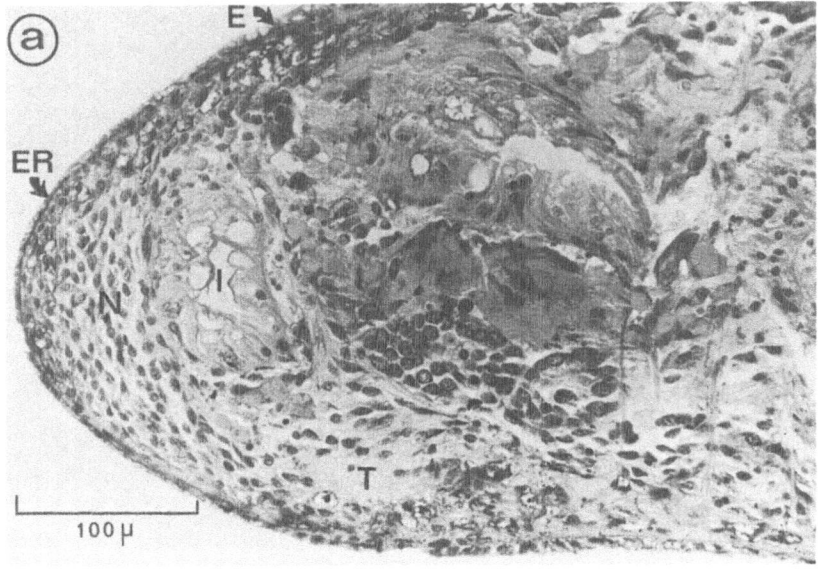

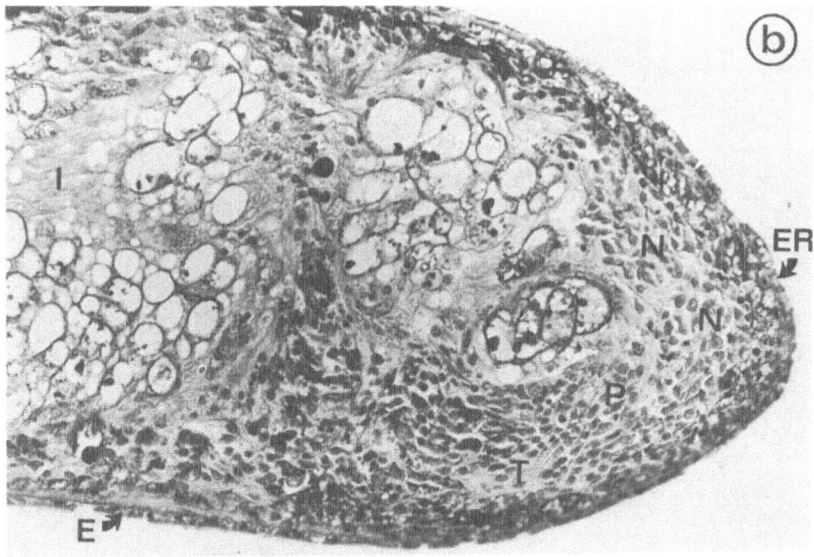

Fig. 2. Coupes parasagittales d'un blastème antérieur (a) et d'un blastème postérieur (b), 10 jours après une section pratiquée au milieu de la région prépharyngienne de *Crenobia alpina*; régénération à 10°C. *E*, épiderme ancient; *ER*, épiderme régénéré; *I*, intestin; *N*, néoblastes; *P*, ébauche du pharynx en régénération; *T*, tronc nerveux ventral. De Beauchamp, hémalun-éosine.

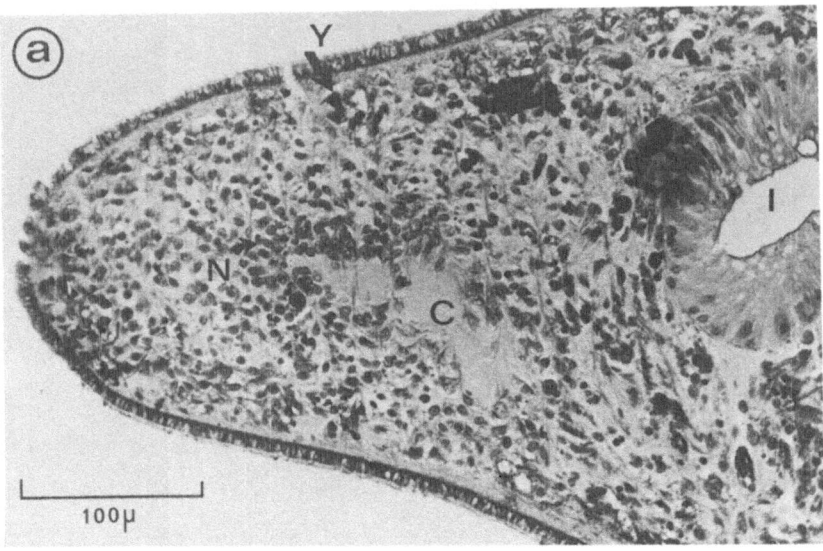

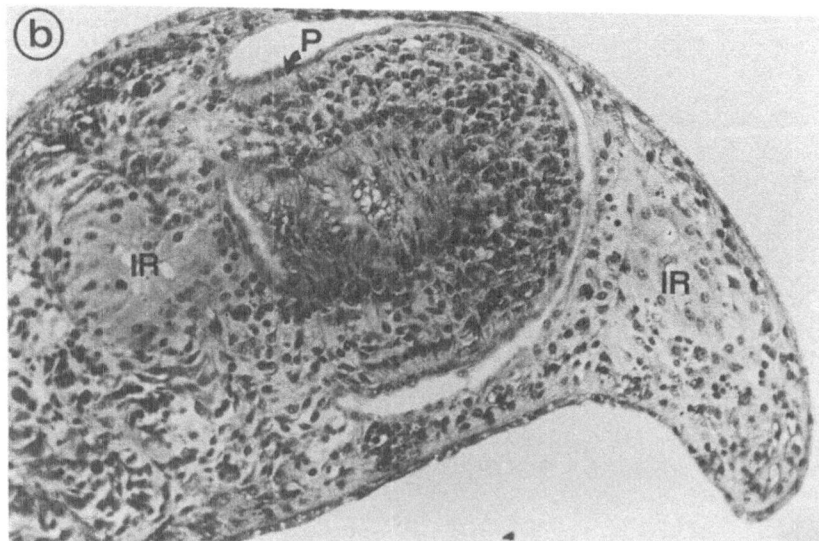

Fig. 3. Coupes parasagittales d'un blastème antérieur (a) et d'un blastème postérieur (b), 20 jours après une section pratiquée au milieu de la région prépharyngienne de *Crenobia alpina*, régénération à 10°C. *C*, cerveau en voie de régénération, entouré de nombreux néoblastes *N*; *I*, intestin ancien; *IR*, intestin régénéré; *P*, pharynx régénéré; *Y*, œil régénéré. De Beauchamp, hémalun-éosine.

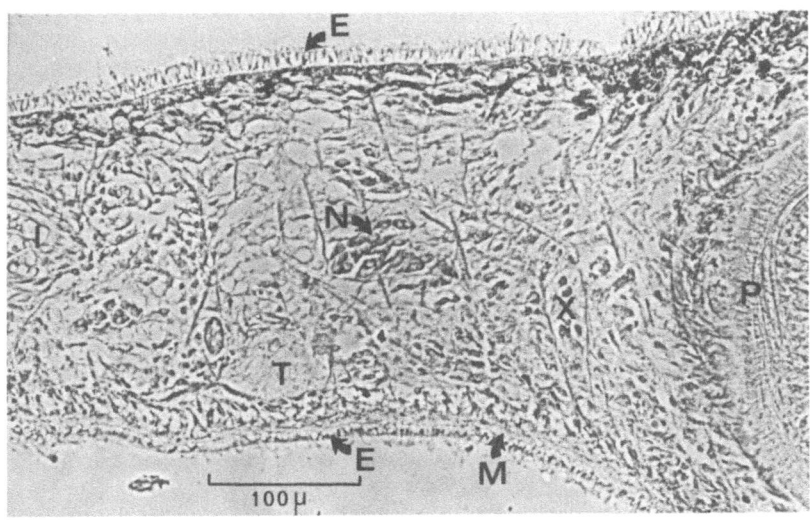

Fig. 4. Portion d'une coupe transversale de *Crenobia alpina*. *E*, épiderme; *I*, intestin; *M*, muscles sous-épidermiques; *N*, groupe de néoblastes; *P*, pharynx; *T*, tronc nerveux ventral; *X*, parenchyme traversé par des fibres musculaires dorso-ventrales. Alcool acétique, vert de méthyle-pyronine, contraste de phase.

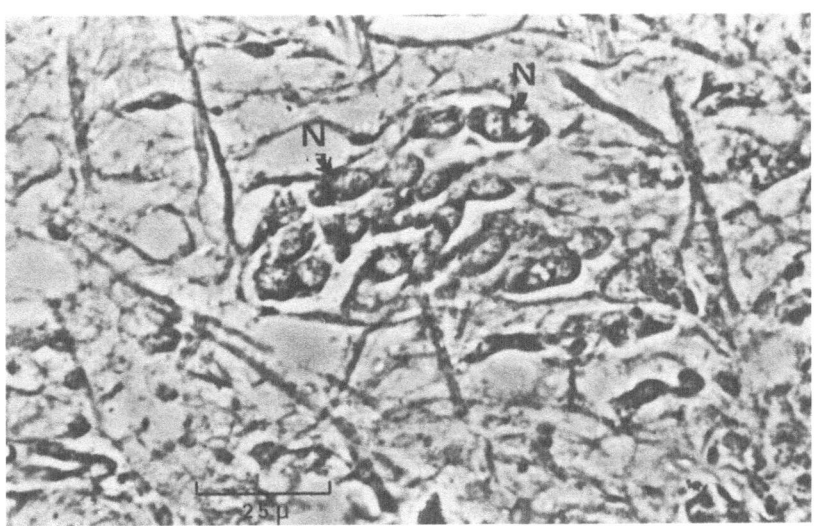

Fig. 5. Groupe de néoblastes *N* dans le parenchyme de *Crenobia alpina*. Alcool acétique, vert de méthyle-pyronine, contraste de phase.

On trouve dans le parenchyme des Planaires de nombreuses petites cellules ressemblant beaucoup par tous leurs caractères à celles du blastème (Fig. 4 et 5). Ces cellules sont capables de migrer sur de longues distances jusqu'au niveau d'une lésion. Elles sont totipotentes. La démonstration en a été faite principalement par Wolff et Dubois (1947 et 1948), Dubois (1949b) et Lender et Gabriel (1965).

Pouvoir de migration et totipotence des néoblastes

L'expérience cruciale consiste à irradier aux rayons X (5.000 r) la région prépharyngienne d'une Planaire *Dugesia lugubris*, détruisant ainsi tous les néoblastes dans cette zone (Fig. 6). Si l'irradiation n'est pas suivie d'une amputation, toute la région prépharyngienne se nécrose peu à peu et s'élimine jusqu'au niveau de la région saine, qui survit. Si l'on décapite cette Planaire irradiée localement, un court blastème apparaît quelques jours après la section, mais il ne se différencie pas et régresse rapidement. Ce premier blastème voué à la dégénérescence est constitué uniquement par les néoblastes radiolésés, qui se nécrosent dès qu'ils ont achevé leur migration au niveau de la section. Ce n'est que 4 semaines après l'amputation qu'un deuxième blastème se forme et grandit en régénérat définitif, qui comporte bientôt cerveau et yeux. Le délai d'apparition du second blastème est d'autant plus long que la zone irradiée est plus étendue. Il est d'une quinzaine de jours pour une tranche égale au dixième de la longueur du corps. Il atteint une centaine de jours, si huit dixièmes de la longueur du corps ont été irradiés.

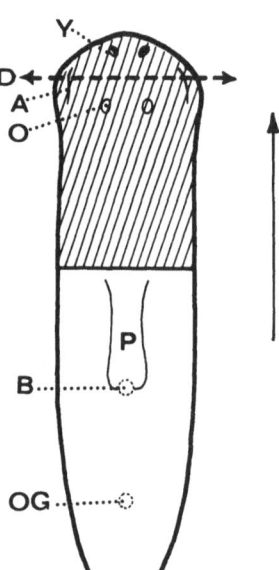

Fig. 6. Irradiation régionale aux rayons X chez *Dugesia lugubris*. La Planaire est décapitée *D*, puis la région prépharyngienne est irradiée (*hachures*). Un régénérat céphalique apparaîtra environ un mois après l'amputation (au lieu de 7 jours chez les témoins non irradiés). La flèche indique le sens de la migration des néoblastes. *A*, auricule; *B*, bouche; *O*, ovaire; *OG*, orifice génital; *P*, pharynx; *Y*, œil. (D'après Dubois, 1949b.)

Ces résultats s'interprètent aisément si l'on admet que des cellules de la région saine migrent à travers la zone irradiée jusqu'à la surface de section. Une autre expérience (Dubois, 1949b) démontre de façon frappante le pouvoir de migration et le rôle des néoblastes dans la régénération. Une *D. lugubris* est irradiée totalement. Tous ses néoblastes sont donc détruits et la Planaire est vouée à la mort. Il suffit cependant d'un fragment de tissu sain, prélevé dans une Planaire non irradiée et implanté vers le milieu de la Planaire irradiée, pour que celle-ci échappe à la nécrose et retrouve sa capacité de régénérer (Fig. 7). Si l'on sectionne sa tête, un bourgeon de régénération se forme au niveau de la section après un délai plus ou moins long, selon la distance qui sépare le greffon de la section. Si l'on a pris soin de greffer un fragment provenant par exemple d'un individu sombre sur une Planaire à pigmentation claire, on constate que la tête régénérée est d'une teinte sombre conforme à celle du greffon.

Sur la base d'expériences négatives, réalisées sur *D. dorotocephala*, Flickinger (1964) a nié la nécéssité de migrations étendues pendant la régénération de la tête. Pour lui, les néoblastes impliqués dans la reconstitution d'une tête sont issus exclusivement de la zone située immédiatement

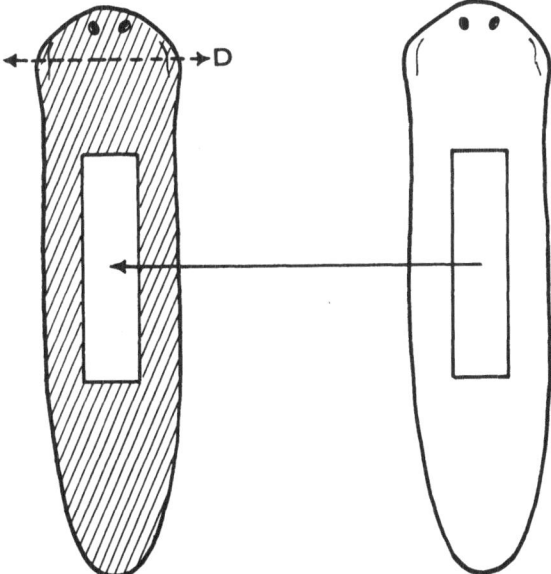

Fig. 7. Transplantation d'un greffon sain (*blanc*) dans un hôte irradié *in toto* aux rayons X (*hachures*). Environ un mois après la décapitation *D* de l'hôte, un régénérat de tête se forme vers l'avant ; il sera pigmenté comme le greffon. En quelques mois, l'hôte entier aura la teinte du donneur. (D'après Dubois, 1949b.)

en arrière de la section. Un greffon provenant d'un donneur préalablement marqué au carbone-14 est implanté dans la région prépharyngienne
d'un hôte non marqué, dont la tête est sectionnée. Les autoradiographies
montrent que le régénérat de tête ne contient pas l'isotope radioactif,
mais que la totalité des cellules greffées reste dans le greffon. Ce résultat
peut s'expliquer aisément, sans qu'il permette pour autant de mettre en
doute le grand pouvoir de migration des néoblastes. En effet, dans les
expériences de Flickinger, l'hôte n'est pas irradié aux rayons X. Ses néoblastes sont donc sains et capables de donner naissance à un régénérat
normal. Ceux qui se trouvent entre le niveau de la section et le greffon
sont peut-être suffisants pour édifier un blastème. Les néoblastes du
greffon sont inutiles et restent inutilisés.

Réalisant selon un schéma expérimental analogue des transplantations
chez *D. gonocephala* (Fig. 8), Lender et Gabriel (1965) ont apporté la
preuve irréfutable du grand pouvoir de migration des néoblastes et de leur
totipotence au cours de la régénération. L'expérience consiste à greffer un
fragment de Planaire saine marquée préalablement à l'uridine tritiée dans
une Planaire hôte irradiée totalement aux rayons X. Le greffon marqué est
prélevé dans la région prépharyngienne et est implanté dans la région
prépharyngienne de l'hôte irradié. La tête de celui-ci est sectionnée 24
heures après la prise du greffon. L'étude histoautoradiographique des
Planaires en voie de régénération permet de suivre la migration des néoblastes sains marqués depuis le greffon jusqu'au niveau de la section. Là
s'édifie le blastème formé exclusivement par les néoblastes radioactifs.
Les Planaires irradiées non greffées et décapitées ne régénèrent pas.

Ainsi la preuve est faite (1) que les néoblastes sains migrent jusqu'au
bord antérieur d'une Planaire irradiée et décapitée, et (2) qu'ils y édifient
tous les tissus du régénérat.

Il existe d'autres preuves de la totipotence des néoblastes. L'observation histologique des zones radiolésées envahies par les néoblastes sains
et celles des bourgeons de régénération permet de suivre la différenciation
progressive des cellules de régénération en divers types de cellules spéciali-
sées: cellules nerveuses, musculaires, épidermiques, etc.

En particulier, dans le cas d'irradiation de la région prépharyngienne,
les ovaires et leur contenu en cellules germinales dégénèrent rapidement
après l'irradiation. L'emplacement des ovaires est alors colonisé graduellement par les néoblastes, qui se transforment en ovogonies, puis en ovocytes
sains. La reconstitution des ovaires se termine 1 à 3 semaines après la
régénération de la tête. Le Ver est alors à nouveau capable de se reproduire.

Le cas de la régénération des testicules n'est pas moins démonstratif.
L'utilisation de rayons X peu pénétrants (7 kV, 15 mA) permet la destruction totale et sélective des testicules, situés dans la région dorsale du corps
chez *D. lugubris* (Fedecka-Bruner, 1964). L'irradiation n'affecte pas les
cellules du parenchyme et de l'épiderme dorsaux. Seuls les testicules
subissent des radiolésions irréversibles se traduisant par des pycnoses.
Leurs cellules germinales sont particulièrement radiosensibles et sont

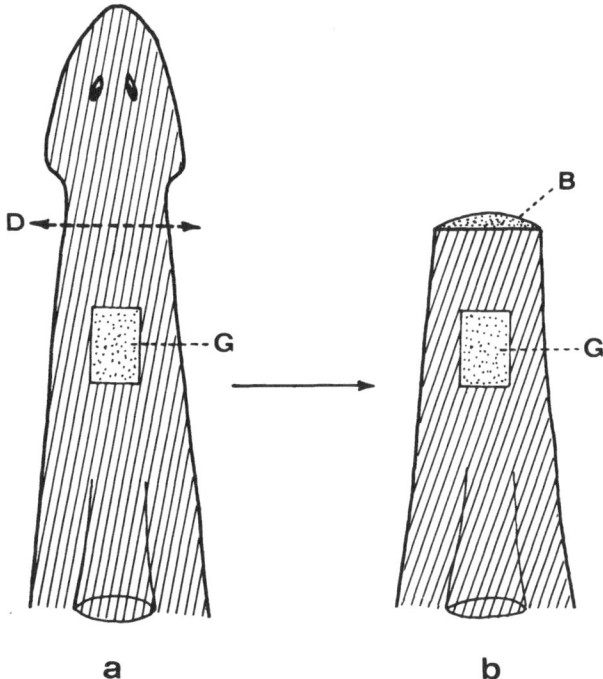

Fig. 8. Démonstration de la migration des néoblastes chez *Dugesia gonocephala*. Un greffon *G* marqué à l'uridine tritiée (*pointillé*) est implanté dans la région prépharyngienne d'une Planaire irradiée *in toto* aux rayons X (*hachures*). Quatre à six jours après la décapitation *D* de l'hôte, un bourgeon de régénération antérieur *B* apparaît, qui est entièrement constitué par les néoblastes radioactifs. (D'après Lender et Gabriel, 1965.)

toutes éliminées 15 jours après l'irradiation. La phase de réparation débute par une accumulation de néoblastes dans la région ventrale de la Planaire, particulièrement autour des troncs nerveux. Ensuite les néoblastes migrent vers la face dorsale du corps, en se glissant le long des fibres musculaires dorso-ventrales. Arrivés à l'emplacement des testicules nécrosés, les néoblastes s'organisent en petits amas, puis se transforment en spermatogonies, qui bientôt entrent en mitose. Les testicules arrivent à maturité entre le vingt-cinquième et le trentième jour après l'irradiation.

Cette expérience est intéressante à plus d'un point de vue. Non seulement elle fournit un exemple de plus de la potentialité des néoblastes, mais elle apporte aussi des précisions sur divers aspects de leur migration. Jusqu'ici, les expériences de Wolff et Dubois (1947) et de Dubois (1949a) avaient montré que les néoblastes étaient capables de migrer vers l'avant, vers l'arrière et latéralement. Les résultats de Fedecka-Bruner (1964) prouvent qu'ils sont également capables de se déplacer suivant un axe dorso-ventral. De plus, on pensait (Stéphan-Dubois, 1965) que seule une incision franche provoque la migration des néoblastes vers le lieu de la blessure. On vient de voir qu'une lésion interne, en l'occurrence la dégéné-rescence des testicules irradiés, est aussi capable de mobiliser et de faire migrer les néoblastes.

Les facteurs qui provoquent cette mobilisation, et plus précisément le moyen de communication entre la lésion et les néoblastes, restent encore inconnus. Il peut s'agir d'une nécro-hormone libérée dans le milieu interne par des cellules lésées ou d'un stimulus transmis par les troncs nerveux.

Propriétés des néoblastes cultivés *in vitro*

C. Sengel (1960 et 1963) a mis au point une technique de culture *in vitro* de blastèmes de régénération de Planaires. Les blastèmes sont prélevés 2 à 3 jours après l'amputation de la tête ou de la queue. A ce stade, ils sont encore indifférenciés et se présentent sous la forme d'une petite languette

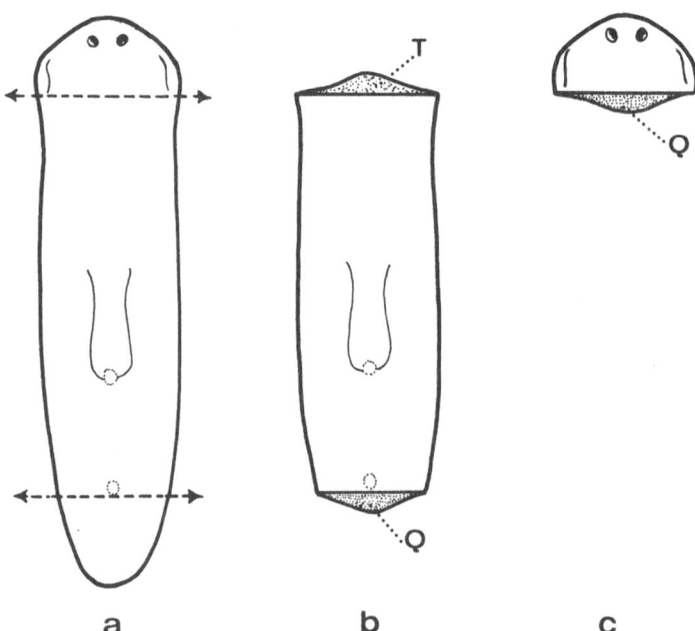

a   b   c

Fig. 9. Formation des blastèmes de régénération antérieur et postérieurs chez *Dugesia lugubris*. (a) Amputation de la tête et de la queue; (b) régénération d'un blastème de tête *T* et d'un blastème de queue *Q* après 2 à 3 jours; (c) régénération d'un blastème de queue *Q*. (D'après C. Sengel, 1960.)

étroite, translucide et non pigmentée (Fig. 9). Les expériences montrent que de tels blastèmes sont pourtant déjà déterminés.

Douze à vingt-quatre heures après l'explantation, l'épiderme cicatriciel recouvre la surface de section et l'explant prend l'aspect d'un petit nodule sphérique, puis s'aplatit en forme de lentille. Il s'agit donc d'une culture organisée de néoblastes contenus dans un sac d'épithélium cicatriciel. Les explants, dans ces conditions, sont incapables de croissance, mais ils se différencient comme ils le feraient *in vivo*: ils forment du pigment le troisième jour, des muscles le quatrième et le cinquième jour, un cerveau et des yeux entre les huitième et quinzième jours de culture (Fig. 10). Mais le résultat le plus important de ces expériences est que les explants se différencient en tête (avec cerveau et yeux) s'ils proviennent d'un blastème antérieur, et en queue (sans cerveau ni yeux) s'ils proviennent d'un blastème postérieur. Ils se différencient donc conformément à leur destinée normale. On peut affirmer qu'au moment où ils sont prélevés, les blastèmes sont déjà entièrement polarisés et déterminés par leur contact avec les tissus anciens. Le fait de les isoler dès le deuxième jour ne change plus rien à leur différenciation ultérieure. De plus il est remarquable de constater qu'il ne se reconstitue pas, à partir d'un blastème isolé, une Planaire entière: un blastème de tête ne forme jamais de queue ni de région pharyngienne, non plus qu'un blastème de queue n'édifie de tête ou de région pharyngienne. Les néoblastes au deuxième jour de régénération ne sont donc plus totipotents. Leur capacité morphogénétique se trouve limitée aux tissus et aux organes que le blastème était appelé à remplacer.

Dans le cas pourtant de l'association *in vitro* d'un blastème de tête avec un ou deux blastèmes de queue (Fig. 11), la fusion et la coopération

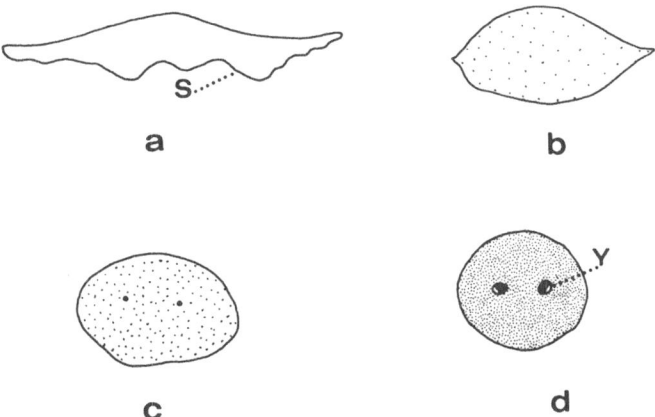

Fig. 10. Différenciation *in vitro* d'un blastème céphalique de *Dugesia lugubris*. (a) Aspect de l'explant au moment de la mise en culture; (b) après 48 à 72 heures; (c) après 8 jours; (d) après 10 à 15 jours. *S*, surface de section; *Y*, œil. (D'après C. Sengel, 1960.)

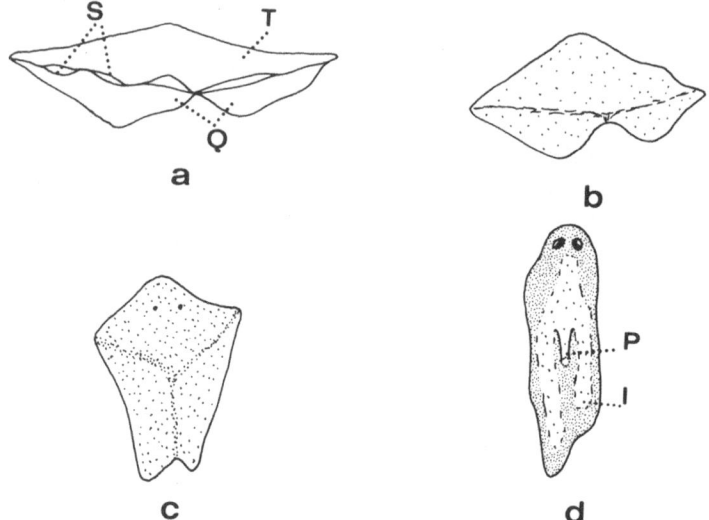

Fig. 11. Culture *in vitro* de l'association d'un blastème céphalique et de deux blastèmes caudaux. (a) Au moment de l'explantation; (b) après 48 à 72 heures; (c) après 8 jours; (d) après 10 à 15 jours. *I*, intestin à trois branches; *P*, pharynx avec sa bouche; *Q*, blastèmes caudaux; *S*, surfaces de section; *T*, blastème céphalique. (D'après C. Sengel, 1960.)

des explants conduisent à la reconstitution d'une petite Planaire entière, pourvue d'un pharynx et d'une bouche ainsi que d'une ébauche de tube digestif à trois branches. Ce résultat ne peut s'expliquer simplement par la quantité plus grande de néoblastes explantés dans ces associations, car la réunion de plusieurs blastèmes de tête ou de plusieurs blastèmes de queue n'aboutit jamais à la reconstitution d'une Planaire entière (Fig. 12).

Propriétés des néoblastes chez une Planaire à pouvoir de régénération limité

Il est évident que la totipotence et le pouvoir de migration des néoblastes ne sont pas des facteurs suffisants pour aboutir à une régénération harmonieuse et réparatrice. Nous venons de voir qu'au cours de la formation du blastème les néoblastes sont déterminés, par la restriction de leurs potentialités, à former les organes qui doivent être remplacés. Les facteurs de cette détermination émanent sans aucun doute des tissus anciens adjacents au bourgeon de régénération. Ces tissus constituent un territoire morphogène qui exerce sur les néoblastes une action inductrice déterminante. Nous aurons l'occasion d'en voir divers exemples dans la deuxième partie de cet article, consacrée aux facteurs de la morphogenèse régénérative (p. 96).

Mais la notion de territoire morphogène est particulièrement bien illustrée par les expériences réalisées sur *Dendrocoelum lacteum*, Planaire à pouvoir de régénération limité (Kolmayer et Stéphan-Dubois, 1960;

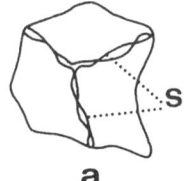
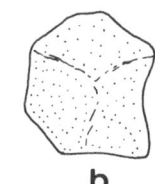

a   b

Fig. 12. Culture *in vitro* de
l'association de trois blastèmes
céphaliques. (a) Au moment de
l'explantation; (b) après 48 à
72 heures; (c) après 8 jours;
(d) après 10 à 15 jours. *S*, surfaces de section; *Y*, yeux.
(D'après C. Sengel, 1960.)

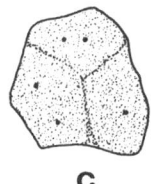
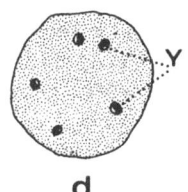

c   d

Stéphan-Dubois et Gilgenkrantz, 1961; et Stéphan-Dubois, 1965). Cette
Planaire ne régénère jamais de tête à partir d'un fragment postérieur
sectionné en arrière de la base du pharynx. Au contraire à partir d'une section
passant juste en arrière des yeux, une tête se reconstitue en 8 jours dans
100% des cas. Si la section est pratiquée au milieu de la région prépharyngienne, la régénération de la tête se réalise avec un retard de 5 jours dans 7
cas sur 10 seulement. Le pouvoir de régénération céphalique n'existe donc
qu'en avant du pharynx et diminue selon un gradient céphalo-caudal le
long de la région prépharyngienne pour devenir nul en arrière de la racine
du pharynx.

Comment expliquer cette inaptitude de la partie postérieure du corps
à régénérer une tête?

Elle n'est pas due à l'absence ou à la densité moindre des néoblastes
dans la région du corps postérieure à la base du pharynx. Des dénombre-
ments de néoblastes, caractérisés par leur colorabilité spécifique au vert de
méthyle-pyronine (Stéphan-Dubois et Kolmayer, 1959; Stéphan-Dubois,
1961; et Brøndsted et Brøndsted, 1961) ont montré que des néoblastes
existent à tous les niveaux du corps, répartis d'une manière sensiblement
uniforme, tout comme chez les Planaires à pouvoir de régénération total
(Pedersen, 1959; et Lender et Gabriel, 1960). Du reste, *D. lacteum* est
capable de régénérer une queue à partir d'une section pratiquée à n'importe
quel niveau. Cette inaptitude n'est pas le fait non plus d'une diminution du
pouvoir morphogène des néoblastes postérieurs, qui auraient perdu la
faculté d'édifier une tête. Les expériences d'irradiation localisée de Kolmayer
et Stéphan-Dubois (1960) en donnent la démonstration.

La région prépharyngienne de *D. lacteum* est irradiée au rayons X, de
sorte que tous les néoblastes de cette région sont détruits. La décapitation

de la Planaire déclenche la migration vers l'avant des néoblastes sains de la région postpharyngienne. Après un long délai, ils parviennent au niveau de la section, y édifient un blastème et reconstituent une tête. Ces néoblastes sont donc totipotents et doués des mêmes propriétés que les néoblastes de Planaires à pouvoir de régénération total.

Les expériences de transplantation combinée à l'irradiation aux rayons X (Stéphan-Dubois et Gilgenkrantz, 1961) donnent l'explication du pouvoir de régénération limité de *D. lacteum*. Un greffon de territoire prépharyngien irradié est implanté dans la région postpharyngienne (Fig. 13). Après la prise de la greffe, la Planaire est sectionnée à travers le greffon. Celui-ci, incapable de régénérer par lui-même, puisque tous ses néoblastes ont été détruits par l'irradiation, est envahi par les néoblastes sains de la zone postpharyngienne. Parvenus au niveau de la section, ils reconstituent une tête. Les tissus de l'hôte se cicatrisent sans régénérer. L'expérience inverse consiste à implanter un fragment postpharyngien sain dans la région prépharyngienne d'un hôte irradié *in toto* et dont on sectionne la

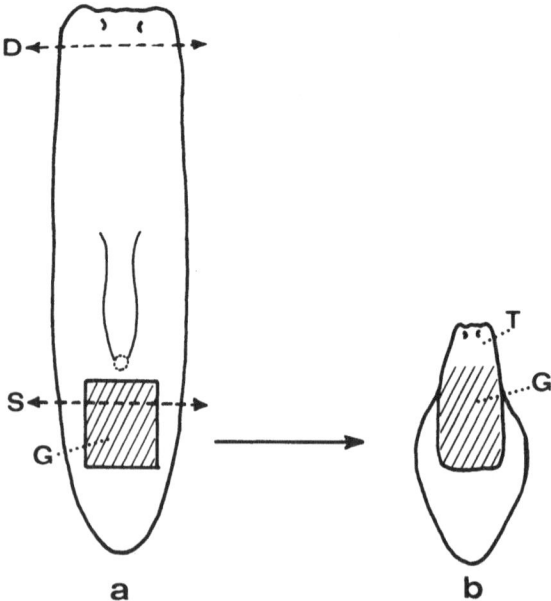

a   b

Fig. 13. Implantation d'un greffon prépharyngien irradié aux rayons X dans la région postpharyngienne d'un hôte sain, chez *Dendrocoelum lacteum*. (a) La Planaire hôte est décapitée *D*; après la prise du greffon *G* irradié (*hachures*), une deuxième section *S* est pratiquée en travers du greffon; (b) le fragment postérieur régénère une tête *T* vers l'avant, après un certain délai. (D'après Stéphan-Dubois et Gilgenkrantz, 1961.)

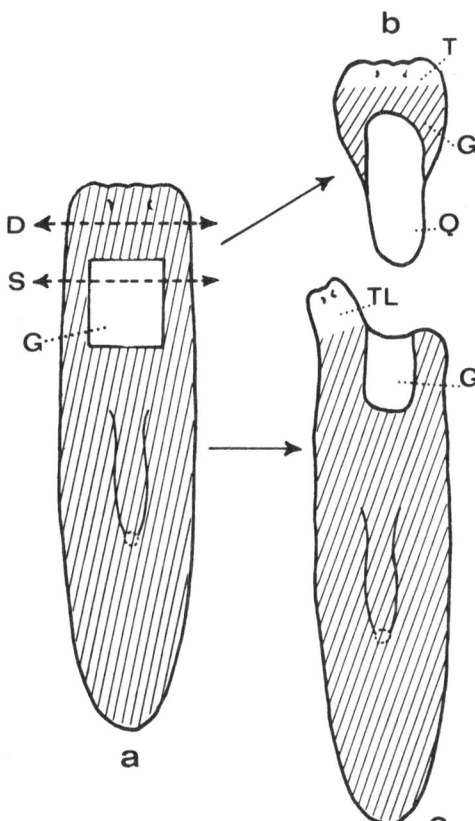

Fig. 14. Implantation d'un greffon postpharyngien sain dans la région prépharyngienne d'un hôte irradié *in toto* aux rayons X chez *Dendrocoelum lacteum*. (a) La planaire hôte irradiée (*hachures*) est décapitée *D*; après la prise du greffon *G*, une deuxième section *S* est pratiquée en travers du greffon; (b) le fragment antérieur régénère une queue *Q* vers l'arrière et, après un certain délai, une tête *T* vers l'avant; (c) le fragment postérieur régénère une tête latérale *TL* vers l'avant. (D'après Stéphan-Dubois et Gilgenkrantz, 1961.)

tête en arrière des yeux (Fig. 14). Une nouvelle section est pratiquée ensuite en travers du greffon. Le fragment antérieur régénère aussitôt une queue vers l'arrière à partir du greffon sain. Par la suite, les tissus irradiés de l'hôte sont colonisés par les néoblastes postpharyngiens sains qui édifient une tête en avant. De même dans le fragment postérieur, les néoblastes sains migrent latéralement hors du greffon et sont capables de régénérer une tête latérale typique.

En conclusion, tous les néoblastes de *D. lacteum* sont totipotents, capables de migrations étendues et possèdent le pouvoir de régénérer une tête, à condition qu'ils soient ou qu'ils parviennent en territoire prépharyngien. La notion de territoire morphogène inducteur est ainsi mise en évidence: seule la région prépharyngienne est capable d'induire la différenciation d'une tête; les territoires postérieurs à la racine du pharynx sont dépourvus de ce pouvoir.

Développement embryonnaire de *Polycelis nigra* et propriétés des néoblastes embryonnaires

Le développement embryonnaire des Planaires présente bon nombre de caractères bien particuliers, qu'il convient de rappeler ici. La plupart des descriptions sont déjà anciennes (voir la bibliographie correspondante dans Le Moigne, 1966). Mais récemment Le Moigne (1963) a fait une étude complète et détaillée du développement de *Polycelis nigra*. Il distingue sept stades embryonnaires de la ponte à l'éclosion.

Stade 1 (Fig. 15): c'est le stade de la multiplication des blastomères. Au moment de la ponte du cocon, celui-ci contient de 5 à 15 oeufs en métaphase de deuxième division de maturation (Lepori, 1949) et de 10 à 20.000 cellules vitellines. Les oeufs sont gynogénétiques, les spermatozoïdes n'ont qu'un rôle d'activation. La régulation chromosomique se fait par dédoublement des chromosomes avant la méiose (Le Moigne, 1962). On trouve deux types de cellules vitellines: les petites formeront le syncytium

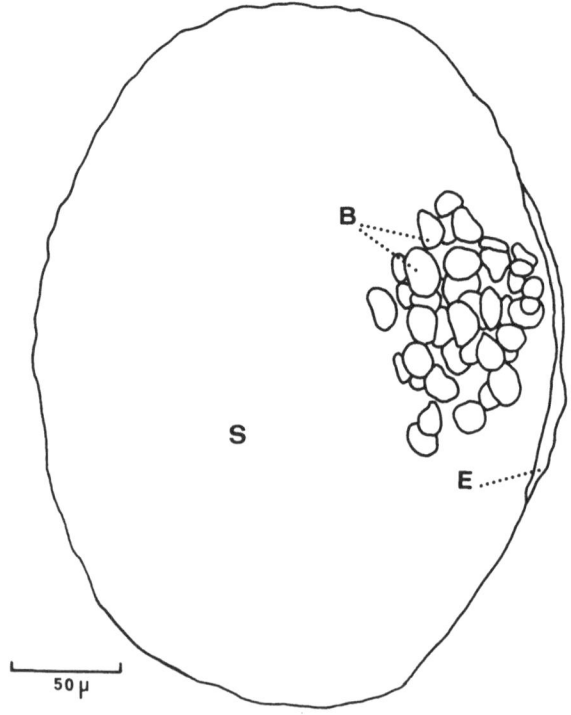

Fig. 15. Embryon de *Polycelis nigra* au stade 1, comptant une soixantaine de blastomères groupés à l'endroit des futurs pharynx et intestin embryonnaires. *B*, blastomères; *E*, épiderme embryonnaire; *S*, syncytium nourricier. (D'après Le Moigne, 1966.)

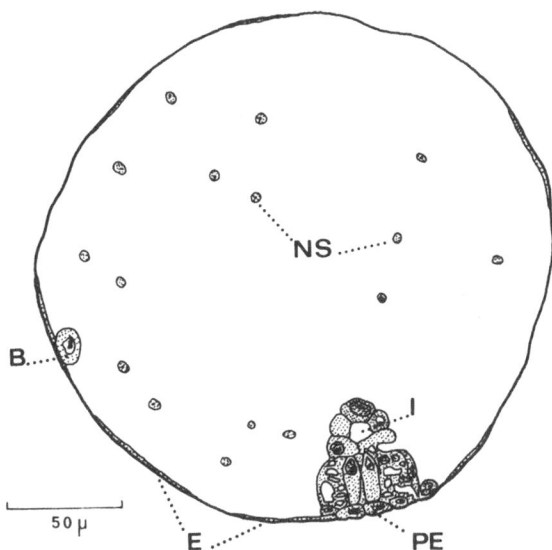

Fig. 16. Coupe sagittale d'un embryon de *Polycelis nigra* au stade 2. Le pharynx et l'intestin embryonnaires sont en voie de différenciation. *B*, blastomère; *E*, épiderme embryonnaire; *I*, intestin embryonnaire; *NS*, noyaux du syncytium nourricier; *PE*, pharynx embryonnaire. (D'après Le Moigne, 1966.)

nourricier; les grandes seront absorbées ultérieurement (stade 3) par l'intestin embryonnaire. La segmentation, irrégulière, aboutit à l'édification d'un embryon d'une cinquantaine de blastomères. A ce moment, s'ébauche un épithélium superficiel, l'épiderme embryonnaire. Les autres blastomères migrent vers le lieu de l'édification des futurs pharynx et intestin embryonnaires.

Au *stade* 2 (Fig. 16), se différencient le pharynx et l'intestin embryonnaires. L'embryon est âgé de 2 à 4 jours, il mesure de 0,2 à 0,3 mm de diamètre. Il comporte de 60 à 70 cellules, dont une dizaine forme l'épiderme. Une vingtaine de cellules s'ordonnent alors à l'emplacement de la future bouche, pour former l'ébauche du pharynx embryonnaire. Quatre cellules constituent celle de l'intestin embryonnaire; quatre autres occupent une position intermédiaire entre le pharynx et l'intestin. De 20 à 30 cellules indifférenciées restent libres dans le syncytium nourricier. L'enveloppe épidermique se complète peu à peu.

Le stade 3 (Fig. 17) est caractérisé par l'absorption des grandes cellules vitellines et la multiplication des cellules embryonnaires. L'embryon, alors âgé de 3 à 7 jours, a pris l'aspect d'une outre ovoïde d'environ 0,8 mm de long. Il est bordé par un épiderme très distendu, dont les cellules ne tardent pas à dégénérer. Le syncytium, repoussé à la périphérie, délimite une

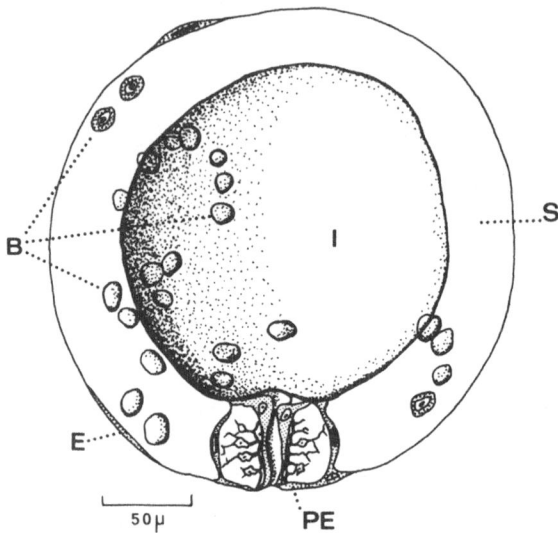

Fig. 17. Embryon de *Polycelis nigra* au début du stade 3,
en coupe médiane. Les blastomères sont répartis dans la
paroi formée par le syncytium nourricier. Les grandes
cellules vitellines absorbées ne sont pas représentées. *B*,
blastomères; *E*, épiderme embryonnaire; *I*, intestin embryonnaire; *PE*, pharynx embryonnaire; *S*, syncytium
nourricier. (D'après Le Moigne, 1966.)

cavité centrale, l'intestin embryonnaire, s'ouvrant à l'extérieur par le
pharynx embryonnaire. C'est par ce dernier que sont absorbées les cellules
vitellines, qui pénètrent ainsi dans la cavité intestinale. Celle-ci se tapisse
progressivement de cellules à caractère phagocytaire. Dès ce stade,
l'embryon acquiert sa polarité dorso-ventrale. La majorité des blastomères
indifférenciés se groupe dans l'hémisphère occupé par le pharynx et
définit ainsi approximativement la face ventrale de l'embryon. Ainsi,
chez un embryon comptant 170 cellules libres, 6 seulement sont situées dans
l'hémisphère dorsal. L'épiderme embryonnaire est remplacé progressivement par de nouvelles cellules superficielles formant l'épiderme définitif.

Au *stade* 4 (Fig. 18), 7 à 8 jours après la ponte, un pharynx transitoire,
formation propre à *Polycelis nigra-tenuis*, vient remplacer le pharynx
embryonnaire. Les cellules du parenchyme embryonnaire, constitué par les
blastomères restés entre l'épiderme et l'épithélium intestinal, se multiplient
activement surtout du côté ventral. Certaines, fusiformes, ressemblent
aux néoblastes.

A partir du *stade* 5 (Fig. 19), l'embryon subit des transformations qui
l'amèneront graduellement à la morphologie adulte. Le système nerveux,
le pharynx définitif et l'intestin triclade se différencient entre le neuvième
et le quatorzième jour après la ponte. L'embryon possède dès lors une polarité

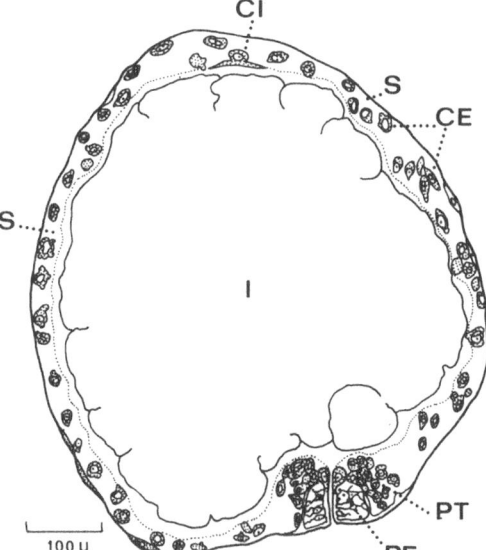

Fig. 18. Coupe sagittale d'un embryon de *Polycelis nigra* au stade 4. Des cellules s'amassent au-dessus et autour du pharynx embryonnaire pour former le pharynx transitoire. *CE*, cellules du parenchyme embryonnaire; *CI*, cellule intestinale en voie de différenciation; *I*, intestin empli de cellules vitellines; *PE*, pharynx embryonnaire; *PT*, pharynx transitoire; *S*, syncytium nourricier; *en pointillé*, limites du syncytium nourricier.
(D'après Le Moigne, 1966.)

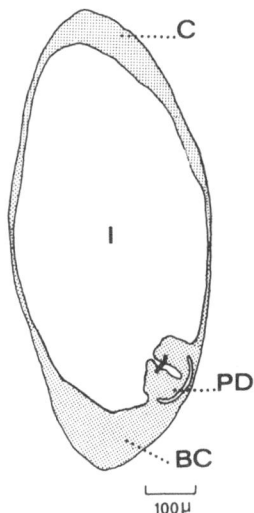

Fig. 19. Embryon de *Polycelis nigra* au stade 5, d'après une coupe longitudinale. *BC*, bourgeon caudal; *C*, cerveau; *I*, intestin; *PD*, pharynx définitif. (D'après Le Moigne, 1966.)

céphalo-caudale. La majorité des cellules est groupée dans une bande germinative ventrale, limitée à l'extérieur par un épiderme continu, à l'intérieur par un épithélium intestinal en formation. Dans cette bande, les cellules prennent l'aspect des néoblastes de l'adulte. A l'avant se différencient les premières cellules nerveuses, qui sont à l'origine des ganglions cérébroïdes. On ne distingue pas encore de fibres nerveuses. Peu après, le pharynx définitif s'édifie, à l'emplacement même du pharynx transitoire (ce qui est exceptionnel chez les Triclades). En arrière se dessine le bourgeon caudal, qui donnera naissance à la région postpharyngienne, où s'organisent les deux branches postérieures de l'intestin. Sur la face dorsale, l'épiderme, constitué de cellules étirées, reste encore très mince. Au fur et à mesure que l'embryon grandit, de nouvelles générations de cellules viennent en surface s'ajouter aux cellules épidermiques précédentes, augmentant ainsi la densité des cellules épithéliales. Ce processus caractéristique de migration centrifuge des cellules du parenchyme profond vers la surface se poursuit pendant toute la vie de la Planaire (Skaer, 1965). Au neuvième jour après la ponte, les rhabdites se développent dans le parenchyme. Les cellules les contenant migrent alors vers la surface et remplacent rapidement toutes les cellules de l'épiderme transitoire. Chez l'adulte, le remplacement continu des cellules épidermiques se réalise de la même façon. En se déplaçant vers la surface, les cellules à rhabdites entraînent avec elles les conduits des glandes unicellulaires.

Au *stade* 6 (Fig. 20), l'embryon possède les traits essentiels de l'animal à l'éclosion. Il mesure alors environ 1,2 mm de long. Les fibres nerveuses se différencient d'avant en arrière dans la bande germinative entre le

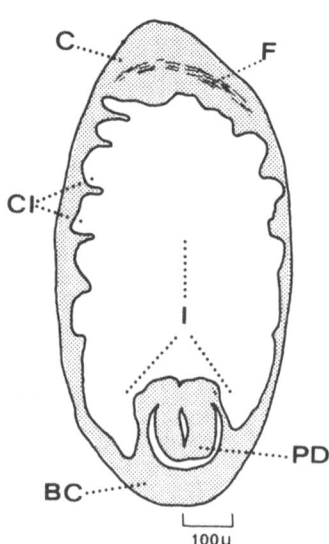

100μ

Fig. 20. Embryon de *Polycelis nigra* au stade 6, d'après une coupe frontale. *BC*, bourgeon caudal; *C*, cerveau; *F*, fibres nerveuses; *I*, intestin à trois branches; *CI*, caeca intestinaux; *PD*, pharynx définitif. (D'après Le Moigne, 1966.)

douzième et le dix-huitième jour. Les premières taches oculaires apparaissent au plus tôt 14 jours après la ponte, au moment où les troncs nerveux ont atteint la moitié de la longueur de l'embryon. Les autres organes parfont leur accroissement et leur différenciation. La ciliature apparaît sur l'épiderme ventral. Dès ce stade, des embryons extraits du cocon sont capables de survivre et peuvent s'alimenter au bout de 24 jours.

Durant le *stade* 7 (Fig. 21), les divers organes prennent leur forme définitive: le cerveau édifie sa commissure transverse, les troncs nerveux longitudinaux atteignent l'extrémité caudale et les yeux se développent d'avant en arrière. Le pharynx acquiert sa structure définitive avec ses diverses assises musculaires. La bouche s'ouvre. Le parenchyme est toujours très riche en cellules indifférenciées à l'aspect de néoblastes. La pigmentation apparaît. Pendant cette dernière période, la Planaire s'aplatit dorso-ventralement. A l'éclosion, elle mesure de 2 à 5 mm de long. Les organes génitaux se forment, aux dépens des néoblastes, quelques semaines ou quelques mois après l'éclosion.

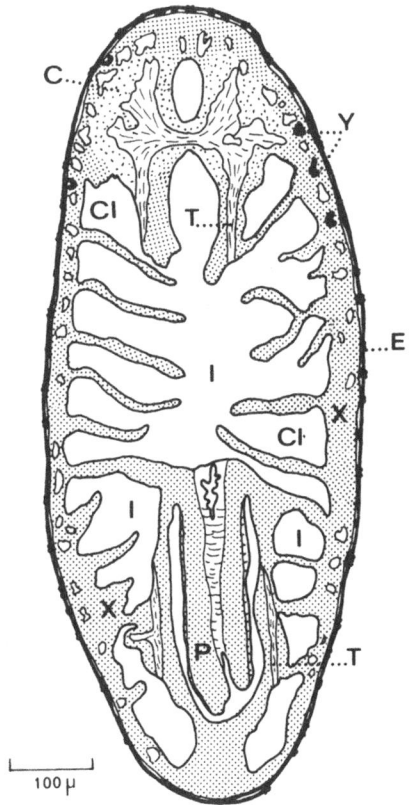

Fig. 21. Coupe frontale d'un embryon de *Polycelis nigra* au stade 7. *C*, cerveau; *CI*, caeca intestinaux; *E*, épiderme avec de nombreux rhabdites; *I*, intestin à trois branches; *P*, pharynx; *T*, troncs nerveux; *X*, parenchyme; *Y*, yeux. (D'après Le Moigne, 1966.)

Ce développement singulier des Planaires Triclades n'a pas encore été soumis à l'investigation expérimentale. On ne sait rien encore des mécanismes de la différenciation des principaux organes. Toujours est-il qu'à aucun moment du développement on ne peut distinguer plusieurs feuillets embryonnaires (nonobstant ces animaux sont classés par les zoologistes dans le grand groupe des Métazoaires Triploblastiques). Tous les organes s'édifient au sein d'une seule masse de blastomères ayant les mêmes propriétés que les néoblastes de l'adulte.

Toutefois, dans ses recherches récentes de régénération chez l'embryon, Le Moigne (1965a et b, et 1966) a montré que, très tôt, il convient de distinguer deux lignées de cellules embryonnaires. L'une comprend les cellules appelées à se différencier immédiatement au cours des stades suivants de l'ontogenèse. L'autre comporte les cellules qui, au stade 5, prennent l'aspect morphologique des néoblastes, restent indifférenciées et conservent, semble-t-il, toutes leurs pòtentialités embryonnaires. Les expériences ont porté seulement sur des embryons des stades 5, 6 et 7 et sur de jeunes *Polycelis nigra-tenuis* à l'éclosion. Aux stades plus précoces, elles sont impraticables. Les animaux sont coupés transversalement en deux parties (Fig. 22). Aux stades 5 et 6, la section isole une partie postérieure dépourvue de troncs nerveux. Au stade 7, les troncs nerveux sont bien développés en arrière de la section.

La capacité de régénération, exprimée par la proportion des individus opérés qui régénèrent au bout de 9 à 19 jours, augmente avec l'âge des embryons. Au stade 5, la mortalité est très élevée: Le Moigne n'a pu obtenir que deux cas de régénération d'une tête et d'un cerveau à partir d'un fragment postérieur. Ce résultat suffit pourtant à démontrer que, dès ce stade, la régénération est possible. Au stade 6, 85 % des fragments antérieurs ont régénéré une queue et un pharynx et 59 % des fragments postérieurs ont régénéré une tête et un cerveau. Au stade 7, les proportions sont respectivement de 91 % pour la régénération postérieure et 59 % pour la régénération antérieure. Enfin chez les jeunes éclos la régénération se réalise dans 100 % des cas.

Les fragments antérieurs, aux stades 6 et 7, semblent donc avoir un pouvoir de régénération plus élevé que les fragments postérieurs. Cela pourrait s'expliquer par l'existence d'un gradient morphogénétique (Brøndsted, 1955), dont le système nerveux, qui se développe d'avant en arrière, pourrait être le support anatomique.

L'étude histologique des fragments d'embryons en cours de régénération et de ceux qui ne régénèrent pas a permis de dégager les faits suivants:

1. La régénération chez l'embryon se fait à partir de cellules ayant les mêmes propriétés que les néoblastes de l'adulte. Celles-ci migrent vers la surface de section pour y édifier le blastème. La formation du blastème est toutefois plus lente que chez l'adulte, et chez certains embryons elle n'a pas lieu.

2. La différenciation des organes se poursuit normalement aussi bien chez les individus qui régénèrent que chez ceux qui ne le font pas. En

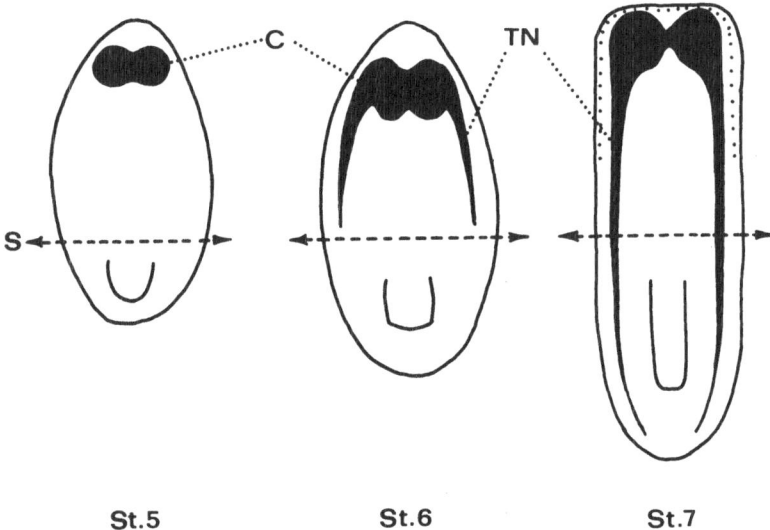

Fig. 22. Section *S* d'embryons de *Polycelis nigra* aux stades 5, 6 et 7. *C*, cerveau;
TN, troncs nerveux. (D'après Le Moigne, 1965c.)

particulier, le système nerveux continue son développement; même dans les parties postérieures qui, au départ, ne possèdent pas d'éléments nerveux, les troncs nerveux se différencient indépendamment du cerveau, qu'il y ait régénération ou non.

Il est possible que le système nerveux joue un rôle important dans la mobilisation des néoblastes. Avant que celui-ci ne soit suffisamment développé, les néoblastes semblent incapables de répondre rapidement au stimulus traumatique. Dès l'instant que les troncs nerveux sont présents, les néoblastes sont mobilisables et capables d'édifier un blastème. Chez l'adulte le système nerveux joue un rôle activateur analogue, comme l'ont montré Lender et Gripon (1962).

Enfin, les expériences d'irradiation aux rayons X, réalisées par Le Moigne (1965b et 1966) sur des embryons à tous les stades du développement, ont mis en évidence la ségrégation progressive des néoblastes au cours de l'ontogenèse. La radiosensibilité des embryons à des doses de 750 à 3.000 r diminue avec l'âge croissant. Aux stades 1 et 2, tous les embryons sont tués. Au stade 3, la mortalité est de 75 %; parmi les survivants, 85 % sont anormaux à l'éclosion. Au stade 4, 75 % des embryons irradiés survivent et, parmi eux, 90 % naissent normaux. Enfin tous les embryons aux stades 5, 6 et 7 éclosent normaux.

Cette variabilité de la radiosensibilité s'explique aisément si l'on admet que les cellules à caractère embryonnaire et à grand pouvoir mitotique sont plus sensibles aux rayons X que les cellules déterminées ou différenciées. Or, la proportion de ces cellules totipotentes diminue évidemment au cours du développement, à mesure que les organes se différencient. Mais il est une catégorie de cellules qui conserve pendant toute le vie de la Planaire ses caractères embryonnaires. Ce sont les néoblastes. L'étude histologique des embryons irradiés montre précisément qu'au moment de l'éclosion la plus grande partie des néoblastes a disparu. Ceux qui restent présentent tous des lésions caractérisées par l'absence de nucléole ou son appauvrissement en acide ribonucléique. L'amputation d'embryons irradiés et parvenus à l'éclosion révèle par ailleurs qu'ils sont incapables de régénérer.

En conclusion, les irradtions totales d'embryons lèsent les néoblastes ou leurs précurseurs en épargnant les cellules en voie de différenciation ou déjà différenciées. C'est pratiquement à partir du stade 4 qu'on réalise cette discrimination. C'est également à ce stade que commencent à se former les principaux organes et qu'on peut reconnaître dans le parenchyme embryonnaire des cellules à l'aspect de néoblastes. C'est donc vraisemblablement pendant cette période du développement que se séparent les deux lignées cellulaires de l'embryon de Planaire: les néoblastes d'une part, les cellules déterminées dans leur différenciation immédiate d'autre part.

LES FACTEURS DE LA MORPHOGENESE REGENERATIVE: INDUCTIONS ET INHIBITIONS

Nous avons vu que les néoblastes, dont nous connaissons à présent les propriétés, sont un élément nécessaire mais non suffisant de la régénération. Des facteurs morphogènes émanant des tissus restants influencent la destinée des néoblastes, en limitant leur pouvoir de différenciation à l'édification des seules parties manquantes.

De nombreuses expériences démontrent l'existence de ces facteurs, tantôt inducteurs, tantôt inhibiteurs, dont l'équilibre aboutit à la reconstitution d'un organisme normal et harmonieusement proportionné. Nous prendrons comme exemple le cas de la régénération de divers organes, dont le mécanisme a été étudié d'une manière approfondie: ce sont la tête, le cerveau, les yeux et le pharynx.

La régénération de la tête et du cerveau

La régénération de la tête à partir d'un fragment postérieur est généralement caractérisée, à l'examen macroscopique, par l'apparition des yeux dans le blastème et, à l'examen histologique, par la différenciation du cerveau.

On ne connaît pas encore le déterminisme de la différenciation de la tête. Mais il est hors de doute que les tissus anciens, la base, jouent un rôle essentiel dans sa régénération. Souvenons-nous que chez *Dendrocoelum lacteum* les territoires postpharyngiens sont incapables d'induire la formation d'une tête (Stéphan-Dubois et Gilgenkrantz, 1961), alors que la région prépharyngienne est dotée de cette faculté. Au demeurant, les expériences de Chevtchenko (1936 et 1937) montrent que le cerveau d'une Planaire exerce une puissante inhibition sur la différenciation d'une tête greffée dans son voisinage. L'inhibition décroît avec la distance du cerveau, et la tête greffée se développe d'autant mieux qu'elle est plus loin.

Notre connaissance des facteurs de la régénération de la tête se limite au rôle joué par divers corps chimiques stimulateurs ou inhibiteurs et à l'action inhibitrice de certains extraits tissulaires sur la différenciation du cerveau.

Kanatani (1957 et 1959) constate que, chez *Dugesia gonocephala*, la régénération de la tête est stimulée par l'insuline (10 et 50 μg/ml, surtout en présence de glucose à 0,1 %) et par l'oestradiol (0,01 et 0,1 μg/ml). Lorsque les animaux sont placés pendant un certain temps après la décapitation dans des solutions de ces substances, la surface occupée par le régénérat en fin d'expérience est plus grande chez les sujets traités que chez les témoins. Mais ces substances n'ont aucun effet sur la date d'apparition des yeux. Il est vraisemblable qu'il s'agisse simplement d'une stimulation trophique de la croissance.

Des résultats plus intéressants ont été obtenus par Coward, Flickinger et Garen (1964) chez *Dugesia dorotocephla*. Chez cette Planaire possédant un gradient physiologique antéropostérieur très prononcé, la fréquence de régénération de tête (*head frequency* de Child, 1941), qui est de 100 % pour une section passant en arrière des yeux, tombe à 50 % pour une section pratiquée dans la région caudale (Brøndsted, 1955). La fréquence de tête traduit la capacité d'un niveau donné du corps à régénérer des structures céphaliques complètes. Elle s'exprime aisément par la durée d'apparition des taches oculaires. Des tronçons transversaux de *D. dorotocephala* sont placés dans diverses solutions nutritives contenant soit du ribonucléate de sodium de levure (0,005 à 0,01 %), soit du sérum de Cheval (5 %), soit des acides aminés (1 %), soit de l'extrait d'embryon de Poulet (1 %). Tous ces milieux augmentent significativement la fréquence de tête et la vitesse de formation des yeux, particulièrement dans les tronçons postérieurs. Concomitamment le taux d'incorporation de CO_2 est accru, montrant que la rapidité de la régénération va de pair avec une activité de synthèse protéique plus élevée.

Divers auteurs (Kanatani, 1958; Flickinger, 1959; Doumain, 1960; et Flickinger et Coward, 1962) ont soumis des Planaires à des inhibiteurs métaboliques variés, qui en général empêchent ou retardent la régénération céphalique. Mais l'effet remarquable de ces substances se traduit par la différenciation fréquente d'hétéromorphoses allant jusqu'à la formation de Planaires bipolaires, régénérant une tête normale en avant et une deuxième tête hétéromorphique à la place de la queue. Les meilleurs résultats ont été obtenus avec la colcémide (déacétylméthylcolchicine), le chloramphénicol et le β-mercaptoéthanol. Ces substances, dont on sait qu'elles dépriment l'activité métabolique des cellules, effacent en quelque sorte le gradient physiologique céphalo-caudal, perturbant par là-même les facteurs du maintien de la polarité antéro-postérieure.

De telles inversions de la polarité n'avaient réussi tout d'abord que sur des Planaires décapitées. Par l'application ingénieuse d'un traitement chimique en deux temps, Flickinger et Coward (1962) ont montré qu'il est pourtant possible de produire une tête hétéromorphique de polarité inverse en la présence de la tête normale. Des *D. dorotocephala* sont placées pendant un jour dans une solution de colcémide (0,02 mg/ml). Après quoi, on isole la région prépharyngienne par une section transversale passant juste en avant du pharynx. La tête de ces régions prépharyngiennes est ensuite plongée dans une solution de colcémide (0,01 à 0,02 mg/ml), la Planaire étant maintenue immobile dans un bloc d'agar (Fig. 23). L'extrémité postérieure du fragment prépharyngien n'est pas en contact direct avec le liquide contenant la colcémide. Il s'établit ainsi un gradient de concentration de l'inhibiteur métabolique; celle-ci est maximum à la tête, minimum au niveau de la section. Après 18 à 48 heures de ce traitement, les Vers sont replacés dans l'eau, où ils édifient bientôt un blastème de régénération postérieur.

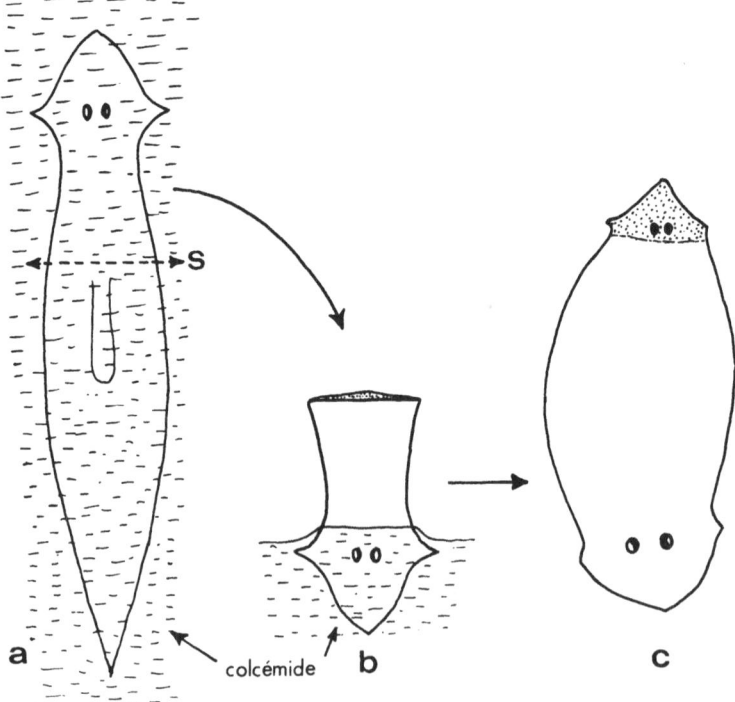

Fig. 23. Régénération d'une tête hétéromorphique de polarité inverse chez *Dugesia dorotocephala*. (a) Traitement préalable de la Planaire entière à la colcémide; (b) après section *S* en avant du pharynx, la région prépharyngienne est incluse dans un bloc d'agar, puis la tête seule est immergée dans une solution de colcémide; (c) le bourgeon de régénération postérieur (*pointillé*) se différencie en une tête avec des yeux, malgré la présence de la tête normale. (D'après Flickinger et Coward, 1962.)

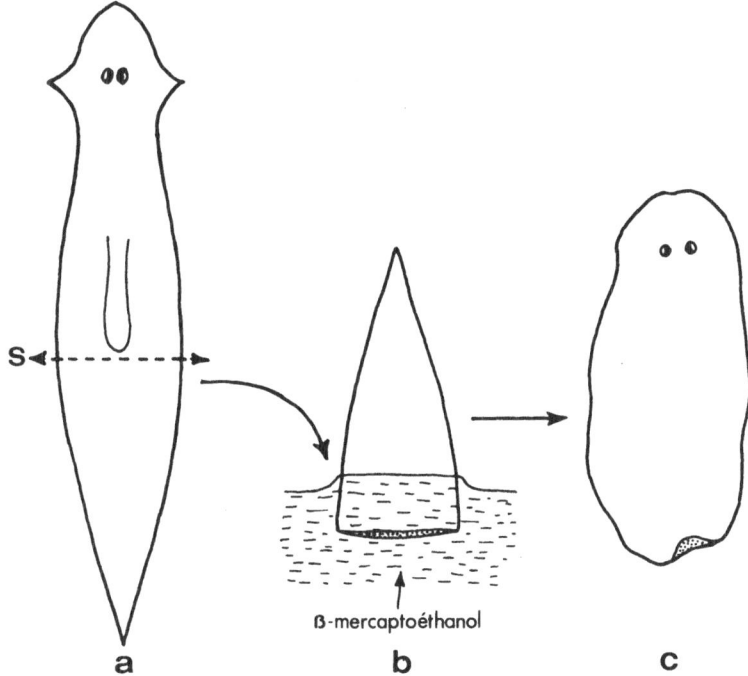

Fig. 24. Différenciation d'une tête de polarité inverse dans une queue, chez *Dugesia dorotocephala*. (a) isolement de la région postpharyngienne par une section *S* en arrière du pharynx; (b) l'extrémité antérieure de la région post-pharyngienne est immergée dans une solution de β-mercaptoéthanol; (c) une tête avec des yeux se différencie dans les tissus de la queue; la régénération antérieure est inhibée. (D'après Flickinger et Coward, 1962.)

Dans 5 cas sur 150 survivants, une tête complète avec cerveau et yeux s'est différenciée à partir de l'extrémité postérieure de tels fragments, et cela malgré la présence de la tête normale.

Une inversion de la polarité peut-être plus spectaculaire encore a été réalisée par les mêmes auteurs dans des fragments postpharyngiens de *D. dorotocephala*. Ces parties postérieures sont incluses dans un bloc d'agar et placées au contact d'une solution d'iodoacétamide (10^{-5} M) pendant 2 jours, ou d'une solution de colcémide (0,02 mg/ml) pendant 18 à 48 heures, ou encore d'une solution de β-mercaptoéthanol (10^{-3} M) pendant 1 jour, de telle façon que seule la surface de section plonge dans le liquide (Fig. 24). Dans 1 cas sur 8 en présence d'iodoacétamide et dans 4 cas sur 36 en présence de β-mercaptoéthanol, la queue de ces fragments a formé des taches oculaires typiques qui signent la différenciation d'une tête hétéromorphique de polarité inverse. La colcémide n'a pas eu d'effet aussi net, mais bon nombre des individus présente des taches oculaires incomplètes et manifeste une bipolarité fonctionnelle. En aucun cas, la tête ne régénère à l'extrémité antérieure. Il est remarquable de constater qu'une tête peut se différencier par morphallaxis pure dans une région du corps qui n'est pas en cours de régénération.

En contrecarrant le gradient physiologique céphalo-caudal, ces substances lèvent l'inhibition qui pèse normalement sur les parties postérieures, leur permettant ainsi de réaliser des potentialités supérieures à leur destinée normale.

Mais dans les conditions normales (en l'absence d'inhibiteurs métaboliques), il est manifeste que la région antérieure d'une Planaire exerce effectivement une puissante action inhibitrice sur la régénération d'une tête et particulièrement sur celle du cerveau (Morgan, 1902, et Chevtchenko, 1937). Un cerveau greffé dans la région antérieure d'une *Dugesia lugubris* décapitée (Lender, 1960) inhibe la régénération d'un cerveau dans le blastème antérieur. Tantôt le cerveau greffé vient occuper l'emplacement où aurait dû se former un cerveau dans le régénérat, tantôt une tête régénère indépendamment du cerveau greffé, mais cette tête ne contient pas ou contient très peu de cellules nerveuses.

Lender (1955a et c, et 1956c) a tenté d'élucider le mécanisme de cette inhibition. Des *D. lugubris* et des *Polycelis nigra* décapitées sont élevées dans de l'eau contenant des broyats de fragments de Planaire. En présence de broyat de tête, les Vers régénèrent une tête avec des yeux, mais le cerveau ne s'y différencie pas ou reste beaucoup plus petit que chez les témoins. Les broyats de queue, par contre, n'exercent aucune inhibition sur la formation du cerveau. La tête des Planaires contient donc une substance diffusible, soluble dans l'eau, capable d'empêcher la régénération d'un cerveau. Cette substance est stable à 60°C (Lender, 1956a). Elle ne présente pas de spécificité zoologique, car un broyat de tête de *D. lugubris* ou de *D. gonocephala* inhibe la régénération du cerveau de *P. nigra*. Une telle substance inhibitrice, émise régulièrement par le cerveau et diffusant dans le corps de la Planaire en concentration graduellement décroissante d'avant en arrière, fournit un support matériel plausible à la théorie des gradients physiologiques de Child (Wolff, 1962).

La régénération des yeux

La régénération des yeux est entièrement sous la dépendance du cerveau. En l'absence de cerveau, les yeux ne régénèrent pas. Les expériences de Wolff et Lender (1950) et de Lender (1950, 1951a et b, et 1952a et b) sur *Polycelis nigra* l'ont démontré.

Si l'on excise simultanément un groupe d'yeux latéraux et le cerveau, la régénération du cerveau se réalise en 4 jours, celle des yeux en 7 jours (Fig. 25). Si, par des excisions répétées tous les 2 jours, on empêche le cerveau de se reconstituer, la régénération des yeux n'a pas lieu. Sept jours après la dernière excision du cerveau, les yeux se régénèrent enfin, chaque fois que le cerveau lui-même a pu se reconstituer. Dans certains cas, celui-ci ne se régénère plus après ces multiples mutilations successives; dans ces conditions les yeux non plus ne se reforment pas.

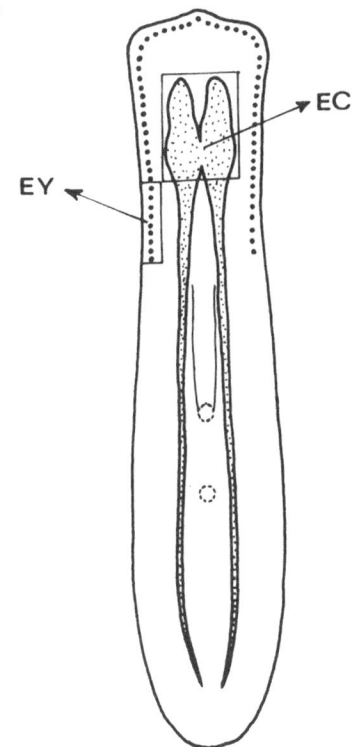

Fig. 25. Excision simultanée du cerveau, *EC*, et d'un groupe d'yeux latéraux postérieurs, *EY*, chez *Polycelis nigra*. *En pointillé*, le système nerveux. (D'après Lender, 1952a.)

Les nerfs optiques ne sont pas nécessaires à la régénération des yeux, car ceux-ci s'édifient avant que les connexions nerveuses ne se rétablissent. Bien plus, les yeux sont capables de régénérer en l'absence de néoblastes issus de la région cérébrale. Après la destruction des néoblastes de la région céphalique par l'irradition aux rayons X, les yeux marginaux postérieurs excisés se reforment à partir de néoblastes sains provenant des zones non irradiées. Ainsi le cerveau exerce sur la régénération des yeux une action inductrice à distance, vraisemblablement par l'intermédiare d'une substance diffusible.

Lender (1950 et 1951a) a mis en évidence l'existence d'un champ inducteur autour du cerveau. Un fragment de bord oculé est prélevé, les yeux en sont excisés; puis il est greffé à différents niveaux de la Planaire, entre le cerveau et la queue. Les yeux régénèrent quand le greffon est placé en avant de la bouche (Fig. 26). Dans la région postpharyngienne ils ne régénèrent pas. Ainsi la zone d'influence du cerveau est limitée à la région prébuccale. La régénération des yeux d'un fragment de bord oculé implanté dans la zone postpharyngienne est pourtant possible si l'on y greffe un cerveau à proximité du bord oculé (Fig. 27).

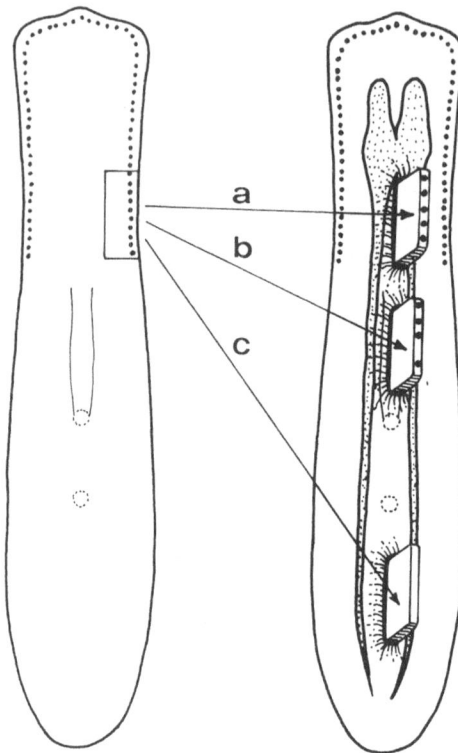

Fig. 26. Greffe d'un bord oculé (dont les yeux sont excisés) dans différentes régions du corps, chez *Polycelis nigra*. (a) Après implantation au voisinage du cerveau, les yeux régénèrent; (b) après implantation dans la région pharyngienne, les yeux régénèrent, mais en plus petit nombre; (c) après implantation dans la région caudale, les yeux ne régénèrent pas. *En pointillé* le système nerveux. (D'après Lender, 1952a.)

L'inducteur des yeux émanant du cerveau n'a pas de spécificité zoologique. Un bord oculé privé de ses yeux, greffé au voisinage du cerveau de *D. lugubris*, régénère ses yeux normalement. De plus, cette transplantation xénoplastique démontre que la présence des nerfs optiques n'est pas nécessaire à la régénération des yeux, car aucune connexion nerveuse ne s'établit entre le cerveau de *D. lugubris* et les yeux de *P. nigra*.

L'analyse du facteur inducteur de la régénération des yeux a été poussée plus loin (Lender, 1954a et b, 1955a et b). Des homogénats de tête de Planaires de diverses espèces (*P. nigra, D. lugubris, D. gonocephala* et *D. lacteum*) provoquent la régénération des yeux de Planaires privées de cerveau. On sait déjà qu'en présence de ces homogénats le cerveau ne régénère pas. Ils contiennent donc à la fois la substance inductrice des yeux et la substance inhibitrice du cerveau. Les homogénats de queue n'ont pas le pouvoir d'induire la régénération des yeux. Toutefois, s'ils sont portés à 60°C pendant quelques minutes, ils acquièrent le même pouvoir inducteur que les homogénats de tête. On peut penser que la substance inductrice est présente dans tout le corps de la Planaire, mais qu'elle est inactive dans la région postpharyngienne.

D'autres données intéressantes sur la morphogenèse des yeux de *P. nigra* ont été apportées par Pentz (1963). Cet auteur s'est surtout attaché au problème de la distribution et de la disposition des yeux le long du bord oculé de cette espèce. En effet, les yeux sont répartis sur le bord frontal et sur les bords latéraux du Ver selon un patron bien défini. Les yeux frontaux et latéraux antérieurs sont petits, ne comportent qu'une ou deux cellules visuelles; l'intervalle qui sépare deux yeux voisins est court. Les yeux latéraux postérieurs sont plus grands, comptent trois cellules visuelles et sont plus écartés les uns des autres. Selon l'hypothèse de Pentz et Seilern-Aspang (1961), chaque œil est entouré d'un champ inhibiteur, empêchant la formation d'un autre œil en-deçà d'une distance minimum. Ce champ est plus étendu autour des grands yeux latéraux postérieurs qu'autour des petits yeux frontaux. Il existe un antagonisme entre la substance inductrice des yeux issue du cerveau et une substance inhibitrice issue des yeux. A proximité du cerveau, dans le bord frontal et latéral antérieur, le champ

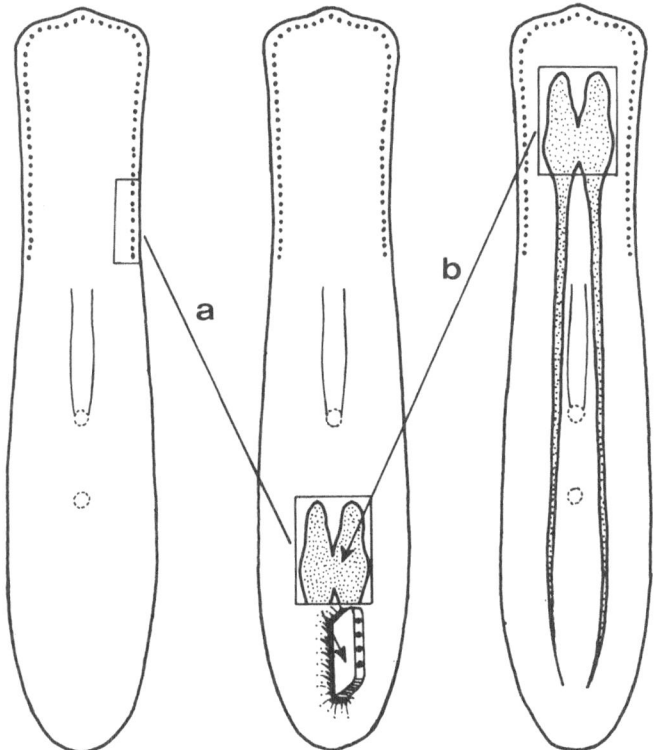

Fig. 27. Double greffe (a) d'un bord oculé (dont les yeux sont excisés) et (b) d'un cerveau dans la région caudale de *Polycelis nigra.* Le bord oculé, placé au voisinage du cerveau greffé, régénère des yeux. *En pointillé,* le système nerveux. (D'après Lender, 1952a.)

inhibiteur des yeux est contrebalancé par l'influence inductrice du cerveau; l'écart entre les yeux est réduit. L'influence du cerveau diminuant d'avant en arrière, le champ inhibiteur périoculaire est d'autant plus grand que les yeux sont plus éloignés du cerveau, et l'intervalle entre les yeux augmente concomitamment. De plus, l'écartement des yeux se répercute sur leur dimension. L'inhibition réciproque qu'exercent l'un sur l'autre deux yeux voisins est plus forte entre yeux serrés, qui restent donc petits, qu'entre yeux espacés, qui peuvent devenir trois fois plus grands.

Cette hypothèse a été vérifiée par Pentz (1963) à l'aide de transplantations de bord oculé chez *P. nigra* (Fig. 28). Un morceau de bord oculé postérieur, privé de ses deux ou trois grands yeux, est implanté à proximité du cerveau. Il régénère de cinq à neuf petits yeux à une seule cellule visuelle. Si les deux yeux du greffon de bord oculé postérieur ne sont pas excisés préalablement, il s'y forme pourtant, au voisinage du cerveau, un à deux petits yeux supplémentaires dans l'intervalle entre les deux grands yeux. Inversement, dans un bord oculé latéral antérieur, privé de ses cinq à sept petits yeux et greffé près de la racine du pharynx, deux à trois grands yeux régénèrent, formés chacun de trois cellules visuelles. Par conséquent, la taille et la densité des yeux ne dépendent pas du territoire où s'effectue la régénération. La compétence et les potentialités du bord oculé sont identiques en tous points depuis l'extrémité antérieure jusqu'au dernier œil latéral postérieur. La dimension et l'écartement des yeux en régénération sont déterminés par la distance séparant le cerveau inducteur du bord oculé.

La régénération du pharynx

L'histogenèse du pharynx en régénération a fait l'objet de nombreux travaux déjà anciens (Thacher, 1902; Stevens, 1907; Steinmann, 1908;

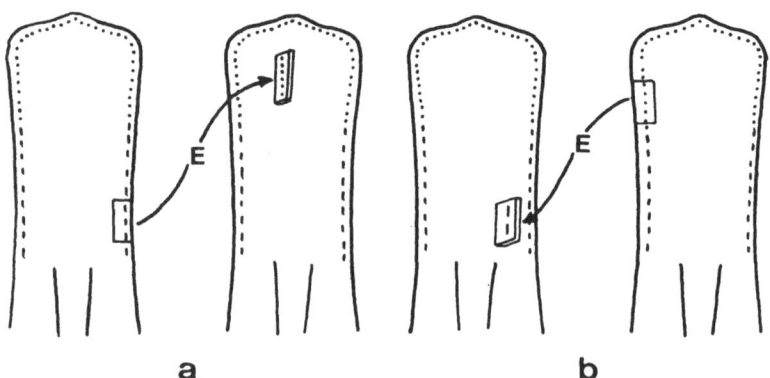

a   b

Fig. 28. Régénération des yeux chez *Polycelis nigra*. Un fragment de bord oculé est transplanté, après extirpation *E* des yeux, de la région latéro-postérieure dans la région du cerveau (a) et de la région latéro-antérieure à proximité de la racine du pharynx (b). (D'après Pentz, 1963.)

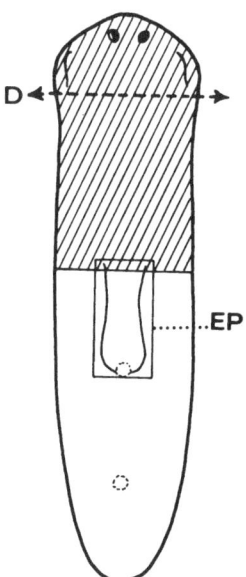

Fig. 29. Excision du pharynx *EP* d'une Planaire *Dugesia lugubris* décapitée *D* et dont la région prépharyngienne a été irradiée aux rayons X (*hachures*). (D'après P. Sengel, 1951b.)

Bartsch, 1923; Bandier, 1936; et Van Asperen, 1946). Elle a été reprise et précisée plus récemment par Kido (1957, 1959, et 1961a et b) et par Gazso (1958). La régénération du pharynx débute toujours par une accumulation de néoblastes, qui se creuse ensuite et délimite ainsi la gaine du pharynx. Enfin intervient un simple phénomène de croissance du pharynx à l'intérieur de sa gaine.

Santos (1929 et 1931) a mis en évidence pour la première fois un mécanisme inducteur de la différenciation chez les Planaires. Ses expériences, réalisées sur *Planaria maculata*, montrent qu'un greffon céphalique, implanté dans la région postpharyngienne, induit un ou deux pharynx supplémentaires à l'intérieur des tissus de l'hôte. La tête greffée joue donc le rôle d'un inducteur, les tissus de l'hôte réagissent en différenciant un pharynx, sans qu'il y ait régénération à proprement parler. Okada et Sugino (1934a et b et 1937) et Okada et Kido (1943) précisent qu'il suffit, chez *Dugesia dorotocephala*, d'un greffon prélevé dans la région prépharyngienne pour obtenir le même résultat.

P. Sengel (1951a et b) s'est demandé si la tête ou la région prépharyngienne est aussi l'inducteur de la régénération *normale* du pharynx. Ses expériences ont été réalisées sur *Dugesia lugubris*. La tête est sectionnée et sa régénération est retardée de 3 à 4 semains par l'irradiation aux rayons X de la région prépharyngienne (Fig. 29). Dans ces conditions, le pharynx est toujours régénéré au plus tard 6 jours après l'excision. Des tranches latérales au pharynx sont découpées dans la Planaire et leur région prépharyngienne est soit sectionnée soit irradiée aux rayons X (Fig. 30). Le pharynx est toujours régénéré entre le neuvième et le dixième jour après l'opération, en l'absence de toute formation céphalique.

Le pharynx avec sa gaine est extrait à l'emporte-pièce. En même temps, la région prépharyngienne est sectionnée au ras de la racine du pharynx. Deux jours après l'opération, une deuxième section entame légèrement la région de la racine du pharynx. Le pharynx est régénéré entre le cinquième et le septième jour après le début de l'expérience, en l'absence de tête, de région prépharyngienne et même de bourgeon de régénération antérieur. Dans une autre expérience, toute la région comprise entre le bord antérieur de la Planaire et un niveau transversal passant juste en arrière de la bouche est irradiée aux rayons X (Fig. 31). La région prépharyngienne est sectionnée au ras de la racine du pharynx et celui-ci est extirpé à l'emporte-pièce. Dans ces conditions, le pharynx est régénéré au cours de la migration des néoblastes sains, qui balaient la région pharyngienne d'arrière en avant. Au bout de 11 jours, le pharynx est présent bien avant que ne se constitue le blastème définitif de régénération antérieur. Le pharynx se régénère donc

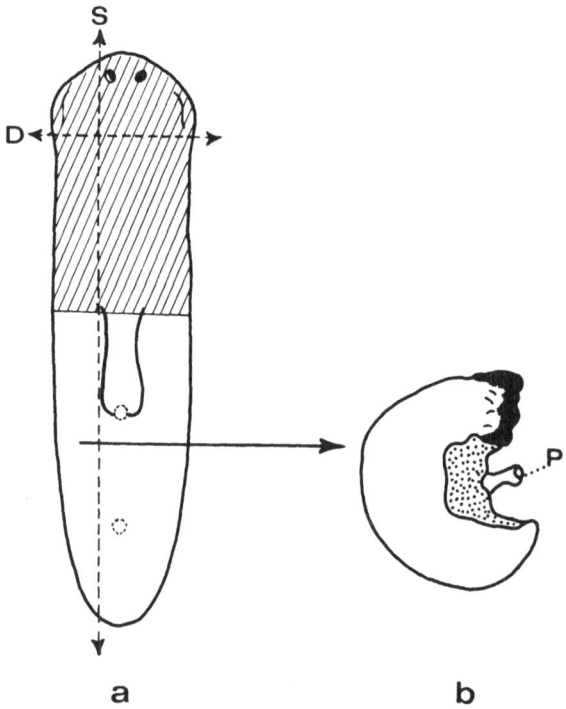

a   b

Fig. 30. Régénération du pharynx dans un morceau latéral de *Dugesia lugubris*. (a) Une section longitudinale *S* est pratiquée dans une Planaire décapitée *D* et dont la région prépharyngienne a été irradiée (*hachures*) aux rayons X; (b) 9 à 10 jours après l'opération, le pharynx *P* est régénéré dans les tissus néoformés (*pointillé*); *en noir*, nécrose du blastème de régénération antérieur. (D'après P. Sengel, 1951 b.)

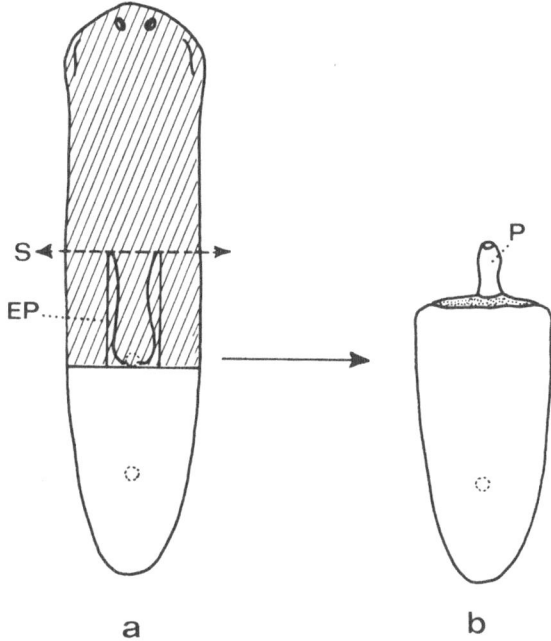

Fig. 31. Régénération du pharynx en l'absence de région prépharyngienne, chez *Dugesia lugubris*. (a) La région prépharyngienne et la région pharyngienne sont irradiées (*hachures*) aux rayons X; une section *S* est ensuite pratiquée en avant du pharynx, qui est excisé *EP* à l'emporte-pièce; (b) le fragment postérieur régénère un pharynx *P*, qui se dévagine par une bouche s'ouvrant en avant. (D'après P. Sengel, 1951b.)

non seulement en l'absence de tête, mais aussi en l'absence de la région prépharyngienne.

Il est légitime de penser que c'est la région pharyngienne elle-même qui possède la faculté de régénérer le pharynx. Deux types d'expériences le démontrent (P. Sengel, 1951b et 1953). Un fragment de région pharyngienne est prélevé latéralement au pharynx, puis implanté dans la région céphalique ou postpharyngienne d'une autre Planaire (Fig. 32). En place, ce transplant régénère un ou deux pharynx, sans donner lieu à la formation d'une tête. Un greffon céphalique contenant le cerveau est implanté dans la région postpharyngienne (Fig. 33). Au bout de 3 à 4 semaines, on constate que ce greffon a induit un ou plusieurs pharynx supplémentaires dans les tissus de l'hôte. Le greffon est ensuite excisé. Le ou les pharynx supplémentaires sont extraits à leur tour. Au bout de 8 à 15 jours, ils sont régénérés avec leur polarité propre, que celle-ci soit concordante ou inverse par rapport à l'hôte. On peut extirper le ou les pharynx supplémentaires régénérés une deuxième fois; ils se régénèrent à nouveau (Fig. 34).

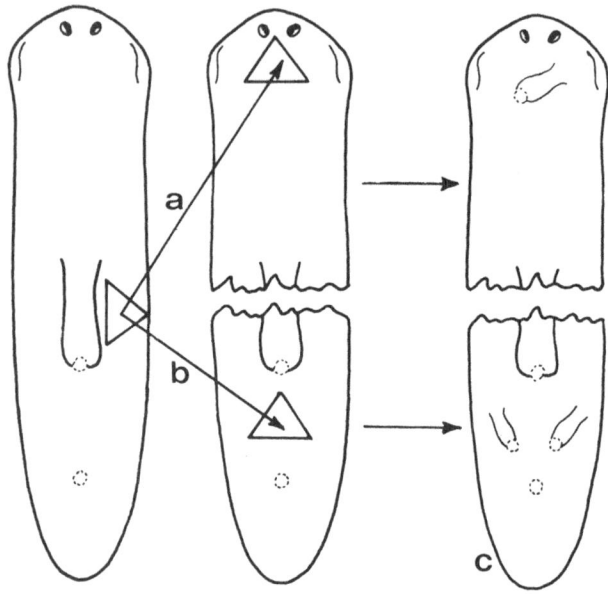

Fig. 32. Transplantation d'un fragment de région pharyngienne, soit dans la tête (a), soit dans la queue (b), chez *Dugesia
lugubris*. (c) Le greffon en place régénère un ou deux pharynx
dont la bouche s'ouvre sur la face ventrale. (D'après P. Sengel,
1951 b.)

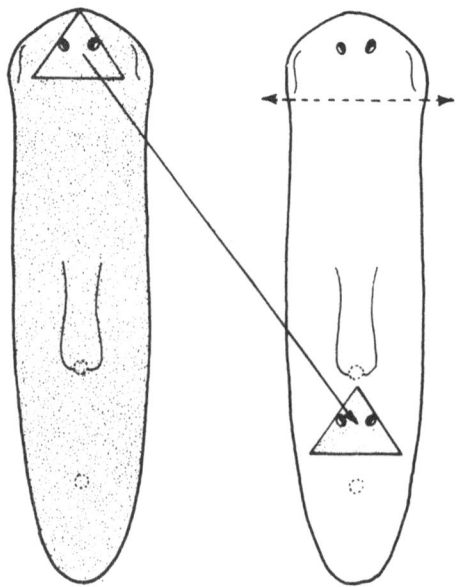

Fig. 33. Implantation d'un greffon
céphalique, comprenant le cerveau
et les yeux, dans la région postpharyngienne, chez *Dugesia lugubris*.
En grisé, le donneur, de pigmentation sombre; *en blanc*, l'hôte
de pigmentation claire. (D'après
P. Sengel, 1951 b.)

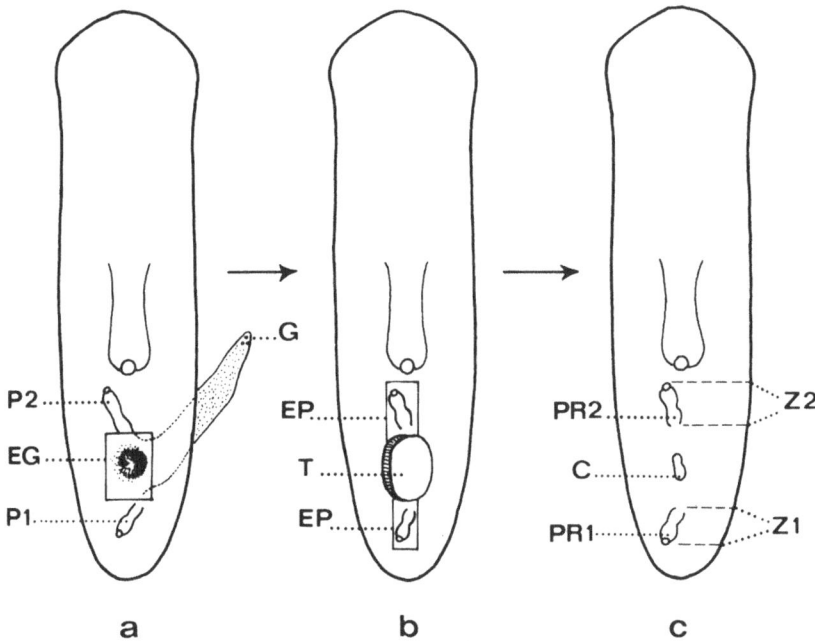

Fig. 34. Exemple d'induction de deux zones pharyngiennes par un greffon céphalique implanté dans la région postpharyngienne de *Dugesia lugubris*. (a) La Planaire hôte, en vue ventrale, 4 semaines après l'implantation du greffon; le greffon est excisé; (b) le lendemain de l'excision du greffon, les deux pharynx supplémentaires sont extirpés à leur tour; (c) 8 à 15 jours plus tard, les deux pharynx supplémentaires ont régénéré. *C*, cicatrice laissée par l'excision du greffon; *EG*, excision du greffon; *EP*, excision des pharynx supplémentaires; *G*, greffon formant une excroissance sur la face dorsale de l'hôet. *P*1, pharynx supplémentaire, de polarité concordante; *P*2, pharynx supplémentaire de polarité inverse; *PR*1, pharynx supplémentaire régénéré, de polarité concordante; *PR*2, pharynx supplémentaire régénéré, de polarité inverse; *T*, trou après l'excision du greffon; *Z*1, zone pharyngienne induite, de polarité concordante; *Z*2, zone pharyngienne induite, de polarité inverse. (D'après P. Sengel, 1953.)

Ce résultat peut s'interpréter de la façon suivante: le greffon céphalique, après sa différenciation propre, détermine dans l'hôte l'apparition d'une région pharyngienne supplémentaire. Après l'ablation du greffon, la nouvelle région pharyngienne conserve sa propriété de *zone pharyngienne*; si l'on extrait son pharynx, elle le régénère en lui conservant sa polarité primitive. On aboutit à la notion d'*induction de zone*. Une région dominante, la tête, induit une zone dominée, la région pharyngienne. La nouvelle zone pharyngienne induit elle-même la différenciation de son pharynx. En ce qui concerne la régénération du pharynx, la zone pharyngienne apparaît donc comme une région parfaitement indépendante des formations céphalique et prépharyngienne.

Selon Wolff (1953), on assiste à une véritable cascade d'inductions: la tête greffée induit d'abord une zone prépharyngienne, celle-ci induit une zone pharyngienne, et la zone pharyngienne différencie un pharynx. Il est possible que de multiples inductions de zone se succèdent quand la Planaire doit reconstituer une grande partie de son corps. Il convient vraisemblablement de distinguer l'induction de zone de celle d'organes précis comme les yeux.

Les cas d'induction de deux ou plusieurs pharynx supplémentaires par un greffon céphalique sont relativement rares. Lorsque cela se produit, les pharynx sont toujours éloignés les uns des autres, souvent orientés différemment, les uns en polarité conforme, les autres en polarité inverse par rapport à l'hôte. On est tenté de penser que chacun de ces pharynx supplémentaires occupe le centre d'une zone pharyngienne propre; il y aurait eu induction de deux ou plusieurs zones pharyngiennes distinctes. Par ailleurs, il semble que la régénération de plusieurs pharynx à l'intérieur d'une même zone pharyngienne n'ait jamais été observée. Existe-t-il donc pour le pharynx, comme pour le cerveau, une substance inhibitrice produite par le pharynx (ou par la zone pharyngienne) et qui empêcherait la différenciation d'autres pharynx à son voisinage?

Les expériences récentes de C. Ziller-Sengel (1965) montrent que tel est vraisemblablement le cas, chez *D. lugubris*, *D. tigrina* et *P. nigra*. Des Planaires, dont le pharynx a été extirpé à l'emporte-pièce, sont mises à régénérer dans de l'eau contenant de l'extrait de région céphalique, de région pharyngienne ou de région caudale. Les extraits de tête et de queue ont plutôt une action stimulatrice sur la régénération du pharynx. L'extrait de région pharyngienne a une nette action retardatrice, sans toutefois empêcher la régénération du pharynx. Fait curieux, l'effet inhibiteur le plus puissant est obtenu avec des extraits de région pharyngienne dont le pharynx a été supprimé préalablement. On aurait pu penser qu'une zone pharyngienne privée de pharynx produirait plutôt des substances inductrices de la régénération du pharynx. L'interprétation de ce fait semble devoir attendre le résultat d'autres expériences, où serait testée l'action inhibitrice ou stimulatrice de la zone pharyngienne à divers stades de la régénération de son pharynx. Il est fort possible qu'une région pharyngienne, selon l'état physiologique où elle se trouve, ait tantôt un effet stimulant, tantôt un effet inhibiteur sur la régénération du pharynx.

CONCLUSIONS

De l'ensemble des faits que nous venons de rapporter se dégagent les conditions nécessaires à la morphogenèse régénératrice.

Il faut tout d'abord qu'existent dans l'organisme des cellules douées du pouvoir de se différencier pour former les tissus et les organes qui doivent être remplacés. Pour remplir leur rôle, ces cellules doivent être capables de se déplacer pour se rendre sur le lieu de la lésion.

L'existence de telles cellules, les néoblastes, a été amplement démontrée chez les Planaires. Elles sont effectivement totipotentes et douées d'un grand pouvoir de migration. Elles apparaissent très tôt au cours du développement embryonnaire. Elles constituent une lignée de cellules distinctes de celles qui se différencient progressivement. Elles conservent toutes leurs propriétés de cellules embryonnaires pendant toute la vie de l'organisme. Dès l'instant de cette ségrégation, elles sont capables chez l'embryon déjà de donner naissance à un régénérat. Le mécanisme qui déclenche leur migration est encore inconnu.

Arrivés à l'emplacement de la lésion, les néoblastes, en se multipliant, édifient un blastème de régénération. A l'intérieur du blastème, qui s'allonge progressivement en un régénérat, les néoblastes se différencient pour donner naissance dans un ordre très précis aux divers organes régénérés. Dès le deuxième jour après la lésion, les néoblastes du blastème ont perdu leur totipotence. Ils sont déterminés à ne former que les tissus et les organes manquants.

La détermination du blastème et la différenciation successive des organes sont le résultat de multiples actions inductrices et inhibitrices émanant soit des tissus anciens, soit des organes en voie de régénération. La succession ordonnée dans le temps de ces inductions et de ces inhibitions au cours de la morphogenèse régénératrice aboutit finalement à la reconstitution d'un organisme harmonieux.

De nombreuses inductions et inhibitions spécifiques ont été mises en évidence. Le cerveau induit la différenciation des yeux par l'intermédiaire d'une substance diffusible. Les yeux restreignent le développement d'yeux voisins. La tête induit une zone prépharyngienne et pharyngienne, mais elle inhibe la différenciation du cerveau. A son tour, la zone pharyngienne induit la formation d'un pharynx.

Selon l'hypothèse de Wolff (1962), la doctrine des gradients physiologiques de Child (1920) pourrait être expliquée biochimiquement par la diffusion de substances inductrices ou inhibitrices le long de l'axe céphalocaudal de la Planaire.

En définitive, il semble que le pouvoir de régénération des Planaires réside dans deux propriétés essentielles de leurs tissus. La première est l'existence dans le parenchyme de cellules qui conservent indéfiniment leurs potentialités embryonnaires: ce sont les néoblastes. La seconde est la production par les divers territoires de l'organisme de facteurs morphogènes, substances inductrices ou inhibitrices, qui déterminent et règlent la qualité de la différenciation des néoblastes.

REFERENCES

Bandier, J., Histologische Untersuchungen über die Regeneration von Landplanarien. *Roux' Arch. Entwicklungsmech. Organ.,* 1936, **135**, 316.
Bartsch, O., Die Histogenese der Planarienregenerate. *Roux' Arch. Entwicklungsmech. Organ.,* 1923, **99**, 187.
Brøndsted, H. V., Planarian regeneration. *Biol. Rev.,* 1955, **30**, 65.

Brøndsted, A. et H. V. Brøndsted, Number of neoblasts in the intact body of *Euplanaria torva* and *Dendrocoelum lacteum. J. Embryol. Exp. Morphol.*, 1961, **9**, 167.

Chevtchenko, N. M., Polaritätsumkehrung bei der Unterdrückung der dominierenden Regionen bei Planaria. *Zool. Anz.*, 1936, **115**, 232.

Chevtchenko, N. M., Die Wechselwirkung von Teilen von verschiedener physiologischer Activität bei Planarien. *Biol. Zhur.*, 1937, **6**, 581.

Child, C. M., Some considerations concerning the nature and origin of physiological gradients. *Biol. Bull.*, 1920, **39**, 147.

Child, C. M., *Patterns and Problems of Development.* Chicago: *University of Chicago Press.* 1941.

Coward, S. J., R. A. Flickinger et E. Garen, The effect of nutrient media upon head frequency in regenerating Planaria. *Biol. Bull.*, 1964, **126**, 345.

Doumain, J., Action de la colchicine et de la démécolcine sur la régénération de la Planaire *Dugesia subtentaculata* (Drap.). *C. R. Soc. Biol.*, 1960, **154**, 1453.

Dubois, F., Potentialités histogénétiques des néoblastes chez *Euplanaria lugubris*, et facteurs qui déclenchent leur migration. *Ann. Biol.*, 1949a, **26**, 693.

Dubois, F., Contribution à l'étude de la migration des cellules de régénération chez les Planaires dulcicoles. *Bull. Biol. Fr. Belg.*, 1949b, **83**, 213.

Fedecka-Bruner, B., Radiodestruction des testicules suivie de régénération chez la Planaire *Dugesia lugubris. C. R. Acad. Sci.*, 1964, **258**, 3353.

Flickinger, R. A., A gradient of protein synthesis in Planaria and reversal of axial polarity of regenerates. *Growth*, 1959, **23**, 251.

Flickinger, R. A., Isotopic evidence for a local origin of blastema cells in regenerating Planaria. *Exp. Cell Res.*, 1964, **34**, 403.

Flickinger, R. A. et S. J. Coward, The induction of cephalic differentiation in regenerating *Dugesia dorotocephala* in the presence of the normal head and in unwounded tails. *Devel. Biol.*, 1962, **5**, 179.

Gazso, L. R., Contribution à l'étude de la régénération du pharynx du *Dendrocoelum lacteum* et de la *Dugesia lugubris. Acta Biol. Acad. Sci. Hung.*, 1958, **8**, 263.

Kanatani, H., Effect of insulin on the head regeneration in the Planarian, *Dugesia gonocephala. J. Fac. Sci. Univ. Tokyo*, 1957, **8**, 11.

Kanatani, H., Formation of bipolar heads induced by demecolcine in the planarian. *Dugesia gonocephala. J. Fac. Sci. Univ. Tokyo*, 1958, **8**, 253.

Kanatani, H., Effect of estradiol on head regeneration in the planarian, *Dugesia gonocephala. J. Fac. Sci. Univ. Tokyo*, 1959, **8**, 439.

Kido, T., Remarks on the so-called induction of the pharynx in Planaria. *Sci. Rep. Kenazawa Univ.*, 1957, **5**, 49.

Kido, T., Location of the new pharynx in regenerating piece of planaria. *Sci. Rep. Kenazawa Univ.*, 1959, **6**, 77.

Kido, T., Studies on the pharynx regeneration in planarian *Dugesia gonocephala*, I: Histological observation in transected pieces. *Sci. Rep. Kanazawa Univ.*, 1961a, **7**, 107.

Kido, T., Studies on the pharynx regeneration in planarian *Dugesia gonocephala*, II: Histological observation in the abnormal regenerates produced experimentally. *Sci. Rep. Kanazawa Univ.*, 1961b, **7**, 125.

Kolmayer, S. et F. Stéphan-Dubois, Néoblastes et limitation du pouvoir de régénération céphalique chez la Planaire *Dendrocoelum lacteum. J. Embryol. Exp. Morphol.*, 1960, **8**, 376.

Le Moigne, A., Etude de formules chromosomiques de quelques *Polycelis* (Turbellariés, Triclades) de la région parisienne. *Bull. Soc. Zool. Fr.*, 1962, **87**, 259.

Le Moigne, A., Etude du développement embryonnaire de *Polycelis nigra* (Turbellarié, Triclade). *Bull. Soc. Zool. Fr.*, 1963, **88**, 403.

Le Moigne, A., Sur la régénération et la différenciation de fragments d'embryons de *Polycelis nigra tenuis* (Turbellarié, Triclade). *C. R. Soc. Biol.*, 1965a, **159**, 54.

Le Moigne, A., Effets des irradiations aux rayons X sur le développement embryonnaire et le pouvoir de régénération à l'éclosion de *Polycelis nigra* (Turbellarié, Triclade). *C. R. Acad. Sci.*, 1965b, **260**, 4627.

Le Moigne, A., Mise en évidence d'un pouvoir de régénération chez l'embryon de *Polycelis nigra* (Turbellarié, Triclade). *Bull. Soc. Zool. Fr.*, 1965c, **90**, 355.

Le Moigne, A., Etude du développement embryonnaire et recherches sur les cellules de régénération chez l'embryon de la Planaire *Polycelis nigra* (Turbellarié, Triclade). *J. Embryol. Exp. Morphol.*, 1966, **15**, 39.

Lender, T., Démonstration du rôle organisateur du cerveau dans la régénération des yeux de la Planaire *Polycelis nigra* par la méthode des greffes. *C. R. Soc. Biol.*, 1950, **144**, 1407.

Lender, T., Sur les propriétés et l'étendue du champ d'organisation du cerveau dans la régénération des yeux de la Planaire *Polycelis nigra*. *C. R. Soc. Biol.*, 1951a, **145**, 1211.

Lender, T., Sur les capacités inductrices de l'organisateur des yeux dans la régénération de la Planaire *Polycelis nigra* (Ehr.). Action du cerveau en voie de dégénérescence et en greffes hétéroplastiques. *C. R. Soc. Biol.*, 1951b, **145**, 1378.

Lender, T., Le rôle inducteur du cerveau dans la régénération des yeux d'une Planaire d'eau douce. *Bull. Biol. Fr. Belg.*, 1952a, **86**, 140.

Lender, T., Le rôle inducteur du cerveau dans la régénération des yeux d'une Planaire d'eau douce. *Ann. Biol.*, 1952b, **28**, 191.

Lender, T., Sur la régénération des yeux de la Planaire *Polycelis nigra* en présence de broyats de la région antérieure du corps. *C. R. Acad. Sci.*, 1954a, **238**, 1742.

Lender, T., Sur l'activité inductrice de la région antérieure du corps dans la régénération des yeux de la Planaire *Polycelis nigra*: Activité de broyats frais traités à la chaleur. *C. R. Soc. Biol.*, 1954b, **148**, 1859.

Lender, T., Mise en évidence et propriétés de l'organisine de la régénération des yeux chez la Planaire *Polycelis nigra*. *Rev. Suisse Zool.*, 1955a, **62**, 268.

Lender, T., Sur quelques propriétés de l'organisine de la régénération des yeux de la Planaire *Polycelis nigra*. *C. R. Acad. Sci.*, 1955b, **240**, 1726.

Lender, T., Sur l'inhibition de la régénération du cerveau de la Planaire *Polycelis nigra*. *C. R. Acad. Sci.*, 1955c, **241**, 1863.

Lender, T., L'inhibition de la régénération du cerveau des Planaires *Polycelis nigra* (Ehr.) et *Dugesia lugubris* (O. Schm.) en présence de broyats de tête et de queue. *Bull. Soc. Zool. Fr.*, 1956a, **81**, 192.

Lender, T., Recherches expérimentales sur la nature et les propriétés de l'inducteur de la régénération des yeux de la Planaire *Polycelis nigra*. *J. Embryol. Exp. Morphol.*, 1956b, **4**, 196.

Lender, T., Analyse des phénomènes d'induction et d'inhibition dans la régénération des Planaires. *Ann. Biol.*, 1956c, **32**, 457.

Lender, T., L'inhibition spécifique de la différenciation du cerveau des Planaires d'eau douce en régénération. *J. Embryol. Exp. Morphol.*, 1960, **8**, 291.

Lender, T., Factors in morphogenesis of regenerating fresh-water Planaria. *Advances in Morphogenesis*, Vol. 2, 1962, 305, Academic Press, N.Y.

Lender, T. et A. Gabriel, Etude histochimique des néoblastes de *Dugesia lugubris* (Turbellarié, Triclade) avant et pendant la régénération. *Bull. Soc. Zool. Fr.*, 1960, **85**, 100.

Lender, T. et A. Gabriel, Les néoblastes marqués par l'uridine tritiée migrent et édifient le blastème de régénération des Planaires d'eau douce. *C. R. Acad. Sci.*, 1965, **260**, 4095.

Lender, T. et P. Gripon, La régénération des yeux et du cerveau de *Dugesia lugubris* en présence de deux troncs nerveux inégaux. *Bull. Soc. Zool. Fr.*, 1962, **87**, 387.

Lepori, N. G., Ricerche sulla ovogenesi e sulla fecondazione nella Planaria *Polycelis nigra* Ehr. con particolare riguardo all'ufficio del nucleo spermatico. *Caryologia*, 1949, **1**, 280.

Morgan, T., Regeneration in planarians. *Roux' Arch. Entwicklungsmech. Organ.*, 1900, **10**, 58.

Morgan, T., Growth and regeneration in *Planaria lugubris*. *Roux' Arch. Entwicklungsmech. Organ.*, 1902, **13**, 179.

Okada, Y. K. et T. Kido, Further experiments on transplantation in Planaria. *J. Fac. Sci. Univ. Tokyo*, 1943, **6**, 1.

Okada, Y. K. et H. Sugino, Transplantation experiments in *Planaria gonocephala*, I. *Proc. Imper. Acad. Japan*, 1934a, **10**, 37.

Okada, Y. K. et H. Sugino, Transplantation experiments in *Planaria gonocephala*, II. *Proc. Imper. Acad. Japan*, 1934b, **10**, 107.

Okada, Y. K. et H. Sugino, Transplantation experiments in *Planaria gonocephala* (Dugès). *Japan. J. Zool.*, 1937, **7**, 373.

Pedersen, K. J., Cytological studies on the planarian neoblast. *Z. Zellf.*, 1959, **50**, 799.

Pentz, S., Über die Augendifferenzierung bei *Polycelis nigra* Ehrb. *Roux' Arch. Entwicklungsmech. Organ.*, 1963, **154**, 495.

Pentz, S. et F. Seilern-Aspang, Die Entstehung des Augenmusters bei *Polycelis nigra* durch Wechselwirkung zwischen dem Augenhemmfeld und der Augeninduktion durch das Gehirn. *Roux' Arch. Entwicklungsmech. Organ.*, 1961, **153**, 75.

Santos, F. V., Studies on transplantation in Planaria. *Biol. Bull.*, 1929, **57**, 188.

Santos, F. V., Studies on transplantation in *Planaria dorotocephala* and *Planaria maculata*. *Physiol. Zool.*, 1931, **4**, 111.

Sengel, C., Culture *in vitro* de blastèmes de régénération de Planaires. *J. Embryol. Exp. Morphol.*, 1960, **8**, 468.

Sengel, C., Culture *in vitro* de blastèmes de régénération de la Planaire *Dugesia lugubris*. *Ann. Epiphyties*, 1963, **14**, 173.

Sengel, P., Sur les conditions de la régénération normale du pharynx chez la Planaire *Dugesia* (= *Euplanaria*) *lugubris* O. Schm. *C. R. Soc. Biol.*, 1951a, **145**, 1381.

Sengel, P., Sur les conditions de la régénération normale du pharynx chez la Planaire d'eau douce *Dugesia lugubris* O. Schm. *Bull. Biol.*, 1951b, **85**, 377.

Sengel, P., Sur l'induction d'une zone pharyngienne chez la Planaire d'eau douce *Dugesia lugubris* O. Schm. *Arch. Anat. Microsc. Morphol. Exp.*, 1953, **42**, 57.

Skaer, R. J., The origin and continuous replacement of epidermal cells in the Planarian *Polycelis tenuis*. *J. Embryol. Exp. Morphol.*, 1965, **13**, 129.

Steinmann, P., Untersuchungen über das Verhalten des Verdauungssystems bei der Regeneration der Tricladen. *Roux' Arch. Entwicklungsmech. Organ.*, 1908, **25**, 523.

Stéphan-Dubois, F., Les cellules de régénération chez la Planaire *Dendrocoelum lacteum*. *Bull. Soc. Zool. Fr.*, 1961, **86**, 172.

Stéphan-Dubois, F., Les néoblastes dans la régénération chez les Planaires, *Regeneration in Animals and Related Problems: An International Symposium*, edited by V. Kiortsis and H. A. L. Trampusch. Amsterdam: North-Holland Publishing Co., 1965, p. 112.

Stéphan-Dubois, F. et F. Gilgenkrantz, Transplantation et régénération chez la Planaire *Dendrocoelum lacteum*. *J. Embryol. Exp. Morphol.*, 1961, **9**, 642.

Stéphan-Dubois, F. et S. Kolmayer, La migration et la différenciation des cellules de régénération chez la Planaire *Dendrocoelum lacteum*. *C. R. Soc. Biol.*, 1959, **153**, 1856.

Stevens, N. M., A histological study of regeneration in *Planaria simplicissima, Planaria maculata* and *Planaria morgani*. *Roux' Arch. Entwicklungsmech. Organ.*, 1907, **24**, 350.

Thacher, H. F., The regeneration of the pharynx in *Planaria maculata*. *Amer. Natur.*, 1902, **36**, 638.

Van Asperen, K., Pharynx regeneration in postpharyngeal fragments of *Polycelis nigra*. *Proc. Kon. Ned. Ak. Wet.*, 1946, **49**, 1083.

Wolff, Et., Les phénomènes d'induction dans la régénération des Planaires d'eau douce. *Rev. Suisse Zool.*, 1953, **60**, 540.

Wolff, Et., Recent researches on the regeneration of Planaria, *in Regeneration, 20th Growth Symposium, Society for the study of development and growth*. New York: Ronald Press Co., 1962, p. 53.

Wolff, Et. et F. Dubois, Sur une méthode d'irradiation localisée permettant de mettre en évidence la migration des cellules de régénération chez les Planaires. *C. R. Soc. Biol.*, 1947, **141**, 903.

Wolff, Et. et F. Dubois, Sur la migration des cellules de régénération chez les Planaires. *Rev. Suisse Zool.*, 1948, **55**, 218.

Wolff, Et. et T. Lender, Sur le rôle organisateur du cerveau dans la régénération des yeux chez une Planaire d'eau douce. *C. R. Acad. Sci.*, 1950a, **230**, 2238.

Wolff, Et. et T. Lender, Sur le déterminisme de la régénération des yeux chez une Planaire d'eau douce *Polycelis nigra. C. R. Soc. Biol.*, 1950b, **144**, 1213.

Wolff, Et. et T. Lender, Les néoblastes et les phénomènes d'induction et d'inhibition dans la régénération des Planaires. *Ann. Biol.*, 1962, **1**, 500.

Wolff, Et., T. Lender et C. Ziller-Sengel, Le rôle de facteurs auto-inhibiteurs dans la régénération des Planaires. (Une interprétation nouvelle de la théorie des gradients physiologiques de Child). *Rev. Suisse Zool.*, 1964, **71**, 75.

Ziller-Sengel, C., Inhibition de la régénération du pharynx chez les Planaires, *Regeneration in Animals and Related Problems: An International Symposium*, edited by V. Kiortis and H. A. L. Trampusch. Amsterdam: North-Holland Publishing Co., 1965, p. 193.

Chapter 9

Aspects of Planarian Biology and Behavior

Marie M. Jenkins

Department of Biology
Madison College
Harrisonburg, Virginia

Planarians are most interesting creatures to study, and I think part of their fascination, for me at least, is their unpredictability. For example, if I decide after weeks of study that they are likely to do a certain thing under certain conditions, I promptly find their ideas on the subject are quite different. McConnell's story of their climbing on an electrode to avoid shock is typical (1965).

You know, from Dr. Kenk's excellent resumé, that numerous species and races of fresh-water planarians are available for study, and it follows that one would encounter variations not only among individuals, but also among races and species. Keeping this in mind, I shall include other groups as much as possible, but I shall deal primarily with the polytypic species, *Dugesia dorotocephala*, common in springs and spring brooks in the United States, since these are the animals with which I have done most of my work. The aspects of planarian biology and behavior which I propose to discuss briefly are: nutrition, respiration and metabolism, excretion and osmoregulation, innate reactions, and reproductive activities.

NUTRITION

Planarians are essentially carnivorous, although many are, at times, scavengers, and cannibalism is not uncommon. The digestive tract of planarians has but one opening, which must serve both for the entrance of food and the ejection of waste. When a planarian feeds, a long, muscular tube, the pharynx, is protruded through the mouth, a small pore on the midventral surface of the animal. Fully extended, the pharynx may be over half as long as the planarian itself. Food is ingested through the orifice of the pharynx and not, strictly speaking, through the mouth. A nerve plexus within this tube enables the isolated pharynx to carry out the usual activities, including actual ingestion of food (Hyman, 1951).

A planarian coming into contact with food juices diffusing into the water exhibits a characteristic behavior (Pearl, 1903). The worm pauses if moving, or bestirs itself if resting, then raises its head and swings it slowly about as if testing the direction from which the juices are coming. Having ascertained the direction, the worm turns and glides directly toward the food, then crawls over it, gripping the food by more or less wrapping itself about it and, at the same time, covering it with sticky slime. If the food is not too large, the planarian will also attach itself firmly to the substrate by means of exuded adhesive secretions. The pharynx is immediately protruded, makes exploratory movements, then is inserted into the food. If the planarian is feeding on a small invertebrate, the pharynx moves about within the body of the prey, sucking out all soft parts. Even when organisms as small as 300 μ in diameter are fed upon, the pharynx may be elongated and inserted (Jennings, 1962). This is confirmed by Reynoldson and Young (1963), who found that only 10 triclads, out of 1030 examined (four species), had ingested small, whole organisms. However, many planarians will feed avidly on whole *Daphnia*, or water fleas, which average 2 to 3 mm in length, and Forrest (1963) has had success in feeding brine shrimp. My impulse here is either to agree with McConnell (1965) that some planarians cannot read instructions or else to assume that planarians that ingest tiny, whole crustaceans are so hungry they are not particular. A planarian feeding on liver may get enough tough connective tissue into its pharynx that it can drag a chunk of liver much larger than itself a distance of several centimeters. If both liver and worm are placed out of water on a smooth surface and the observer holds the liver steady with a pair of forceps, the worm will soon contract violently and expel the mass, then crawl free, leaving behind it a slender rod of liver tissue molded to fit the inside of the pharyngeal tube.

Pearl (1903) stated that planarians are not attracted to food in the absence of chemical stimuli, nor is the pharynx protruded unless the mouth is in direct contact with the food itself. Without doubt, chemical stimuli, in the form of juices diffusing from damaged prey, play a major role in attracting planarians to food (Reynoldson and Young, 1963), but some individual planarians are as unpredictable in feeding reactions as in any other type of activity. On numerous occasions, in planarians starved for several days, I have seen the pharynx protruded and moved about, with the tip opening at times to form a tiny funnel, when no food of any kind had been placed in the fingerbowl.

Planarians may also be attracted to intact prey, from which no juices are diffusing, if the prey produces some type of disturbance in the water. Reynoldson and Young (1963) attached 25 intact invertebrates to the bottom of a container with smears of petroleum jelly, then added 20 planarians to the dish. The planarians promptly attacked and ate about 80 % of the struggling animals. By contrast, only 4 to 12 % of intact, nonstruggling prey were ingested by controls.

Jennings (1957) stated that planarians produce slime networks which serve as snares that entrap and entangle small invertebrates. The planarians do not lie in wait for the prey but are attracted by the disturbance created by the captured animal struggling to free itself. Animals larger than planarians may be immobilized in this way and used as food. Jennings added that there is no evidence that the mucus is toxic, for prey rescued after complete immobilization showed no ill effects.

It is interesting to note, in connection with the question of the toxicity of planarian mucus, that homogenates of *Dugesia dorotocephala* (Booth, *unpublished*) did not inhibit the growth of several species of bacteria examined, but, on the contrary, seemed to give a slight stimulation to growth in *Bacillus subtilis*. Damselfly and dragonfly nymphs appear to be the only predators of any importance (Young and Reynoldson, 1965). Young *Dugesia*, however, may be attacked and consumed by the protozoan *Dileptus* if the concentration of the latter is high (Brown and Jenkins, 1962).

Although planarians, if hungry enough, may feed on any food available, not all foods are equally nutritious. A series of studies made around forty years ago showed that planarians thrive upon raw liver and can be maintained indefinitely on this as an exclusive diet (Wulzen, 1927). Thymus, adrenal, kidney, and brain cortex are also good (Wulzen, 1923), but, if tissues used for food are taken from animals on deficient diets, the planarians exhibit pathological abnormalities (Wulzen and Bahrs, 1936). Refrigeration of food causes a slight deterioration (Bahrs, 1929), as does perfusion (Bahrs and Wulzen, 1932). An unknown growth substance is apparently washed out by the latter procedure.

There is some evidence that planarians may absorb nutrients through the epidermis. Osborne and Miller (1962) reported a definite penetration of peroxidase enzyme through the intact epidermis, which seemed to be taking place via the canals left by extrusion of rhabdites.

For many years it has been accepted that the breaking up of food into particles and their consequent ingestion through the pharyngeal tube was a purely mechanical process, but Jennings (1962) has provided convincing evidence that some degree of proteolysis occurs before ingestion. This appears to be the basis for the planarian habit of inserting the pharynx into the prey even when the latter is less than half a millimeter in diameter.

As soon as the food enters the gut, phagocytosis begins, and the greater part of digestion occurs intracellularly in food vacuoles, although some enzymes are secreted into the gut lumen during the first few hours after feeding (Jennings, 1962). Depending on the size and condition of the worm, as well as the availability of food, the intestines are filled after about 1 to $1\frac{1}{2}$ hours. Phagocytosis is essentially completed about 8 hours after feeding, and digestion of one feeding requires about 5 days (Willier, Hyman, and Rifenburgh, 1925). Food particles too large for phagocytosis, which survive the initial discharge of enzymes, persist unchanged and are eventually expelled from the gut in the process of defecation (Jennings, 1962). When such occurs, the entire body of the worm contracts three or four times, then the intestinal debris is shot out of the pharynx with surprising force. If worms that have been eating liver are handled, blood may be "belched" and discolor the water for a distance of 2 to 3 cm.

Food reserves in planarians appear to consist largely of fat. Willier, Hyman, and Rifenburgh (1925) found in planarians, some days after feeding, an abundance of fat in phagocytic cells, and also granules which gave a protein reaction. Both fat and protein diminished during starvation, although considerable quantities of the former remained, even after several weeks. von Brand (1936) reported in two European species of planarians the cyclical storage of small quantities of polysaccharides which reached a peak during autumn. These stores were utilized during the first week of starvation. Jennings (1957) confirmed the occurrence of much fat and some glycogen in gut cells but added both are to be found also in mesenchyme cells. He was of the opinion that gonads, mesenchyme, and general body tissues constituted the major protein reserve and concluded (1962) that protein granules in gut cells do not serve as protein reserve but are the source of the enzymes responsible for intraluminal digestion during the first few hours after feeding.

RESPIRATION AND METABOLISM

Basal metabolism, medically speaking, can be expressed as respiratory rate at minimal activity, and metabolic rate in planarians is commonly given as rate of oxygen uptake. Planarians have no specialized structures for gaseous exchange, and it is assumed that the process takes place by simple diffusion, through the body surface. In an early series of studies Hyman, using the Winkler method with asexual *Dugesia dorotocephala*, found the rate of oxygen consumption to be 0.2 to 0.3 cc per gram per hour, depending on size, interpreted as age, and on degree of starvation (1919a and c and 1920). Large (old) worms consumed less oxygen per unit weight than small (young) ones. She found that, shortly after feeding, oxygen consumption reached a maximum which was followed within a few hours by a marked fall in respiratory rate. If the planarians were not fed, the pronounced decline in oxygen uptake continued for 6 to 8 days, to be succeeded by a third phase characterized by a very slight and continuing decline. This more or less constant level of metabolic activity has been reported to extend from 1 to 3 weeks in *Dugesia dorotocephala* (Hyman, 1919a and 1920) to as long as 6 weeks in the race formerly known as *Planaria agilis* (Allen, 1919). Sexual *Dugesia dorotocephala* show these same variations in oxygen uptake, but, since the sexual specimens are much larger, the numerical values are considerably lower, the range being 0.12 to 0.16 cc per gram per hour, wet weight (Jenkins, 1960). Hyman found that respiratory rate was increased during the later stages of starvation (1920) and during regeneration (1919b). According to Lund (1921), oxygen concentration becomes a limiting factor when the concentration falls below about one-third saturation.

According to Child's theory (1915), there exists in axiate organisms a gradation in rate of physiological processes along the axis, the chief factor in correlation being antero-posterior dominance, i.e., each level dominates the region behind and is dominated by that in front. Child's views were supported by Hyman's work on respiratory gradients in planarians (1923), but later work has supplied evidence both for and against the idea of metabolic gradients. Løvtrup (1953), using the Cartesian diver technique and basing respiratory measurements on total nitrogen rather than on wet weight, found no evidence to support Child's theory, nor did Pedersen (1956), but the latter suggested this might be due to the fact that the total nitrogen content of the worm is much reduced during starvation. However, no data are available to show whether reduction of nitrogen content and of wet weight are comparable.

Aspects of cellular respiration in flatworms have received little attention. Hublé and Van Grembergen (1949), using *Planaria torva* and *Polycelis nigra*, found that respiration of intact cells (finely divided tissue pulp) was as high as 350 μl per hour per gram of fresh tissue at a pH of 7.1 to 7.4. Little change was noted over a wide range of pH, from 3.7 to 10.5. These authors found tissue respiration in planarians to be completely independent of oxygen pressure but agreed that Hyman's and Lund's findings were due to their use of entire animals, since gaseous diffusion is quickly reduced at low oxygen pressures as the distance of inner cells from the oxygen source is increased. They found a typical cytochrome oxidase system to be present which was subject to KCN inhibition and CO poisoning in the dark. Mengebier and Jenkins (1964), using asexual *Dugesia dorotocephala*, also reported evidence of the presence of the cytochrome oxidase system but found that maximum succinoxidase activity occurred at a pH of 8.3 rather than at the usually accepted optimum of pH 7.4. They suggested this was correlated with the ability of planarians to convert their external medium to a constant pH which, in their laboratory, was found to be 7.9 to 8.0.

Hammen and Lum, who have studied carbon dioxide fixation in a number of invertebrates, reported that several flatworms, including *Dugesia*, differ from other invertebrates studied in that they incorporate labeled carbon initially into malic acid (1962). Hammen (1964) suggested that the function of carbon dioxide fixation in animals may be to provide materials for amino acid biosynthesis from Kreb's cycle acids, since it is known that this process can occur.

Flickinger (1959), using *Dugesia tigrina* incubated with $C^{14}O_2$ or glycine C^{14}, found both chemicals were incorporated into proteins by the worms along an axial antero-posterior gradient. Treatment of the worms with an inhibitor of protein synthesis flattened the gradient and lowered the rate of incorporation. Coward and Flickinger (1965) reported, in intact planarians, a gradient of incorporation of nucleic acid precursors, estimates of synthetic capacities being based on relative amounts of uptake. Best and his associates (1965), however, were able to demonstrate little or no incorporation of amino acids and nucleotides in solution, but a large incorporation was obtained when the same chemicals were ingested in the form of particulate food. They reported for *Dugesia dorotocephala* a protein turnover of at least 46% in 7 weeks.

EXCRETION AND OSMOREGULATION

The excretory system of turbellarians is a protonephridial system consisting of delicate tubules composed of a one-layered epithelium. These tubules give off numerous branches which form a network, each branch terminating in a flame bulb (Hyman, 1951). The morphology of the excretory system is not completely known for any planarian (Hyman, 1956), but evidence points to the view that the chief function of the protonephridial system, and the only function of the flame bulbs, is regulation of water content (Hyman, 1951). Steinbach (1962) found that *Dugesia tigrina* was able to concentrate Na, K, and Cl ions and showed remarkable ability to maintain constant internal water and salt concentrations under several conditions of salt concentration. Evidence indicated the worms were freely permeable to water, and that movement of ions into and out of the animals was dependent on the properties of the worm itself and not on the magnitude of a diffusion gradient.

Nothing is known of the form in which nitrogenous waste is excreted in planarians, but, since protein food is stored chiefly as fat, a process of deamination must occur, with consequent formation of ammonia or a related compound. Hyman (1951) thought it probable that turbellarians convert nigrogenous wastes into insoluble granules which are then stored permanently in the mesenchyme and in pigment cells. Skaer (1961), in a study of *Polycelis nigra*, found that rhabdites consisted largely of combined arginine and purine and suggested that these structures were organelles of purine excretion. Rhabdites, which are soft, rodlike extrusions from epidermal glands, swell when placed in water, and form a semifluid, gelatinous layer which invests the body of the planarian. Although it is accepted that a primary function of this material is that of acting as a "fluid cuticle" to protect the epidermis (Jennings, 1957), rhabdites may possibly play a dual role and also serve as an excretory product.

INNATE BEHAVIOR

Behavior has been described (H. S. Jennings, 1962) as being merely a collective name for the most obvious and most easily studied processes of the organism. In the case of planarians, I am not sure whether or not their responses are always obvious, but I am sure they are not always easily studied. One of the earliest and most complete investigations of planarian behavior was that of Pearl (1903), who attempted to analyze planarian behavior into component factors. Although not all of his conclusions are valid in the light of later studies, his paper, together with Hyman's volume II (1951) of her series on *Invertebrates*, are perhaps the two most generally helpful references available for those investigating planarian behavior.

Movements

Pearl reported that planarians made use of two kinds of locomotor movements: gliding and crawling. Quite small planarians, not over 2 to

3 mm in length, appear to move entirely by ciliary gliding in which the ventral cilia beat against the thin layer of mucus which is constantly secreted by the animal. The ciliary waves appear to be independent of the nervous system. As the animals increase in size, gliding is still the characteristic type of movement, but ciliary action is supplemented more and more by imperceptible muscular waves that pass from the anterior end backward. In triclads, the waves are mostly monotaxic, i.e., they pass across the whole width of the animal. Relative roles played by cilia and by muscular waves in locomotion can be determined by treating the worms with lithium chloride, which paralyzes ciliary action, and with magnesium chloride, which paralyzes the muscles (Hyman, 1951). As the worm glides, the head may be raised and lowered or moved from side to side, and the auricles may also be extended and moved as if the worm were testing its environment. There is no sinuous bending of the body in any way that contributes to propulsion. Speed of gliding in minutes per second is greater in larger animals, but there is little difference in species. Gliding backward apparently does not occur (Pearl, 1903).

Crawling, a more hasty type of locomotion, is induced when the worm is stimulated. This consists of large, obvious muscular contractions and expansions, in which the head may be stretched forward and attached to the substrate and the body then pulled after it. The animal may crawl either forward or backward, depending on the stimulation. Crawling movements usually subside in a short time, unless repeated stimulation occurs, and the worm reverts to characteristic gliding. Movement can occur in the absence of the brain, as is shown by the fact that recently cut sections are capable of independent locomotion.

Adhesive mucus is continually produced at the body margins of planarians. These mucous deposits constitute the characteristic slime trails left by planarians on the substrate and may provide temporary anchorage during contraction. As the worm moves, the mucus may be drawn into strands 2 to 3 cm long, which may stretch, in nature, between stones or, in the laboratory, across corners of a container. Such strands constitute the food snares mentioned above and may be demonstrated by flooding the culture dishes with weak eosin. Under suboptimal conditions, the number of strands is greatly increased. I have seen planarians in lake water containing fine silt (after a rain), which could not be filtered out, construct three-dimensional webs, or nets, which were clearly visible outlined by deposited silt. The worms crawled about on these sticky threads and, indeed, appeared to prefer them to the container surface.

Swimming does not occur in planarians, although they may glide upside down on the surface film of the water. This movement is slow, because the film is too elastic to offer much support. A planarian can support itself on the surface film by only the tip of the tail and may often do so, waving the rest of the body about below as if seeking the substrate. If an object to which the planarian may attach is not encountered, the worm may construct a slime thread and dangle downward as it gradually lowers itself to the bottom.

Correlated with movement in planarians is their ability to change length. In general, increase and decrease in length is a result of contraction of circular and of longitudinal muscles respectively, and flattening is produced by contraction of dorso-ventral strands. Recent studies have shown that change of shape involves not only muscle fibers but also regularly disposed, inextensible fibers in the epidermal basement membrane, which are arranged in the form of a lattice of right- and left-hand spirals. In much the same way that a garden trellis may be lengthened or shortened, the planarian may become longer or shorter as a change occurs in the angle between intersecting elements of the fiber lattice (Clark and Cowey, 1958). It follows that the position of relaxation, which is that of greatest stability, is not coincident with the condition of greatest length. This substantiates Pearl's observations that a relaxed planarian is not one that is fully extended and gliding (1903).

Activity and Rest

Planarians may undergo periods of activity and rest, but whether or not these periods are cyclic has not been determined. Many observers report that planarians are more active at night, after 2 or more hours of darkness (Pearl, 1903). Onset of light will cause resting planarians to glide about, but they usually come to rest within a short time in the most shadowed area of the container. If the worm is disturbed, it will move again, but the periods of activity become progressively shorter and shorter. Planarians at rest usually form groups or collections, 1 to 2 in. in diameter, in which the worms may be so closely packed together that overlapping occurs.

Response to Mechanical Stimuli

Planarians are extremely sensitive to mechanical disturbances such as vibration or contact. Even so minor a disturbance as breaking the water-surface film with a needle may cause nearby planarians to contract momentarily. Pearl's analysis of reactions to mechanical stimuli, with special reference to the head, is as follows: A positive response begins with a momentary pause, followed by a longitudinal extension of the anterior end of the body and a turning toward the stimulus. The anterior end is then raised and the worm moves toward the stimulus. A negative response consists of a pause followed first by longitudinal contraction of the anterior end, then by the worm turning and moving from the stimulus. Pearl found that, in general, positive responses are given to stimuli of low intensity, while to stimuli of high intensity negative responses are made. Positive reactions are not exhibited by postpharyngeal regions, nor by prepharyngeal regions in the absence of a brain. Pearl characterized an indifferent response as a local contraction followed by continued movement in the same direction.

Marked or persistent disturbances induce an excited condition in the animals, and they will not then give the finer responses. Positive responses to weak stimuli seem to be of an investigative nature. If the stimulus is increased in intensity so that it becomes noxious to the planarian, the latter will then turn away, repeatedly if necessary. However, if a noxious stimulus is continued, the planarian may cast all discretion to the winds, so to speak, and make a wild jump toward the offending source of stimulation. One gets the impression that the planarian has decided that, if normal avoidance measures won't work, rushing the enemy might. Alternatively, the planarian may simply contract and refuse to budge, even with stimulations so strong that death follows.

Response to Contact

Planarians are markedly thigmotactic, the dorsal surface being negatively and the ventral surface positively thigmotactic. The strength of the latter response is shown particularly in the characteristic "righting reaction," in which a planarian, placed dorsal surface downward, will quickly twist itself into a spiral so that the ventral surface of the head is in contact with the substrate. Mucous secretions assist the head in attaching to the substrate, then the worm glides forward, unwinding the spiral as it progresses. The original head turn may be either to the right or to the left, so that the spiral unwinds clockwise or counterclockwise. This depends on the relation of the somewhat convex dorsal surface to the bottom of the container (Pearl, 1903). Only a few seconds is required for the righting reaction, depending on the size and activity of the animal.

Although a planarian will move away from a light object placed on its dorsal surface, it will not turn over or give a "reverse righting reaction" to avoid such. Due to the negative thigmotaxis of the dorsal surface, planarians are less likely to come to rest in a situation where both surfaces are in contact, but resting planarians will allow other planarians to rest on them. Also, I have found groups of planarians congregated on the bottom of small stones, partially buried, under conditions where the sand had to be in contact with their dorsal surfaces, and I have had planarians under observation in a Syracuse watch glass, which contained a thick layer of paraffin, crawl into the narrow crevice between paraffin and glass. In the latter case, the negative thigmotactic response may have been overcome by the stronger negative response of planarians to light.

Response to Water

Rheotaxis, or response to water currents, is not pronounced in planarians inhabiting still waters, such as the common *Dugesia tigrina*. *Dugesia dorotocephala*, however, a spring-dwelling species, shows a positive rheotaxis by traveling upstream in nature, or by crawling up into continuous-flow aerators in laboratory aquaria. In general, planarians exhibit a positive response to weak currents, which is probably part of the food-catching mechanism, since the animals are attracted to intact prey creating a water disturbance. Strong currents usually elicit a pronounced contraction of the body and increased adhesion to the substrate. Although rheoreceptors are thought to be distributed over the entire body surface, highest sensitivity is found in the auricular groove or auricular sense organ. This area is generously supplied with nerve cells and bears short cilia that are probably sensory bristles (Hyman, 1951).

Pearl (1903) stated that no true hydrotaxis, in the sense of movement toward water, was to be found in planarians. He found no evidence that planarians would voluntarily leave the water, but, if they were taken from the water and placed on a dry surface, they would curl up and tuck under the head, thus exposing as little surface as possible. At short intervals the animals would straighten out and extend the head, but, if no water was encountered, they would curl up again. A short exposure to dryness will result in death, although, under such conditions, a profuse production of slime, which absorbs water, may protect the animal for a time. Water can be lost in an atmosphere of 100% humidity, but recovery will normally occur if water loss does not exceed 45% of the fresh weight (Hyman, 1951).

Contrary to Pearl's observations, it is not uncommon for at least some species of fresh-water triclads to leave the water and crawl upon a wet surface. I have noticed this repeatedly in Dugesia dorotocephala in the laboratory. In fact, it is such a common occurrence that one of our three code symbols for reporting mortality is "DB" which means "crawled out of water and dried on bowl." Crawling out of water may also occur in nature. I have seen members of this same species, a half-dozen or more at a time, crawl out of a spring on to the wet sand and wander about for a few minutes before returning to the water, their black bodies glistening in the sunlight. Planarians also, if strongly provoked, may protrude a part of the body from the water and leave it there until it dries completely. On one occasion I placed a large, sexual specimen in a Syracuse watch glass filled with water and containing numerous specimens of the ciliate, Dileptus. The ciliates attacked the planarian, assaulting it with lashes of their whiplike proboscisces. The toxin discharged at each lash caused the planarian to cringe noticeably, and, apparently in desperation, it stuck its sensitive head up out of the water where the ciliates could not reach it and left it there until the head dried. The planarian then wandered off, minus a head, but otherwise not much worse off.

Response to Light

In general, the entire body surface of a planarian is sensitive to light. Nondirectional light is stimulating in that it causes increased activity in the form of random movements and also prompts resting worms to begin moving. A negative response occurs when a more lighted area is reached, and planarians eventually come to rest in the most darkened area of the container (Taliaferro, 1917). This preference for shadowed areas can be strikingly shown by placing an opaque strip, such as an electric light cord, over the container so that a distinct shadow line is cast across the bottom. The planarians will congregate in the shadow, accurately outlining it. The sense organs responsible for this generalized phobic reaction have not been identified (Hyman, 1951).

A fairly precise negative response is given to directed light, mediated by the anterior end only, primarily by the eyes. The location of the bending of the body, when the head is turned away, depends on the intensity of the light. The more intense the illumination, the more posteriorly the point of bending is located. It is not, however, located farther back than the pharynx (Taliaferro, 1918). Planarians without eyes will move away from a light source, but with less precision and more slowly than those with eyes (Parker and Burnett, 1900).

There is a wide range of sensitivity to light among planarians. The average threshold illumination for green light in *Dendrocoelum lacteum* was found by Marriott (1958) to be 39,000 quanta per eyespot per second. Particularly sensitive animals responded one log unit below this. Marriott reported that the threshold of the most sensitive animals was of the same order as the illumination received from a clear, moonless night sky. Planarians are insensitive to infrared light, but are relatively much more sensitive than man to ultraviolet. The spectral sensitivity curve is roughly similar to that of rod vision in the human eye without a lens (Pirenne and Marriott, 1955). That these responses are mediated by the eyes was demonstrated by performing the same experiments on animals from which the eyes had been removed. The power of responding was lost but returned when the eyespots regenerated.

Although the directed response of planarians is mediated by the eyespots, eyeless planarians are sensitive both to ultraviolet radiation and to white light. In the white, eyeless flatworm from Mammoth Cave, *Sphalloplana percaeca*, the sensitivity was found to be so great that the worms writhed violently upon a 30-second exposure to sunlight and were killed by a 2-minute exposure (Buchanan, 1936). It is possible this sensitivity is due to the presence of some chemical. Krugelis-Macrae (1956) reported finding in planarians uroporphyrin which, like other porphyrins, is strongly photoactive and is efficient in absorbing light around 4,000 Å. Although the presence of porphyrin is not uncommon in various inert animal tissues, in the planarian the chemical is distributed rather generally and is not stored in inert tissues. Krugelis-Macrae suggested that possibly the light sensitivity of planarians might be due, in some degree, to the presence of porphyrins.

Response to Electric Current

Sensitivity of planarians to electric current was investigated by Pearl (1903). He found, in contrast to other behavior he studied, planarian reactions to constant direct current were not clear cut. The worms turned incompletely toward the cathode but rarely oriented themselves along the direction of the current. If they did line up, the posterior end contracted, then relaxed when the current was broken. This did not occur when the animals were at right angles to the current. Furthermore, he found the worms became wholly or partially paralyzed. He thought the current affected the oriented muscle fibers directly and that neither sense organs nor cilia were stimulated.

Hyman and Bellamy (1922) found that planarians oriented themselves in an electric current with the head toward the cathode and suggested that planarian orientation in an electric current was related to the direction of bioelectric currents present within the animal. Hyman (1932) proposed that orientation took place by means of the regular neuromotor or ciliary mechanism of the animal, but that the posture assumed was controlled by internal electric currents.

More recent studies of the responses of planarians to electric current have been made by Viaud, using *Dugesia lugubris*. The resistance of the planarian body to the passage of an electric current varies between 10^5 and 10^6 ohms. When a planarian is exposed to a direct current, it orients itself in such a way that its body offers the least resistance to the passage of the current. Under normal conditions, when the planarian's head is toward the cathode, resistance is least and conductance is greatest. Viaud used a series of planarians of different sizes, ranging in length from 6 to 16 mm, and tested their threshold responses (simple contraction) in terms of conductance, with the head pointed first toward the anode then toward the cathode. He found the difference of these two values increased linearly with the size of the planarian per unit of surface cross section (1951). Since the current passes through the planarian body more easily when the head is toward the cathode, Viaud concluded that the planarian body is constituted of a series of polarized elements, or dipoles, oriented in the same direction and able to develop an EMF. He was of the opinion that the presence of a nervous system was not necessary for a reaction to be produced, basing this on the fact that cut sections or an isolated pharynx will also show cathodic orientation (1952).

In studying the EMF developed in the planarian, which opposed the flow of current, Viaud used a series of values ranging from 10 mv to over 4,000 mv. He found the variations in the opposing EMF to be a function of the intensity of the current. Below 1,000 mv, the planarian EMF was proportional to the intensity of the current. Above this value, the development of the opposing EMF diminished little by little, reaching a maximum at 2,000 mv. After this the opposing EMF dropped to zero at a point between 3,000 and 4,000 mv. At the maximum, the EMF produced by the planarians was 118 mv when the head was toward the cathode and 85 mv when toward the anode. The planarian EMF varied with the size of the animal according to an approximately parabolic function but was still greater in cathodally oriented worms (Viaud, 1954a). Viaud also reported there was a lag in the development by the planarians of the opposing EMF and, furthermore, several minutes were required after the current was cut off for the planarians to return to their normal state of feeble polarization (1954b).

The taxic response of planarians to electric current, i.e., orientation toward the cathode, can be reversed to some degree by exposure of the worms to polarized light and to a greater degree by exposure to ultraviolet light. When both polarized and ultraviolet light are used together, a still greater percent of inversion is obtained (Viaud, 1955). Viaud termed such an orientation toward the anode, produced by polarized or ultraviolet light, "true galvanotropism," but added that similar reactions, produced merely by fatiguing the animals, were "pseudotropisms" (1957). Viaud summed up his findings by stating that the bases of galvanotropism were to be found in three areas, one superimposed upon the other: electric anisotropy of the animal on a physicochemical plane; anisotropy of excitation, on a physiological plane; and the tendency of the worm to orient itself toward the maximum source of excitation, on a psychological plane (1959).

Response to Geophysical Stimuli

Planarians not only respond to such obvious stimuli as contact, light, and electric current, but are also capable of very precise rhythmic responses to pervasive geophysical factors. These responses have been demonstrated by Brown and his associates who have studied metabolic fluctuations of numerous organisms in an automatically recording respirometer, completely sealed away for several months from expected outside influences. They have shown beyond reasonable doubt that living things, even while in so-called constant conditions, have access to outside information as to the time of day (or position of the sun), time of lunar day (or position of the moon), time of lunar month, and also time of year. It is obvious that responses to subtle geophysical fields, if they occurred, would be rhythmic, for all geophysical forces we know about display periodisms related especially to relative motions of the earth, moon, sun, and stars. It is also obvious that, if the living system is responding to natural fields, it must exhibit an extraordinary sensitivity for extremely weak fields. This is not unexpected when we consider, e.g., that the responses of living things to temperature changes cover only a very short range out of the hundreds of degrees of change that are possible.

Brown has found in planarians of the genus *Dugesia*, as in other organisms studied, a sensitivity to very weak magnetic fields of the order of strength of the earth's own. In testing the responses, the worms were placed in position and induced to crawl in different compass directions over a polar coordinate grid. That worms could distinguish compass directions was shown by the fact that, when started north or south, they veered clockwise as they crawled, but, when headed east or west, they veered counterclockwise. As the apparatus was rotated to various other geographical directions, the responses of the worms changed. Their ability to resolve small differences in orientation seemed to approach that of a magnetic compass needle (Brown, 1965). During winter months a clear monthly cycle was evident, with maximum counterclockwise turning on the day of new moon and maximum clockwise turning at the time of full moon. During summer months a semimonthly cycle was followed, with maximum clockwise turning near the times of both new and full moon. Periods of transition occurred during the fall and spring. In an experimental field 20 times as strong as the earth's, the response was abolished. The cycle was repeated, with essentially no change, during 3 years of study (Brown, 1962b).

Brown has also shown that *Dugesia* can respond to very weak electrostatic fields. The planarians can detect a change, while in water, of 2 v/cm in electrostatic gradient in the surrounding air. The nature of the response to exactly the same field change, at a given time of day, varied with the geographic orientation of the worms. The responses of northward- and southward-bound worms were the opposite of one another; the responses of east- and west-directed worms were very small. The response for each of the compass directions in the afternoon was exactly opposite to the response given in the morning hours (1962c). The degree of sensitivity demonstrated by the experiments exceeds by many times that which would be necessary for planarians to become aware, e.g., of a thundercloud several miles away (1962a).

The sensitivity of planarians to very weak gamma fields, less than 10 times the natural ambient background, has also been demonstrated by Brown (1963). The worms veered away from the gamma source whether it was placed on the right or on the left, and the strength of the response was quantitatively related to the strength of the field. The response was also related to geographical spatial orientation, and semimonthly, semiannual, and annual variations were found. No correlation was shown with concurrent temperature variations.

Few other studies of rhythmic variations in planarian reactions are available, but May and Birukow (1966) report that fluctuations in planarian responses to light are more marked during the lunar month than during the solar day.

REPRODUCTIVE ACTIVITY

Planarians may reproduce sexually or asexually or by a combination of both methods. Variations exist not only among species, but from population to population within a species. There are also demonstrable individual variations.

Asexual Reproduction

Asexual reproduction is common in many fresh-water planarians, especially in the genera *Dugesia* and *Phagocata*. Planarians propagating by this means undergo transverse fission, i.e., they divide into two pieces, usually in a plane just posterior to the mouth, and each piece thereafter regenerates the missing parts. In North American planarians, a worm about to undergo fission shows no morphological changes prior to the event. When the worm reaches a suitable size and other factors are favorable, the rear end suddenly adheres firmly to the substrate as the animal glides about, while the anterior end continues an attempt to advance. After a few anterior-end struggles the tail may release its hold, and the worm continue unperturbed. If this occurs, the process is soon repeated one or more times and, sooner or later, as the anterior end struggles to move forward, the rear end does not release its hold and the planarian pulls in two. The head then continues with its usual activities, but the strongly contracted tail remains extremely sluggish for 2 or 3 days. Curtis (1902) believed the division was accomplished by a constriction of the circular muscles. He considered the process to be a spontaneous and perfectly normal process of multiplication.

Immediately after fission each wound surface contracts strongly, forming a shallow, concave, dark crescent. In the head piece, as new tissue develops, the tail first becomes roundly blunt but soon assumes a normally pointed shape. In the tail piece, a small, unpigmented head soon forms, tiny eyespots appear, and internal rearrangements occur. After 5 days it is often difficult to identify positively either head or tail fission products.

Sexual Reproduction

In many planarians a complete and complex set of sexual organs develops. These include, since sexual animals are hermaphroditic, a pair of ovaries just behind the eyes, paired oviducts, separate yolk glands, and testes scattered throughout the body. Both male and female ducts open eventually into a cavity, termed an *atrium*, which, in turn, opens to the outside by means of a genital pore. A cavity for temporary storage of sperm is found in many species. Copulation is mutual, and cross-fertilization occurs. An exception to the latter statement is found in *Cura foremanii*. Anderson and Johann (1958) reported that in four stocks of this species, of diverse geographical origin, in which individuals were isolated from birth, sexual reproduction involving only self-fertilization occurred. There are no reports of copulation having been witnessed in this species, nor is any evidence known of their reproduction by fission.

Physiological Races

Although many studies of planarian reproduction have been published, the relative importance of the two types of propagation in the life histories of the animals is not understood, nor are the factors influencing the appearance of either the one type or the other. Many have the impression that planarians, in general, have a full set of reproductive organs throughout their lives, but that the animals apparently are not convinced of their potential and simply divide when the spirit moves them. On the whole, this is a false assumption. Kenk (1937, 1940, and 1941) and others have confirmed an earlier observation of Curtis (1902) that *Dugesia tigrina* occurs in at least two physiological races, a sexual and an asexual one. The latter, according to observations over a period of several years, reproduces exclusively by fission. Temperature and nutrition influence the rate of fission but do not induce sexuality. The sexual race develops sexual organs and deposits cocoons during the colder months of the year, but, when the breeding season is over, the sex organs degenerate and fission occurs during the warmer months. Such an alternating cycle may be repeated several times during the lifetime of the individual worm. This type of life cycle, with an alternation of periods of sexual and asexual activity, is found in a number of species of several genera.

Reproduction in *Dugesia dorotocephala*

For many years *Dugesia dorotocephala* was thought to propagate in nature almost exclusively by fission (Hyman, 1939, and Kenk, 1935), although some specimens kept in laboratory cultures at low temperatures occasionally became sexual. Some years ago Harley Brown and I were fortunate enough to find a race of this species in Oklahoma, which remained sexual during all months of the year both in nature and in the laboratory. We have published several papers on sexual activities and behavior in this race, and, although much of my present research is quite incomplete, I should like to report on some of the work that has been done in so far as it is applicable here.

The Oklahoma *Dugesia dorotocephala* adults grow to be quite large, averaging close to 30 mm in length, although 40 mm is not uncommon. The average weight of 100 specimens, weighed five at a time, was 52 mg per individual. These worms become sexual, on an average, in 4 to 6 months at a temperature of 18 to 20°C and remain sexual thereafter, without fission, and with continuous cocoon production. I have at present several isolated pairs of sexual worms, many of which have produced 1 to 3 cocoons every 4 to 5 days since the onset of sexuality some time ago.

Cocoon production does vary somewhat from month to month, as can be seen in Table 1, but there is no seasonal trend. The only influence I have been able to note thus far is temperature. In worms subjected to rising temperature, cocoon production dips downward practically to zero, while worms at a constant temperature continue regular production of cocoons. A temperature rise is not followed by loss of sex organs nor by a return to fission. The worms remain sexual and either die when the temperature is high or apparently recover fully when a lower temperature is reestablished.

Data in Table 2 show that there is no seasonal trend in the production of young. In other words, not only are cocoons produced continually throughout the year, but a high per cent of cocoons are viable all months of the year. It is noticeable, however, that more young are produced per cocoon in older worms. Whether this is due to maturity of the parents, to older worms having more stored sperm, since sperm remain viable for several weeks after copulation, or to some other factor is unknown. At present there are no data on how long these worms remain sexual. I have

TABLE 1

Monthly cocoon production in four groups of *Dugesia dorotocephala*, in terms of number of cocoons produced per worm per month. $S = F_1$ generation of interbred siblings, all of which emerged from one cocoon. $S\text{-}S = F_2$ generation, crossbred; the two generations became sexual without prior fission. $FS\text{-}S = F_2$ generation, crossbred; the F_2 generation became sexual without prior fission, and the F_1 generation fissioned once before onset of sexuality. $FS\text{-}FS = F_2$ generation, crossbred; in both generations one fission occurred prior to sexual maturity. Dates given indicate time of appearance of first cocoons.

	Jan.	Feb.	Mar.	Apr.	May	June	July	Aug.	Sept.	Oct.	Nov.	Dec.
S (1964)	...	...	...	...	...	...	...	1.5	0.6	1.2	1.2	0.9
	2.6	1.7	1.1	1.6	2.1	0.9	1.1	2.2	2.2	1.2	2.8	2.5
	1.7	1.2	1.6	1.1	1.6							
S-S (1965)	...	...	...	...	...	...	1.4	1.2	1.6	2.0	2.8	2.0
	1.6	2.3	2.4	2.6	3.8							
FS-S (1965)	...	...	...	...	0.5	0.9	1.2	0.6	1.8	1.9	2.7	2.6
	2.3	1.9	1.8	1.7	2.6							
FS-FS (1965)	...	...	...	...	...	6.5	3.7	2.8	3.0	2.9	3.9	3.4
	3.9	3.3	3.4	2.6	4.5							

a number of specimens over $2\frac{1}{2}$ years old that are still producing viable cocoons.

Copulation in the laboratory can occur at any time of day or night, but I have found it is more likely to occur within the first half hour or so after worms kept in darkness are exposed to light. The entire process of copulation requires about 6 to 7 minutes. The heads of the copulants face in approximately the same direction, both adhering firmly to the container. The posterior halves are raised slightly, and the ventral surfaces moved gently against each other for perhaps 2 minutes, as if each worm were seeking the genital pore of the other. When actual copulation occurs, a marked dent or dimple appears on the dorsal surface of each copulant, above the gonopore. During this time the animals are quite contracted, completely immobile, and indifferent to ordinary stimuli. After about 3 minutes the worms gradually relax and separate. Copulation occurs repeatedly throughout the adult life of the individual (Jenkins and Brown, 1964).

Sperm received during copulation soon pass up the oviducts to paired sperm receptacles near the ovaries. Eggs are fertilized as they pass down the oviducts, and separate yolk cells are added from the yolk glands (Hyman, 1951). The mass of fertilized eggs and yolk cells is formed into a cocoon in the atrium, with a delicate, slender stalk, several millimeters long, attached at one side. Cocoon deposition may occur within a couple of days after copulation. When the cocoon is deposited, the stalk is first extruded and attached to the substrate (Jenkins, 1966), then the cocoon is "squeezed" through the gonopore. One to four hours are required for the complete process. During this time a pale area can be seen on the dorsal surface of the worm in the region of the gonopore, and the worm is quite constricted at this point (Jenkins and Brown, 1963). Planarians depositing cocoons are usually indifferent to normal stimuli, such as moderate light or other worms crawling over them. If food is offered, they may ignore it or may move to it and eat. In the latter case, the cocoon may not be deposited until the meal is finished, or it may be "popped out" without being attached to the substrate.

Newly deposited cocoons are creamy white but soon turn red, then black. Cocoons, especially fresh ones, apparently are a delicious morsel, for hungry planarians may devour the contents, sometimes beginning eagerly even before the cocoon is completely deposited. At a temperature of 18 to 20°C the cocoons hatch in 3½ to 4½ weeks. From one cocoon 1 to

TABLE 2

Monthly production of living young in four groups of *Dugesia doroto-cephala* in terms of number of offspring produced per worm per month. For explanation of symbols, see Table 1.

	Jan.	Feb.	Mar.	Apr.	May	June	July	Aug.	Sept.	Oct.	Nov.	Dec.
S (1964)	...	...	...	...	...	...	...	4.2	5.9	8.6	6.6	4.9
	10.5	19.9	5.0	5.0	9.1	4.0	12.1	21.1	25.4	15.8	29.5	28.2
	25.5	19.6	10.9	13.6	13.5							
S-S (1965)	...	...	...	...	...	...	4.8	0.6	6.2	7.2	14.9	10.3
	8.8	10.9	10.8	11.3	8.2							
FS-S (1965)	...	...	...	...	0.8	1.2	2.9	2.3	8.6	10.2	15.5	12.2
	12.1	9.7	9.7	9.7	10.0							
FS-FS (1965)	...	...	...	...	...	12.5	21.5	15.9	17.7	21.3	22.1	18.8
	21.9	18.4	21.3	15.7	30.2							

TABLE 3

Variations in size of offspring produced from cocoons deposited by one group of sexual *Dugesia dorotocephala* during four 2-month periods representing respectively: onset of sexuality (9–10/64 = September–October, 1964), 6 months later, 1 year later, and 1½ years later. Number of adult worms = 8. Number of offspring during first 22 months = 2,379. Numbers in columns show per cent of juveniles of size indicated. $T = 1$–2 mm; $S = 3$–4 mm; $M = 5$–6 mm; $L = 7$–10 mm; $O = $ all others, including juveniles of larger size or those that have fissioned or have become sexual.

Per cent of planarians in each size group

Date	At emergence				Two months later					Four months later				
	T	S	M	L	T	S	M	L	O	T	S	M	L	O
9–10/64	23.6	62.7	13.7	...	56.3	39.5	2.1	2.1	...	50.0	16.6	16.6	16.6	...
3–4/65	11.6	44.2	40.4	3.8	3.9	28.2	17.9	20.5	29.5	...	3.9	3.9	9.8	82.4
9–10/65	27.0	69.0	4.0	...	7.1	23.8	26.2	4.8	38.1	...	...	...	...	100.0
3–4/66	16.3	72.5	11.2	...	12.0	60.5	23.9	3.0	0.6	(No data yet)				
Total	21.6	65.5	12.4	0.5	15.8	45.4	19.7	7.2	11.9	4.1	4.1	4.1	6.9	80.8

35 baby planarians may emerge. As a usual rule, the more babies there are per cocoon, the smaller the average size of the juveniles.

The size of immature or nonsexual planarians varies greatly, not only at the time of emergence from the cocoon, but for months thereafter. Data in Table 3 show the great majority of newly emerged young are 3 to 4 mm long, but lengths of 1 to 2 mm and 5 to 6 mm are common. Greater lengths, up to 12 mm, are rare but do occur. In a number of studies by various investigators, the size of planarians has been correlated with age, but data in Table 3 show such is not necessarily the case, at least in this race. At 6 months of age, when many worms are already sexual and have attained a length of over 2 cm, a certain proportion may still be very small. In this example the size of the worms at 6 months of age is related, to some extent, to the age of the parents. Worms hatched from the earliest cocoons produced failed to grow as well as worms hatched later when the parents were more mature.

In this race, mortality is high. In the four groups shown in Table 4, the highest mortality by far was found in the F_1 interbred group. Parents of all juveniles in this group were siblings, hatched from one cocoon. In the other three groups, which are the F_2 generations of crossbred parents, the highest mortality occurred, during the year's study, in worms whose parents and grandparents became sexual without previous fission (S-S), and least mortality was found in the group whose parents and grandparents fissioned once before becoming sexual (FS-FS).

Although fission occurs in this race, the per cent of worms that fission is always low. Data in Table 4 show that, particularly in the crossbred groups, only about one-third to one-fourth as many worms fission as become sexual without fission. In most planarians fission is a method of reproduction, but in this race it appears to be a most ineffective means of propagation. Planarians that fission may indeed fission again—one lone planarian, born of sexual parents, fissioned 14 times in 10 months—but, sooner or later, these planarians become sexual or die. In a study of the fate of offspring of 35 cocoons, it was found that fissions were at

TABLE 4

Comparison of per cent of mortality, fission, and onset of sexuality without prior fission, in four groups of *Dugesia dorotocephala*. For explanation of symbols, see Table 1.

	S	S-S	FS-S	FS-FS
Mortality	83.1	62.9	61.7	41.5
Fission	4.5	8.1	6.1	12.2
Sexual	3.0	27.9	29.8	45.8
Remaining	9.4	1.1	2.4	0.5

all numerous in only 12 of the 35 groups. Results of the study of these 12 are presented in Table 5. One fission product lived 150 weeks after the date of emergence, and another lived 94 weeks, but the great majority died during the first year. Several of those that became sexual are still alive and are producing viable cocoons.

TABLE 5

Total number of fissions and fate of individuals in planarians hatched from 12 cocoons of *Dugesia dorotocephala*. Offspring from these 12 showed the greatest number of fissions in 35 groups studied. H = number hatched from each cocoon; F = number of original fissions; F-L = number of later fissions; S = number that became sexual; D = number that died. D-1, D-2, and D-3 indicate number of deaths that occurred during the first, second, or third year after date of hatching.

Cocoon	H	Fate of original planarians			Fate of fissioned planarians				
		F	S	D-1	F-L	S	D-1	D-2	D-3
No. 1	16	1	..	15	5	..	7	..	..
2	18	6	2	10	6	5	3	8	..
3	8	7	..	1	25	3	14	18	1
4	12	7	..	5	15	..	21	8	..
5	10	7	..	3	71	3	39	38	..
6	16	2	6	8	5	..	8	..	..
7	14	5	..	9	12	5	13	4	..
8	29	10	..	19	15	3	32	..	..
9	11	6	..	5	..	4	8	..	..
10	11	2	..	9	11	..	11	4	..
11	18	7	..	11	21	2	34	..	..
12	13	5	..	8	9	..	12	7	..
Total	176	65	8	103	195	25	202	87	1

Referring again to Table 4, we find that in the *FS-FS* group not only is mortality much lower than in the other groups, but both fissioning and onset of sexuality without fission are much higher. Although data show that fissioning *per se* is an inefficient means of reproduction in this race, there is the possibility that a correlation exists between occurrence of fissioning and strength and vigor in mature animals. Experiments in progress at present may shed further light on this problem.

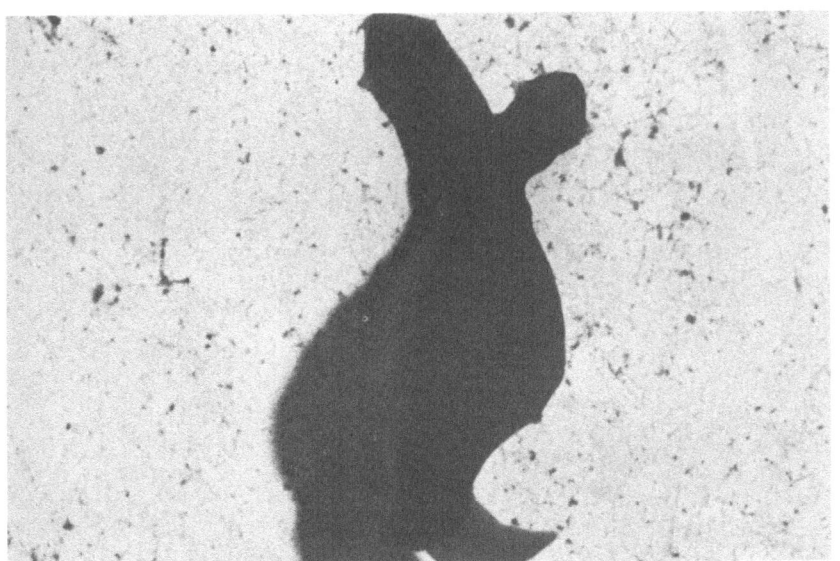

Fig. 1. *Dugesia dorotocephala* in copulation. The posterior ends of the planarians are raised slightly from the substrate and adhere closely in the region of the gonopore. The worms are contracted and immobile during this time, although the heads may move about slowly and assume various angles relative to each other.

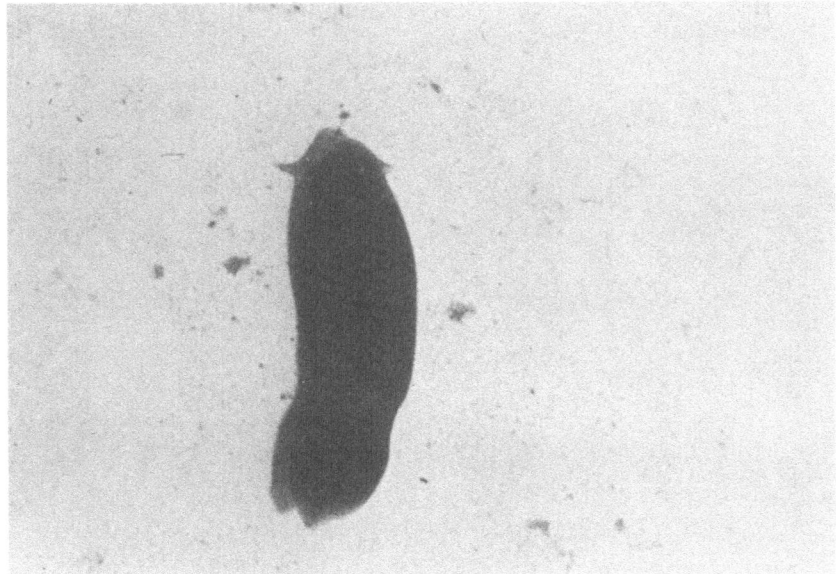

Fig. 2. *Dugesia dorotocephala* ovipositing, dorsal view. An hour or more is required for cocoon deposition. The entire planarian is somewhat contracted, with a pronounced constriction in the region of the gonopore. A dorsal hump, which is due to the contained cocoon, can be seen as a grayish-white spot in the constricted area.

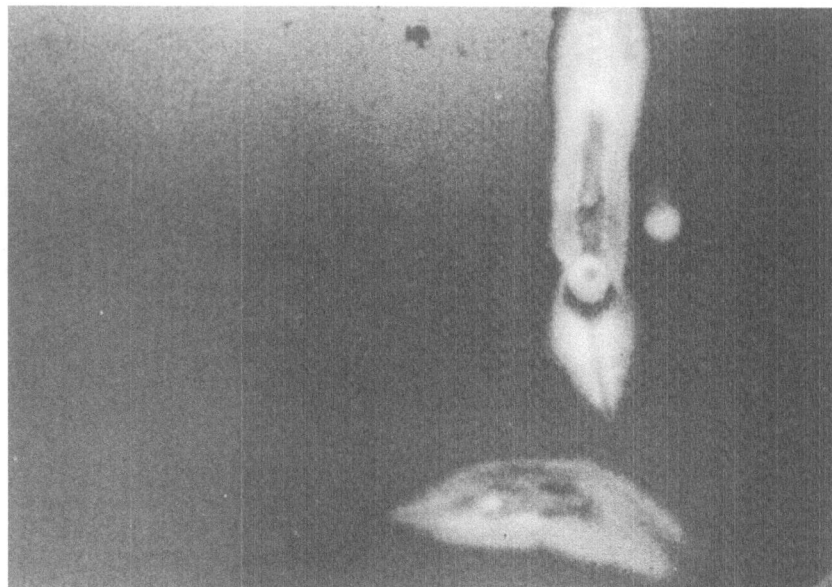

Fig. 3. *Dugesia dorotocephala* ovipositing, ventral view. A white cocoon, just extruded, can be seen under the larger worm. The spot in the center of the cocoon is the attachment disc which is connected to the cocoon by a slender thread. In the smaller worm, only a small portion of the cocoon is visible through the slightly enlarged genital opening.

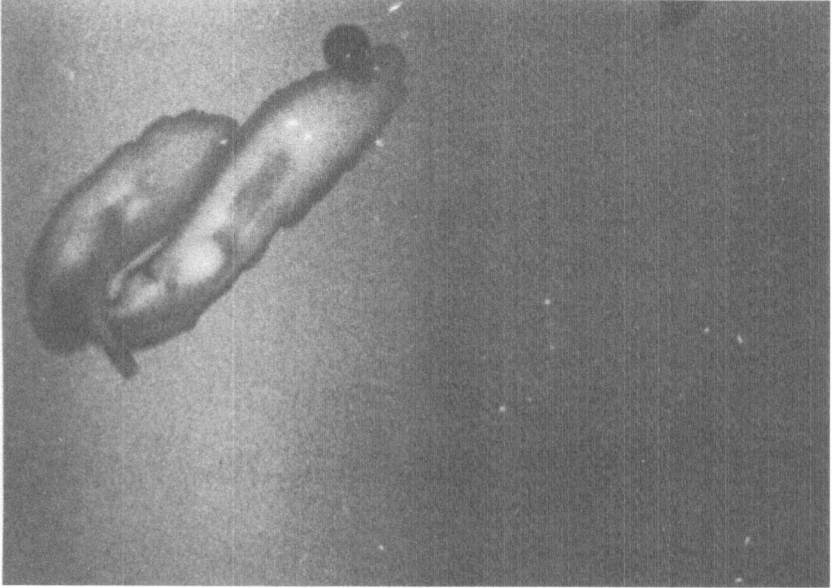

Fig. 4. Cannibalism during oviposition in *Dugesia dorotocephala*. The protruded pharynx of a hungry planarian can be seen attempting to reach a newly deposited cocoon, even before the worm that deposited the cocoon has released its hold on the glass and moved on.

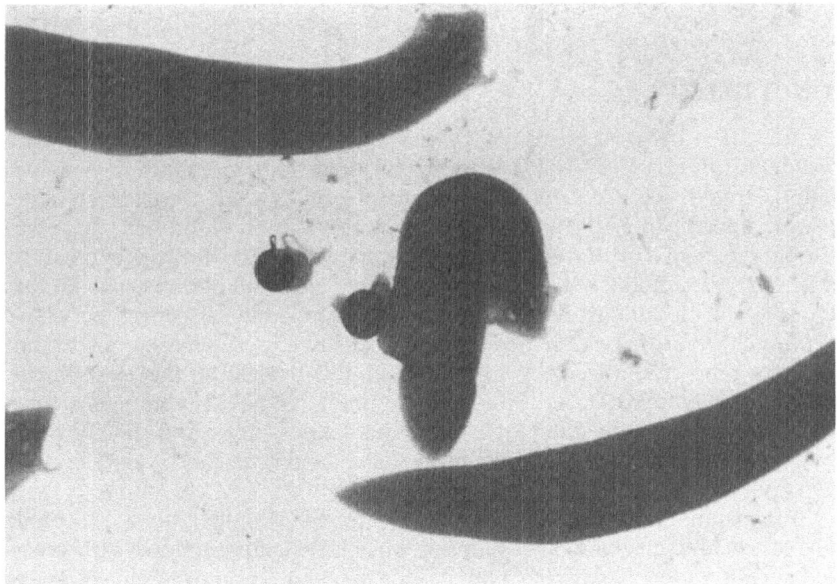

Fig. 5. *Dugesia dorotocephala* ingesting the contents of a newly deposited cocoon. The worm bent in a U shape in the photograph devoured eggs and yolk cells in a very few minutes and left only a thin husk of shell. Nearby planarians appear to be attracted to the feeding area, but they rarely attempt to eat the contents of cocoons several or more hours old, owing, presumably, to the hardened covering.

Fig. 6. Newly hatched *Dugesia dorotocephala*. The juveniles are transparent, but lightly pigmented, at emergence. They vary in length from 1 to over 10 mm. On an average, 12 to 15 worms develop in one cocoon, although the known range is from 1 to 35.

CONCLUSION

As the various aspects of planarian physiology and behavior are considered, in relation to the responses of planarians to stimuli, it becomes obvious that there are numerous innate responses that are characteristic, to some degree, of all groups. These responses may be modified, however, by either external or internal changes, or by both. Allowance must be made for the influence on behavior of such controllable external factors as degree of temperature or amount of light, and also for pervasive geophysical factors which are not subject to experimental control but which can be minimized by a change in the design of the experiment. Too, the physiological condition of the animal must be taken into consideration. Hunger, sexual state, age, previous experience, and many other physiological variables will inevitably have a greater or lesser effect on any observed reactions at any given time.

Planarians may seem, superficially, to be simple animals, capable only of set and mechanical responses, but closer acquaintance with cross-eyed, fresh-water flatworms will convince even an indifferent observer that planarians do not always obey rules, are often unpredictable, and are likely to exhibit a high degree of individuality. Perhaps we can sum it up by saying that planarians are living creatures and, as such, have planarian personalities.

REFERENCES

Allen, G. D., Quantitative studies on the rate of respiratory metabolism in planaria, II: The rate of oxygen consumption during starvation, feeding, growth and regeneration in relation to the method of susceptibility to potassium cyanide as a measure of rate of metabolism. *Amer. J. Physiol.*, 1919, **49**, 420.

Anderson, J. M., and J. C. Johann, Some aspects of reproductive biology in the fresh-water triclad turbellarian, *Cura foremanii. Biol. Bull.*, 1958, **115**, 375.

Bahrs, A. M., The effect of refrigeration upon the growth-promoting power of rabbit tissues for planarian worms. *Physiol. Zool.*, 1929, **2**, 491.

Bahrs, A. M., and R. Wulzen, Variations in the growth-promoting power of kidney for planarian worms. *Physiol. Zool.*, 1932, **5**, 198.

Best, J. B., R. Rosenvold, J. Souders, and C. Wade, Studies on the incorporation of isotopically labeled nucleotides and amino acids in planaria. *J. Exp. Zool.*, 1965, **159**, 397.

Booth, B., An analysis of filtrate obtained from homogenized *Dugesia dorotocephala* for indications of antibiotic activity. Research report, undergraduate, 1966. *Unpublished*.

von Brand, Th., Studies on the carbohydrate metabolism in planarians. *Physiol. Zool.*, 1936, **9**, 530.

Brown, F. A., Jr., Extrinsic rhythmicality: A reference frame for biological rhythms under so-called constant conditions. *Ann. New York Acad. Sci.*, 1962a, **98**, 775.

Brown, F. A., Jr., Responses of the planarian, *Dugesia*, and the protozoan, *Paramecium*, to very weak horizontal magnetic fields. *Biol. Bull.*, 1962b, **123**, 264.

Brown, F. A., Jr., Response of the planarian, *Dugesia*, to very weak horizontal electrostatic fields. *Biol. Bull.*, 1962c, **123**, 282.

Brown, F. A., Jr., An orientational response to weak gamma radiation. *Biol. Bull.*, 1963, **125**, 206.

Brown, F. A., Jr., A unified theory for biological rhythms, in J. Aschoff, ed., *Circadian Clocks*. Amsterdam: North Holland Publishing Co., 1965, p. 252.

Brown, H. P., and M. M. Jenkins, A protozoon (*Dileptus*; Ciliata) predatory upon metazoa. *Science*, 1962, **136**, 710.

Buchanan, J. W., Notes on an American cave flatworm, *Sphalloplana percaeca* (Packard). *Ecology*, 1936, **17**, 194.

Child, C. M., *Individuality in Organisms*. Chicago, Ill.: University of Chicago Press, 1915.

Clark, R. B., and J. B. Cowey, Factors controlling the change of shape of certain nemertean and turbellarian worms. *J. Exp. Biol.*, 1958, **35**, 731.

Coward, S. J., and R. A. Flickinger, Axial patterns of protein and nucleic acid syntheses in intact and regenerating planaria. *Growth*, 1965, **29**, 151.

Curtis, W. C., The life history, the normal fission and the reproductive organs of *Planaria maculata*. *Proc. Boston Soc. Nat. Hist.*, 1902, **30**, 515.

Flickinger, R. A., A gradient of protein synthesis in planaria and reversal of axial polarity of regenerates. *Growth*, 1959, **23**, 251.

Forrest, H., Observing planarians feeding on brine shrimp. *Turtox News*, 1963, **41**, 34.

Hammen, C. S., Carbon dioxide fixation in marine invertebrates: a review. *Proc. Symposium Exp. Marine Ecol.*, *Publ. No.* 2, University of Rhode Island Graduate School of Oceanography, 1964.

Hammen, C. S., and S. C. Lum, Carbon dioxide fixation in marine invertebrates, III: The main pathway in flatworms. *J. Biol. Chem.*, 1962, **237**, 2419.

Hublé, J., and G. van Grembergen, Contribution expérimentale à la biochimie de la respiration chez les Turbellariés. *Enzymologia*, 1949, **13**, 268.

Hyman, L. H., Physiological studies on planaria, I: Oxygen consumption in relation to feeding and starvation. *Amer. J. Physiol.*, 1919a, **49**, 377.

Hyman, L. H., Physiological studies on planaria, II: Oxygen consumption in relation to regeneration. *Ibid.*, 1919b, **50**, 67.

Hyman, L. H., Physiological studies on planaria, III: Oxygen consumption in relation to age (size) differences. *Biol. Bull.*, 1919c, **37**, 388.

Hyman, L. H., Physiological studies on planaria, IV: A further study of oxygen consumption during starvation. *Amer. J. Physiol.*, 1920, **53**, 399.

Hyman, L. H., Physiological studies on planaria, V: Oxygen consumption of pieces with respect to length, level, and time after section. *J. Exp. Zool.*, 1923, **37**, 47.

Hyman, L. H., Studies on the correlation between metabolic gradients, electrical gradients, and galvanotaxis, II: Galvanotaxis of the brown hydra and some nonfissioning planarians. *Physiol. Zool.*, 1932, **5**, 185.

Hyman, L. H., North American Triclad Turbellaria, IX: The priority of *Dugesia* Girard 1850 over *Euplanaria* Hesse 1897 with notes on American species of *Dugesia*. *Trans. Amer. Microsc. Soc.*, 1939, **58**, 264.

Hyman, L. H., *The Invertebrates*, Vol. II: *Platyhelminthes and Rhynchocoela*. New York: McGraw-Hill Book Company, 1951.

Hyman, L. H., Textbook planarians and the reality. *Amer. Biol. Teacher.*, 1956, **18**, 124.

Hyman, L. H., and A. W. Bellamy, Studies on the correlation between metabolic gradients, electrical gradients, and galvanotaxis. *Biol. Bull.*, 1922, **43**, 313.

Jenkins, M. M., Respiration rates in planarians, I: The use of the Warburg respirometer in determining oxygen consumption. *Proc. Okla. Acad. Sci.*, 1960, **40**, 35.

Jenkins, M. M., Note on stalk formation in cocoons of *Dugesia dorotocephala* (Woodworth) 1897. *Trans. Amer. Microsc. Soc.*, 1966, 85, 158.

Jenkins, M. M., and H. P. Brown, Cocoon production in *Dugesia dorotocephala* (Woodworth) 1897. *Ibid.*, 1963, **82**, 167.

Jenkins, M. M., and H. P. Brown, Copulatory activity and behavior in the planarian *Dugesia dorotocephala* (Woodworth) 1897. *Ibid.*, 1964, **83**, 32.

Jennings, H. S., *Behavior of the Lower Organisms*. Bloomington, Ind.: Indiana University Press, 1962, p. 339.

Jennings, J. B., Studies on feeding, digestion, and food storage in free-living flatworms (Platyhelminthes: Turbellaria). *Biol. Bull.*, 1957, **112**, 63.

Jennings, J. B., Further studies on feeding and digestion in triclad Turbellaria. *Ibid.*, 1962, **123**, 571.

Kenk, R., Studies on Virginian triclads. *J. E. Mitchell Soc.*, 1935, **51**, 79.

Kenk, R., Sexual and asexual reproduction in *Euplanaria tigrina* (Girard). *Biol. Bull.*, 1937, **73**, 280.

Kenk, R., The reproduction of *Dugesia tigrina* (Girard). *Amer. Nat.*, 1940, **74**, 471.

Kenk, R., Induction of sexuality in the asexual form of *Dugesia tigrina* (Girard). *J. Exp. Zool.*, 1941, **87**, 55.

Krugelis-Macrae, E., The occurrence of porphyrin in the planarian. *Biol. Bull.*, 1956, **110**, 69.

Løvtrup, E., Studies on planarian respiration. *J. Exp. Zool.*, 1953, **124**, 427.

Lund, E. J., Oxygen concentration as a limiting factor in the respiratory metabolism of *Planaria agilis. Biol. Bull.*, 1921, **41**, 203.

Marriott, F. H. C., The absolute light-sensitivity and spectral threshold curve of the aquatic flatworm *Dendrocoelum lacteum. J. Physiol.*, 1958, **143**, 369.

May, E., and G. Birukow, Lunarperiodische Schwankungen des Lichtpräferendums bei der Bachplanarie. *Naturwissenschaften*, 1966, **53**, 182.

McConnell, J. V., *A Manual of Psychological Experimentation on Planarians.* Ann Arbor, Mich.: Planarian Press, 1965.

Mengebier, W. L., and M. M. Jenkins, Succinoxidase activity in homogenates of *Dugesia dorotocephala. Biol. Bull.*, 1964, **127**, 317.

Osborne, P. J., and A. T. Miller, Jr., Uptake and intracellular digestion of protein (peroxidase) in planarians. *Biol. Bull.*, 1962, **123**, 589.

Parker, G. H., and F. L. Burnett, The reactions of planarians, with and without eyes, to light. *Amer. J. Physiol.*, 1900, **4**, 373.

Pearl, R., The movements and reactions of fresh-water planarians: a study in animal behavior. *Quar. J. Microsc. Sci.*, 1903, **46**, 509.

Pedersen, K. J., On the oxygen consumption of *Planaria vitta* during starvation, the early phase of regeneration and asexual reproduction. *J. Exp. Zool.*, 1956, **131**, 123.

Pirenne, M. H., and F. H. C. Marriott, Light sensitivity of the aquatic flatworm *Dendrocoelum lacteum. Nature*, 1955, **175**, 642.

Reynoldson, T. B., and J. O. Young, The food of four species of lake-dwelling triclads. *J. Anim. Ecol.*, 1963, **32**, 175.

Skaer, R. J., Some aspects of the cytology of *Polycelis nigra. Quar. J. Microsc. Sci.*, 1961, **102**, 295.

Steinbach, H. B., Ionic and water balance of planarians. *Biol. Bull.*, 1962, **122**, 310.

Taliaferro, W. H., Orientation to light in Planaria *n. sp.* and the function of the eyes. *Anat. Rec.*, 1917, **11**, 1.

Taliaferro, W. H., Reactions to light in *Planaria maculata* with special reference to the function and structure of the eyes. *J. Exp. Zool.*, 1918, **31**, 59.

Viaud, G., Anisotropie électrique et variations de la conductance en fonction de la taille chez *Planaria lugubris* O. Sch. *C. R. Soc. Biol.*, 1951, **146**, 489.

Viaud, G., Galvanotropisme, anisotropie d'excitation et anisotropie électrique chez une Planaire (*Planaria = Dugesia lugubris* O. Schm.). *J. Physiol.* (Fr.), 1952, **44**, 343.

Viaud, G., Etude quantitative de la force électro-motrice d'opposition produite par les Planaires (*Planaria = Dugesia lugubris* O. Schm.) en réponse à une excitation électrique due à un courant continu. *C. R. Soc. Biol.*, 1954a, **148**, 2068.

Viaud, G., Mise en évidence d'une force électro-motrice d'opposition produite par les Planaires en réponse à une excitation électrique due à un courant continu. *Arch. Anat. Hist. Embryol.*, 1954b, **37**, 145.

Viaud, G., Inversion du sens du galvanotropisme de *Planaria* (= *Dugesia*) *lugubris* O. Schm. par l'action de radiations visibles et ultraviolettes polarisées. *C. R. Soc. Biol.*, 1955, **149**, 2221.

Viaud, G., Recherches nouvelles sur le galvanotropisme des animaux. *Actes XVe Congrès International Psych., Bruxelles*, 1957, pp. 254–255.

Viaud, G., Le galvanotropisme animal sous son nouvel aspect. *Strasbourg Medical*, 1959, **10**, 605.

Willier, B. H., L. H. Hyman, and S. A. Rifenburgh, A histochemical study of intra-
cellular digestion in triclad flatworms. *J. Morph. Physiol.*, 1925, **40**, 299.
Wulzen, R., A study in the nutrition of an invertebrate, *Planaria maculata. Univ. of Calif.
Publ.*, 1923, **5**(15), 175.
Wulzen, R., The nutrition of planarian worms. *Science*, 1927, **65**, 331.
Wulzen, R., and A. M. Bahrs, A common factor in planarian and mammalian nutrition.
Physiol. Zool., 1936, **9**, 508.
Young, J. O., and T. B. Reynoldson, A laboratory study of predation on lake-dwelling
triclads. *Hydrobiologia*, 1965, **26**, 307.

Chapter 10

The Neuroanatomy of the Planarian Brain
and Some Functional Implications[1]

Jay Boyd Best

Department of Physiology and Biophysics
Colorado State University
Fort Collins, Colorado

The planarians have been used extensively for studies on regeneration, morphogenesis, and, in the last decade, for provocative experiments on the chemical and cellular bases of memory. Yet, in spite of this interest in certain selected classes of phenomena which the planarians manifest, the anatomical and physiological substratum underlying these has been only imperfectly studied or known. This gap in the critical information needed to decipher fully this, now reasonably numerous, body of experiments has rendered it less valuable than it has the potential to be.

Over the last two years we have at Colorado State University been devoting a considerable effort toward clarifying the histological structure of *Dugesia dorotocephala*. Much of this attention has been directed toward the nervous system of this creature. These studies have involved a fairly systematic approach starting with the optically visible histological structure in serial paraffin sections and working into progressively higher magnifications to study the ultrastructure with electron microscopy.

We have had a considerable amount of success in these efforts due in no small part to the technical skill of Jacqueline Souders in the preparation of the optical sections and Michio Morita in the preparation of the electron micrographs. By working back and forth between the lower magnifications of optical microscopy and the higher magnifications obtainable with electron microscopy, and also through the use of several different staining procedures, we now know considerably more than we did a couple of years ago, especially of the nerve and muscle systems.

One will notice that all of the pictures of the planarian nervous system shown in classic works, such as Libby Hyman's treatise on the invertebrates, are sketches or line drawing reconstructions. These are a credit to the powers of observation of these earlier workers for, as we found in our initial

[1] A portion of the research described was suggested by PHS grant No. MH07603 and NASA grant No. NSG-625.

studies, it was difficult to obtain paraffin sections that really displayed the structure of this nervous system very clearly. Some of the reasons for this will be returned to later when I discuss our electron micrographic studies.

We tried using some of the Cajal, i.e., silver, staining procedures such as Teichii Betchaku had employed successfully with *Planarian vittae*, but we were entirely unsuccessful with these. Inasmuch as the theory underlying the silver stains is not well understood, it is difficult to say just where we went wrong, but we were unable to stain nervous tissue specifically using these methods. We stained everything or nothing and eventually gave up trying to use them.

Curiously enough we found a very good nervous system stain for optical sections when we were trying to do something else. Michio Morita, who had come to work in my laboratory at Colorado State University in 1964, had prepared some good electron micrographs of planarian brain. In these micrographs we were able to see what appeared to be neuro-secretory granules. At the suggestion of Dr. Dorothy Billenstein, who has worked in the area of neurosecretion, we tried staining paraffin sections with the chromealumhaematoxylin and aldehyde-fuchsin techniques to see if we could obtain the distinctive staining characteristics of neuro-secretory granules. Both of these stains, as it turned out, displayed the nervous system so nicely that even a freshman zoology student would have no trouble in delineating it.

It will be of interest first, by way of general orientation, to take a histological tour through the planarian by examining several transverse paraffin sections made at different levels along the length of the animal. The drawing (Fig. 1) shows the various levels at which the sections were made.

The first section (Fig. 2), taken at level I, shows the epidermal layer of cells with the basilar membrane lying just internal to it. It seems quite likely that this basilar membrane constitutes a relatively impermeable barrier to the diffusion of materials into the planarian. In the "wings" of the section, i.e., the regions just interior to the lateral margins of the planarian, can be seen the tissue regions generally referred to as *mesenchymal tissue*. This has a kind of lattice, i.e., reticular, structure. Occupying most of the central regions of the section is the prominent digestive cavity. Although several branches appear in this section, this apparent multiplicity really arises from its complex involutions. There are really, at this level of the cut, only two main branches. The phagocytic cells of the digestive

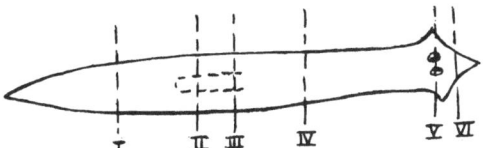

Fig. 1. Levels of transverse histological sections
of planaria shown in Figs. 2–7.

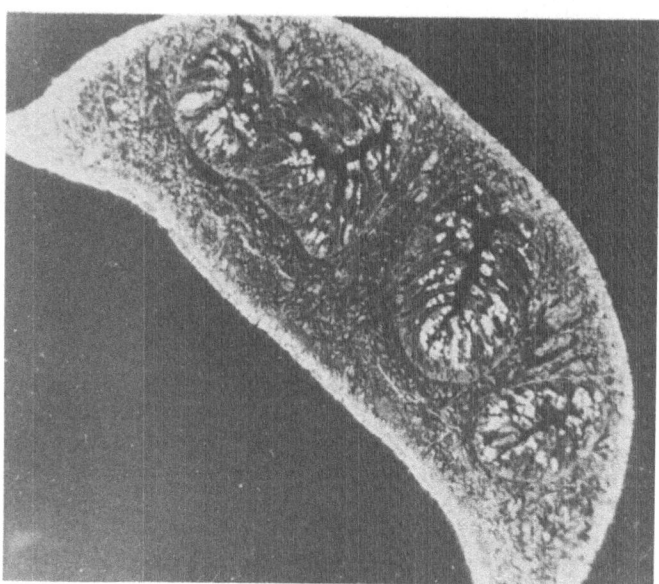

Fig. 2. Transverse section of the planarian *Dugesia dorotocephala*
taken behind pharynx at level I.

cavity are clearly visible and these can be seen to contain reddish[2] staining
droplets which are probably partially digested food materials and the
enzymes which are digesting them. Also apparent is a basement membrane
which seems to serve as an anchor for the phagocytic cells and separates
them from the mesodermal regions. This basement membrane appears
to be interposed as an enveloping barrier around the digestive cavity and
phagocytic cells so that these are probably more or less isolated from the
compartment comprising the milieu of most of the other cells of the animals.
The two main longitudinal nerve cords, stained bluish-gray, can be observed
just a little inside the epidermal basement membrane of the ventral surface.
    The section in Fig. 3 is taken at level II and cuts through the pharynx
of the planarian. This structure is thick walled, and the turquoise-staining
layers of muscle are very evident. The various lamina contain circular,
longitudinal, and radially oriented bundles of muscle fibers. Although it
may be difficult to see in this picture, the pharynx also contains a fine
plexus of nerve fibers. This musculature in the walls of the pharynx is by
far the heaviest found anywhere in the planarian and accords with its
observed functional capability of tearing out relatively large pieces of meat

    [2] The slides of the optical sections which were shown at the meetings were in color
which enables one to see the structures more clearly. Unfortunately these cannot be used
in this published account.

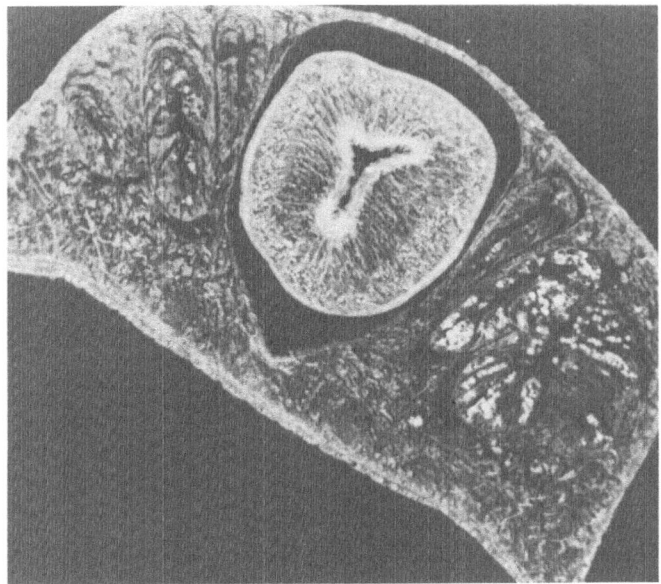

Fig. 3. Transverse section of the planarian *D. dorotocephala* taken through the pharynx at level II.

from subdued prey. The various orientations of its musculature account for the marked ability of this organ to extend, contract, and change its dimensions in general. Note the cavity into which the pharynx can recess when contracted.

In Fig. 4 the mooring attachment of the pharynx is shown. The lumen of the pharynx of course opens into the digestive cavity of the planarian and is the sole passageway of connection between the digestive cavity and the outside.

A transverse section at level IV, anterior to the point of attachment of the pharynx, can be seen in Fig. 5. At this level the digestive cavity has only one main branch, although, because of its complex involutions, there appears in the section to be more. The same general features of epidermal layer, epidermal basement membrane, mesenchymal tissue, phagocytic cells of the digestive cavity, and ventral nerve cords which were evident in Fig. 2 can be seen in this section also.

Figure 6 shows a transverse section at level V, just about through the region of the optic organs and the auricles. The cut at this level shows the two lobes of the brain and a portion of the optic tract connecting one of the eyes to the brain. The neuropil of the brain as well as the optic tract are stained a grayish-blue color, as were the ventral nerve cords in the previous sections. The cells clustering about the amorphous-appearing region of

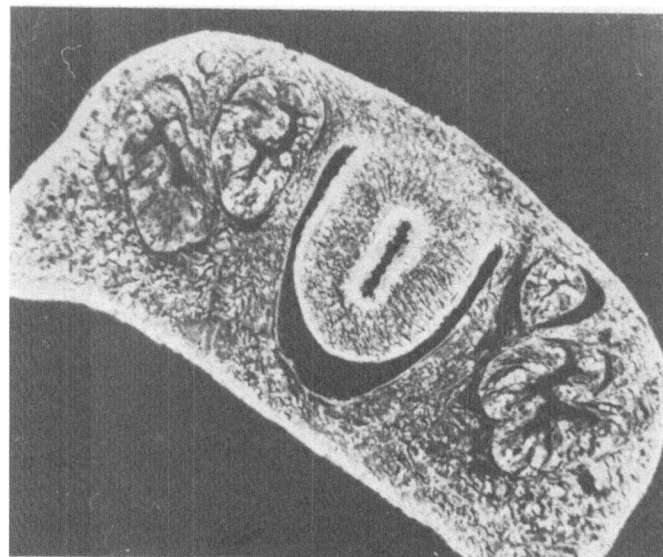

Fig. 4. Transverse section of the planarian *D. dorotocephala* taken through the pharynx at level III.

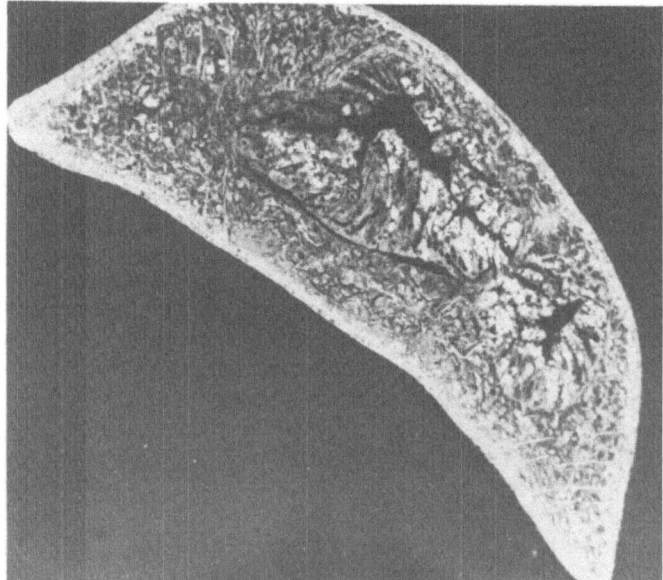

Fig. 5. Transverse section of the planarian *D. dorotocephala* taken anterior to the pharynx at level IV.

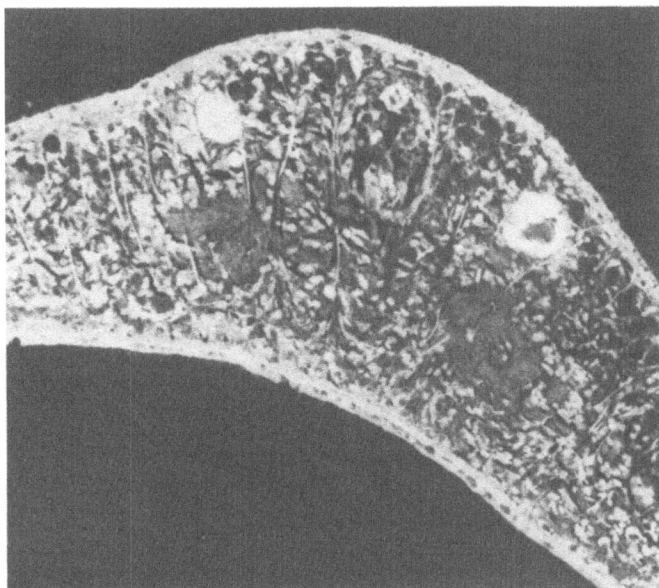

Fig. 6. Transverse section of the planarian *D. dorotocephala* taken through the eye spots at level V. This section shows two lobes of the brain as well as the rostral extremity of the digestive cavity.

the neuropil are the nerve cell bodies. It should be remarked at this point that the reason the region of the neuropil appears amorphous, aside from a slightly evident fibrous texture, is that it is mostly comprised of nerve fibers that are less than a wavelength of blue light in diameter, which are tightly compacted together. This is evident from the electron micrographs which I shall discuss later. The barely discernible fibrous texture probably arises from the small portion of the fibers that are just about a wavelength of blue light in diameter. The digestive cavity can be seen between the two lobes of the brain. The brain is a kind of hairpin-shaped structure and, at least in *Dugesia dorotocephala*, the digestive cavity extends rostrally to just beyond the level of the eyes. This farthermost extension of the digestive cavity thus nestles in the loop of the hairpin-shaped brain.

The transverse section of Fig. 7 is at level VI anterior to the optic organs and auricles. At this level the neuropils of the two lobes of the brain appear fused. The reason for this is that the plane of the cut at this level is passing through the region of the bend of the brain. Both in this slide and the previous one, one can see a ventrodorsally oriented palisade of scarlet-stained muscle fibers. A number of these pass uninterrupted through the brain. While the resolution of these pictures is not good enough to see it, one can under oil immersion just perceive, at the limit of resolution, a

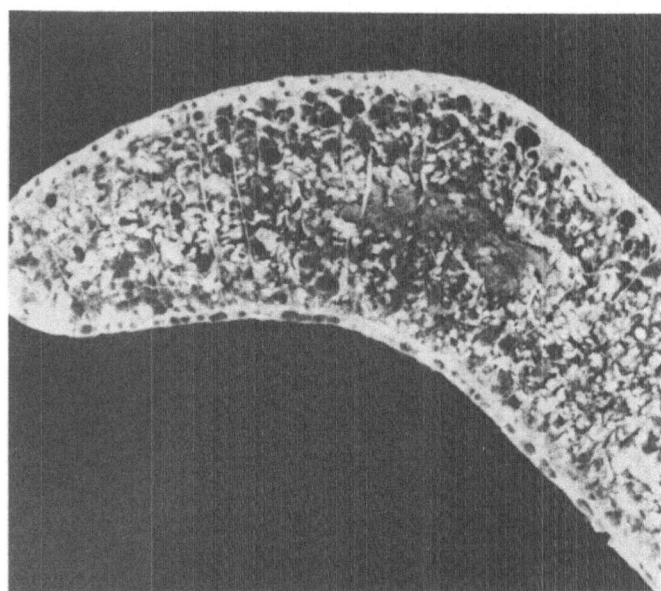

Fig. 7. Transverse section of the planarian *D. dorotocephala* taken anterior to the eye spots at level VI. This section shows the brain as a single structure.

fine basketwork of nerve fibers enveloping these muscle fibers. These muscle fibers are not striated, but on the other hand neither is their morphology strictly like vertebrate smooth muscle.

It is worth commenting briefly on the probable function significance of these muscle fibers of the head and their close proximity to the brain. If one watches a planarian, such as *Dugesia dorotocephala*, under a dissecting microscope, while it is engaged in the capture of a small active prey such as brine shrimp, one sees that its head is a prehensile structure capable of complex, well coordinated, grasping motions. The delicacy with which it can grasp a rapidly moving brine shrimp, or one partially ensnared by mucous threads, almost gives the illusion of a hand inside a mitten. Such movements are almost certainly being executed by this head muscle system with its direct innervation from the neuropil of the brain.

Having now become oriented in regard to the optically visible histological anatomy of the planarian and its nervous system, let us now look at the fine structure of its nervous system.

Figure 8 shows an electron micrographic montage of a horizontal, i.e., in the plane of the animal, section of the planarian brain in the boundary region between neuropil and the cortex, containing the nerve cell bodies.

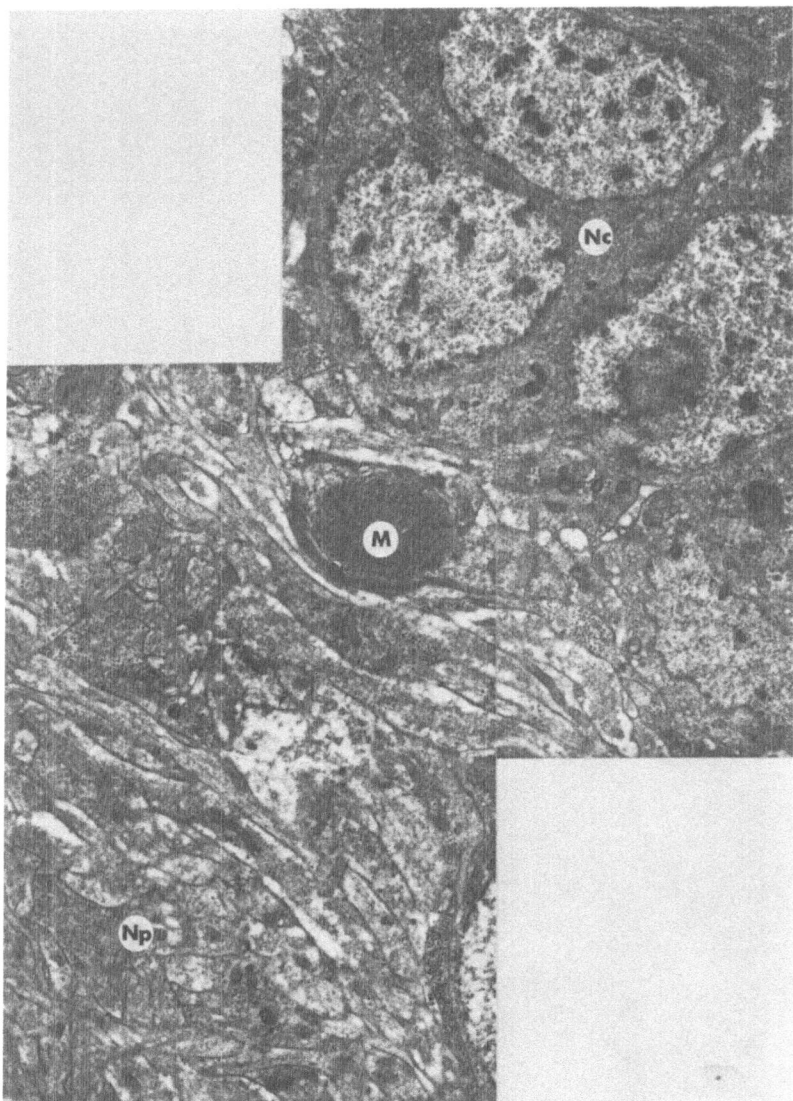

Fig. 8. Electron micrographic montage of a horizontal section of the planarian brain showing a region of approximately 180 μ^2 in the boundary between the neuropil (Np) and the region containing the nerve cell bodies (Nc). A cross section of a muscle fiber (M) can be seen in the center.

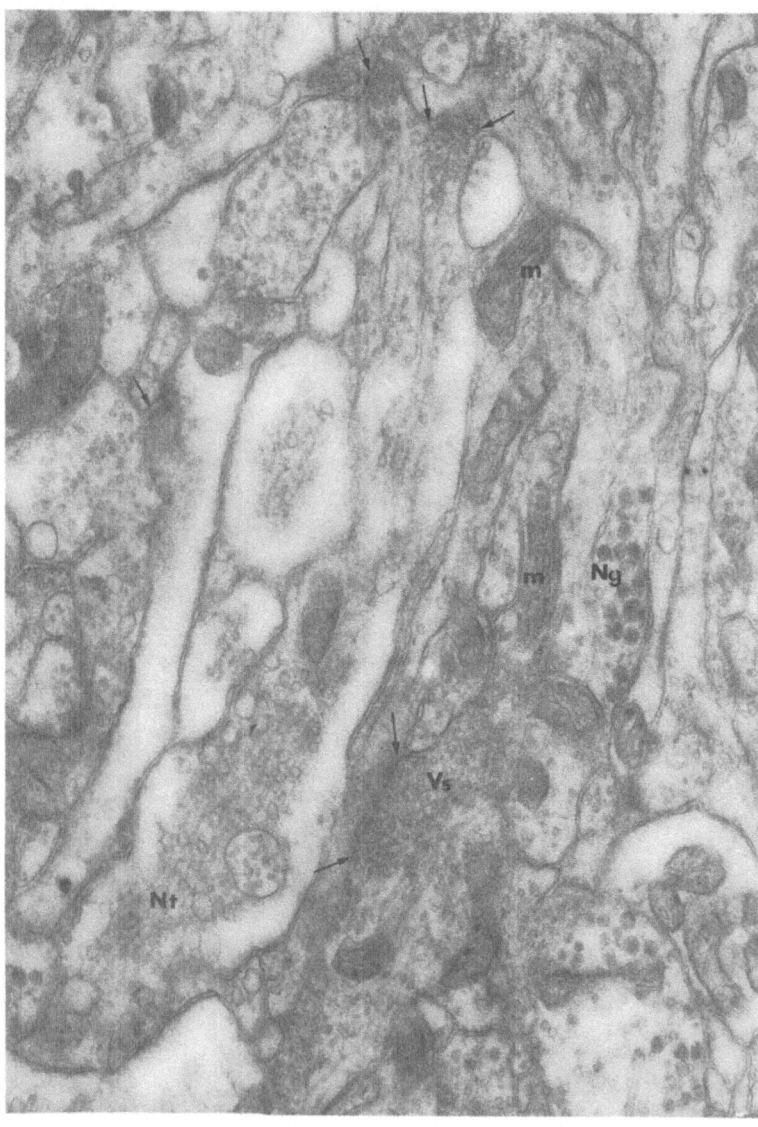

Fig. 9. Electron micrograph of a region of neuropil magnified approximately
40,000 × showing synaptic junctions (arrows), neurosecretory granules (*Ng*),
neurotubules (*Nt*), synaptic vesicles (*Vs*), and mitochondria (*m*).

The montage spans an area about 20 by 12 μ in extent. By way of reference, this region would be about three human red blood corpuscles in length and two in width. The neuropil can be seen to be a compact mass of unmyelinated nerve fibers ranging from about 0.2 to 0.4 μ in diameter. Most of the fibers are less than 0.35 μ in diameter. The mottled, more or less ovoid-shaped, objects about 2 μ wide and 3 or 4 μ in length are the nuclei of the nerve cells. It is evident from this electron micrograph, as from the optical sections, that the neuropil does not contain any cell bodies. In some regions of the montage, between the cell bodies of the cortical region, the nerve fibers are organized into tracts and the plane of the cut is very nearly parallel to these fibers of the tract. Note the relatively scanty cytoplasm of the cell bodies, so that the nucleus occupies a large portion of the volume of the nerve cell body.

Distributed through the cytoplasm of the nerve cell bodies as well as through the fibers of the neuropil, you will note a considerable number of the sausage-shaped mitochondria. Perhaps not evident in the reproductions but clear upon close examination of the original photographs, are the characteristic cristae in these mitochondria. On the whole, these mito-chondria are much smaller than those which one would find in a tissue such as dog heart, although they are probably about average for the lower invertebrates. These relative sizes are in accord with the established role of the mitochondria as the primary, if not the exclusive, site of the Krebs-cycle respiratory system. One would anticipate that tissues with large respiratory rates would have larger mitochondria than those with smaller respiratory rates.

Here and there in the neuropil one can see nerve fibers containing clusters of round electron-dense granules looking somewhat like bird shot. These granules are neurosecretory granules, and the fibers containing them are processes of neurosecretory cells. I shall have more to say about these below.

One can see the cross section of several muscle fibers in this montage. These muscle fibers are part of the system of ventrodorsally oriented muscle fibers that I pointed out previously in the optical sections. Since this electron micrographic section is in the plane of the animal, it is oriented perpendicularly to the axis of the muscle fibers. Hence all these fibers are seen in cross section in this electron micrograph.

Aside from these muscle fibers, Morita and I believe we can distinguish three major kinds of cells in the brain of the planarian *Dugesia doroto-cephala*. These are neurons, neurosecretory cells, and neural accessory cells. These latter cells, the neural accessory cells, appear to be analogs to the glia of higher animals. The relative dimensions of this nervous system to those of the vertebrates' nervous system are worth emphasizing. The entire area shown in this montage is only about 180 μ^2. The diameter of a single human ventral horn-cell soma typically runs about 40 to 50 μ, so its cross-sectional area would average about 2,000 μ^2. Thus, the entire region covered by this montage is only about one-tenth the cross-sectional

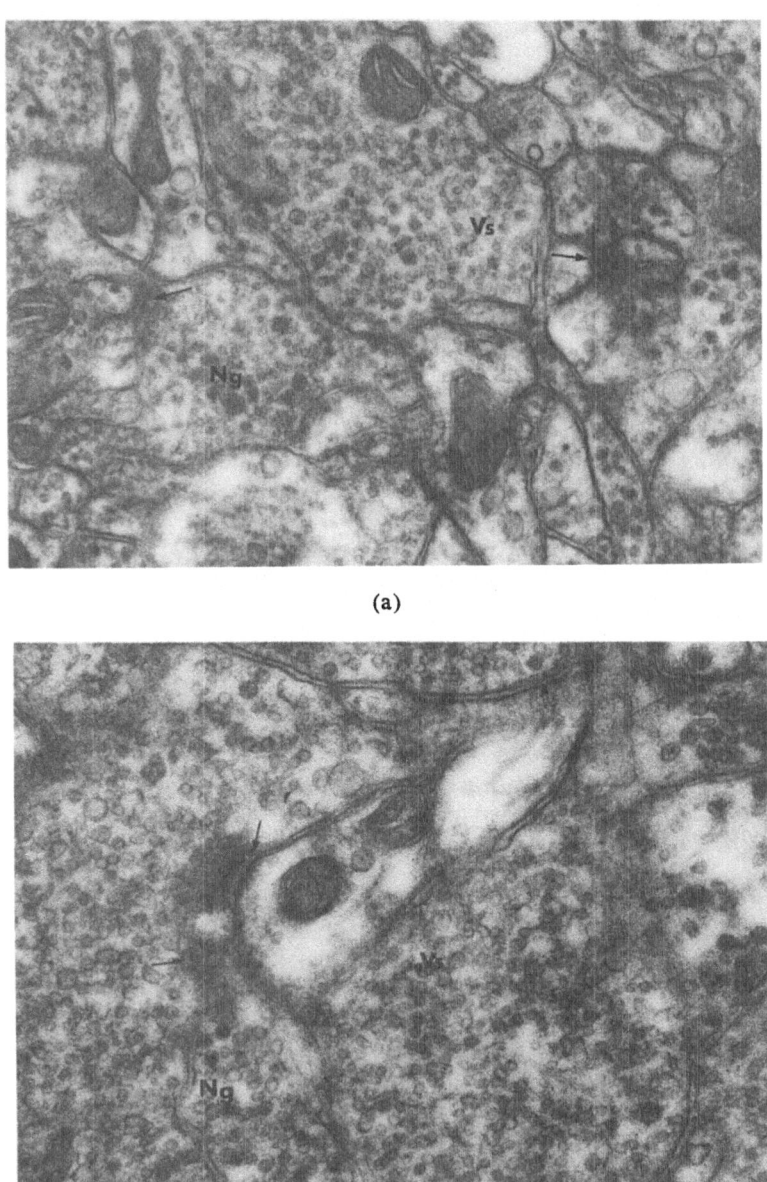

(a)

(b)

Fig. 10. Electron micrograph of a region of neuropil of the brain of *D. doroto-cephala* magnified approximately 60,000 × showing details of synaptic junctions (arrows) and synaptic vesicles (*Vs*). Cristae can be seen clearly in two of the mitochondria of the upper picture (a) and in one of the lower picture (b).

area of the soma of a human ventral horn cell. A portion of the planarian brain containing hundreds of neurons with their associated fiber tracts and connection systems could be placed inside the soma of such a single ventral horn cell. The planarian brain is thus microminiaturized, so that one might anticipate a somewhat larger cybernetic complexity than its small size would appear to warrant.

Having become acquainted with the general topography of the planarian brain, let us now examine in Fig. 9 a region of the neuropil at higher magnification. Several interesting aspects of the fine structure can be seen in this section. First are the synaptic junctions, indicated by arrows, with characteristic distributions of synaptic vesicles on the presynaptic side. Second are the neurotubules *Nt*. Third is an aggregation of neurosecretory granules *Ng*. In close relation to the group of neurotubules that have been labeled is a multivesicular body. These multivesicular bodies are puzzling objects with a very characteristic and easily recognizable morphology, but their function is at the moment completely unknown.

Two different kinds of synapses are observable in this slide. Those at the top are both formed by the terminus of a neural process synapsing with another neural process, while the one at the lower center is formed by a fusion of the lateral membranes of two parallel nerve fibers. The functional significance of these two different morphological types of synapses is not clear. It is worth mentioning at this point that all of the synaptic junctions we have found in the brain have been in the neuropil and are either axoaxonic or axodendritic in type. We have not observed any axosomatic synapses. Functionally, this would indicate that the nerve cells of the cortex make all contact with one another by sending processes into the neuropil and synapsing there.

The clumping of the neurotubules and the clear space between them and the membrane of the nerve fiber is probably artifact. Most workers feel that the natural distribution of the neurotubules is one in which they are more or less uniformly distributed through the cross section of the axon.

The next figure (Fig. 10) shows two synapses at higher magnification that were stained with uranium to display their fine structure better. Figure 10b shows a synapse (arrow) with its typical morphological characteristics. Notice the aggregation of electron-dense material in the cytoplasm immediately neighboring the point at which the membranes make their synaptic juncture. Also there is the characteristic intrasynaptic space, a lightly staining, clearly demarcated region about 150 Å wide. The relatively dense aggregation of synaptic vesicles on the presynaptic side distinguishes it from the postsynaptic side containing very few such vesicles. The close proximity of a mitochondrion to the synapse is also characteristic. Figure 10 shows a synapse with a complex set of invaginations and "fingers" about the region of the synaptic junction. The function of these complex structural arrangements is not clear, but similar ones have been observed about the synapses in the nervous systems of other species as well.

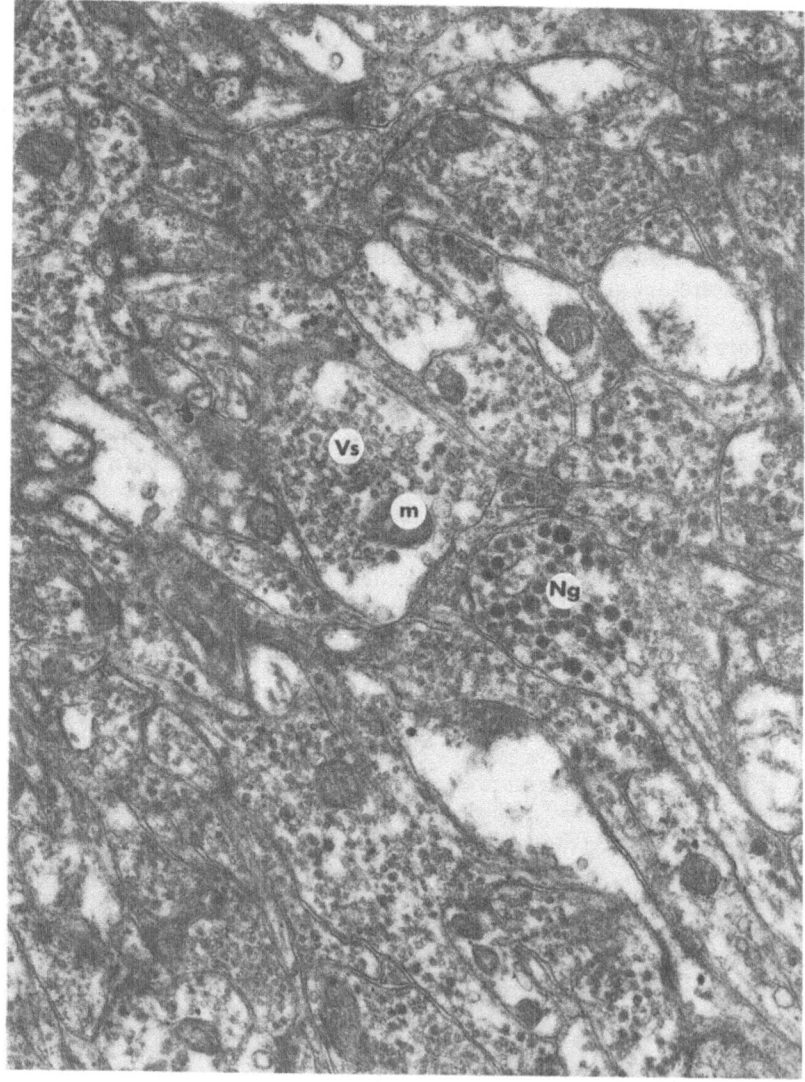

Fig. 11. Electron micrograph of a region of neuropil of the brain of *D. dorotocephala* magnified approximately 40,000 × showing a cluster of neurosecretory granules (*Ng*) in a terminal bulb.

Figure 11 shows another region of neuropil of the grain of *Dugesia dorotocephala*. In this section one can see an enlargement, or swelling, on the end of a neural process which contains a cluster of neurosecretory granules. Neurotubules running down through the axoplasm of the nerve fiber into the neurosecretory cells of this terminal bulb can also be observed.

Michio Morita and I believe that we have elucidated the major aspects of the formation and transport of these planarian neurosecretory granules. I say "believe" because of the unavoidable element of uncertainty that is attendant to any study in which a sequence of events is reconstructed from electron micrographs which one believes are showing one different stages of that sequence. The neurosecretory granules are almost certainly made in the soma of the neurosecretory cells and then migrate out along the inside of the nerve fibers which the neurosecretory cells extend into the neuropil. The evidence for this is that one never finds partially formed neurosecretory granules in the nerve fibers of the planarian brain. If they were formed in the axons or dendrites, one would expect to find some partially formed neurosecretory granules in such nerve fibers. Such is not the case. The neurosecretory granules of the nerve fibers are either present or not present. Nor do we ever see them outside the cell, only inside.

We do, however, find abundant evidence of their manufacture in the cytoplasm of the soma of the neurosecretory cells in the cortex of the planarian brain. There appear to be two major kinds of configurations for the soma of neurosecretory cells. One of these, which can be seen in Fig. 12, contains a rather well-defined endoplasmic reticulum in the outer regions of the cytoplasm and an inner or central region that is largely devoid of endoplasmic reticulum. One or more Golgi complexes can usually be observed in the portions of this central region just interior to the outer zone of endoplasmic reticulum. These are quite evident in this electron micrograph. Few organelles appear in the outer region with the endoplasmic reticulum, but a reasonable number of such organelles, e.g., mitochondria and multivesicular bodies, can be observed in the central region along with some partially formed neurosecretory granules. These partially formed neurosecretory granules do not yet have their full complement of electron-dense material. At one edge of this electron micrograph one can see a portion of the nucleus and the region of moderately heavy electron density bordering the nuclear membrane. The region covered by this picture is about 4 by 6 μ.

Figure 13 shows Golgi complexes at four different stages during their activity and existence. In Fig. 13a and b the Golgi apparatus appears as a bundle of more or less tubular laminae with accumulations of electron-dense material in certain regions. The ends of these Golgi tubes can be observed to be breaking off vesicular structures containing a moderate amount of electron-dense material. In c and d of Fig. 13 the Golgi complex can be seen breaking up, and, as it passes into dissolution, a number of partially formed neurosecretory granules are appearing in its place.

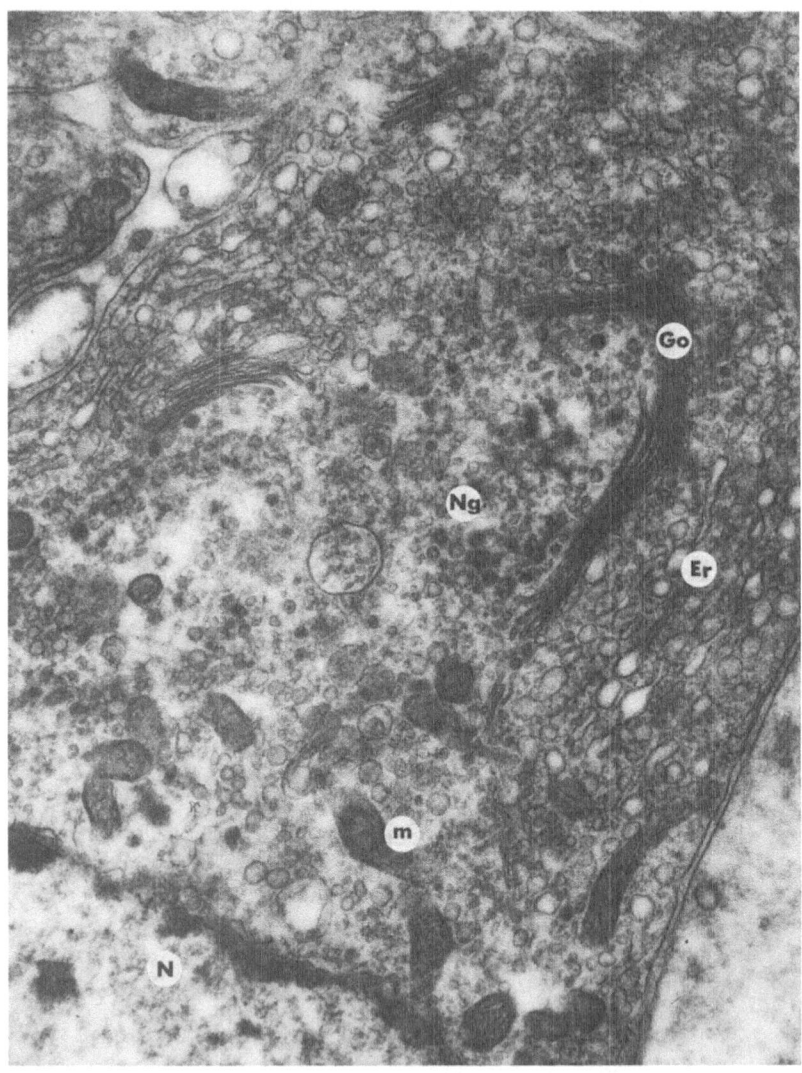

Fig. 12. Electron micrograph of the soma of a neurosecretory cell in the cortex of the brain of the planarian *D. dorotocephala* showing Golgi apparatus (*Go*) and endoplasmic reticulum (*Er*).

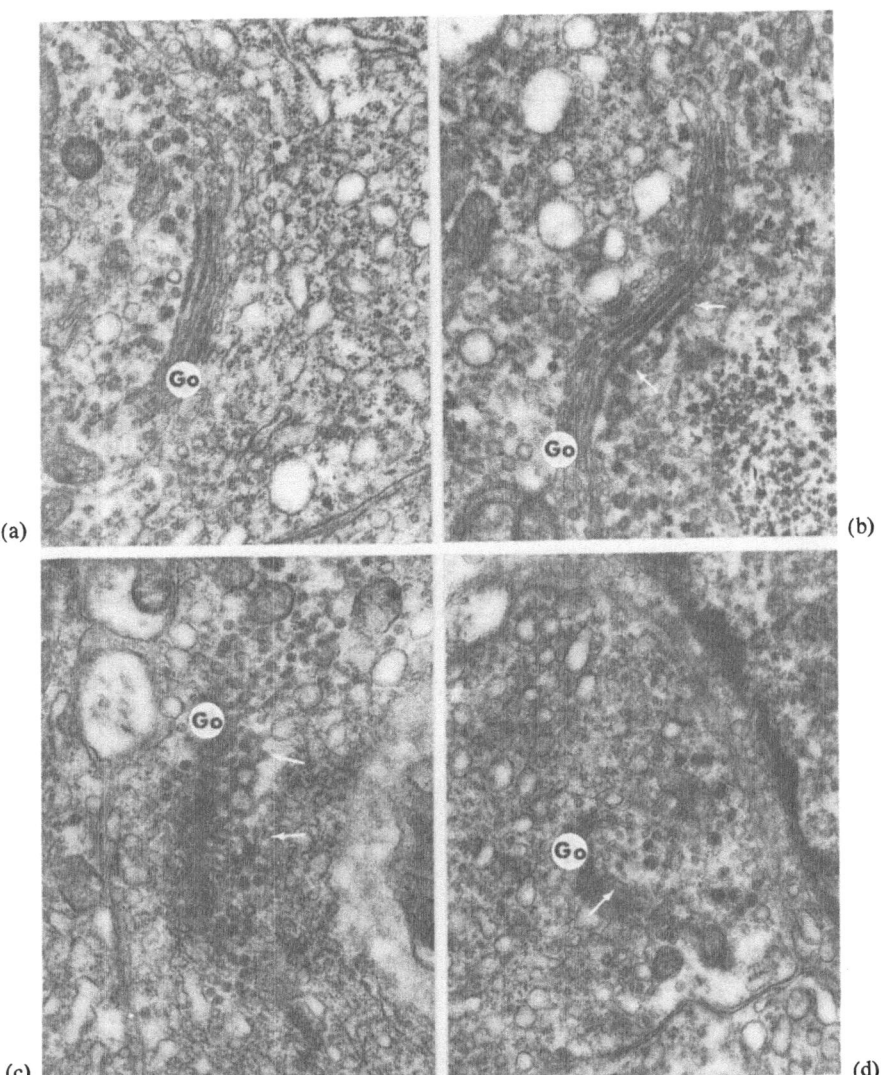

Fig. 13. Electron micrographs of cytoplasm of the soma of neurosecretory cells in the cortex of the brain of the planarian *D. dorotocephala* showing various stages in the Golgi apparatus (*Go*) and its manufacture of neurosecretory granules.

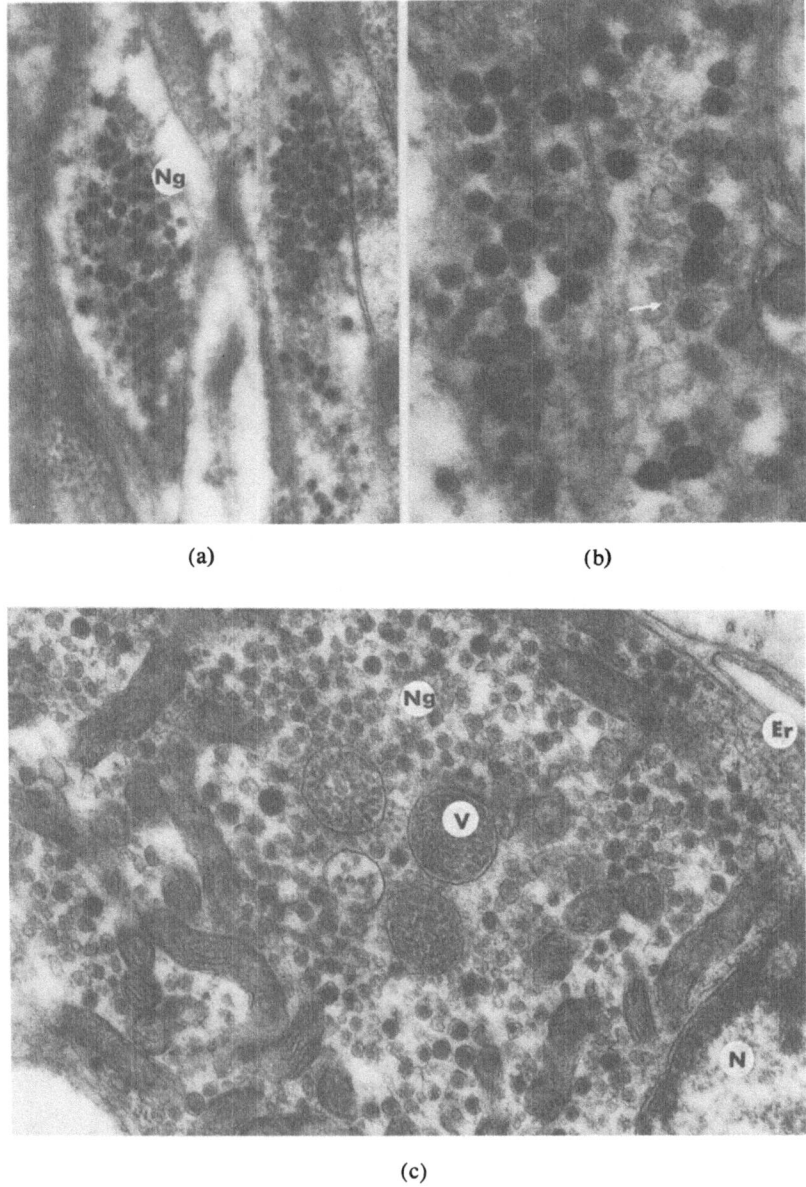

(a)   (b)

(c)

Fig. 14. Electron micrographs of selected regions of the brain of the planarian *D. dorotocephala*. Top two pictures (a, b) show clusters of neurosecretory granules in fibers of neuropil. Bottom picture (c) shows cytoplasm of neurosecretory cell soma in phase 2 of activity cycle in which it contains large numbers of neurosecretory granules, mitochondria, and multivesicular bodies (*V*).

In Fig. 14 we can see clusters of completed neurosecretory granules in the fibers of the neuropil. Also in this picture we can see the cytoplasm of the soma of a different form of neurosecretory cell. This form of neurosecretory cell, unlike the first that I showed, has very little endoplasmic reticulum and no Golgi apparatus. There is no division into an outer and central region, and the cytoplasm contains a large number of subcellular organelles such as mitochondria, multivesicular bodies, and neurosecretory granules distributed throughout. The two forms of neurosecretory cells which we have seen in these slides are probably two different stages in the activity cycle of the neurosecretory-cell soma associated with the manufacture of the neurosecretory granules. Thus, in the first form, when the neurosecretory granules are being manufactured, the Golgi complex and endoplasmic reticulum are prominent and probably involved in this manufacture. In the second form, when the neurosecretory granules are completed and awaiting transport out through the cellular extensions into the neuropil, the Golgi complex and endoplasmic reticulum have disappeared and the cytoplasm is loaded with organelles, such as mitochondria, multivesicular bodies, and neurosecretory granules.

Let me, at this point, recapitulate and elaborate upon the probable role of the Golgi complex in the production of the neurosecretory granules.

The sharp demarcation boundary of the neurosecretory granules stems from the fact that they are enclosed by a bounding membrane. This feature is not a unique characteristic of neurosecretory granules in planarians but seems to be a fairly general aspect of the morphology of secretory granules, e.g., pancreatic zymogen granules. F. Sjostrand has suggested that the role of the Golgi complex is the manufacture of the membrane envelope of secretory granules. This suggestion seems to apply to our study on planarian neurosecretory granules.

Scharrer has suggested that the predecessors of the dense neurosecretory material are manufactured in the ergastoplasm (Nissl substance) and passed on to the neighborhood of the Golgi complex. In Fig. 12, which contains phase 1 of the neurosecretory-cell activity cycle, one can see the rows and clusters of ribosomes in the endoplasmic reticulum. One can see clouds of electron-dense material both in these regions and in the neighborhood of the Golgi complex. The Nissl substance is, of course, RNA, probably mostly ribosomal RNA. Those regions of the Golgi tubes which intersect the clouds of dense material are themselves electron dense, as though concentrating electron-dense material. The vesicles, which can be seen budding off the ends of the Golgi tubes and in the later stages forming from the dissolution of the Golgi tubes, contain dense material, but not as much as mature neurosecretory granules. The granules seen in the cytoplasm of phase 2 of the neurosecretory cell are nearly, but not quite, as dense as those neurosecretory granules observed in the neural fibers of the neuropil.

These observations would all accord with the mechanism postulated by Sjostrand for the formation of pancreatic zymogen granules. The membranes of the Golgi tubes concentrate electron-dense secretory material which is originally manufactured by the RNA of the endoplasmic reticulum. These Golgi-tube membranes subsequently swell and bud off to make a little sac, and the side of the membrane that formed the exterior of the Golgi tube is still outside the sac and the side that formed the interior is still inside. This closed membrane envelope, or vesicle, then continues to concentrate the dense material from the surrounding cytoplasm until it finally acquires the electron density characteristic of a mature neurosecretory granule.

The physiological role of these neurosecretory granules for the planarian and for the functioning of its brain is still not clear, although a couple might in a speculative way be suggested. Lender[3] has shown the role of the brain as a cephalic organizer in planarians. The substance or substances contained in the neurosecretory granules might be the mediator of this effect. Secondly, the interesting work of McConnell and his co-workers have shown that modifications in behavior produced in light-shock conditioning situations can be transmitted from "trained" planarians to "untrained" ones by feeding the trained to the untrained subjects. One might imagine the substances contained in the neurosecretory granules to be the mediators of this effect.

Irrespective of the precise role of these neurosecretory granules, however, several things are evident from these studies. The first is that we need no longer pretend that the brain of the planarian is an unknown "black box" about whose internal organization we know nothing. Second, from the small sizes of the fibers it is clear that any preconceptions of the capabilities or limitations of the planarian brain which were based exclusively upon light microscopy must be essentially speculative in character. All but the gross organization of it is simply too small to have been resolved with a light microscope. Third, it appears as though the basic components, the "nuts and bolts and parts," are essentially similar to those found in the brains of higher animals. There are synaptic junctures. These synaptic junctions show the essential morphological characteristics of the synapses of higher animals. There are axons. These axons have neurotubules. There are neurosecretory cells which make and transport neurosecretory granules. The process by which these neurosecretory granules are manufactured involves the Golgi complex, as it does in higher animals. There are neural accessory cells which appear to be analogous to the neuroglia of the higher animals.

Until one knows precisely what structures are really involved in those modifications we call learning, one cannot, from anatomical knowledge, definitely assert or deny whether a particular nervous system can learn and, if so, how much. But on the other hand there does not appear to be any compelling reason, based upon the neuroanatomy or neural ultrastructure, to draw a line between planarians and higher animals except in size and

---

[3] The work of Lender is reviewed by Sengel in this volume.

quantity. All the basic parts seem to be there. There would seem to be no particular reason, based on neuroanatomical grounds, to assume that the nervous system of a planarian is not subject to the modifications involved in learning. Unless, of course, one wishes to contend that a brain comprised of 10^{11} of unmodifiable elements is modifiable, while one made of 10^3 of such elements is not.

Chapter 11

Some Characteristics of the Light-Evoked Electrical Response of the Planarian Eyecup[1,2]

H. Mack Brown[3]

V.A. Hospital and Department of Neurology
University of Utah College of Medicine
Salt Lake City, Utah

Early studies of planarian behavior correlated the structure of the animal's eyecup with its negative phototaxis (Parker and Burnett, 1900, and Taliaferro, 1920). More recent studies have attributed experimental augmentation of planarian phototaxic behavior to learning (Corning and John, 1961; McConnell, Jacobson, and Kimble, 1959; and Thompson and McConnell, 1955) or other factors (Brown and Beck, 1964; Brown, Dustman, and Beck, 1966a, and b; and Halas, James, and Knutson, 1962). These studies have not contributed knowledge of the direct effects of light upon the nervous system of planaria. It is the purpose of this communication to describe briefly some characteristics of the light-evoked, extracellular, slow potentials recorded from the planarian eyecup with conventional microelectrode techniques.

1. The ocellar potential (OP) evoked by a brief (800 μsec), intense (5000 Lux) flash of light is a negative monophasic potential of long duration (0.7 to 1 second). The minimum latency of the OP is approximately 35 msec.

2. Systematic changes in the OP occurred under different conditions of light adaptation. (1) When intense, paired, identical flashes were presented, the response to the second flash increased as a simple exponential function of the time in the dark (interflash interval) with a time constant of approximately 30 seconds. (2) The amplitude of the OP varied inversely with the logarithm of background illuminance. (3) When the planarian ocellus was progressively light adapted with a train of brief but dim light flashes, the OP decreased exponentially. (4) Adaptation by a train of bright flashes of light caused the OP to behave in an anomolous fashion. The

[1] A summary of reports that have appeared elsewhere (Brown and Ogden, 1965, and Brown and Ogden, 1966). A full description of this work is in preparation.

[2] This project was aided by NIH Grant No. NB-04135.

[3] VA Post-Doctoral Research Associate.

first flash in such a series elicited the maximal response; the second flash elicited the smallest response. Subsequent flashes (three to seven) elicited responses that grew by approximately equal increments until the steady state was attained. This result is not readily understood in terms of conventional photopigment kinetics.

3. Temperature affected both the latency and amplitude of the OP. As the temperature was raised from 15 to 27°C, the latency decreased ($Q_{10} = 1.5$) and amplitude increased ($Q_{10} = 1.75$). At higher temperatures OP amplitude diminished, although the latency continued to decrease. These changes were not reversible if the temperature was raised above 30 to 32°C. The potential was abolished altogether when the temperature was raised to 45°C.

These results indicate that the photoreceptors of the primitive organized planarian eye behave in most respects like the photoreceptors of higher forms.

REFERENCES

Brown, H. M., and E. C. Beck, Does learning in Planaria survive regeneration? *Fed. Proceed.*, 1964, **23**, 254.

Brown, H. M., R. E. Dustman, and E. C. Beck, Experimental procedures that modify light response frequency of regenerated Planaria. *Physiol. Behav.*, 1966a, **1**, 245.

Brown, H. M., R. E. Dustman, and E. C. Beck, Sensitization in planaria. *Physiol. Behav.*, 1966b, **1**, 305.

Brown, H. M., and T. E. Ogden, Some characteristics of electrical activity evoked in the eyecup of the Planarian *Dugesia tigrina*. *The Physiologist*, 1965, **8**, 124.

Brown, H. M., and T. E. Ogden, Electrical response of the planarian eyecup in different states of light adaptation. *The Physiologist*, 1966, **9**, 145.

Corning, W. C., and E. R. John, Effect of ribonuclease on retention of conditioned response in regenerated planarians. *Science*, 1961, **134**, 1363.

Halas, E. S., R. L. James, and C. A. Knutson, An attempt at classical conditioning in the planarian. *J. Comp. Physiol. Psychol.*, 1962, **55**, 969.

McConnell, J. V., A. L. Jacobson, and D. P. Kimble, The effects of regeneration upon retention of a conditioned response in the planarian. *J. Comp. Physiol. Psychol.*, 1959, **52**, 1.

Parker, G. H., and F. L. Burnett, The reactions of planaria with and without eyes to light. *Amer. J. Physiol.*, 1900, **4**, 373.

Taliaferro, W. H., Reactions to light in *Planaria maculata* with special reference to the structure of the eyes. *J. Exp. Zool.*, 1920, **31**, 59.

Thompson, R., and J. McConnell, Classical conditioning in the planarian *Dugesia dorotocephala*. *J. Comp. Physiol. Psychol.*, 1955, **48**, 65.

Chapter 12

Tracer Studies of the Uptake of Organic Compounds by Planarians[1]

Edward L. Bennett, Marie Hebert, and
Ann M. Hughes

Lawrence Radiation Laboratory
University of California
Berkeley, California

INTRODUCTION

For this symposium it is superfluous to review either the numerous theories of memory or the interest that has existed, and for many still exists, in planarians as a useful tool in studies of chemical processes which occur in learning. Because of this interest in planarians, a few years ago we attempted to train planarians reliably and reproducibly. At the same time we studied the uptake of a number of organic compounds by planarians. Our objective was not to investigate the intermediary metabolism of these compounds in detail, but rather to determine in a general way (1) the relative efficiency with which different classes of organic compounds were utilized, (2) the variation within a class of the degree of utilization, and (3) some variables which influence uptake. The parameters affecting uptake which we studied included the external concentration of the organic compounds and the time of contact. In addition, the rate of loss of the label when the planarian was removed from the tracer solution and the distribution of the utilized chemical between "soluble" and "insoluble" compounds were determined. The retention and distribution of the label when transferred by cannibalism were also investigated. Radioautographic studies were carried out to confirm the uptake of the tracer compounds and to investigate their distribution among cells of the planarian.

Our ultimate objective, which we did not attain, was to demonstrate that labeled macromolecules transferred by ingestion or injection would subsequently reach the neural tissue of the recipient planarian. Many of us would probably be much more receptive to the notion that memory can be transferred if it could be conclusively shown that neural material from the donor was transferred to neural tissue of the recipient. The recent electron

[1]This research was supported by the U.S. Atomic Energy Commission.

166

microscope studies of the neural structure of planarians by Morita and Best (1965), which are included in this volume, illustrate some of the problems to be surmounted before transfer of neural components to neural tissue can be demonstrated in planarians.

In our studies, we compared the uptake of (1) purines, pyrimidines, and nucleosides, (2) amino acids, (3) fatty acids, and (4) carbohydrates. Of these compounds, fatty acids are taken up most extensively. Butyrate-2-^{14}C and valerate-2-^{14}C are taken up more rapidly than the lower fatty acids and form at least four or five major compounds and about ten minor compounds. Propionate-2-^{14}C, which is also extensively utilized by planarians, is lost only slowly and is transferred to, and retained by, cannibals. Of the nucleic acid precursors tested, adenine and cytosine exhibited the greatest uptake. Adenine was converted to 5'-adenylic acid and its derivatives and to nucleic acid adenine and guanine. The amino acids and carbohydrates, as well as inorganic phosphate, were taken up comparatively slowly.

METHODS

Typically, the planarians *Dugesia dorotocephala* or *D. tigrina* were placed in a radioactive solution contained in a 10- or 20-ml beaker and stored in a dark cupboard. The "concentration" of the planarians was three to four per milliliter of solution, and the concentration of the radioactive compound varied from 1.0 to 10 μM/ml. A local spring water (Alhambra) was used, and the pH was adjusted to 7 after the compound was dissolved.

The amino acids, organic acids (as sodium salts), glucose, and sucrose had been prepared in our laboratory. Each compound had approximately 0.5 to 1 mCi/mM and was used without further dilution. Purity was checked prior to use by paper chromatography and radioautography. The valine sample had 10 to 15% of an unidentified impurity present. The nucleic acid precursors were commercial samples and were diluted when used at 2.5 μM/ml to approximately 3 mCi/mM for the ^{14}C-containing compounds or 20 mCi/mM for the ^{3}H-containing materials. At 10 μM/ml, the specific activities were further reduced by a factor of 2 by addition of appropriate concentrations of carrier. Concentrations were then determined by spectrophotometric methods. About 40% of the radioactivity of the adenine-^{3}H sample was volatile when a solution was evaporated to dryness. Data obtained with this sample were corrected accordingly. No similar check was made of the other ^{3}H-containing samples. Prior to analysis, a planarian was removed from the radioactive solution and passed through three successive rinses of water in a spot plate. The following basic fractionation procedures were followed:

Trichloroacetic Acid (TCA) Method

An individual planarian was homogenized in 500 μl of distilled water by use of a micro Teflon-glass homogenizer. After 25 μl of the homogenate was removed for a determination of the total radioactivity present, 100 μl of 50% TCA was added to the remainder of the homogenate, and the mixture was centrifuged. The precipitate was washed twice with 500 μl of 10% TCA, and the activity of the combined supernatants (TCA-soluble fraction) was determined by counting a 100-μl aliquot. In some experiments, the TCA was removed by continuous ether extraction, and the activity in the aqueous phase was redetermined. The residue in the homogenizer was extracted twice with 500 μl of 95% ethanol (this step was eliminated in some preliminary experiments) and then dissolved by warming overnight at 37°C with 50 μl of 1 N KOH. The KOH solution was then diluted to 500 μl and heated to 80°C for 2 to 3 hours. Aliquot portions were taken of the ethanol extracts and the KOH-hydrolyzed residue for determination of radioactivity. An approximate measure of the size of the worm was usually made by determining the protein content of an aliquot of the KOH-soluble residue by means of the Folin-Wu color reaction (Lowry et al., 1951). The results were expressed in terms of mμM uptake per 100 μg protein or 1 mg planarian (a planarian was assumed to be 10% protein, and those used normally weighed 2 to 4 mg).

Ethanol Extractions

In a later series of experiments, the fractionation procedure was patterned after that commonly used in photosynthesis experiments. The planarian was homogenized and extracted twice with 500 μl of 80% ethanol and 20% H_2O at 50°C, then twice with 500 μl of 30% ethanol and 70% H_2O. In one series of experiments acetone-water mixtures were also used with similar results. The residue was dissolved by warming with 50 μl of 1 N KOH or by heating overnight at 80°C with 100 μl of 6 N HCl. In either case, the solution was diluted to 500 μl with H_2O prior to sampling. No protein determinations were made after acid hydrolysis.

Scintillation counting (Tri-Carb, Packard Instrument Co.) was used to determine radioactivity. The scintillator solution consisted of 5.0 g of 2,5-diphenyl oxazole (PPO), 100 mg of 1,4-bis-[2-(5-phenyloxazolyl)] benzene (POPOP), 50 g naphthalene, 400 ml of p-dioxane, 250 ml of ethanol, diluted to 1 liter with toluene. Normally, samples were counted shortly after addition to the scintillator solution in order to minimize a small (maximum 10%) decrease in activity due to settling of particulate material in some samples. The basic protein hydrolysates were neutralized in the counting vial by the addition of HCl. The efficiency was routinely determined by addition of toluene-[14]C (or toluene-^{3}H in the experiments with tritiated compounds) and was normally 60 to 65% for ^{14}C and 10% for [3]H.

Extracts from the planarians were chromatographed two-dimensionally on Whatman No. 1 filter paper in phenol and water and then in 40% butanol, 25% propionic acid and 35% H_2O (wt %). Radioactive compounds were located by radioautography with Kodak X-ray film, and the distribution of radioactivity was estimated by counting the principal radioactive areas with a Mylar-window GM tube. Carrier amino acids were located by spraying with ninhydrin. In experiments with nucleic-acid precursors, 60% propanol, 30% ammonium hydroxide and 10% H_2O (by volume) was substituted for phenol-H_2O, and the location of carrier compounds was determined under ultraviolet light.

To test for radioactivity transfer by cannibalism, planarians were labeled for 2 days in experiment 1 with either adenine-^{3}H with a specific activity of 12 mCi/mM or propionate-2-^{14}C with a specific activity of 0.8 mCi/mM and a concentration of 10 μM/ml. It was estimated that the cannibalized planarians contained 400,000 dpm of ^{3}H or 150,000 dpm of ^{14}C. In experiment 2, labeling was carried out for 3 days with propionate- -2-^{14}C, and then the donor planarians were placed in water for 4 days.

For the cannibal experiments, radioactive *D. tigrina* were cut transversely into four approximately equal segments, and each segment was cannibalized by an unlabeled, hungry *D. tigrina* in a 3-in. petri dish containing 25 ml of water. Only an approximate estimate of the radioactivity fed could be made by determination of the radioactivity in similar uncut worms at the time of feeding. The time to attack and ingest the victim was observed and was typically about 10 minutes. In about 15% of the cases, the cannibal refused to eat and had to be replaced.

For the histological studies, planarians were first rinsed in several changes of water, as had been done for the chemical fractionation procedure. They were then fixed in 10% neutral formalin, embedded in paraffin, and serial sections 5 μ thick were cut. After being placed on slides, the planarian sections were deparaffinized and hydrated. Slides were then dipped in Eastman NTB$_2$ emulsion, air-dried, and stored in light-tight boxes in the refrigerator for 3 to 4 weeks. At the end of this time, the radio-autographs were developed in Kodak D$_{19}$ for 6 minutes, rinsed briefly in distilled water, and fixed in sodium thiosulfate. Slides were washed for 1 hour in running distilled water, then air-dried overnight. The standard hematoxylin-eosin method was used for poststaining the sections.

RESULTS

Uptake of Nucleic-Acid Precursors

One of the first experiments undertaken investigated the effect of concentration and time on the uptake of several nucleic-acid precursors: adenine-8-^{14}C, cytosine-^{3}H, and thymine-2-^{14}C. The results for adenine, shown in Fig. 1, clearly indicate that within the range investigated uptake was approximately proportional to the external concentration and was also time-dependent. Adenine taken up by the planarians was retained for long periods; the total radioactivity decreased by only one-half over a 3-week period, even though the planarians had been fed brine shrimp every 3 or 4 days after removal from the adenine solution (Fig. 2).

The uptake of several nucleic-acid components is compared in Fig. 3. In terms of total uptake, adenine is taken up 3 to 5 times more than the pyrimidine bases uracil-2-^{14}C and cytosine. The relative uptake of the

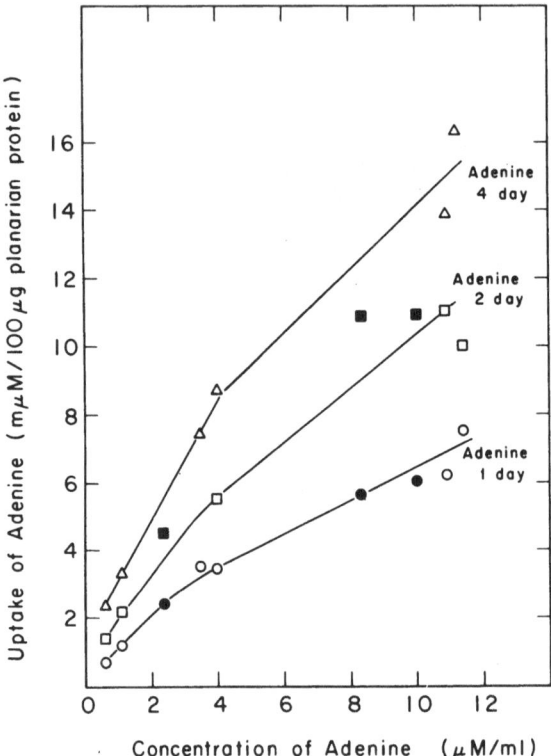

Fig. 1. Effects of concentration and time on the total
uptake of adenine-8-^{14}C by *D. tigrina*. The concentrations of adenine in exp. 1 were 0.6, 1.1, 3.5, 4.0, 10.9, and
11.4 μM/ml; in exp. 2 concentrations were 2.4, 8.3, and
10.0 μM/ml. The radioactive concentrations of adenine-
^{14}C used was between 5 and 15 μCi/ml, and that of
adenine-^{3}H was 50 μCi/ml. Data were obtained from
one planarian for each point in exp. 1, and two planarians for each point in exp. 2. Results are expressed
in millimicromolar units of compound incorporated per
100 μg of planarian protein.

nucleosides—uridine-2-^{14}C, cytidine-^{3}H, and adenosine-^{3}H—is even less.
In another experiment, the uptake of thymidine was also shown to be much
less than that of adenine.

So far in this discussion, we have emphasized total uptake and retention only, but in these and other experiments the planarians were fraction-
ated into a trichloracetic acid-soluble (TCA-soluble) fraction and a trichloracetic acid-insoluble (TCA-insoluble) fraction. The results were summarized in millimicromolar units of precursor incorporated in the TCA-
insoluble fraction, presumably RNA and DNA. The percentage of the total

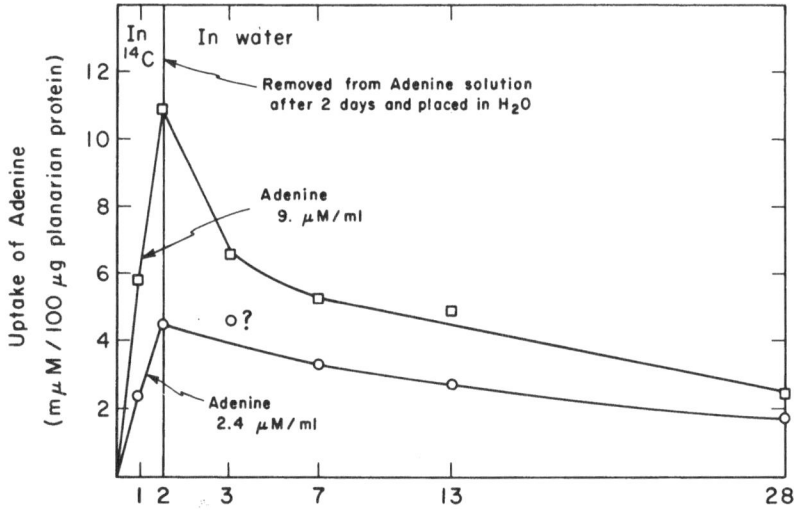

Fig. 2. Loss of radioactivity after removal of *D. tigrina* from an adenine solution.

activity in the planarians which is incorporated in this fraction is given in Table 1. When expressed in this manner it appears that cytidine and adenine have comparable efficiencies as precursors of nucleic acid in planarians when the external concentration is about 2.5 μM/ml. The superiority of cytosine as a precursor of nucleic acid is even more obvious when we compare the data obtained at an external concentration of 10 μM/ml (Fig. 4). This apparent discrepancy in total incorporation is merely a reflection of the size of the soluble nucleotide pool. It is well known that in higher mammals the soluble adenylic acid pool is much larger than that of other nucleotides. Indeed, we have shown that in planarians more than 50 % of the TCA-soluble radioactive material after adenine uptake can be accounted for as AMP and its derivatives, together with small quantities of the guanylic acid derivatives. Much of the remainder is hypoxanthine and adenine. Total adenine incorporation in the insoluble fraction continues to increase as precursors for nucleic acids are drawn from the soluble pool. On the other hand, the radioactivity from cytosine in nucleic acids decreases more rapidly than that from adenine. This is a reflection of the small pool size of soluble cytosine derivatives as compared with the large pool size of the soluble adenine compounds.

Of the nucleic-acid precursors studied, cytosine appears to be the best for obtaining the maximum proportion of label in the nucleic acids, and the minimum possible in soluble nucleotides and other low-molecular-weight compounds. These low-molecular-weight compounds would only confuse the interpretation of the fate of RNA during studies of cannibalism or regeneration.

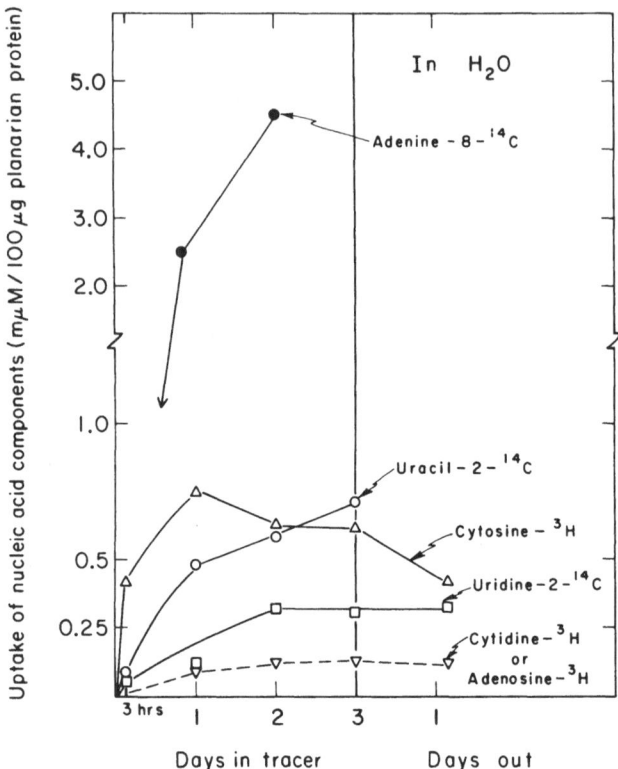

Fig. 3. Fifteen *D. dorotocephala* were placed in a 10-ml beaker containing 10.4 to 10.9 μM of the radioactive compound dissolved in 4 ml of spring water, and the pH was adjusted to 7.0 with NaOH. Approximately 30 μCi of a ^{14}C-labeled compound or 200 μCi of a ^{3}H-labeled compound was used. Two planarians were sampled at each time interval. The uptake of cytidine and adenosine was nearly equivalent, and only one line has been placed on the graph for these two compounds.

Uptake of Amino Acids

The uptake of five amino acids—glycine-2-^{14}C, DL-serine-3-^{14}C, DL-alanine-2-^{14}C, L-leucine-3-^{14}C, and L-valine-4,4'-^{14}C—was investigated. The uptake of each of these compounds increased with time up to 2 days and was 1 mμM/100 μg of protein at 1 day in the TCA-soluble fraction when the external concentration was 10 μM/ml (Fig. 5). This value increased at 2 days to 1.5 to 2.0 mμM/100 μg planarian protein. Depending somewhat on the time chosen for comparison, the decreasing order of uptake is alanine, serine, glycine, leucine, and valine.

The uptake into TCA-insoluble material is shown in Fig. 6. The uptake of serine, glycine, alanine, and leucine is in the range from 1.0 to 1.5 mμM/ 100 μg of planarian protein at 2 days. Valine is taken up less efficiently. After removal of the external radioactive substances, the fraction incorporated into TCA-precipitable material increased as the activity in the acid-soluble pool decreased.

Another experiment indicated that the uptake of glycine was approximately proportional to the external concentration in the range from 1.0 to 10 μM/ml (Fig. 7). No further investigations were made of the types of compounds into which the amino acids were incorporated.

TABLE 1

Comparison of Uptake and Retention of Nucleic-Acid Precursors in Cold 10 % TCA-Insoluble Material

Results are expressed in the millimicromolar-unit equivalent of radioactivity found in the 10 % TCA-insoluble fraction per 100 μg of planarian protein and the percentage of the total uptake which this represents. Total uptake data are given in Fig. 3. Adenine and cytosine data are from several experiments which have been combined.

Time in radio-active solution	Adenine-8-^{14}C		Cytosine-^{3}H		Uracil-2-^{14}C	
	TCA-insol.	% Total in TCA-insol.	TCA-insol.	% Total in TCA-insol.	TCA-insol.	% Total in TCA-insol.
3 Hours	...	..	0.12	28	0.002	2
1 Day	0.1	4	0.25	33	0.03	6
2 Days	0.16	4	0.37	59	0.03	5
3 Days	...	..	0.38	62	0.07	10
3 Days In						
+ 1 Day Out	...	..	0.32	76		
2 Days In						
+ 3 Days Out	0.66	14	0.40	69		
+ 7 Days Out	0.64	19	0.32	82		
+ 13 Days Out	0.87	32	0.29	83		
+ 28 Days Out	0.49	31	0.10	83		
	Adenosine-^{3}H		Cytidine-^{3}H		Uridine-2-^{14}C	
3 Hours	0.006	17	0.005	16	0.009	15
1 Day	0.03	25	0.025	18	0.027	23
2 Days	0.05	45	0.09	36	0.07	22
3 Days	0.06	43	0.04	29	0.08	26
3 Days In						
+ 1 Day Out	0.05	48	0.05	29	0.12	36

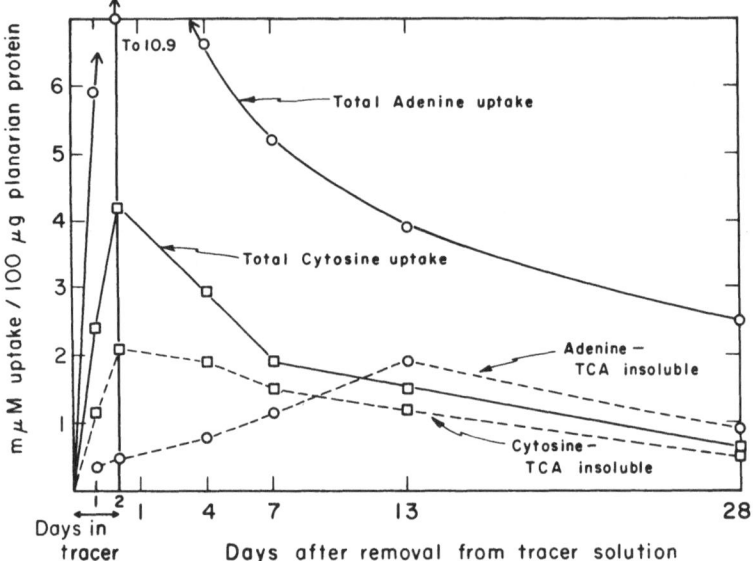

Fig. 4. The uptake and retention of adenine and cytosine by *D. dorotocephala* in terms of total uptake and uptake into the TCA-insoluble (nucleic acid) fraction.

Uptake of Fatty Acids

Of the compounds tested, the fatty acids were taken up most extensively and rapidly. The uptake increased with increasing molecular weight from formate through valerate (Fig. 8). The distribution of radioactivity in planarians kept in formate-^{14}C, propionate-2-^{14}C, and valerate-2-^{14}C was very similar. Approximately 50 to 60% of the radioactivity was TCA-soluble, 10 to 20% of the radioactivity was extracted by ethanol, and approximately 25 to 35% of activity was both TCA- and ethanol-insoluble. From planarians maintained in acetate-2-^{14}C, 40% of the activity was extracted by cold TCA, 25% by ethanol, and the remainder was insoluble. The distribution of radioactivity did not change markedly up to 1 week after removal of the planarians from the radioactive solutions.

Some preliminary investigation was made of the type of compounds into which these acids were incorporated. Only 10 to 20% of the TCA-extractable material was subsequently removed by continuous liquid-liquid extraction with ether. This indicates that only a small amount of the activity was present as fatty acids. Inasmuch as the planarians were much more radioactive after soaking in butyrate or valerate, more compounds could be detected by radioautography after paper chromatography of the extracts than from the lower fatty acids. Figure 9 shows a radioautograph obtained from the TCA-soluble fraction of planarians after soaking in

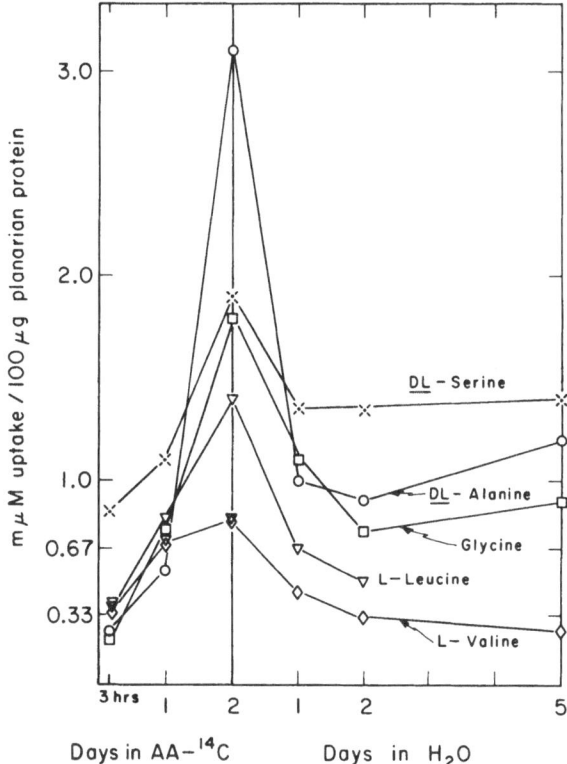

Fig. 5. The uptake and retention of glycene-2-^{14}C, DL-alanine-2-^{14}C, 1-valine-4-^{14}C, and L-leucine-3-^{14}C in 10% TCA-soluble material by *D. tigrina*. The planarians were in the radioactive amino acid solution (10 μM/ml; sp. act. approximately 0.5 μCi/μM) for periods up to 2 days, after which they were placed in water for periods up to 5 days. Data for the TCA-soluble fraction are given in Fig. 6.

valerate for 3 days. Glutamic acid, glutamine, aspartic acid, and alanine appear to be the main radioactive compounds formed. The large amount of radioactivity remaining at the origin yielded primarily glucose after hydrolysis with HCl. Similar patterns, with not as many minor spots detectable, were obtained from "acetate," "propionate," and "butyrate" worms.

Uptake of Glucose and Sucrose

Glucose-U-^{14}C and sucrose-U-^{14}C were taken up about equally but were relatively little utilized by *D. tigrina*. The incorporation from glucose into the 10% TCA-soluble fraction was about 1 mμM/mg of planarian, or

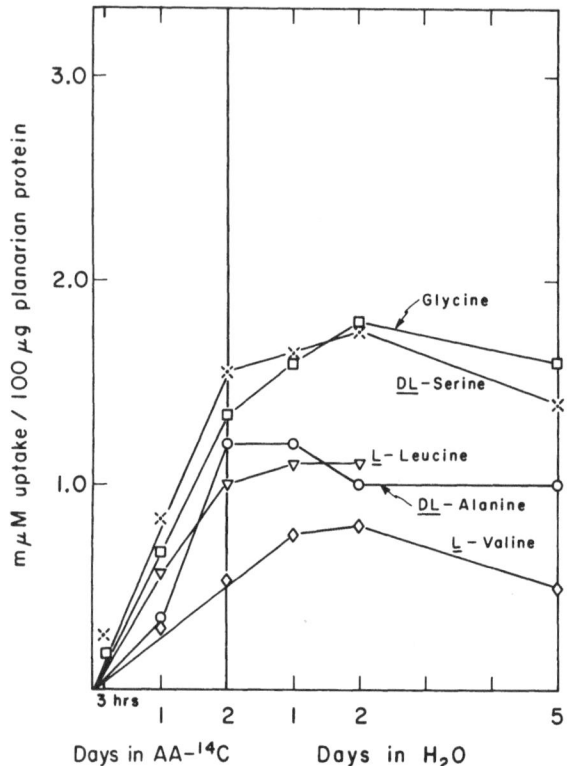

Fig. 6. The uptake and retention of amino acids in the 10%
TCA-insoluble material by *D. tigrina*. Data for TCA-
soluble fraction are given in Fig. 5.

about one-fourth that found with acetate; the incorporation into the
fraction subsequently extracted by ethanol was less than 0.5 mμM/mg,
and less than 1 mμM/mg was TCA-insoluble. Sucrose gave about four times
as much incorporation into the TCA-soluble fraction, but only slightly
more incorporation than glucose into the ethanol-soluble and the residue
fractions.

Uptake of Phosphate

Only one experiment was done to investigate the uptake of phosphate.
In this experiment phosphate uptake was found to increase with external
concentration up to 0.01 M, at which concentration the uptake was about
1.4 mμM/mg of planarian. About 90% of the radioactivity was extracted by
cold 10% TCA, and the remainder of the activity was about equally
divided between the ethanol and acetone extractions and the TCA-insoluble

fraction. Seven days after removal from the external phosphate, only about 50% of the radioactivity was extracted by cold TCA, 20 to 25% by the organic solvents, and the remainder was in the residue.

Of all the compounds tested, PO_4^{3-} was taken up the least, as shown in Fig. 10, which summarizes the total uptake in 2 days of each of the compounds investigated when supplied at a concentration of 0.01 M.

Transfer of Radioactivity by Cannibalism

When donor planarians, labeled with either adenine-^{3}H or propionate-2-^{14}C, were cannibalized by D. tigrina, a significant proportion of the ingested radioactivity was incorporated into the cannibal and retained for long periods (Table 2). In the case of propionate, the radioactivity was distributed in a similar manner in the recipient to that in the donor, i.e., about 20% was extracted by 80% ethanol, 20% by 30% ethanol, and the remainder was ethanol-insoluble. These proportions did not change markedly with time after ingestion, nor did they appear to differ depending upon the portion (head, midsection, or tail) of the victim which had been cannibalized. In the case of cannibals eating adenine-labeled planarians, 40% of the radioactivity was TCA-insoluble as compared with 5 or 10% in the donors. Again, this fraction did not appear to change markedly with time.

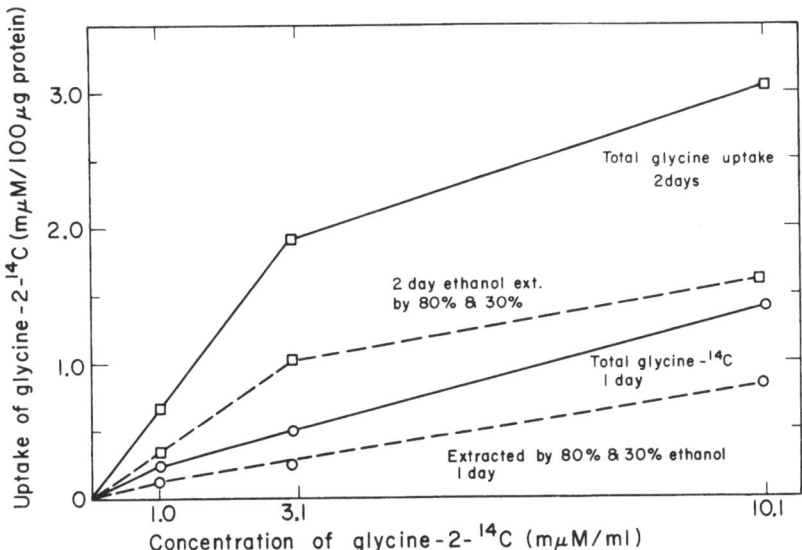

Fig. 7. The uptake of glycine-2-^{14}C by D. tigrina as a function of concentration and time. In this experiment, the ethanol fractionation procedure was used, and the fraction extracted is indicated by the dashed lines.

TABLE 2

Radioactivity Transferred by Feeding Radioactive Planarians

| Days after feeding | "Propionate" planarians | | "Adenine" planarians |
	Exp. 1 dpm/four cannibals	Exp. 2 dpm/four cannibals	Exp. 1 dpm/four cannibals
1	71,000		256,000
2		43,000	
3	44,000		203,000
4		32,000	
7		20,000	
8	39,000		205,000
15	40,000		86,000

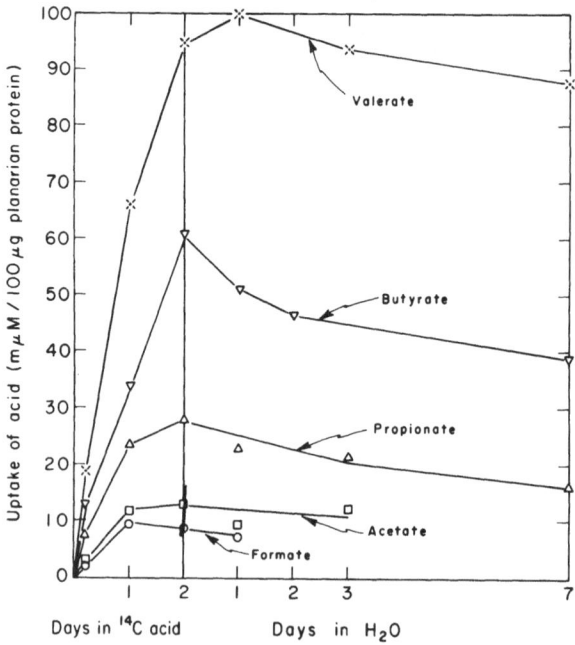

Fig. 8. Rate of uptake and retention of five labeled fatty acids from 0.01 M solutions, pH 7, by *D. tigrina*. The substrates were all alpha labeled (except for formate), and 20 to 50 μCi of each was used in each 4 ml of solution. After 2 days in the radioactive solution, the planarians were removed to fresh bowls of water.

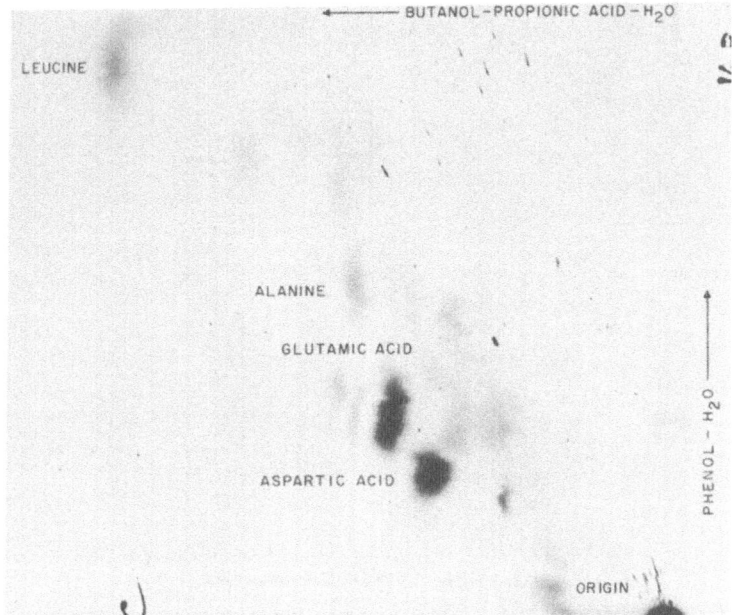

Fig. 9. Radioautograph of 10% TCA-extractable material from planarians soaked in 0.01 M valerate for 3 days. The TCA was first removed by continuous liquid-liquid extraction, and the aqueous phase was placed on Whatman No. 1 filter paper and chromatographed in phenol-water and then butanol-propionic acid water. Identified compounds include aspartic acid, glutamic acid, glutamine, and alanine. Analysis of the radioactive origin by hydrolysis in HCl and rechromatography yielded largely glucose.

In other experiments carried out at the same time with propionate-fed worms, the rate of disappearance of label from worms transected and allowed to regenerate did not differ markedly from that of planarians removed from the radioactive solution, i.e., the sum of the halves approximated the whole.

Radioautographic Studies

Radioautographic studies of planarians grown in adenine and propionate showed, in general, a rather uniform distribution of label. There were essentially no silver grains above background on the surface of the worms, which indicated that the activity found by the chemical analyses was not merely adsorbed on the planarian.

Though the concentration of silver grains, representing radioactivity, was relatively uniform, the highest concentration was over the cells of the digestive system and the lowest was over the nervous system (Figs. 11 through 13).

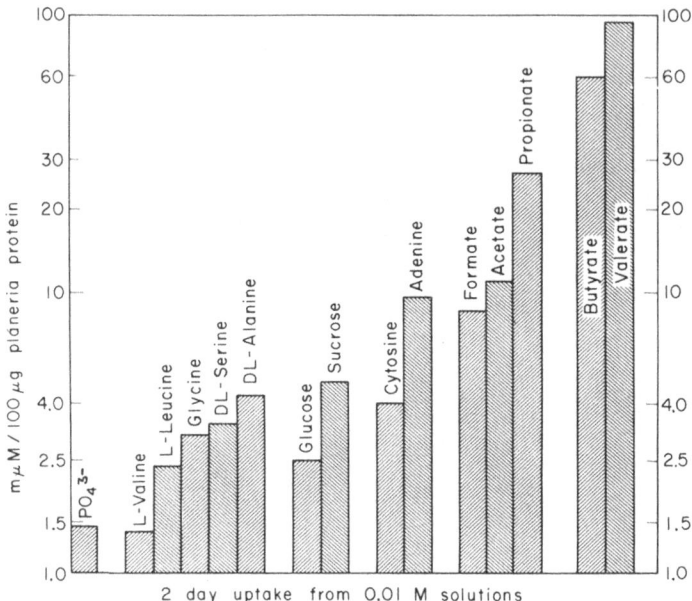

Fig. 10. Summary of total uptake of phosphate, sugars, amino acids, nucleic acid bases, and fatty acids from 0.01 M solutions in 2 days by planarians. Uptake is expressed in terms of millimicromolar-unit equivalent of radioactivity/100 μg planarian protein.

DISCUSSION

At the time we started our studies of the radioactive uptake of various organic compounds by planarians, we were aware only of the study by Flickinger (1963) to establish the axial metabolic gradient when the flatworm was exposed to glycine or to $^{14}CO_2$ for 2 hours. Hammen and Lum (1962) also in short-time experiments, showed the rapid conversion of propionate-2-^{14}C into alpha-ketoglutaric, succinic, fumaric, malic, citric, isocitric, lactic, and mesoxalic acids and the formation of aspartic and glutamic acids and threonine from $^{14}CO_2$. More recently, Best, Rosenvold, Souders, and Wade (1965) have investigated the long-term uptake of leucine-^{14}C and uracil-^{14}C by feeding appropriately labeled yeast to the planarians over a period of several weeks to several months. Although each of these investigations was able to demonstrate incorporation of radioactivity into the planarians, in all cases the amount of radioactivity incorporated was very low. For example, the specific activity of the leucine-^{14}C fed to the planarians by Best *et al.* was 5.5 μCi/mM, and it resulted in planarians labeled to the extent of 2.0 μCi/mM after 7 weeks of feeding. Best *et al.* report that an individual planarian (size not stated, but estimated to be about 1.5 mg from the protein analysis given) contains 0.02 μM of leucine. On this basis, each planarian contained 88 dpm in the protein fraction. In other experiments, Best *et al.* attempted to label planarians by soaking in amino acid solutions containing 0.02 μCi leucine-^{14}C/ml of unstated specific activity. Under these conditions, it is not surprising that uptake could not be readily demonstrated.

In our experiments, using L-leucine with a specific activity of 1 mCi/mM and 0.01 M concentration, the activity found in the TCA-insoluble fraction was over 5000 dpm/100 μg of planarian protein after 2 days of soaking. Using 100 μM and 100 μCi of leucine in 10 ml of water and 4 worms/ml, 40 planarians could be labeled to this extent. However, L-leucine is among the least efficient compounds that we tested for labeling planarians. The fatty acids, in particular, were much more effective as precursors. For example, when valerate is supplied at an external concentration of 0.01 M or 10 μM/ml, within 2 days the activity inside the planarian is equivalent to that of a 0.1 M valerate solution. In other experiments, not summarized in this report, planarians were maintained for periods up to 4 weeks in acetate or propionate solutions (0.01 M), and for periods up to 15 days in 0.003 M adenine solutions. In these long-term experiments, the total radioactivity tended to saturate or plateau after approximately 4 to 7 days. A practical difficulty with long-term experiments, particularly those utilizing amino

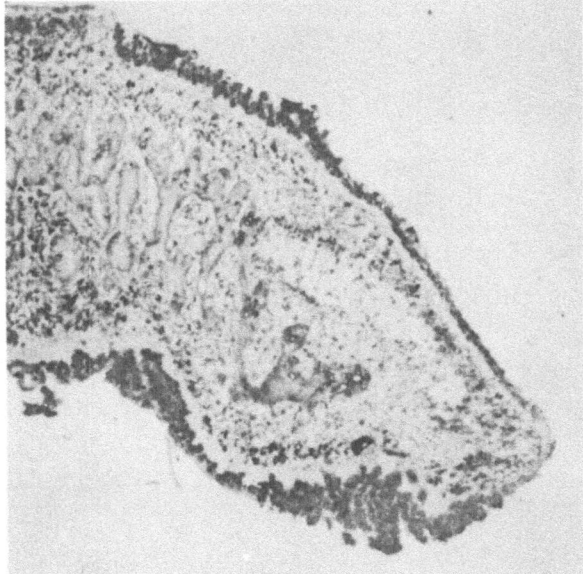

Fig. 11. Low-power photomicrograph of a longitudinal section of the anterior end of a *D. tigrina*. The relatively unstained area is neural tissue, and the digestive tract is the treelike area in the center.

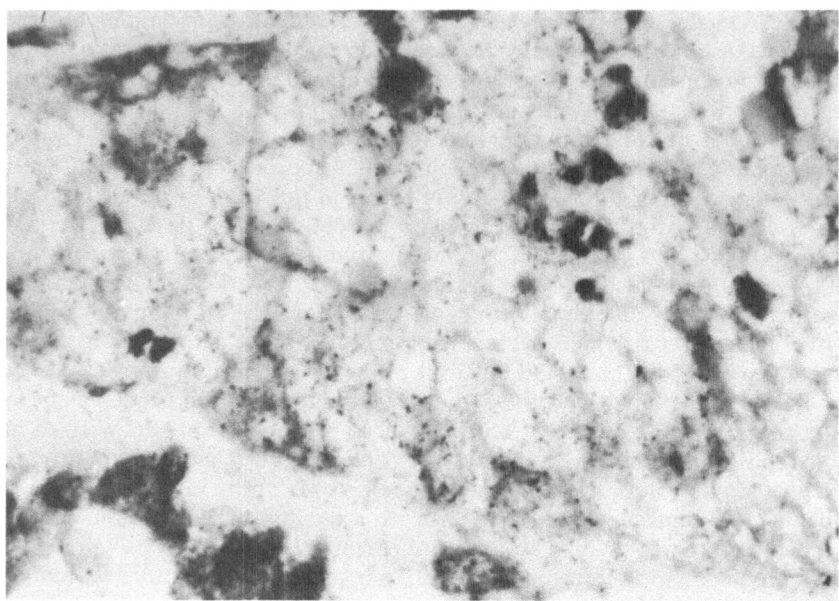

Fig. 12. Radioautograph of the digestive tract from a planarian maintained in propionate-
^{14}C for 3 days.

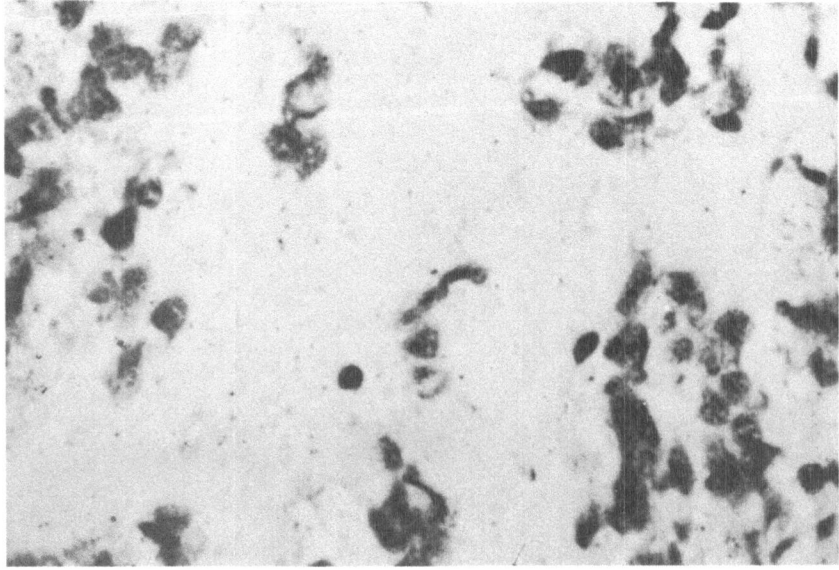

Fig. 13. Radioautograph of the neural tissue of the same animal as in Fig. 12. Note the
relative absence of silver grains.

acids, was bacterial contaminated of the solutions. This became noticeable after 2 or 3 days. To minimize this contamination, we generally transferred the planarians to fresh radioactive solutions every 2 or 3 days. In addition, in several experiments, the addition of 6 mg penicillin G and 6 mg of streptomycin-dihydrostreptomycin per 100 ml of solution reduced but did not eliminate bacterial growth.

Best *et al.* have expressed concern about the radiation dose to which planarians are exposed in soaking experiments. We believe it is more proper to take the average energy of the ^{14}C beta (0.055 MeV) rather than the maximum energy when calculating the radiation dosage received. On this basis, a solution containing 10 μCi/ml would result in a dosage of about 50 rep over a 2-day period. When the calculated uptake is 10 mμM/100 μg of planarian protein, the "concentration" of radioactivity inside the planarians equals the external concentration. If ^3H-labeled compounds are used, generally higher radioactive concentrations are employed to compensate for the decreased counting efficiency encountered. However, the average energy from tritium is about one-tenth that from carbon, so it is unlikely that excessive radiation is received by the planarians in a typical experiment.

In our search to find a method by which a large proportion of the radioactivity could be incorporated into high-molecular-weight compounds, cytosine appears to be the most efficacious compound. After several days in a cytosine solution and then 7 days out, about 75% of the radioactivity of planarians was not extracted by cold TCA. More work should be done with cytosine-^{14}C so that the compounds into which cytosine is incorporated can be more readily studied by chromatographic and radioautographic techniques.

REFERENCES

Best, J. B., R. Rosenvold, J. Souders, and C. Wade, Studies on the incorporation of isotopically labeled nucleotides and amino acids in Planaria. *J. Exp. Zool.*, 1965, **159**, 397.

Flickinger, R. A., The site of incorporation of ^{14}C-labeled CO_2 and glycine in the planarian *Dugesia dorotocephala*. *Exp. Cell Res.*, 1963, **30**, 605.

Hammen, C. S., and S. C. Lum, Carbon dioxide fixation in marine invertebrates, III: The main pathway in flatworms. *J. Biol. Chem.*, 1962, **237**, 2419.

Lowry, O. H., N. H. Rosenbrough, A. L. Farr, and R. J. Randall, Protein measurement with the folin phenol reagent. *J. Biol. Chem.*, 1951, **193**, 265.

Morita, M., and J. B. Best, Electron microscopic studies on Planaria, II: Fine structure of the neurosecretory system in the planarian *Dugesia dorotocephala*. *J. Ultrastruct. Res.*, 1965, **13**, 396.

Discussion

P. Wells: Could Dr. Kenk identify planarians solely on the gross morphological characteristics without histological sections? Very often, when one prepares sections, the gonads are not developed anyway. Also, is there a manual that you can recommend for identification?

R. Kenk: If one can observe live animals in good physiological condition when they are gliding quietly, then there are some good clues for the identification of the species, such as the outline of the head, presence or absence of auricles, position of the eyes, etc., provided that the planarians are from a well-investigated geographic area. In a few years I hope to have a manual published to help in this identification. The latest reviews now available are Libbie Hyman's treatments in R. W. Pennak's *Fresh-Water Invertebrates* (1953) and in H. B. Ward and G. C. Whipple's *Fresh-Water Biology* (1959).

H. Lenhoff: I was pleased to hear Dr. Kenk emphasize the importance of controlling temperature and pH when experimenting with planaria. In my presentation I will expand on this view and will point out that, in addition to being affected by pH and temperature, aquatic invertebrates can be extremely sensitive to the kinds and concentrations of environmental ions. All of these factors affect the animals and may either alter their behavior or modify the degree of their responses. Ever since becoming aware of such effects, I always see a red flag when I learn that an experiment was conducted with the animals having been kept in either pond water or in "conditioned" tap water.

Similarly, another "environmental" factor that I have become aware of through my experiments using hydra is the presence in the medium of metabolic products produced by the experimental animals themselves. The concentration of these products can vary depending upon the number of animals in the experimental container, their location in it, the feeding schedule, temperature, and even the locomotory activities of the animals.

One other factor that may lead to variable results in experiments with aquatic invertebrates is bacterial contamination. I have thought about this a great deal in my work with hydra, and I would like to make a few suggestions that might make it possible to raise bacteria-free planaria. Assuming that planaria embryos within a newly laid capsule are free of bacterial contamination, then it should be possible to hatch germ-free worms by first treating the capsule with some germicidal agent, such as Merthiolate, and then placing the capsule in a sterile solution until the worms hatch.

Successful application of the foregoing technique has been applied by Provasoli (*Biol. Bull.*, 1959, **117**, 347) to cysts of the brine shrimp *Artemia* *salina.* The germ-free shrimp derived by Provasoli's method are excellent food for planaria. Thus, with a controlled environment, germ-free planaria, and germ-free *Artemia*, it should be possible to eliminate many of the environmental variables that often can have major effects on "naked" aquatic invertebrates.

Regarding the chemistry of Planaria, there is no reason to assume that Planaria's composition is similar to that of all other metazoans. To the contrary; from my experience with hydra carotenoids (Krinsky and Lenhoff, *Comp. Biochem. Physiol.*, 1965, **16**, 189) and nematocyst "collagens" (Blanquet and Lenhoff, *Science*, 1966, **154**, 152), I have learned to expect peculiarities when analyzing for the first time animals that have not been subject to much chemical study.

A few years ago I analyzed an acid hydrolysate of the amino acids composing the capsule walls of planaria cocoons. To my surprise, I observed a spot on the chromatogram which indicated the presence of alpha-amino adipic acid. If this finding can be confirmed by other tests for alpha-amino adipic acid, then, to my knowledge, it would be the first case of this amino acid being present in an animal protein. Furthermore, we might expect a special codon for alpha-amino adipic acid in planaria nucleic acid. Perhaps analysis of other cellular components of planaria may lead to other surprises.

M. Clay: It has amazed me that there has been as much consistency as has appeared in the literature concerned with planarian behavior. Species differences can be marked, and I have frequently run across planarian researchers who think they have one species but in actuality have another.

S. Coward: I wonder if, in regeneration, we can rule out differentiation of parenchymal cells rather than neoblast cells—or a combination of both? I was thinking of a paper by Chandebois in which she showed neoblasts breaking down in "amitotic divisions" and with various other nuclear abnormalities. She theorized that the neoblasts donated their contents to the parenchymal cells and that these differentiated.

W. Corning: A paper of McWhinnie and Gleason (1957) presents evidence of mitosis in what they called "parenchymal amoebocytes." Interference with this mitotic activity with colchicine in sectioned animals (*Dugesia dorotocephala*) prevented regeneration.

J. Best: We have never seen a mitotic figure in *D. dorotocephala* either in the blastema or elsewhere in the system.[1] Another point concerns irradiation effects on regeneration. If we irradiate with dosages which are sufficient

[1]Since this symposium, Dr. Best reports that he has observed mitotic figures in planarians. This was achieved by a method developed by Thomas Slaga and Professor T. S. Campbell of the Department of Biology of Steubenville College, Steubenville, Arkansas. In this method regenerating planarians are treated first with phytohemagglutinin, then with colchicine. The blastema is then removed, stained with aceto orcein, and observed as a squash preparation. A large number of mitotic figures can be seen in such preparations. These mitotic figures, arrested in metaphase by the colchicine, are sufficiently clear to do chromosome counts on them. This corrects the assertion made at the symposium that mitotic figures were never observed in planarians.

to inhibit regeneration, we still see what we think are neoblasts. This contradicts what E. Wolff (1962) reported in a recent review in which the phrase "completely depopulated" was used. There are still cells there.

With respect to uridine tagging, we have found it difficult to tag planarians by immersing them in uracil-C^{14}, tritiated uracil, tritiated thymidine, or tritiated thymine. I am wondering what techniques Dr. Sengel used.

P. Sengel: The animals were immersed prior to the experiments. I am not sure of the details because I was not involved in the experiment.

H. Lenhoff: Perhaps those looking for mitotic figures in planaria are unable to find them because they are using worms which do not have many cells dividing. Such was the case with hydras. If you look at sections of hydras that were fed the previous day, you see few mitotic figures. Dr. R. Campbell, now of the University of California at Irvine, reasoned that, if cell divisions are to occur, they would most likely be found soon after the animals had eaten. Accordingly, he often found a tenfold increase in mitotic activity 4 to 6 hours following the ingestion of a meal by hydras. Similarly, perhaps planaria having a significant number of cells undergoing cell division may be obtained by looking for the appearance of mitotic figures at varying intervals following the ingestion of food by the worms.

Campbell was also able to get labeled thymidine into the nuclei of dividing cells by microinjecting the thymidine into the gut of hydras while they were digesting a meal of brine shrimp. Another way of getting the thymidine into the hydra cells was devised by Dr. Chandler Fulton of Brandeis University (*personal communication*). Fulton first labeled a developing chick embryo with thymidine and then fed the highly labeled tissue to the hydra.

I have had success in labeling both hydras and planaria by this method of prelabeling the food. While at the Carnegie Institution of Washington (Lenhoff, *Carnegie Institution of Washington Year Book*, 1958, **57**, 157), a mouse was made radioactive by injecting it with a hydrolysate of S^{35}-labeled yeast. Some labeled mouse liver was fed to planaria and hydras. Sections of the planaria were made and radioautographs ·showed heavy labeling (see Figs. A and B). This method of labeling planaria by feeding them prelabeled food should apply to labeling the animal with virtually any isotope.

J. Best: We tried what Dr. Lenhoff suggests by first tagging *E. coli* DNA, and then taking the extracted DNA and mixing it with liver and feeding it to planarians. The planarians show a big increase in count, but in a short time you find that the tag has become PCA-soluble and then the tag is lost from the animal. This occurred whether planarians were regenerating or not.

One other point about using tritiated thymidine. If the tritiated thymidine has been sitting around in aqueous solution for any period of time, there may be some hydrogen exchange. You can get some incorporation by hydrogen exchange—not very high incorporation, but some.

S. Coward: I have seen very few mitotic figures in *D. dorotocephala* that are regenerating. As far as labeling with tritiated thymidine, we have been able to get incorporation simply by immersion. These were low-level incorporations, but in the mitotic figures we did see, the chromosomes were labeled.

W. Corning: Henry Quastler and Eberhard Fetz were involved in planarian-tagging studies at Brookhaven. They found that, when they were using tritiated cytidine, they could not get labeling when they fed the worm labeled materials or when they immersed the animal; they were successful only when they injected the tag directly into the body cavity. Also, in some work I was involved in at Brookhaven, we found high incorporation of P^{32} in RNA could be achieved by simply immersing the planarians in a solution with the isotope.

H. Lenhoff: Will the brain-inducing material from one species of Planaria induce eye differentiation in another species of flatworm? What are the chemical and physical properties of the inducer?

P. Sengel: The inductor of eye regeneration is not species-specific as it will induce the regeneration of eyes in other species. Lender (1955a and b) has

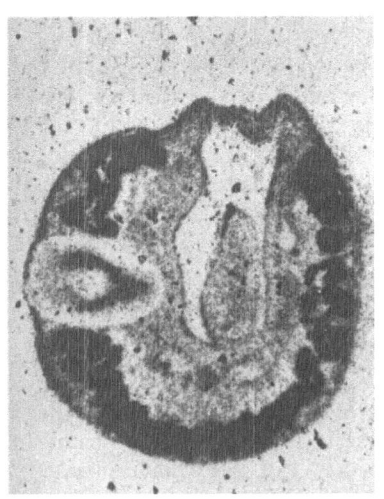

Fig. A. Radioautograph of a section of "coiled" planaria that has been fed S^{35}-labeled mouse lung 18 hours before it was killed (reprinted from Plate 5 of the *Carnegie Institution of Washington Year Book* 57, with permission of the Carnegie Institution of Washington).

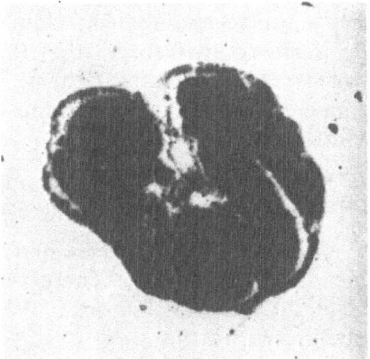

Fig. B. Radioautograph of a cross section of a planaria that had been fed S^{35}-labeled mouse lung 36 hours before it was killed (reprinted from Plate 5 of the *Carnegie Institution of Washington Year Book* 57, with permission of the Carnegie Institution of Washington).

shown that the substance is not stable at boiling. There has not been much characterization of this inductive substance as yet.

M. Cohen: Dr. Sengel mentioned the culture work of his sister in which she combined three bits of head tissue. Was the brain three times as large, and was the brain a normal one?

P. Sengel: The brain was not, I believe, three times as large, but it was larger and it was organized. The limiting factor in attempting to grow a "super brain" would be the epidermis. It is not strong enough to hold large masses together.

M. Cohen: I was wondering if Dr. Best felt he had evidence of dendrites in the planarian, a branching from the soma membrane, as we have in the vertebrate.

J. Best: I'm assuming that a nerve cell process that makes contact with another nerve cell process has to be either an axon or a dendrite. But, if you want to know whether I can differentiate between an axon and a dendrite in the neuropil without looking at the synapse, the answer is no. The only way you can tell is by looking at a synapse and determining the presynaptic and postsynaptic elements. Now as for the dendritic processes, there do not appear to be multiple processes emerging directly from the cell body. The one process does, however, branch considerably once it has left the cell body and entered the neuropil.

J. McConnell: I have a couple of rather minor points to add to Dr. Jenkin's presentation. First, Dr. Best has evidence of diurnal activity cycles in planaria (Best, 1960). In fact it was his study of the diurnal rhythms that got me started on cannibalism. Concerning slime toxicity, there is some evidence by Barnwell in Georgia that the land planarians seem to paralyze earthworms when they attack them. Finally, in contrast to what Hyman claims, I have evidence that planarians are omnivores; they will eat plants as well as meat.

R. Kenk: You sometimes find unpigmented planarians which are completely green having just eaten some algae.

J. Best: I find my species pretty exacting, even to the point of rejecting meat that isn't fresh. There is one thing in Dr. Jenkins' review that is generally accepted but does not appear to be true; this is the statement that planarians do not store polysaccharides. We have data that indicate the opposite. We found that some of the material which was stained by the PAS method could be removed from the histological sections by pretreatment of the section with saliva, a convenient source of amylase.

D. Jensen: From Dr. Brown's comments on dark adaptation, it is apparent that the intertrial interval is quite important. The usual procedure when running planarians in a classical conditioning situation is to nudge the animal with a brush to get him running in one direction. This allows the possibility of the intertrial intervals being controlled by the experimenter, i.e., he makes the decision when to poke the animal and when not to. If this procedure is used, there is a greater time between trials and this could bring about an apparent increase in responsivity by allowing longer adaptation times.

H. Brown: The procedure initiated by Thompson and McConnell in 1955, which has been adhered to subsequently by most other investigators, makes it virtually impossible to obtain a consistent intertrial interval. The investigator must wait until the planarian resumes gliding following a shock stimulus. The time that the animal requires to resume gliding is highly variable and is usually greater at the end of a series of light-shock trials than it is at the beginning of the series. When one considers the intensity of light employed in most conditioning investigations (up to 10,000 lux) and the time required for the planarian eye to adapt to dark following a brief (800 μsec) flash of the same intensity used in these experiments (Brown, this volume), there is no question that the light receptors are in varying states of sensitivity throughout the conditioning procedure.

J. McConnell: Dr. Brown has told us a great deal about what happens when light stimulates the eye. Before the animal responds, however, there is a rather interesting nervous system that must mediate this input from the eye. He speaks, e.g., of shortening the duration of the light stimulus. I suggest that, if the CS [conditioned stimulus] is too short, you will find that the animal does not respond to it. The information has to be present for a minimum period of time before the system can process it. If you calculate the probability that an animal will respond as a function of the length of time of the CS, you find that the probability goes up the longer the CS is on. From what Brown says, you would expect responses to go down, but just the opposite occurs. You not only have to worry about what to do at the receptor side but also what to do with the response side. I think that's why many of us have chosen a longer CS simply because we have the rest of the animal to consider.

H. Brown: My only point is that it would be worthwhile to make a systematic investigation of the exposure of light necessary to elicit a reliable behavioral response from planaria, yet cause a minimum change in the state of light adaptation. This would allow the effects of conditioning parameters to be determined independently of the effects of light adaptation and would be a preliminary step in the determination of the "processing time" necessary for a behavioral response. As yet we have no evidence for the time that either event requires. However, as I indicated in my presentation, there is electrophysiological evidence to suggest that the planarian photoreceptor obeys the reciprocity law (although not ideally) for stimulus durations up to 50 to 60 msec; the amplitude of the action potential recorded from the planarian eyespot becomes greater as stimuli with increased durations (up to 50 to 60 msec) and constant intensity are presented. There is no augmentation of the response when the stimulus duration exceeds this value. This gives us some estimate of the limitations imposed by the photoreceptors and suggests that stimuli considerably more brief than the commonly used 3-second stimulus might be just as efficacious for behavioral studies. This alteration in procedure would have the advantage of not introducing the effects of variables that are not being directly studied, such as light and dark adaptation.

W. Corning: It is frequently necessary to present the cue for a certain length of time in order for the information to get processed by central mechanisms. In the Limulus (Corning, 1965), we found on a number of trials that it takes up to 10 seconds of light stimulation to get certain reflexes going. Now, if you're recording the information being processed by the primary receptor, you find that the receptor has started to adapt after a second or two. However, the animal uses the extra information because, if we reduce the duration of our light stimulus to fit photoreceptor characteristics, we do not observe the visual reflexes. So I would agree with McConnell's comments.

J. McConnell: It is probably wiser to start off by letting the animal tell you what intensity it is going to respond to. This can be accomplished by first presenting the subjects with extinction trials and adjusting the light intensity until you reach a given response rate. With this technique, you will probably get greater replicability from experiment to experiment than if you go through rather elaborate physical measurements of the photoreceptor processes.

J. Best: I wonder if Dr. Brown has any evidence of the spectral sensitivity of the planarian eye. The reason I ask is that we run our worms under red

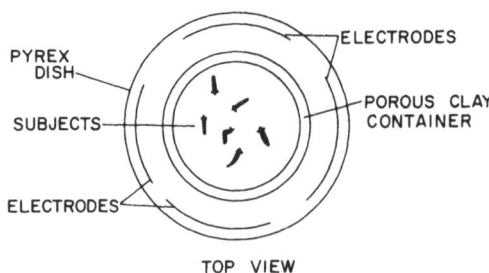

TOP VIEW

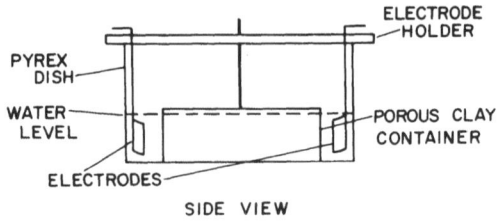

SIDE VIEW

Fig. C. Training apparatus for planarians (Corning and Freed, unpublished).

lighting conditions when we wish to observe them in conditions of darkness, and we really don't know what we are doing to the photoreceptors. Our rationale, of course, was based on the assumption that red light is not visible to them.

H. Brown: Yes, we have some preliminary data to indicate that the peak sensitivity is at approximately 510 mμ. There also appears to be a small peak in the blue region of the spectrum (460 mμ). Sensitivity falls off very rapidly in the red region of the spectrum.

W. Corning: One final comment about the programming of trials. We did succeed in devising a fairly automatic apparatus at Brookhaven which kept the animals oriented in particular ways with respect to the electrodes and delivered trials independent of the experimenter. (Details of these procedures may be found in BNL Report No. 981 which was prepared by Dr. S. Freed; see also Fig. C.) Essentially, we used a porous clay wall to keep the animals from climbing on the electrodes. The polarity of the shock was alternated to prevent any orientation of the animals, and the entire trough was covered with a plastic shield to minimize heat effects. We were able to get fairly clean conditioning results even with mass training.

D. Jensen: We have developed a similar apparatus which is basically a circular maze mounted on a turntable so that I can keep the animals equidistant from the electrodes. I also have a dissecting microscope mounted over the apparatus so that I can easily follow what the animal is doing.

S. Coward: I have one comment concerning starvation and its effect upon uptake of isotopes. I think one generally finds that the longer they are starved, the more readily uptake occurs.

J. Best: By just letting planarians sit in L-leucine for periods up to 3 months, we did not get any uptake and this was with regenerating, nonregenerating, fed, and starved animals. We did get incorporation if we preincorporated it into yeast, acetone-dried the yeast, and mixed it with dried liver powder. We did have some success with uracil when we would preincorporate the uracil into a mutant of *E. coli* that had a uracil requirement and then feed this cellular content from *E. coli*. This did seem to follow the distribution of RNA in the flatworm.

F. Crawford: On the basis of our studies with planarians I cannot tell whether the animals are responding to the onset or offset of light. In our operant situations they continue, for the most part, to circle within the apparatus. Vattano and Hullett reported a study (1964) in which they varied the intertrial interval for forward, backward, and simultaneous conditioning. In forward conditioning they used two groups, one with a 2-second CS duration and the other with a 3-second duration. Their forward conditioning groups were significantly greater in number of conditioned responses than a pseudoconditioning group, and the group with the 3-second CS duration demonstrated the highest number of responses. I presently am using extinction as a measure in my experiments and I use a 3-second duration, because it has been my observation that this is the duration within which the animal typically responds.

E. Halas: You might have better incorporation of thymidine if you block any other way of the cell's getting thymidine. This means blocking thymidilic synthetase with methopterin.

J. Best: The whole advantage of the tracer method is to get turnover measures in cells with a minimum of disturbance and to introduce deliberately such a drastic disturbance seems to violate the purpose of the technique.

REFERENCES

Best, J. B., Diurnal cycles and cannibalism in planaria. *Science*, 1960, **131**, 1884.

Corning, W. C., Habituation in the horseshoe crab. Paper delivered at Midwestern Psychological Association, Chicago, May, 1965 (paper available on request).

Lender, T., Mise en évidence et propriétés de l'organisime de la régénération des yeux chez la Planaire *Polycelis nigra. Rev. Suisse Zool.*, 1955a, **62**, 268.

Lender, T., Sur quelques propriétés de l'organisime de la régénération des yeux de la Planaire *Polycelis nigra. C. R. Acad. Sci.*, 1955b, **240**, 1726.

McWhinnie, M. A., and M. M. Gleason, Histological changes in regenerating pieces of *Dugesia dorotocephala* treated with colchicine. *Biol. Bull.*, 1957, **11**, 371.

Vattano, F. J., and J. W. Hullett, Learning in planarians as a function of interstimulus interval. *Psychon. Sci.*, 1964, **1**, 331.

Wolff, E., Recent researches on the regeneration of planarians, *in* D. Rudnick, ed., *Regeneration*. New York: Ronald Press Co., 1962.

Planarian Research

Demonstrations of Learning and Problems of Interpretation

Chapter 13

Classical Conditioning and the Planarian[1]

Allan L. Jacobson

Department of Psychology
University of California
Los Angeles, California

In several previous papers (1963, 1965, and 1966), I have compiled and attempted to evaluate the evidence for learning in planarians. During the last few years, several particularly convincing demonstrations have all but laid to rest the notion advocated by at least a few scientists that, because of its simple organization, the planarian must be entirely ineducable. This position, however reasonable on a priori grounds, must eventually yield to empirical evidence. And, as Warren has stated, "Overall, the evidence suggests very strongly that learning occurs in flatworms" (1965, p. 100).

I shall concern myself today only with the topic of classical conditioning in planarians. This type of training has assumed a certain prominence in that the first demonstrations of savings after regeneration (McConnell, Jacobson, and Kimble, 1959), of cannibalistic transfer (McConnell, 1962), and of "transfer by injection" (Fried and Horowitz, 1964, and Zelman, Kabat, Jacobson, and McConnell, 1963) all employed classical conditioning. Furthermore, this technique will undoubtedly be used in future research on biochemical correlates of learning in planarians, since it permits both training of animals en masse and a high degree of control over total stimulation (Jacobson, Fried, and Horowitz, 1966a and b).

Even though Thompson and McConnell's (1955) report of classical conditioning in planarians is now more than 10 years old and even though considerable effort has been expended both in defense of and in attack upon their interpretation, the question of the validity of classical conditioning in planarians has certainly not been resolved to everyone's satisfaction. My thesis today is that much of the confusion in the situation has resulted from three sources: (1) incorrect interpretations of data and failure to describe the literature accurately, (2) misuse of certain concepts and terms pertinent to conditioning, and (3) ambiguities and issues relevant to the conception and definition of classical conditioning in general. I shall

[1]The research described in this paper was supported by University of California Research Grant 2068.

attempt to clarify these issues and to present new data from my laboratory which will, I think, contribute to their resolution.

MEASURES OF CONDITIONING

Several years ago, Baxter and Kimmel (1963) reported some interesting observations: Planarians given paired light and shock showed a progressive increase in responding to light, whereas a control group given unpaired light and shock showed a decline in responding to light. When tested 24 hours later in extinction, however, the groups did not differ significantly. These findings have led some scientists to reach conclusions which are simply incorrect. Jensen, for example, basing his conclusions largely on Baxter and Kimmel's failure to find extinction differences, claimed "... the recent work on classical conditioning in Planaria is another instance of pseudo-learning based upon confounding, ignoring certain aspects of the data, and special pleading" (1965, p. 15). James and Halas (1964) also have opposed the notion of conditioning in planarians, largely on the basis of their own failure to find extinction differences.

As I have pointed out elsewhere (Jacobson, 1965, p. 77), the Jensen and the James and Halas position is clearly refuted by both logic and data. Acquisition and extinction are different measures of the "learning process." If I may quote from myself, "It is not unusual for different measures of learning to 'disagree' (Kimble, 1961), and the acquisition measure is commonly accepted as standard in classical conditioning (Kimble, 1961; Pavlov, 1927)" (Jacobson, 1965, p. 77).

Aside from this rather straightforward logical point, Jensen and James and Halas ignored the fact that other investigators (Barnes and Katzung, 1963; Corning and John, 1961; Griffard, 1963; and Vattano and Hullett, 1964) *have* found distinct differences in extinction between classical conditioning groups and appropriate control groups. I shall present some new data later which also illustrate extinction differences due to experimental treatment. A comparison of these several successful extinction experiments with the two negative reports mentioned above reveals an interesting discrepancy: In all cases where differences were observed, rather high levels of responding were achieved during acquisition; in both cases where differences were not observed, only low levels of responding were achieved during acquisition.

Seen in this light, the problem disappears. There is ample evidence of conditioning in planarians, in either acquisition or extinction measure.

BENNETT AND CALVIN ON CLASSICAL CONDITIONING IN PLANARIANS

Recently, Bennett and Calvin (1964) reported a supposed "failure to train planarians reliably." This paper, although published in a journal of limited circulation, has received widespread attention and has cast some doubt upon the entire planarian venture. A meeting such as the present one would seem a perfect occasion to consider the validity and significance of Bennett and Calvin's conclusions. Since Reeva Jacobson Kimble and I gathered some of the data reported in that paper, I feel particularly qualified to conduct this examination. In keeping with the theme of this paper, I shall consider only those data pertinent to classical conditioning. McConnell (1965a and b) has commented admirably and at length upon the entire Bennett and Calvin report.

First, let me quote the conclusion of the Bennett and Calvin paper: "Unsuccessful efforts to demonstrate learning in planarians by classical conditioning, light habituation, maze-learning, operant conditioning, and food reward methods are described. Until more adequate and more useful methods and descriptions of methods to train planarians are available, this animal appears to have little utility for proposed studies of the possible biochemical bases of learning." Let us now examine some of the data upon which these rather strong assertions are based.

My Fig. 1 presents in graphic form some of the data from Bennett and Calvin's Table 1. The increase in responding from initial to terminal level is seen to be regular and highly significant. This high level of responding was maintained during the 7 additional trial sets reported, but, since some animals were not run past set 13, I have terminated the graph at this point. On trial set 1, the range of responses was 2 to 7 for 10 animals. On trial set 13, one animal scored 7; the remaining nine animals scored from 12 to 20. Bennett and Calvin concluded, ". . . an increase in the average response rate from 20 to 60 % of the total number of trials was observed, but this was significantly less than the preset criterion of 92 %".

My Fig. 2 presents in graphic form some of the data from Bennett and Calvin's Table 2. In this case, 20 trials per day were given, and the incidence of responding increased regularly from 9.7 to 63.5 % in the first seven trial sets. Of the 13 animals tested, 9 reached a preset criterion of 18 responses on any 20 consecutive trials, and the other 4 animals all approached this criterion level.

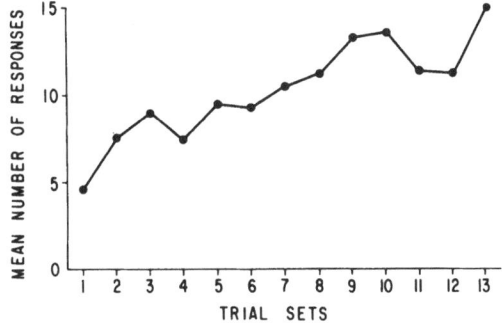

Fig. 1. Mean number of responses per set of 25 trials. (Data from Bennett and Calvin, 1964, p. 6.)

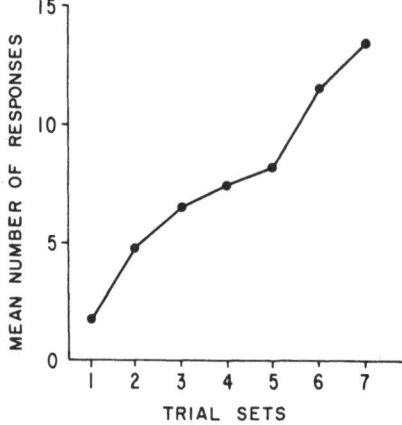

Fig. 2. Mean number of responses per set of 20 trials. Subject 3, which reached criterion on day 7, was assigned a criterion score for that day's set. (Data from Bennett and Calvin, 1964, p. 7.)

In the face of these data, we can only conclude with McConnell that "... 'success' (apparently) means something quite different to Bennett and Calvin from what it means to most other scientists in this field" (1966, p. 119). A noted biochemist admitted to me that he and his friends had accepted Bennett and Calvin's conclusions without troubling to consult the data in question. It is unfortunate but true that the inadequate interpretations which Bennett and Calvin present will probably obscure more valid questions about planarian conditioning for some time to come.

THE NEUTRALITY OF THE CONDITIONED STIMULUS

It has been asserted (e.g., Halas, James, and Knutson, 1962, and Halas, James, and Stone, 1961) that classical conditioning cannot be considered a valid phenomenon in planarians because the typical conditioned stimulus (light) is not initially neutral, i.e., the light initially evokes some responses, although admittedly at a low level. Let us examine this viewpoint more carefully.

Customarily, introductory psychology texts state that the conditioned stimulus in classical conditioning is originally neutral. Most of them then go on to point out that the stimulus is not "absolutely" neutral, since it typically evokes an orienting reaction, but it is neutral with respect to the response to be conditioned. The introductory texts, however, fail to consider the logical and practical implications of this position. In the first place, if we adhere to this conception strictly, we must disqualify as classical conditioning most activities that are currently pursued in the name of classical conditioning. The conditioned stimulus is hardly neutral, e.g., in the conditioning of galvanic skin response (GSR), eyelid closure, heart rate changes, and vasomotor reactions. Logically, then, since we all agree that our terms and concepts should be consistent, we should be sacrificing a great deal if we were to use the neutrality criterion to reject planarian conditioning.

The folly of such a position becomes even more apparent when we recognize the oversimplification inherent in the concept of neutral stimulus. Is a stimulus ever truly neutral? Even if it does not evoke a measurable reaction, how do we know that our measuring instruments are sensitive enough? Furthermore, assuming a stimulus could be shown to be absolutely neutral, it should be recognized that the introduction of the unconditioned stimulus into the situation might cancel this neutrality even in the absence of specific conditioning.

It is precisely for these reasons that American psychologists have not been impressed by the conditioned-stimulus neutrality concept and that they continue their GSR and heart-rate conditioning work. We should require some powerful arguments before we categorically reject an enormous and informative body of data. The concept of conditioned-stimulus neutrality as a necessary component of conditioning is arbitrary and not particularly useful. Classical conditioning is best defined as response changes occurring as a result of certain operations. These changes are distinguished by definition from response changes produced by other, control operations, which I shall discuss shortly. Nowhere is it implicit in this conception that the initial level of response to the conditioned stimulus be zero. What is required is that the initial probability of response be altered by appropriate operations.

I should like to illustrate the contention of this section with some of my group's recent data on planarians. A number of animals were trained with light as the conditioned stimulus. Initially, each animal received 30 trials of light alone. During these adaptation trials, the number of responses ranged from 0 to 2 for different animals. In subsequent training, all subjects showed marked conditioning, and there were no apparent differences between the animals which initially scored 0 and those which initially scored 1 or 2, although for the former group the conditioned stimulus was initially neutral, technically speaking, while for the latter group it was not.

PSEUDOCONDITIONING

I should now like to clarify two terms which have been used very carelessly in discussions of planarian learning, *pseudoconditioning* and *sensitization*. In some contexts, these terms would seem to have been applied to any data that look like learning but occur in a planarian. Despite this pejorative usage and despite the diversity of usages in the literature, I think we can usefully distinguish these terms from each other and from classical conditioning. In doing so, we can perhaps arrive at generally satisfactory criteria for recognizing classical conditioning in the planarian, or in any other organism.

In the first place, a distinction should be maintained between two common usages of the term pseudoconditioning: One usage applies the label to certain operations, procedures, and stimulus arrangements; the other usage refers to outcomes, effects, and results. Thus it is entirely possible that a pseudoconditioning control group will show no pseudoconditioning at all, just as an extinction procedure may yield no decrement in responding.

We may also distinguish a narrow from a broad application of the pseudoconditioning label, and this distinction will apply to pseudoconditioning as both procedure and effect. Let us imagine a classical aversive conditioning situation, in which the experimental group receives the typical forward conditioned-unconditioned stimulus pairing. By a narrow usage of the term, either of the following would constitute pseudoconditioning controls: (1) subjects given randomly interspersed presentations of a conditioned and unconditioned stimulus, or (2) subjects given an entire series of unconditioned stimuli, followed by a test series of conditioned stimuli. Any augmentation of responding to the conditioned stimulus which occurred as a result of these treatments would be defined as a pseudoconditioning effect. This usage corresponds to Hilgard & Marquis' definition of pseudoconditioning: "The strengthening of a response to a previously neutral stimulus through the repeated elicitation of the response by another stimulus, without paired presentation of the two stimuli" (1940, p. 348). Similarly, Gormezano states: "When a noxious UCS [unconditioned stimulus] is presented one or several times prior to presentation of the CS [conditioned stimulus], the procedure will frequently result in the occurrence of a response that is apparently indistinguishable from a CR [conditioned response]. Traditionally such a procedure is called pseudoconditioning and the resulting responses are treated as separate from those acquired by classical conditioning *per se* because of their occurrence in the absence of previous CS-UCS pairings" (1966, p. 389).

Some writers treat the concept of pseudoconditioning more broadly, and it is this usage which has been prevalent in the planarian literature. Thus Beecroft asserts, "A pseudoconditioning effect may be defined as an inflation of response strength, usually temporary, that is attributable to some procedure other than pairing CS and UCS in a forward temporal order" (1966, pp. 15 to 16). By this definition, pseudoconditioning becomes more inclusive, by encompassing the procedures and effects of backward and strictly simultaneous conditioning.

Given these two conceptions of pseudoconditioning, several points deserve emphasis. First, it is apparent that the term pseudoconditioning is purely descriptive. Even by the narrow conception, the two procedures described above differ importantly in that one permits adaptation to the conditioned stimulus, whereas the other does not. Thus the two procedures might well, and sometimes do, yield differing results (e.g., Harris, 1943). A more serious problem is that pseudoconditioning effects, when observed, are little understood. Kimble (1961, pp. 60 to 64) has discussed this topic in some detail and has lamented the paucity of evidence on the subject. Wickens and Wickens (1942), among others, have interpreted pseudoconditioning as a special case of true conditioning, and their research bears out this contention. May (1949) classified explanatory hypotheses about pseudoconditioning effects as "associative" or "nonassociative." In the former class he included explanations of primary stimulus generalization, conditioned trace stimuli, or situational conditioning. Nonassociative explanations included sensitization, response dominance, generalized motor set, and expectancy.

When we consider the broader conception of pseudoconditioning, the complexities are increased enormously. Backward conditioning, in particular, requires separate analysis. Although this procedure typically leads to little conditioning (Kimble, 1961), recent Russian research (described by Asratyan, 1961) indicates that in certain cases enhanced responding may be produced by backward conditioning procedures. Similarly, the results of strictly simultaneous presentation of conditioned and unconditioned stimuli might require a different interpretation than the results of unconditioned stimulus only or unpaired conditioned and unconditioned stimulus presentations.

To summarize the foregoing, I should say that pseudoconditioning is an ambiguous and poorly understood concept. It would be dangerous to assume that we know a great deal about the subject. Our task is greatly simplified, however, if our primary interest is not so much in interpreting pseudoconditioning effects, but rather in distinguishing "true," forward, classical conditioning effects on the one hand from pseudoconditioning effects broadly conceived, on the other. If we set out to distinguish true from pseudoconditioning effects, then, regardless of the explanation of pseudoconditioning, two outcomes can answer our question. In one case, we may observe enhanced responding to the conditioned stimulus with a forward conditioning procedure, but not with backward, simultaneous, or random procedures. In the other case, enhanced responding may occur in one or more of the control groups, but to a significantly lesser extent than in the forward group.

Of the possible controls, the most stringent, in the opinion of many psychologists, is the strictly simultaneous procedure, in which the onset of conditioned and unconditioned stimuli occurs simultaneously, and test trials of conditioned stimulus only are used to assess effects. As Beecroft has written, "Since the power of any pseudoconditioning control should be its ability to detect no conditioning when there is none, the simultaneous procedure appears to be the best pseudoconditioning control" (1966, p. 19). It should be repeated that response enhancement produced by a simultaneous conditioning procedure may merit analysis in its own right. The point is that, if differences do occur between forward and simultaneous groups, we have a powerful way of ruling out pseudoconditioning. Also of great value are random and backward procedures. The unconditioned-stimulus-only procedure suffers from the shortcoming that exposure to the conditioned stimulus is not equated to that for the forward group; nonetheless, this procedure can provide interesting data for analysis of pseudoconditioning proper.

Several examples from the recent conditioning literature will be helpful here. In some cases, little difficulty is encountered in distinguishing pseudo-conditioning controls from the forward conditioning group. Gormezano obtained clear-cut differences between forward and random conditioned-unconditioned stimulus groups in nictitating membrane work (Gormezano, Schneiderman, Deaux, and Fuentes, 1962). Bitterman (1964) found no effects of random conditioned-unconditioned or simultaneous conditioned-unconditioned stimulus presentation in his goldfish conditioning. And Kamin (1963), in one of his studies employing conditioned suppression as a monitor of the conditioned emotional response, observed an extremely clear difference between the results of forward and backward conditioning procedures.

Other response systems do not always fare this well. Shmavonian (1959) reported conditioning of vasomotor responses in humans, but Stolz (1965) failed to find systematic differences in vasomotor response between a forward conditioning group and a control group given randomly interspersed conditioned and unconditioned stimuli. Stolz concluded, "... previous experiments purporting to demonstrate vasomotor conditioning should be reexamined, for the earlier results may be artifactual ..." (1965, p. 182).

The analysis of GSR conditioning has similarly been unclear in some cases. A few studies are of particular interest here, although they tell only a portion of the story on this subject. Champion and Jones (1961) tested three groups of humans on test trials during training: a forward conditioning group, a backward group, and a random conditioned-unconditioned stimulus group. The forward group differed significantly from the random group, but the performance of the backward group was intermediate. According to Champion and Jones, "The general conclusion to be drawn is that pseudoconditioning of the GSR is a basically different process from that occurring in forward and backward conditioning" (1961, p. 60). They believe that the response enhancement in the backward group may represent some degree of true conditioning.

On the other hand, Prokasy, Hall, and Fawcett (1962) found somewhat different results in another study on GSR. Various groups were given forward conditioning, backward conditioning, random conditioned-unconditioned stimulus, conditioned stimulus only, or unconditioned stimulus only, and all were subsequently tested in a series of extinction trials. On the first extinction test trial, the forward group differed significantly only from the conditioned-stimulus-only group. Over the course of extinction trials, the forward group was clearly superior to the conditioned-stimulus-only, backward, and unconditioned-stimulus-only groups but was less easy to distinguish from the random groups. Although there are several procedural differences between the experiments of Champion and Jones and of Prokasy et al., it is nonetheless apparent that assessment of pseudo-conditioning effects in GSR conditioning can be quite tricky. It would be helpful to know the extent to which discrepancies in these experiments are due to test procedures. An experiment which gathered data during both acquisition and extinction with the same subjects would help to resolve this question. One study of this sort (Prokasy and Ebel, 1964) has indicated that pseudoconditioning effects in GSR can be distinguished from true conditioning effects on the basis of several measures in both acquisition and extinction. Kimmel (1959) has contributed an exceedingly interesting observation. When a weak conditioned stimulus was used, forward conditioning was significantly superior to random conditioned-unconditioned stimulus presentation; but this difference did not occur when stronger conditioned stimuli were employed.

SENSITIZATION

Like pseudoconditioning, sensitization has suffered a variety of usages, sometimes being considered more inclusive than pseudoconditioning, sometimes synonymous with it, and sometimes distinct from it. The last usage is the most historically accurate and the most useful. First, it is clear that sensitization refers to an effect rather than a procedure. In fact, the procedure for obtaining sensitization would appear to be identical to that for obtaining classical conditioning. Kimble defines sensitization as "the augmentation of the response to the conditioned stimulus through a conditioning procedure" (1961, p. 64). This could as easily, of course, be a definition of true conditioning. To understand the distinction, we must examine the development of the concept.

As Kimble relates (1961, pp. 64 and 65), sensitization was first formulated in connection with knee-jerk, eyelid, and startle-response conditioning and has been used since then primarily in eyelid-conditioning work. Responses initially evoked by the conditioned stimulus are called alpha responses. Supposedly, in the response systems described above, forward conditioning procedures can enhance alpha responses, and these augmented alpha responses can be distinguished from true conditioned responses, primarily on the basis of latency. According to Gormezano, "Since alpha responses are usually of shorter latency than CR's, both responses can be observed and scored if a sufficiently long CS-UCS interval is employed" (1966, pp. 388 and 389). In order to distinguish sensitization from pseudoconditioning, we must require that the augmentation of responses to the conditioned stimulus occur only in forward-conditioning procedures. Hilgard and Marquis make this same point when they state that the augmented response "... is merely a form of pseudoconditioning, unless it can be demonstrated that the sensitization is a function of the repetition of conditioned and unconditioned stimuli in precise relationship" (1940, p. 42).

As I have indicated, alpha conditioning or sensitization has been elaborated primarily for eyelid conditioning. In the last few years, however, attempts have been made to distinguish alpha responses from conditioned responses in GSR conditioning. Stewart, Stern, Winokur, and Fredman

(1961) believe that, as in eyelid conditioning, latency measures can be used in GSR work to separate the two types of responses. Short latency responses, they claim, represent sensitization effects, while a longer-latency second response can be considered a true conditioned response. This argument leads them to conclude that previous work on GSR conditioning "has dealt with the adaptation and recovery of unconditioned responses rather than conditioning of responses" (1961, p. 66). A recent study by Leonard and Winokur (1963) tends to support Stewart *et al.*'s distinction: When differential reinforcement was applied to two tones, equal numbers of short-latency responses occurred to the two tones, but significantly more long-latency responses occurred to the positive than to the negative tone.

Stewart *et al.*'s analysis of GSR conditioning has been criticized rather severely and cogently by Lockhart and Grings (1963) and by Kimmel (1964), who see inadequacies in Stewart *et al.*'s distinction between the functional characteristics of short- and long-latency responses. For example, Lockhart and Grings point out that the first response can show discriminative properties, especially when magnitude of response (rather than frequency) is used as a measure. This clearly suggests that the response represents more than generalized, indiscriminate sensitization. Also, the first and second responses correlate to a very high degree ($r = 0.90$). Kimmel presents evidence suggesting, ". . . the true conditioned response identified by Stewart *et al.* may actually be a part of the earlier response to the CS which they reject as a sensitized UCR . . ." (1964, p. 165).

A very basic point is involved here. The most compelling reason for distinguishing between short- and long-latency responses would seem to be any functional characteristics that differentiate the two responses. If functional differences do not exist, i.e., if both responses exhibit the properties of conditioned responses, the distinction could well be considered arbitrary and trivial. The question becomes: How can we best experimentally distinguish true conditioned responses from sensitized responses? Lockhart and Grings (1963), Kimmel (1964), and Gormezano (1966) all argue for the use of a differential reinforcement technique to distinguish sensitization from true conditioning. That is, one stimulus would be paired with the unconditioned stimulus, and the other stimulus in the same series would not. Conditioning would be evidenced by ". . . a difference in responding favoring the positive stimulus." Lockhart and Grings put the matter this way: "The magnitude of response to the nonreinforced stimulus is an estimate of sensitization effects, while the magnitude of response to the reinforced stimulus contains both sensitization and conditioning components. The difference between the two response magnitudes, then, may be taken as a measure of conditioning independent of sensitization" (1963, p. 563). The same argument, of course, applies to frequency data.

This line of reasoning, upon which all these writers converge, proposes that sensitization is a nonspecific phenomenon. I have contended that this must be the case for the term to have any value. It would be possible, if one were so inclined, to try to rescue the term, given a discriminative response, by speaking of "stimulus-specific sensitization"; but at this point the term would seem to have lost all meaning as a concept distinct from conditioning. There is, then, reason to question the applicability of the sensitization concept to situations other than eyelid conditioning. Whatever the eventual outcome of this issue, however, the differential reinforcement procedure would appear to provide the definitive method for demonstrating true classical conditioning effects.

PSEUDOCONDITIONING AND THE PLANARIAN

Essentially all the types of pseudoconditioning procedures have been employed with planarians, although, as in studies of other systems, rarely more than one in a given experiment. Thompson and McConnell (1955) found, using interspersed test trials, that responsiveness to light declined slightly when repeated shocks were administered. Baxter and Kimmel (1963) equated the number of exposures to conditioned and unconditioned stimuli in a "random" group with those in a group given paired light and shock stimuli, and they obtained a clear differentiation in acquisition in favor of the paired group. Fried, Horowitz, and I (1966a) used a similar procedure and found an even more pronounced differentiation (see Fig. 3). Subsequent chemical work in this experiment verified the difference between the two groups, although I mention this primarily as an incidental observation.

Griffard (1963) observed a clear difference between groups given conditioned-unconditioned stimulus pairings and those given unconditioned-stimulus-only presentations. There was some suggestion of enhanced responding in the pseudoconditioning group [see Griffard's (1963) Table 2, p. 599], but the amount was not significant and the level fell far short of that for the conditioning group. Crawford, King, and Siebert (1965) found that both a continuous-reinforcement group and a partial-reinforcement group were clearly distinguishable from two unpaired conditioned-unconditioned stimulus control groups. The former groups showed an incremental learning curve, whereas the latter groups showed no increase in response to light. Extinction levels differed accordingly, and it is interesting that the partial-reinforcement group extinguished more slowly than the continuous-reinforcement group.

Recent studies from two laboratories have provided data on several types of pseudoconditioning groups run concurrently. Vattano and Hullett (1964) compared the performances in a series of extinction trials of two forward conditioning groups, two backward conditioning groups, one simultaneous group, and one random conditioned-unconditioned stimulus group. For longitudinal contractions, the forward conditioning groups proved to be significantly superior to all other groups. This was not the case for turns. Since contraction is the unconditioned response to shock, whereas turning is the typical alpha response to light, Vattano and Hullett have concluded, ". . . the contraction response is a more valid measure of

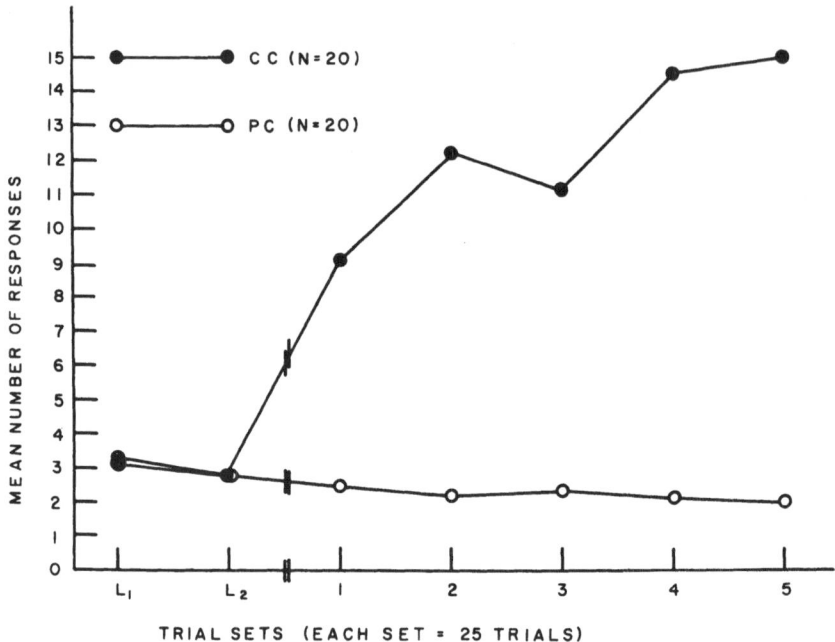

Fig. 3. Mean number of responses for classical conditioning group (solid circles) and random conditioned-unconditioned stimulus group (open circles). On the abscissa, L_1 and L_2 are trial sets in which light alone was presented, and 1 to 5 represent training sets. (From Jacobson, Fried, and Horowitz, 1966a.)

conditioning than is the cephalic turn" (1964, p. 332). My co-workers and I have for some time concentrated on the contraction response for the same reason (Jacobson *et al.*, 1966a and b), and Griffard (1963) also required contractions to occur for a positive response to be scored. This argument does not apply, of course, when the unconditioned stimulus evokes turning rather than contraction (Griffard and Peirce, 1964).

Recently, Horowitz, Fried, and I undertook to replicate the Vattano and Hullett findings, using a blind testing procedure. In our first experiment, four groups were trained: forward conditioning *FC*, simultaneous conditioning *SC*, backward conditioning *BC*, and light only *CS*. The animals were then tested with light only for 25 trials. Figure 4 shows the levels of responding by the several groups when contractions were scored as responses. The means were 4.7, 1.3, 1.6, and 1.5 for groups *FC*, *SC*, *BC*, and *CS*, respectively. This highly significant result ($F = 24.4$; $df = 3/75$; $p < 0.001$) clearly reflects a difference between the *FC* group on the one hand, and the several control groups, on the other. As Fig. 5 shows, the data for turns show little, if any, systematic variation ($F = 1.14$). Thus our findings corroborate quite nicely those of Vattano and Hullett. We

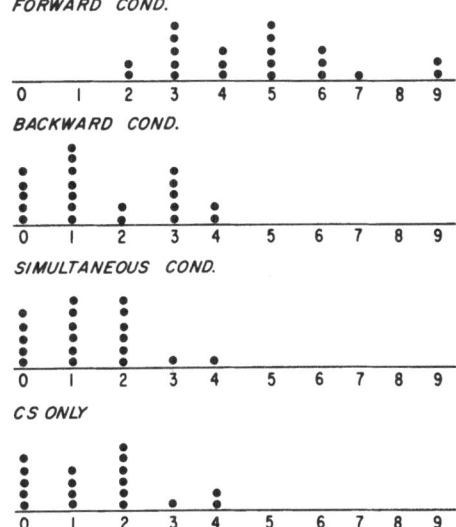

Fig. 4. Distribution of scores for the classical conditioning group and the three control groups. Each circle represents the number of contractions made by a given subject on the test series of 25 light presentations.

have also observed that of 12 planarians given 30 presentations of light alone, all 12 made more turns than contractions. No animal in this test made more than 2 contractions over the course of 30 trials. (These worms were subsequently conditioned in an experiment which I shall describe shortly.)

Since the contraction response is appropriate to the unconditioned stimulus we employ, whereas the turning response is appropriate to light,

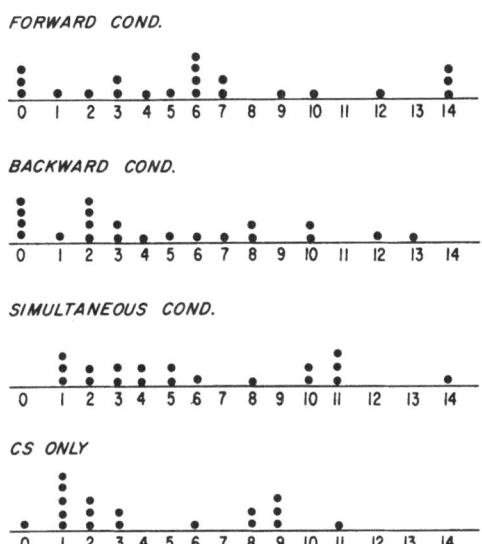

Fig. 5. Distribution of scores for the classical conditioning group and the three control groups. Each circle represents the number of turns made by a given subject on the test series of 25 light presentations.

we have chosen to define only contractions as conditioned responses in our work. I might add also that in this respect light is more desirable than vibration as a conditioned stimulus for planarians, since the dominant alpha response to vibration is contraction. Of the same 12 animals mentioned above, 11 made more contractions than turns to vibration (the twelfth animal responded an equal number of times to both). Although, as Kimmel (1964) has claimed, conditioning can still be demonstrated when the conditioned and unconditioned stimuli elicit qualitatively similar responses prior to training, nevertheless, given a choice, one would probably prefer a stimulus which evokes an alpha response dissimilar from the unconditioned response.

This analysis of light and vibration is supported by several further experiments which Fried, Horowitz, and I have performed. We compared the extinction responding of animals given forward conditioning, simultaneous conditioning, or conditioned stimulus only. In four different experiments we used either of two intensities of light (75 versus 150 watt) and either of two intensities of vibration ("low" versus "high") as conditioned stimuli. All testing was, as usual, conducted in blind fashion.

Figure 6 shows the results of the low-intensity light experiment. The overall differences between groups are highly significant ($p < 0.001$), and it is clear that the forward conditioning group is superior to either of the control groups. Figure 7 shows corresponding data for the high-intensity light experiment. The results are equally clear-cut here, and it is noteworthy that in neither this nor in the low-intensity experiment is there any indication that the simultaneous procedure affects responsiveness to the conditioned stimulus.

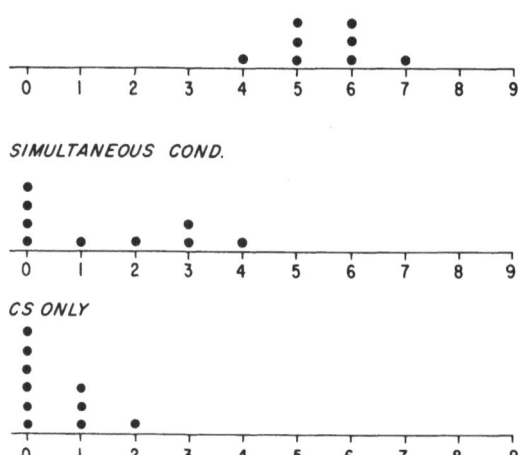

Fig. 6. Distribution of scores when low-intensity light was used as the conditioned stimulus. Each animal received 20 test trials.

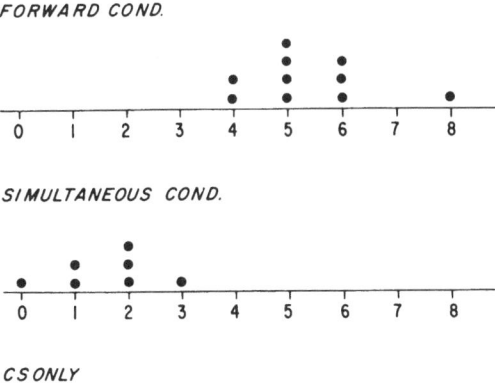

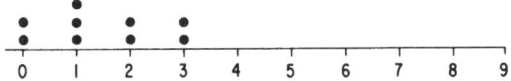

Fig. 7. Distribution of scores when high-intensity light was used as the conditioned stimulus.

The situation for vibration is more complicated. Figure 8 presents the data for high-intensity vibration. The over all groups comparison is highly significant, and a finer analysis shows that the forward group differs from the simultaneous group, which in turn differs significantly from the conditioned-stimulus-only group (Duncan's new multiple-range test). The major difference is a conditioning effect, but there is some response enhancement, whether true or pseudoconditioning, in the simultaneous-presentation condition. Figure 9 shows that, with a lower intensity of vibration, any systematic variation there is becomes quite difficult to interpret. The

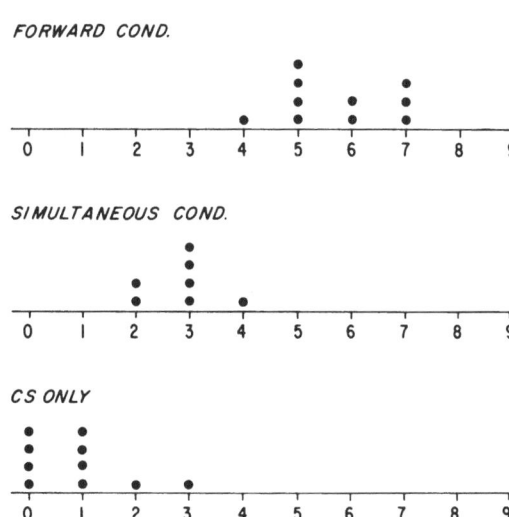

Fig. 8. Distribution of scores when high-intensity vibration was used as the conditioned stimulus.

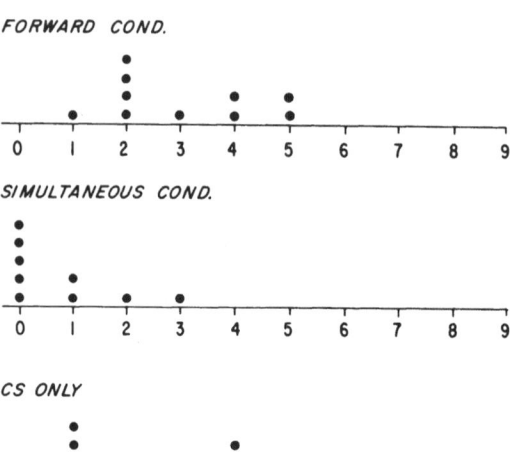

FORWARD COND.

SIMULTANEOUS COND.

CS ONLY

Fig. 9. Distribution of scores when low-intensity vibration was used as the conditioned stimulus.

overall F is significant ($p < 0.01$), and the forward group differs significantly from the simultaneous group, but the conditioned-stimulus-only group is unaccountably intermediate.

It is possible that these results reflect only the properties of the particular levels of stimuli we chose. More likely, though, since we used behavioral criteria in selecting intensities, our data indicate that light is a better conditioned stimulus than vibration for planarians. Still, vibration can be a perfectly suitable conditioned stimulus, as the results of the high-intensity experiment show.

SENSITIZATION AND THE PLANARIAN

Several reports of differential conditioning in planarians have appeared in the last 2 years. Griffard and Peirce (1964) successfully demonstrated conditioned discrimination by using directional shock as the unconditioned stimulus. Light preceded a forced turn in one direction and vibration preceded a forced turn in the other direction. With continued training, subjects turned more and more frequently to the side appropriate to the conditioned stimulus delivered.

Kimmel and Harrell (1964), Fantl and Nevin (1965), and my own group have all used a discrimination paradigm which, in contrast to the Griffard and Peirce experiment, involves only one type of unconditioned response. Two stimuli, light and vibration, were employed in all these studies, one regularly paired, and the other unpaired, with shock. Kimmel and Harrell obtained differential responding when vibration was the positive stimulus but not when light was positive. They pointed out that the vibration may well have been a stronger stimulus than the light. Fantl and Nevin (1964)

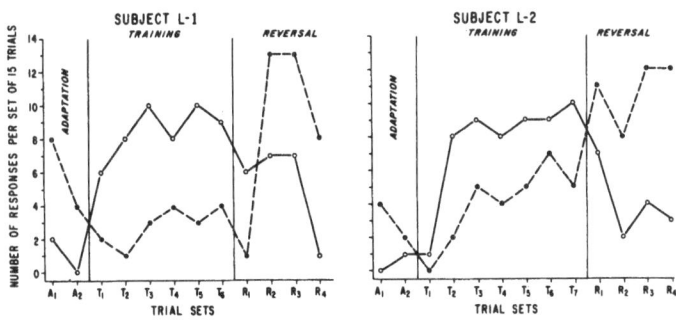

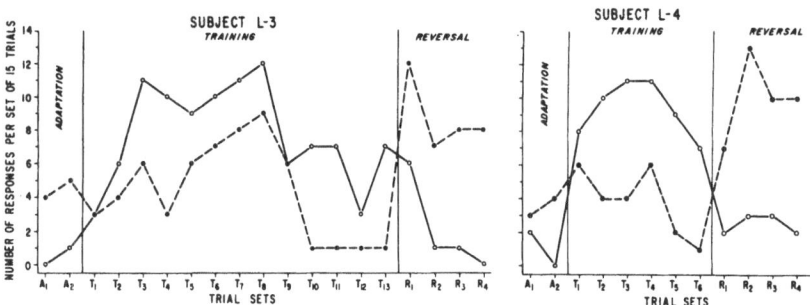

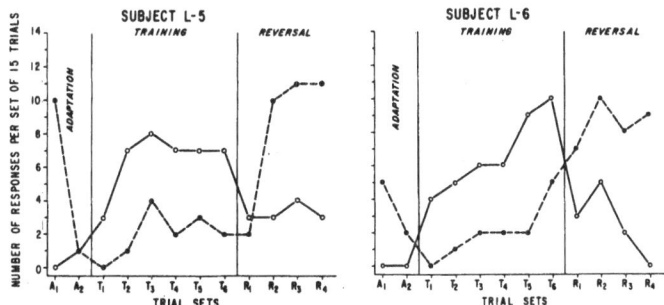

Fig. 10. Responses to light (continuous line) and to vibration (dashed line) for subjects
L-1 to L-6. Each trial set consisted of 15 presentations of light and 15 presentations of
vibration in a fixed, interspersed sequence. During adaptation (trial sets A_1 and A_2),
no shock was given; during training (T_1 to T_n), shock was paired with light; during
reversal (R_1 to R_4), shock was paired with vibration.

seem to have overcome this problem, since they established differential responding in most of their subjects, with either light or vibration as the positive stimulus.

Fried, Horowitz, and I have conducted the following experiment: Initially, worms were adapted to vibration and to light, and their responses were recorded. Then shock was regularly paired with one of the stimuli but not with the other. Training was continued until differential responding had been maintained for several trial sets. At that point, the shock was paired with the previously negative stimulus and not with the previously positive stimulus. Figure 10 shows the individual records of six subjects (Ss) for which light was initially positive. Figure 11 is a composite record for these six Ss. During the first six training sets, these Ss responded significantly more to light than to vibration ($p < 0.001$), and during the four reversal sets they responded significantly more to vibration than to light ($p < 0.001$). Figure 12 shows the individual records of six Ss for which vibration was initially positive. Figure 13 is a composite record for these 6 Ss. Here too, responding to the positive stimulus was significantly greater during both training ($p < 0.001$) and reversal ($p < 0.001$). Details of these new experiments from my laboratory will be presented elsewhere.

We conducted a further experiment in which animals were trained en masse with the differential conditioning procedure. For one group, vibration was positive, and for the other group, light. We then coded the animals and gave each one a blind test series of interspersed lights and vibrations, with no shock. Of 14 Ss for which vibration was positive, 1 responded equally to the two stimuli, and the other 13 all responded more to vibration than to light (sign test, $p < 0.001$). Of 14 Ss for which light

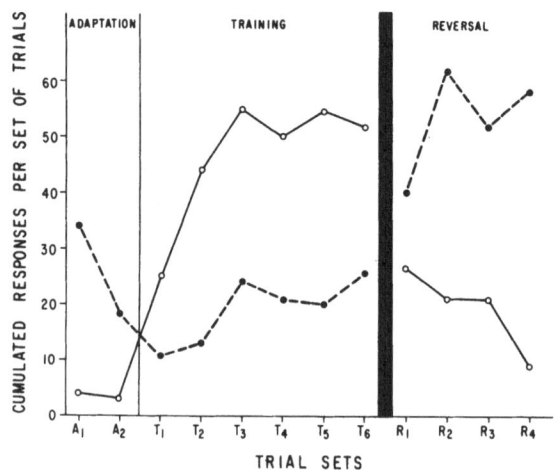

Fig. 11. Responses to light (continuous line) and to vibration (dashed line) for subjects L-1 to L-6 combined. (Training sets beyond T_6 are omitted.)

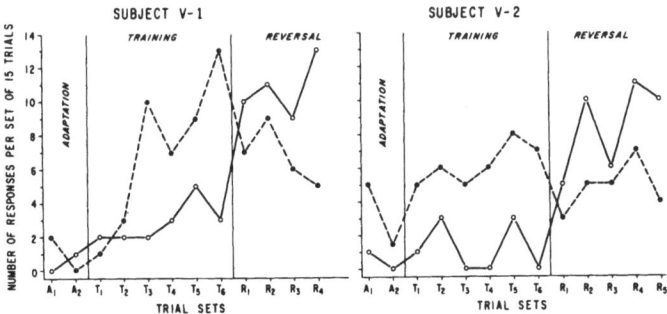

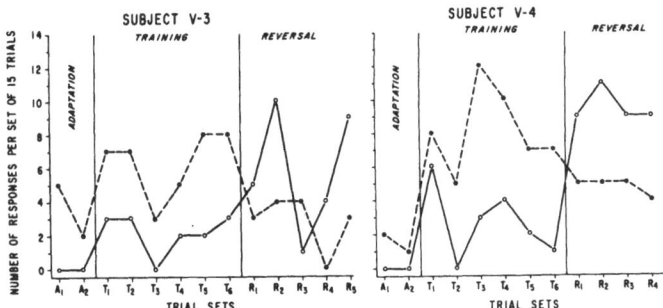

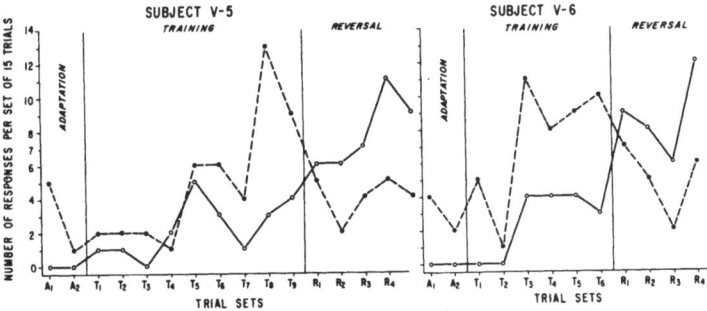

Fig. 12. Responses to light (continuous line) and to vibration (dashed line) for subjects *V*-1 to *V*-6. Each trial set consisted of 15 presentations of light and 15 presentations of vibration in a fixed, interspersed sequence. During adaptation (trial sets A_1 and A_2), no shock was given; during training (T_1 to T_n), shock was paired with vibration; during reversal (R_1 to R_n), shock was paired with light.

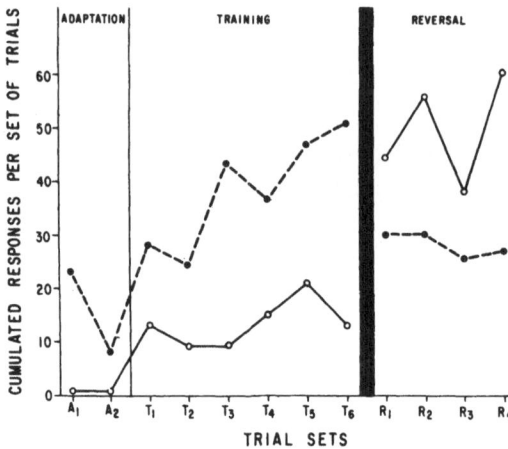

Fig. 13. Responses to light (continuous line) and to vibration
(dashed line) for subjects V-1 to
V-6 combined. Training sets
beyond T_6 and reversal sets beyond R_4 are omitted.

was positive, 2 responded equally to the two stimuli, and 10 of the remaining
12 responded more to light than to vibration (sign test, $p < 0.02$). Differences were small in the light-positive group, since vibration proved to
be a stronger stimulus than light. Despite the fact that total stimulation
was equated for the two groups, the vibration-positive group responded
significantly more to vibration than did the light-positive group ($p < 0.01$),
and, conversely, the light-positive group responded significantly more to
light than did the vibration-positive group ($p < 0.001$).

In a final test of the differential training technique, we assessed reliability of judgment of responses. Two worms were given adaptation and
training trials, one of them with light, and the other with vibration, as
the positive stimulus. Each response was scored independently by two
observers. Agreement ranged from 28/30 to 30/30 on the eight trial sets.
Over all the observers agreed on 230 out of 240 judgments, i.e., on 96%
of the trials.

CONCLUDING REMARKS

To recapitulate, I have discussed and attempted to clarify the conceptions of pseudoconditioning and sensitization. The analysis indicated
that backward, randomly interspersed, and simultaneous stimulus arrangements provide the most stringent controls for pseudoconditioning, and
that the differential conditioning procedure provides the tightest safeguard
against sensitization effects. A review of the literature on classical conditioning in planarians, as well as new data which my co-workers and I
have gathered, reveal that: (1) pseudoconditioning effects have typically
been minimal and easily distinguishable from forward-conditioning
effects; (2) differential conditioning techniques have been shown to produce
appropriate discriminative responding in several different experiments.

It appears reasonable to assert that by the strictest criteria, classical conditioning has been conclusively demonstrated in planarians. More extensive usage of the control procedures discussed in this paper should aid the interpretation of future investigations of the chemical and behavioral properties of conditioning in planarians.

REFERENCES

Asratyan, E. A., Some aspects of the elaboration of conditioned connections and formation of their properties, *in* J. F. Delafresnaye, ed., *Brain Mechanisms and Learning*. Oxford: Blackwell, 1961.

Barnes, C. D., and B. G. Katzung, Stimulus polarity and conditioning in planaria. *Science*, 1963, **141**, 728.

Baxter, R., and H. D. Kimmel, Conditioning and extinction in the planarian. *Amer. J. Psychol.*, 1963, **76**, 665.

Beecroft, R. S., *Classical Conditioning*. Goleta, Calif.: Psychonomic Press, 1966.

Bennett, E. L., and M. Calvin, Failure to train planarians reliably. *Neurosci. Res. Program Bull.*, 1964, **2**, 3.

Bitterman, M. E., Classical conditioning in the goldfish as a function of the CS-UCS interval. *J. Comp. Physiol. Psychol.*, 1964, **58**, 359.

Champion, R. A., and Joan E. Jones, Forward, backward, and pseudoconditioning of the GSR. *J. Exp. Psychol.*, 1961, **62**, 58.

Corning, W. C., and E. R. John, Effect of ribonuclease on retention of conditioned response in regenerating planarians. *Science*, 1961, **134**, 1363.

Crawford, F. T., F. J. King, and Lynne E. Siebert, Amino acid analysis of planarians following conditioning. *Psychonom. Sci.*, 1965, **2**, 49.

Fantl, Stephanie, and J. A. Nevin, Classical discrimination in planarians. *Worm Runner's Digest*, 1965, **7**, 32.

Fried, C., and S. D. Horowitz, Contraction—a leaRNAble response? *Worm Runner's Digest*, 1964, **6**, 3.

Gormezano, I., Classical Conditioning, *in* J. B. Sidowski, ed., *Experimental Methods and Instrumentation in Psychology*. New York: McGraw-Hill Book Company, 1966.

Gormezano, I., N. Schneiderman, E. Deaux, and I. Fuentes, Nictitating membrane: classical conditioning and extinction in the albino rabbit. *Science*, 1962, **138**, 33.

Griffard, C. D., Classical conditioning of the planarian, *Phagocata gracilis*, to water flow. *J. Comp. Physiol. Psychol.*, 1963, **56**, 597.

Griffard, C. D., and J. T. Peirce, Conditioned discrimination in the planarian. *Science*, 1964, **144**, 1472.

Halas, E. S., R. L. James, and C. S. Knutson, An attempt at classical conditioning in the planarian, *J. Comp. Physiol. Psychol.*, 1962, **55**, 969.

Halas, E. S., R. L. James, and L. A. Stone, Types of responses elicited in planaria by light. *J. Comp. Physiol. Psychol.*, 1961, **54**, 302.

Harris, J. D., Studies on nonassociative factors inherent in conditioning. *Comp. Psychol. Monogr.*, 1943, **18**, 93.

Hilgard, E. R., and D. G. Marquis, *Conditioning and Learning*. New York: Appleton-Century-Crofts, 1940.

Jacobson, A. L., Learning in flatworms and annelids. *Psychol. Bull.*, 1963, **60**, 74.

Jacobson, A. L., Learning in planarians: current status. *Anim. Behav.*, *Suppl.*, 1965, **13**, 76.

Jacobson, A. L., Passion and planarians: an editorial. *Worm Runner's Digest*, 1966, **8**, 5.

Jacobson, A. L., C. Fried, and S. D. Horowitz, Planarians and memory, I: Transfer of learning by injection of ribonucleic acid. *Nature*, 1966a, **209**, 599.

Jacobson, A. L., C. Fried, and S. D. Horowitz, Planarians and memory, II: The influence of prior extinction on the RNA transfer effect. *Nature*, 1966b, **209**, 601.

James, R. L., and E. S. Halas, No difference in extinction behavior in planaria following various types and amounts of training. *Psychol. Rec.*, 1964, **14**, 1.

Jensen, D. D., Paramecia, planaria, and pseudolearning. *Anim. Behav. Suppl.*, 1965, **13**, 9.

Kamin, L. J., Backward conditioning and the conditioned emotional response. *J. Comp. Physiol. Psychol.*, 1963, **56**, 517.

Kimble, G. A., *Hilgard & Marquis' Conditioning and Learning*. New York: Appleton-Century-Crofts, 1961.

Kimmel, H. D., Amount of conditioning and intensity of conditioned stimulus. *J. Exp. Psychol.*, 1959, **58**, 283.

Kimmel, H. D., Further analysis of GSR conditioning: a reply to Stewart, Stern, Winokur, and Fredman. *Psychol. Rev.*, 1964, **71**, 160.

Kimmel, H. D., and Virginia L. Harrell, Differential conditioning in the planarian. *Psychonom. Sci.*, 1964, **1**, 227.

Leonard, C., and G. Winokur, Conditioning versus sensitization in the galvanic skin response. *J. Comp. Physiol. Psychol.*, 1963, **56**, 169.

Lockhart, R. A., and W. W. Grings, Comments on "An analysis of GSR conditioning." *Psychol. Rev.*, 1963, **70**, 562.

May, M. A., An interpretation of pseudo-conditioning. *Psychol. Rev.*, 1949, **56**, 177.

McConnell, J. V., Memory transfer through cannibalism in planarians. *J. Neuropsychiat.*, *Suppl.* 1, 1962, **3**, 542.

McConnell, J. V., Editorial. *Worm Runner's Digest*, 1965, **7**, 1.

McConnell, J. V., Comparative physiology: learning in invertebrates. *Ann. Rev. Physiol.*, 1966, **28**, 107.

McConnell, J. V., A. L. Jacobson, and D. P. Kimble, The effects of regeneration upon retention of a conditioned response in the planarian. *J. Comp. Physiol. Psychol.*, 1959, **52**, 1.

Pavlov, I. P., *Conditioned Reflexes* (Transl. by G. V. Anrep), London: Oxford University Press, 1927.

Prokasy, W. F., and H. C. Ebel, GSR conditioning and sensitization as a function of intertrial interval. *J. Exp. Psychol.*, 1964, **67**, 113.

Prokasy, W. F., J. G. Hall, and J. T. Fawcett, Adaptation, sensitization, forward and backward conditioning, and pseudoconditioning of the GSR. *Psychol. Rep.*, 1962, **10**, 103.

Shmavonian, B. M., Methodological study of vasomotor conditioning in human subjects. *J. Comp. Physiol. Psychol.*, 1959, **52**, 315.

Stewart, M. A., J. A. Stern, G. Winokur, and S. Fredman, An analysis of GSR conditioning. *Psychol. Rev.*, 1961, **68**, 60.

Stolz, Stephanie B., Vasomotor response in human subjects: conditioning and pseudo-conditioning. *Psychonom. Sci.*, 1965, **2**, 181.

Thompson, R., and J. V. McConnell, Classical conditioning in the planarian, *Dugesia dorotocephala. J. Comp. Physiol. Psychol.*, 1955, **48**, 65.

Vattano, F. J., and J. W. Hullett, Learning in planarians as a function of interstimulus interval. *Psychonom. Sci.*, 1964, **1**, 331.

Warren, J. M., The comparative psychology of learning. *Ann. Rev. Psychol.*, 1965, **16**, 95.

Wickens, D. D., and Carol D. Wickens, Some factors related to pseudo-conditioning. *J. Exp. Psychol.*, 1942, **31**, 518.

Zelman, A., L. Kabat, Reeva Jacobson, and J. V. McConnell, Transfer of training through injection of "conditioned" RNA into untrained planarians. *Worm Runner's Digest*, 1963, **5**, 14.

Chapter 14

Specific Factors Influencing Planarian Behavior

James V. McConnell

Mental Health Research Institute
The University of Michigan
Ann Arbor, Michigan

Let me begin by stating what I believe to be a terribly important rule of experimental psychology: In at least 90% of the studies published in psychological journals, the most important factors influencing the animals' behavior were those factors the investigators did not measure, did not control, did not mention in their written reports, and probably did not even know about. The rest of this paper is an attempt to prove the validity of this particular point with specific reference to the planarian literature, although I shall begin by talking about rats.

When I was a graduate student at the University of Texas, long before I had ever seen a flatworm, I spent several years training white rats. In those days, the rat laboratory at Texas was housed in a converted mansion appropriately enough called Doom House. It got that titillating name not because of the multitude of rats that died there, by the way, but because it was originally built for a well-known Texas citizen named Colonel Doom. In summer, the temperature inside the house rose to as high as 110°F during the day, so we all ran our rats at night when the mercury dropped to a refreshing 85 to 90°F. During the winter, of course, we ran our rats during the daytime, because the house was not well heated and it got rather chilly at night. I rather imagine that both the change in temperature from season to season and the fact that we ran our animals at different times of the day affected our results. But, since we seldom bothered to repeat our own experiments, much less those of rat-runners elsewhere, we paid little attention to such matters. We never bothered to perform the experiments "blind," either, but perhaps that is a different story entirely. At any rate, in the worm world today, we are all quite aware that temperature and time of day can profoundly influence the behavior of planarians. Van Deventer and Ratner (1964), e.g., showed that planarians are considerably more responsive to photic stimulation (in a light-shock conditioning paradigm) when the water in which they are tested is at 90°F than when it is at 70°F. Joel Cohen (1965) has shown that the speed with which planarians learn a simple T maze is affected by whether the animals are run at night

217

or in the morning, and Kiki Roe (1963) has reported differences between the behavior patterns of worms trained in a hexagonal maze in the morning and their patterns in the afternoon. In my own laboratory we have for several years now attempted to maintain our animals at a constant 70°F during training, not an easy task even in an air-conditioned building. We do find, though, that during light-shock conditioning, the temperature of the water in which the animals are being trained varies less than 2°F across any half-hour training session. Also, we record faithfully the time of day our animals are being run and do our best to train all our animals either late in the morning or early in the afternoon to hold the time variable as constant as possible.

But other external variables are just as important as time of day and temperature. A great many animals, planarians included, are physiologically different from one season to another, yet how many of us record the time of the year during which the animals were run? My colleague at the Mental Health Research Institute at Michigan, Bernie Agranoff, tells me that the goldfish he is working with show tremendous seasonal variations in their ability to learn a simple escape problem. If goldfish have good and bad months, shouldn't one expect the same with planarians? Jay Boyd Best has evidence that the time of the lunar month is an important determiner of learning ability in planarians, and Marie Jenkins has been measuring the effects of barometric pressure on planarian behavior. In my laboratory, we often get the impression that planarians have "good" days and "bad" days, but I suspect that few people other than Jay and Marie have bothered to measure systematically these perhaps critical variables. We do know, however, that the spontaneous activity planarians show in the conditioning trough we use is somehow correlated with the response rate they show to the conditioning stimulus on any given day. For example, recently George Mpitsos and I were interested in testing the effects of the presence or absence of planarian slime on the conditioned response rates of the *Dugesia dorotocephala* we typically use in such experiments. We gave our animals 15 light-shock training trials per day with an intertrial interval of 60 seconds; 30 seconds after each light-shock presentation, we called a "blank trial," a period of 3 seconds during which we did not stimulate the animal in any way but did record any spontaneous response it might have made. The use of such blank trials enables us to gain some notion of the animals' random activity during training. Ignore for the moment the intergroup differences in Fig. 1—we shall have more to say about them later on—but do look at the data obtained on the seventh day of training, when the animals had all reached trial 105. On this date there was a sharp drop in the spontaneous activity of the animals (as shown during the blank trials); there is a corresponding drop in the "percent response" during conditioning trials for both groups of animals. Now, as the rest of the data make clear, the changes in spontaneous activity of the planarians cannot be invoked as an explanation of all the behavioral changes the animals show during training, but it is clear that whatever depressed the activity during the

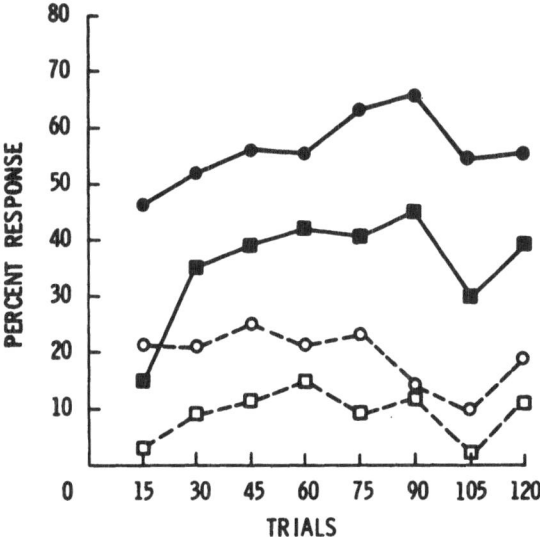

Fig. 1. Performance curves for 24 planarians trained in clean troughs (circle symbols) and for 24 planarians trained in slimed troughs (square symbols). The top two curves show responses made during conditioning trials, and the bottom two curves show responses made during "blank trials."

blank trials on day 7 also depressed the level of the conditioned response on the same day.

I am personally convinced that the chemical composition of the water in which the animals are kept is also very important; hence for the past several years we have kept all of our animals in a precisely defined artificial medium. Sadly enough, judging from what I read in the literature, few of the people active in the field have bothered to control (much less measure and report) most of these highly important external factors that surely influence the behavior of their subjects.

Data such as those discussed above should suggest strongly to any experimenter that, if he is running several groups of animals in an experiment and cannot train all of the animals in all groups at the same time, he absolutely should not train all the subjects in the experimental group at one time and days or weeks later train all the animals in one or more control groups. Instead, he should take one or more subjects from each group he is going to run and train them all at the same time. After they are all finished, he should replicate the entire procedure until he has a large enough sample to be able to determine whether there are statistical differences among the groups.

Another obviously important variable is the species one uses. No, that is not so—species cannot be an "obviously" important variable since several investigators simply do not bother to mention what species they used. For example, in a recent paper Walker and Milton (1966) fail to report the species they used in a "cannibalism transfer" study but do report that they have never been able to get any of their planarians to cannibalize living tissue. In fact, in order to get their "cannibals" to eat, these authors first had to dehydrate the bodies of their trained donors for several hours. Now, anyone who has ever seen hungry *D. dorotocephala* or *D. tigrina* at work will realize that Walker and Milton were not using members of this genus in their research, yet all of the other cannibalism studies (with which Walker and Milton compare their own work) have employed Dugesians as subjects. Ignoring the fact that the chemical composition of the "memory molecule" (if there be such) might well have changed radically during the dehydration process, what are we to make of this research by Walker and Milton, save that they would surely have been better advised to import some *D. dorotocephala* if they wished their work to be comparable to the rest of the studies in the literature?

But the behavioral differences between even such closely related species as *D. tigrina* and *D. dorotocephala* are quite marked, as I have pointed out elsewhere long ago (1965). The importance of these differences becomes clear when one realizes that the Raglands (1965) found it impossible to condition a specific race of *D. tigrina*, although the *D. dorotocephala* they had could be trained quite readily. Interestingly enough, feeding untrained (living) *D. dorotocephala* to the *D. tigrina* did not help at all, but, if trained *D. dorotocephala* were fed to the *D. tigrina*, the latter were at last able to learn.

We must recognize too that there are sometimes very wide behavioral differences within species. *D. dorotocephala* taken from Buckhorn Springs in Oklahoma are often quite different from *D. dorotocephala* taken elsewhere; and Marie Jenkins has evidence that sexual specimens show different learning abilities than asexual specimens of the same species. Yet how many of us bother to report whether our animals were sexually developed or not? For that matter, I wonder how many of us even know whether we are using sexual or asexual specimens? Nor is it sufficient in determining the species involved to rely on the judgment of the biological supply houses, for it is my experience that they are very inconsistent in their labeling and quite often give out the wrong information unintentionally. At any rate, we should all remember that different species can be expected to behave in different ways; just because brand X worms don't learn, or don't show the expected cannibalistic transfer of training, doesn't mean at all that brand Y worms won't show the phenomenon in spades.

There are a number of other "organismic" variables that many worm runners fail to control that should be mentioned here. Several experimenters have found that regenerating animals behave differently than do animals that have not recently undergone fission, although there is considerable lack of agreement as to what these differences are. Van Deventer and Ratner (1964) report an increased sensitivity to light in regenerating heads and tails, a finding repeated by Brown, Dustman, and Beck (1966). It might be noted, however, that Van Deventer and Ratner found the tails more sensitized than the heads, while Brown, Dustman, and Beck found the opposite. In both these studies, although no attempt was made to condition the animals, the authors conclude that the early findings by McConnell, Jacobson, and Kimble (1959) that regenerating heads and tails showed retention of a conditioned response was an artifact and that regeneration merely sensitized the animals to the conditioning stimulus, light. Brown, Dustman, and Beck, however, fail to mention one of the control groups that Allan Jacobson, Dan Kimble, and I ran in that early study, namely, our "regeneration control" animals that were cut in half prior to being trained; far from showing sensitization to the light, these animals showed significantly slower learning than did animals that first were trained and then were cut in half, a finding that we have repeated several times since. Corning and John (1961) likewise failed to find any significant evidence of sensitization in their control animals that were cut in half prior to training, and Westerman (1963) found no evidence of increased sensitization to light in his animals that were cut in half prior to being habituated to light. The Westerman study is particularly pertinent here, for, if cutting a planarian in half sensitized the regenerating portions to light, one might imagine that it would be impossible to demonstrate retention of habituation to light in regenerating pieces of planarians; yet that is just what Westerman showed, and under very carefully controlled conditions. Despite the lack of agreement among experimenters as to what the behavioral effects of regeneration are, however, we might all agree that animals that do fission spontaneously during an experiment should probably be discarded and, just as important, that we should watch our animals for a period of several weeks prior to using them to make sure that they have not fissioned for some time before the experiment begins.

The amount, type, and frequency of food given to animals both prior to the onset of an experiment and during the training itself should be rigidly controlled and the feeding procedure reported in detail in any published account of the work. Likewise, the lighting conditions under which the animals are kept both prior to and during an experiment should be noted. Let me give an illustration of how important a variable this latter one can be. We all know that an animal adapted to complete darkness is likely to be more sensitive to the onset of light than an animal exposed to light for an equal period of time. Suppose we took some animals that had been living in a laboratory with a normal diurnal light-dark cycle; half the animals we begin conditioning using light as the conditioned stimulus; the other half we cut in half and put in the dark to regenerate. As soon as the training animals have reached some criterion, they too are cut in half and put in the dark. At the end of a month, all animals are given test trials. To our surprise, we find that all the animals show a very high response rate to the light. Can we conclude that it was therefore the cutting and regeneration that caused some sort of sensitization? No, of course not, for all the animals had been given rather extensive exposure to absolute darkness for long periods of time, and this dark adaptation might well have made a difference. Yet few authors bother to report whether their animals were housed in darkness, in light, or on a diurnal light-dark cycle during training or regeneration. Indeed, I suspect that much of the discrepancy in the effects of regeneration on light sensitization could be explained if we knew how the animals were housed when they were regenerating. It is my belief, incidentally, that regenerating animals should be kept in normal lighting conditions rather than in darkness, unless one has some pressing experimental reason for doing otherwise.

The type of apparatus used, the experimental paradigm followed, the quality and quantity of the conditioned and the unconditioned stimuli are all critical factors that influence planarian behavior, although you would not always know this from reading the literature. Barnes and Katzung (1963) found that planarians could be classically conditioned using polarized current as the unconditioned stimulus; the planarians showed learning when the shock was always delivered when the animals were heading toward the cathode, but did not learn when they were shocked when heading toward the anode. Roy John has told me informally that he and his colleagues were able to replicate this finding, but in my laboratory we get good learning no matter what the animals' orientation toward the electrodes is, although other aspects of our training situation vary considerably from the paradigm followed by Barnes and Katzung. For instance, I believe their animals were smaller and of a different species than the ones we typically use; they ran their animals much earlier in the morning than we ever do; and they kept their animals in almost complete darkness between trial sets, whereas we typically house our animals in the light. Some experimenters, such as Allan Jacobson (1966), have reported excellent results using alternating current rather than direct current as the unconditioned stimulus; we never had much luck with alternating current until very recently when we copied Jacobson's procedure and apparatus as exactly as possible. Perhaps there's a moral in that, if one cares to look for it.

Whatever type of shock one uses, however, one must control the intensity most carefully. I have long advocated using a behavioral criterion for shock intensity rather than the mere reporting of supposedly precise physical measures of the level of the stimulation used. In short, I believe the worm itself to be a better indicator of the precise intensity of the current than any ammeter, since training apparatuses vary widely from one laboratory to another and 10 volts applied to one training trough might not cause the planarians to react at all, whereas a physically identical 10 volts applied to a different trough might cause the animals to go into convulsions. But, if one uses a slightly suprathreshold amount of shock, just intense enough to cause a good contraction in one's animals, one might legitimately expect one's results to be fairly comparable to those from another laboratory where the same technique was followed, even though the apparatuses used or the water in which the animals were run were slightly different. The same argument could be applied to the intensity of the conditioned stimulus to be employed in a conditioning study. For some time now, we have often followed a procedure first used, I believe, by Corning and John. Prior to the onset of the conditioning proper, all animals are given habituation trials to the conditioned stimulus, or, if a maze is used, the animals are given a number of preference trials in which they are rewarded no matter which arm of the maze they choose. Such a procedure has much to recommend it. To begin with, it allows one to adjust the intensity of the conditioned shock to a point at which one gets but a few unconditioned responses from the bulk of the experimental subjects. Also, if one is going to divide the animals into various groups that get different experimental treatments, one can balance the groups so that their pretreatment response rates are equal. Likewise, such a procedure allows one to discard aberrant animals prior to training, and it must be remembered that one does occasionally find a planarian that, without any prior history of training, seems highly responsive to whatever the conditioning stimulus may be. In the case of maze learning, giving the animals experience in the maze prior to training allows one to take into account the preferences that many animals show for one arm or the other of the maze; one then can make sure that each animal is trained opposite to its original preference. More than this, these pretraining tests allow one to spot gross deficiencies in one's experimental paradigm. For instance, if one finds during pretraining that some of one's groups of naïve planarians begin training with a response rate to the conditioned stimulus that runs as high as 50 to 80%, as was the case in the study reported by James and Halas (1964), one might wish to revise one's experimental situation before continuing the study (McConnell, 1964). Or, if most of the animals show a very strong preference for one particular arm in a T maze, one would surely want to redesign the apparatus.

A great many words have been written over the years about the sensitivity that planarians show toward light, so perhaps now is a good time to comment on this topic. Planarians are generally considered to be photonegative; i.e., they seem to seek out and remain in dark places in preference to brightly lighted areas, and when they cross a sharp light-dark boundary, they often make what appears to be a "startle" reaction. From these two facts, some critics have spun rather wondrous critical attacks against many of the studies demonstrating learning in planarians. Jensen (1965), e.g., has dismissed the maze study of Best and Rubinstein (1962) because in this experiment light was used as a cue. Jensen writes, "According to Pearl, light could produce turning towards or away from the light, depending upon other factors which influence whether the positive or negative reaction is given; this explanation is consistent with the finding by Best and Rubinstein that periods of rejection occur". Jensen's criticism ignores several basic facts. To begin with, the conditions under which

Pearl (1903) was examining planarian behavior were quite different from the conditions that obtained in the Best and Rubinstein experiment, but let us ignore this fact for the moment. More important, Jensen fails to state what the factors were that caused Pearl's planarians to turn toward or away from the light. As I recall matters, Pearl found that planarians were photonegative unless, e.g., they had been severely deprived of food, at which point they became somewhat photopositive. So, if Best and Rubinstein had trained all of their animals to go toward the lighted arm in the maze and had systematically starved them, perhaps one might have expected a shift in preference toward the lighted arm over time. Of course, the reversals Best and Rubinstein found later in the experiment would then be most difficult to explain, but perhaps some other factor could be invoked to handle this fact. But we need not bother with such airy matters, for in fact Best and Rubinstein had appropriate controls that Jensen does not mention: Some of their worms were trained to go to the lighted arm, some to the dark arm. Both groups showed learning, and, under certain specifiable and controllable conditions, both groups also showed reversals. The reversals could be avoided by increasing the intertrial intervals, giving fewer trial sets per day, and allowing the animals access to a large chamber between trials. In fact, the type of photonegative response mentioned by Pearl seems not to be involved at all, and, anyhow, Pearl never claimed that such innate responses might not be modifiable by training. It's also true that Jensen fails to report that others of us have trained planarians to go right or to go left in a maze where light was not a cue, but again let us ignore such matters and continue talking about the responses of planarians to light.

A year or so ago we began working with planarians in a simple T maze with one dark arm, one lighted arm. It is quite true that, if we make one arm jet black and one arm bright white and if we illuminate the maze with a bright light, the planarians show a strong preference for the black arm. However, in a dim light and with light gray and dark gray as the maze colors, their pretraining choices are quite random. Figure 2 shows the results of our giving 300 animals 20 preference trials each; each animal was given 20 trials in which it was rewarded no matter which arm it chose. The reward in this case was being returned to its home aquarium for a period of several minutes. I should say, too, that we actually used two mazes that were mirror images of each other to control for position biases. Oddly enough, one can generate the same bell-shaped curve as that shown in Fig. 2 by giving hundreds of preference trials to a small number of animals. I submit that this is evidence that no strong innate preferences exist in these mazes (a photograph of one such maze is shown in Fig. 3). However, if we take animals that show no strong preferences at the beginning of the experiment and start punishing them by starting them over in the maze whenever they pick the incorrect arm and reward them when they pick the correct arm, the animals rapidly begin to show nonrandom behavior. That is, their choices for the rewarded arm increase dramatically, and within 200 to 300 trials they will almost all reach the criterion of

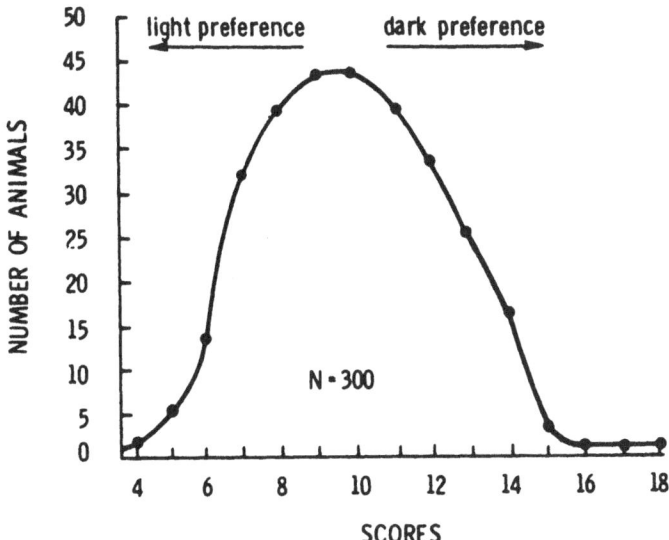

Fig. 2. Preference scores for 300 planarians each given 20 trials in a T maze with one light and one dark arm. A score of 11 means the animal chose the dark arm 11 times out of the 20 trials.

9 correct responses in 10 consecutive trials (for 2 days in a row). If we then allow them to sit for a period of 2 to 3 weeks and retrain them to the same criterion, most of them will regain the criterion in less than 60 trials, a highly significant saving. The learning curves for 24 such animals are

Fig. 3. A simple T maze for planarians.

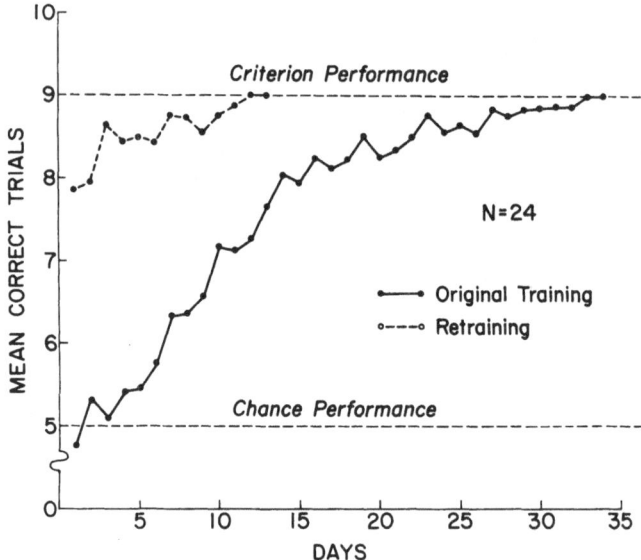

Fig. 4. Learning curves for 24 planarians given 10 trials daily
in a T maze.

shown in Fig. 4. Interestingly enough, I do not believe that any of the
several hundred animals we have trained has ever given us 9 incorrect
responses out of 10 in a day's training. Likewise, those animals given
nondifferential reward (preference trials) for dozens of days of training
simply never reach criterion; indeed, out of thousands of such days of
testing, only twice has a preference animal gone 9 trials out of 10 to a
given arm, and never has this happened two days in a row. Additionally,
we have run animals that have been given random punishment in the maze,
the number of punishments given being the average number we had given
the day before to a matched group of experimental animals. These random-
punishment animals also show completely random-choice behavior, and
not one ever came close to achieving the criterion met by the experimental
animals.

In 1964, Barbara Humphries (1964) ran much the same sort of experiment in a hexagonal maze and found that none of the planarians given
random punishment—in this case, electric shock—showed anything but
random behavior even after they had been put through more than 500
trials, although the experimental animals had all shown clear-cut evidence
of learning by then. Indeed, at the end of 500 random-punishment trials,
these control animals preferred the black arm 51 % of the time, the white
arm 49 % of the time. However, if we then took the random-punishment
animals (after they had been given 500 trials) and started rewarding and

punishing them the way we had the experimental animals, the random-punishment group reached a very high level of correct performance in just 25 training trials, just as one might expect if true learning were involved. Frankly, I do not believe that even a thousand citations from Pearl can explain away these data, and I conclude that, if you design your experiments appropriately, the innate reactions that planarians show to light (or to any other stimulus, for that matter) need not prevent your being able to demonstrate learning in planarians.

There are various factors that influence a planarian's behavior in a testing situation that seem to have to do with the fatigue state of the animal. In my opinion, giving the animals too short an interval between trials or too many trials per day or too many trial sets per week are all conditions that fatigue the animals and cause refractory behavior. Certainly Best and Rubinstein (1962) found this to be the case in their Y-maze study, and Humphries (1964) found, when using a hexagonal maze, that she got the best performance out of her animals by giving them no more than 30 trials per day and training them but 2 days a week. Massing the trials appears to lead to much greater sensitization in classical conditioning studies, too. For instance, anyone who has trained a planarian knows that, if one gives it too many trials a day, the animal will often come to rest and refuse to move unless stimulated, or poked. Few people, however, have bothered to study the behavior of animals that have been poked. We have found that in a classical conditioning paradigm using light as the conditioned and shock as the unconditioned stimulus, a freshly poked animal is not overly reactive if one waits for 10 to 15 seconds after it has been stimulated before beginning the next trial. One of my students, Surreya Dikmen, made an interesting observation about such situations, however: If one begins a trial immediately after poking the animal with a brush, it is more likely to respond if it is a naïve animal than if it is an experimental animal that was at or near criterion before it stopped moving. If one waits for 10 to 15 seconds after the poke before starting the trial, the conditioned animal will usually respond appropriately, whereas the naïve animals hardly ever respond at all. Obviously one could unintentionally elevate the response levels in naïve or control-group animals and decrease the response levels in experimental animals by poking them sufficiently often and by running trials immediately after each poke. So let me adjure all worm runners to wait at least 10 to 15 seconds after each poke before beginning a trial and, more than this, to stop running the animal entirely and give it the rest of the day off if it stops more than two or three times in a given trial set.

Most of the factors that I have mentioned so far are fairly obvious ones to anyone who has studied planarian behavior extensively, although, of course, we have not always done our best in controlling these factors. There is one variable, however, that eluded most of us until recently, and what a critical variable it is! I speak of the slime trails that the planarians leave as they crawl along a maze or a conditioning trough. Back in the bad old days, before we got interested in the chemistry of memory, we never bothered to clean out the troughs or mazes and let considerable slime accumulate in them. When we began attempting to transfer acquired behavioral tendencies by cannibalism or RNA injection, however, we began to worry about those slime trails. Perhaps the animals were leaving each other messages as they went along or perhaps the slime trail left by a trained animal was chemically different from the trail left by a naïve worm. So, to control such possibilities, we started cleaning out the mazes and troughs after each animal had been run. It took us quite some time—a year or so—to discover that planarians are quite sensitive to the presence or absence of slime in their environment, that they much prefer the presence of slime, and that their behavior patterns are markedly aberrant when they are forced to crawl on a freshly cleaned surface.

The first study on slime that we undertook came about more or less by accident. In the summer of 1963, Barbara Humphries had done a series of experiments in the hexagonal maze testing the effects of massed versus spaced training and the effects of the strength of electric shock on the speed of learning (1964). When she attempted to repeat part of this study in 1964, at first she got nowhere. In 1963, the planarians had shown random behavior in the maze until we introduced punishment if they entered the incorrect arms; in 1964, the planarians showed a marked preference for the dark arms and showed little learning at all until they had run at least 200 trials, at which point they slowly began to shift their preference toward the correct arms. The only difference we could find between the experimental paradigms used in 1963 and those in 1964 was that in 1964 we were carefully cleaning out the maze after each animal had run, while in 1963 we had let the slime accumulate. So, in 1964, we carefully slimed up the maze by letting large numbers of naïve animals run through it before attempting to train our experimental animals, and—lo and behold!—when the maze was well slimed, we had no difficulty in reproducing the results we had obtained the summer before. A careful test of the two sets of conditions then followed. When run in a clean maze, the planarians were markedly photonegative; when tested in a slimed maze, the same animals lost their strong photonegativity. Worms given training trials in a slimed maze showed apparent learning in roughly 100 trials or so; worms run in clean mazes preferred the dark arms strongly and overcame this preference only after 200 or more trials. This finding is of some interest, since Bennett and Calvin (1964), in attempting to replicate our 1963 maze study, used clean mazes and reported no learning in their animals after 120 trials. In fact, their animals showed strong photonegativity, and, of course, these authors stopped their study too soon to show learning when using a clean maze. When we ran animals under clean conditions, we got exactly the same results they got; when we ran animals under slimed conditions, the animals learned, and rapidly.

At this point, we began a whole series of studies on the effects of the presence or absence of slime in planarian behavior. In the hexagonal maze, e.g., if we slime just the white portions, the animals tend to stay in the white areas; whereas, if we slime the black arms, they show an even stronger tendency to stay in the black. Planarians show much the same tendency in open field tests, too. George Mpitsos put wax partitions in large soup bowls so that he could let a dozen or so naïve animals provide a layer of slime in half the bowl while the other half remained clean. Then the partition and the slime-providing worms were removed, and a naïve test animal was placed in the bowl and allowed to roam about at will until it came to rest. The vast majority of the animals thus tested spent more time on the slimed half than on the unslimed half of the bowl, and almost all the animals came to rest on the slimed area. Indeed, the planarians showed much the same sort of startle reaction when crossing from slime into a clean area as when crossing from a dark area into one that is brightly lighted. When the animals crossed from a clean area into one containing slime, however, they seldom evinced this startle reaction. The speed of locomotion of the animals on slime is, however, almost exactly the same as when the animals are crawling on a clean surface—about 12 cm/minute. George tested several different species in these open field tests. In general it can be said that most species can recognize slime left by members of their own species and prefer their own species' slime to that of other species. A notable exception is *D. dorotocephala*, which tends to prefer the slime of other species, particularly if well-starved. Indeed, the mere

Fig. 5. A conditioning trough for planarians. The white area is filled with water.

presence of other species' slime is sufficient to elicit pharynx extrusion and food-seeking behavior in these voracious cannibals.

Next, George tested the effects of the presence or absence of slime on classical conditioning (McConnell and Mpitsos, 1965). The subjects were 48 laboratory-hatched *D. dorotocephala* that were 7 to 14 mm in length when tested. The conditioning stimulus was light from a 15-watt bulb suspended 20 cm above the center of the trough; the unconditioned stimulus was d-c shock supplied in the corners of the square trough shown in Fig. 5. A trial consisted of 4 seconds of light, the shock being applied during the final second. Half the animals were run in freshly cleaned troughs, the other half in troughs that had been freshly slimed by a group of slime-providing worms. The slime-providers were never given any training of any kind, and the troughs were freshly cleaned and reslimed before each experimental animal was run. Two animals were run at a time, in troughs housed in separate but adjacent compartments. The experiment was run "blind" in that George was not told which trough was clean, which was slimed. In fact, the behavior of the animals was so markedly different in the two types of trough that I doubt that the study was all that "blind." All 48 animals were given 15 trials a day for 8 days but were run only 3 days a week. At the end of 4 days of training, the "slime-clean" conditions were reversed for 28 animals, while the rest continued to run under the original conditions. The results can be seen in Fig. 6.

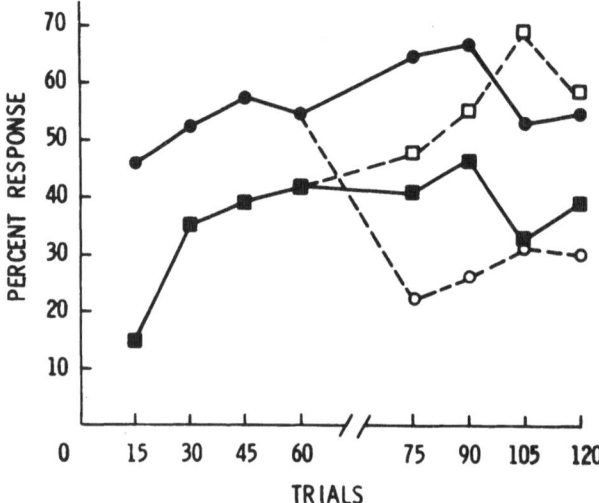

Fig. 6. Performance curves for 24 planarians trained in clean troughs (circle symbols) and for 24 planarians trained in slimed troughs (square symbols). At the end of day 4, half the animals in each group (broken lines) were switched to opposite slime conditions.

As can be seen, the planarians trained in the clean troughs showed a significantly higher initial response rate to the light and reached an asymptote rapidly; when the conditions were reversed, however, they showed a marked drop in response rate but did not fall to the naïve level of responding. The planarians trained in slimed troughs showed the regular increase in responding that we associate with classical conditioning. When switched to clean troughs, they showed a sudden increase in response rate and were actually somewhat superior to the animals trained throughout in the clean troughs.

Incidentally, if you are intrigued by the apparent drop-off in behavior in both groups toward the end of training, I suggest you go back to Fig. 1 of this paper and compare the two graphs. Obviously, any full account of the behavior of planarians in a conditioning trough must include other explanatory factors in addition to slime and the number of reinforced trials the animals have been given at any point in time.

Now, the slime studies are interesting for several reasons. To begin with, they suggest that we can sometimes separate the effects of sensitization from those of conditioning, a point I discuss in another paper in this book. Second, these studies intimate that the effects of sensitization, at least in the case of classical conditioning, may reach asymptote rapidly and remain at that level, while the effects of conditioning are cumulative, a point we might do well to remember. Third, these experiments show that, if one is not careful in designing one's experiment, sensitization may mask any conditioning that might take place and thus the experimenter might be led to rather inaccurate conclusions concerning his results. Oddly enough, most of the experimenters who have had difficulties in achieving clear-cut conditioning in their animals have used clean troughs. Brown, Dustman, and Beck (1966), e.g., carefully cleaned their troughs during the experiment, as did Bennett and Calvin (1964) and several other investigators. I think we are forced to conclude that all results obtained from animals trained in clean apparatus may well have yielded spurious results.

Let me now conclude by drawing a couple of pretty obvious morals. Most of us like to think of ourselves as being animal behaviorists, and we are committed to a scientific point of view that goes something like this: Behavior is controlled by an animal's genetic makeup, its past experience, and the environment in which it finds itself at any given moment. The more we control the organism's genes, its memories, and its present environment, the more we control its present behavior. And unless we do establish firm control over its genes, its past history, and its present stimulation, we cannot hope to·learn very much about that animal. Yet in most of the studies we publish, whether on rats or worms or humans, we make almost no attempt either to report or to control the really relevant variables influencing the subjects' behavior. Somehow, we must ourselves learn to do something about this situation.

Finally, it seems to me that the single most important factor influencing an animal's behavior is the point of view that the investigator running the

study has toward experimental controls. Anyone can run a bad experiment—
it is the easiest thing in the world. But how difficult it is, and how rare a
thing, for someone to take the time to learn about the zoology and the
physiology of the animal before training it, to read the literature carefully,
and to take into account all of the past findings before running a single
beast in the laboratory! And how easy it is, when attempting to replicate
someone else's study, to use a different species because you do not have
the right one at hand; to use a different apparatus, different training
procedures, different stimulus intensities, and different intertrial intervals;
to keep the animals under different conditions during training; and to
ignore such potent variables as slime, the size of the animal, the time of
day, and the time of year the experiment is run; and, then, when the results
are in, to be surprised that one could not replicate the earlier work!
Perhaps what we ought to do at the end of this conference is to draw up
a petition to send to the Great Worm Runner in the Sky asking him to
train us experimenters a little better and to hope that He takes into account
the many critical variables influencing our own behavior when He does so.

REFERENCES

Barnes, C. D., and B. G. Katzung, Stimulus polarity and conditioning in planaria. *Science*, 1963, **141**, 728.

Bennett, E. L., and M. Calvin, Failure to train planarians reliably. *Neurosci. Res. Program Bull.*, 1964, July-Aug.

Best, J. B., and I. Rubenstein, Maze learning and associated behavior in planaria. *J. Comp. Physiol. Psychol.*, 1962, **55**, 560.

Brown, H. M., R. E. Dustman, and E. C. Beck, Experimental procedures that modify light response frequency of regenerated planaria. *Physiol. and Behav.*, 1966, 1(3), 245.

Cohen, J., Diurnal cycles and maze learning in planarians. *Worm Runner's Digest*, 1965, 7(1), 20.

Corning, W. C., and E. R. John, Effect of ribonuclease on retention of conditioned response in regenerated planarians. *Science*, 1961, **134**, 1363.

Humphries, Barbara, and J. V. McConnell, Conditions facilitating learning in planarians. *Worm Runner's Digest*, 1964, 6(1), 52.

Jacobson, A. L., C. Fried, and S. D. Horowitz, Planarians and memory, I: Transfer of learning by injection of ribonucleic acid; II: The influence of prior extinction on the RNA transfer effect. *Nature*, 1966, **209**(5023), 599.

James, R. L., and E. S. Halas, No difference in extinction behavior in planaria following various types and amounts of training. *Psych. Rec.*, 1964, **14**, 1.

Jensen, D. D., Paramecia, planaria, and pseudo-learning. *Animal Behav.*, Suppl. 1, 1965, **13**, 9.

McConnell, J. V., On the turning of worms: a reply to James and Halas. *Psych. Record*, 1964, **14**(1), 13.

McConnell, J. V., ed., *A Manual of Psychological Experimentation on Planarians*. Special publication of *The Worm Runner's Digest*, 1965.

McConnell, J. V., A. L. Jacobson, and D. P. Kimble, The effects of regeneration upon retention of a conditioned response in the planarian. *J. Comp. Physiol. Psychol.*, 1959, **52**, 1.

McConnell, J. V., and G. Mpitsos, Effects of the presence or absence of slime on classical conditioning in planarians. *Amer. Zool.*, 1965, **5**(4), 122.

Pearl, R., The movements and reactions of fresh-water planarians. *J. Micro Sci.*, 1903, **46**, 509.

Ragland, Rae S., and J. B. Ragland, Planaria: interspecific transfer of a conditionability factor through cannibalism. *Psychon. Sci.*, 1965, **3**, 117.

Roe, Kiki, In search of the locus of learning in planarians. *Worm Runner's Digest*, 1963, **5**(2), 16.

Van Deventer, J. M., and S. C. Ratner, Variables affecting the frequency of response of planaria to light. *J. Comp. Physiol. Psychol.*, 1964, **57**, 407.

Walker, D. R., and G. A. Milton, Memory transfer vs. sensitization in cannibal planarians. *Psychon. Sci.*, 1966, **5**, 293.

Westerman, R. A., Somatic inheritance of habituation of responses to light in planarians. *Science*, 1963, **140**, 676.

Chapter 15

Behavioral Modification of Planarians

F. T. Crawford

Department of Psychology
Florida State University
Tallahassee, Florida

In 1955, Thompson and McConnell described an experiment in which flatworms were reported to be classically conditioned. Further studies on regeneration, initiated by McConnell, Jacobson and Kimble (1959), and the role of RNA on the retention of conditioned responses in regenerated planarians (e.g., Corning and John, 1961) have led to a great deal of research on a number of controversial issues. One of these is whether or not planarians are capable of being conditioned; whether they really learn or are instead demonstrating habituation, sensitization, pseudo-conditioning, or something else, owing to improper control of experimental variables. A second question, assuming that planarians can be conditioned, is whether or not retention of the learned response survives regeneration. A third question is whether or not RNA is the means by which retention of the learned response survives regeneration. The present paper is concerned primarily with the first question, although some reference will be made to the other two.

In 1963, Jacobson reviewed the literature for evidence that planarians acquired classically and instrumentally conditioned responses. His review was based upon positive findings in the great majority of studies. These included five studies involving regeneration and two utilizing cannibalism. There were three studies of a negative nature cited by Jacobson (Cummings and Moreland, 1959; Halas, James, and Knutson, 1962; and Halas, James, and Stone, 1961). These were concerned mostly with the problem of light and shock sensitization during classical conditioning. Also cited was the failure of Rice and Lawless (1957) to demonstrate response alternation in a T maze. In overview, it is not difficult to see how Jacobson concluded that there was strong evidence for "learning" in planarians. The world of science, however, is filled with skeptics and cynics. The colorful nature of cannibalism in worms (McConnell, 1962 and 1964), two heads being better than one (Ernhart, 1960 and 1961), retention surviving successive stages of regeneration (McConnell, Jacobson, and Humphries, 1961; and McConnell, Jacobson, and Maynard, 1959), and other such effects invite doubt, and the controversy continues. Since Jacobson's 1963 review, a number of additional studies have appeared reporting negative results, having negative implication, or being interpreted to the effect that the behavioral modification and regeneration effects of worms may not be due to learning. These include Baxter and Kimmel, 1963; Bennett and Calvin, 1964; H. M. Brown, 1964; Brown and Beck, 1964; Cornwell, 1960a; Griffard, 1963; Halas, Mulry, and DeBoer, 1962; Hartry, Keith-Lee, and Morton, 1964; Haynes, Jennings, and Wells, 1965; James and Halas, 1964a and b; Jensen, 1965; Kimmel and Harrell, 1964 and 1966; Pickett, Jennings, and Wells, 1964; Walker, 1966; and Walker and Milton, 1966. Casual observation would seem to indicate that psychologists are the most ready to deny that learning occurs in worms and that the biochemist is the quickest to deny the role of RNA in the "memory" transfer of behavior.

There are studies, however, which have been carried out to investigate specific variables affecting performance. For example, Barnes and Katzung (1963) found that planarians were most easily conditioned when oriented toward the cathode. Similar findings are reported by Halas, Mulry, and DeBoer (1962) and by Shafer and Corman (1963), although these results have not been confirmed by Bennett and Calvin (1964), Best and Elshtain (1966), and by McConnell (in this volume). The effects of light (Best, 1960; DeBold, Thompson, and Landraitis, 1965; Parker and Burnett, 1900; Taliaferro, 1920; Walter, 1908; and Westerman, 1963), slime (McConnell and Mpitsos, 1965), size (H. M. Brown, 1964; VanDeventer and Ratner, 1964), magnetic fields (F. A. Brown, 1962a and b), observer bias (Cordaro and Ison, 1963; and Shinkman and Kornblith, 1965), ultraviolet (H. M. Brown, 1964), shock characteristics (Best and Elshtain, 1966), shock duration (Reynierse, Jensen, and Schroeder, 1966), shock intensity (Baxter and Kimmel, 1963), environmental familiarity (Best and Rubenstein, 1962a), and other factors (F. A. Brown, Jr., 1963; H. M. Brown, 1964; Brown and Park, 1964; McConnell, 1965b and in this volume; and VanDeventer and Ratner, 1964) have been reported. There have been a number of recent articles summarizing various aspects of planarian research (Best, 1963 and 1965; Jacobson, 1965 and in this volume; Jacobson, Fried, and Horowitz, 1966; and McConnell, 1965a and 1966a), but I should like to summarize briefly some of the work which deals with more general aspects of behavioral modification, including some preliminary work and unpublished studies from our laboratory.

Behavioral studies of planarians are said to have begun with van Oye (1920), who trained worms to take an unusual route to food reward. Van Oye had no control groups, but he did conclude that he had observed learning in planarians. Austad (1965) has attempted to improve upon van Oye's procedure, but his results were equivocal and incomplete. Haynes, Jennings, and Wells (1965) also attempted to replicate van Oye's procedure but failed to obtain transfer or facilitation of learning. A later study by Wells, Jennings, and Davis (1966) yielded positive results when the subjects were tested with no food in the maze. McConnell (1965b) also reports successful replication of van Oye's experiment in his laboratory. We have run a number of studies at Florida State University which have attempted to obtain conditioning utilizing food reward. Richard Urbano ran a small number of control and experimental *Dugesia dorotocephala* down a short trough, or runway, to obtain access to a small portion of beef liver in a recessed goal area. We were pleased to find that the experimental worms "ran" significantly faster than the controls. We made some more elaborate runways and employed more control groups and a larger number of worms in an attempt to replicate our original results. When our attempts failed, another student, R. Klingaman, ran two other modifications of the experiment, each intended to be an improvement over the former variations. These studies were also considered failures in that in each instance the experimental groups took consistently longer than the control groups to traverse the runway. An example of this effect is seen in Fig. 1 in another experiment carried out by Larry King. Four groups of *D. dorotocephala* were given eight trials in a runway 1.5 in. long. They were placed in the trough and the time required to glide 0.688 in. into a recessed goal area was recorded. Group 1 was run to food reinforcement, a 0.031 in.^{3} of seared beef liver, one trial a day with a 48-hour intertrial interval. Group 2 was run in the same manner but in addition was fed in the

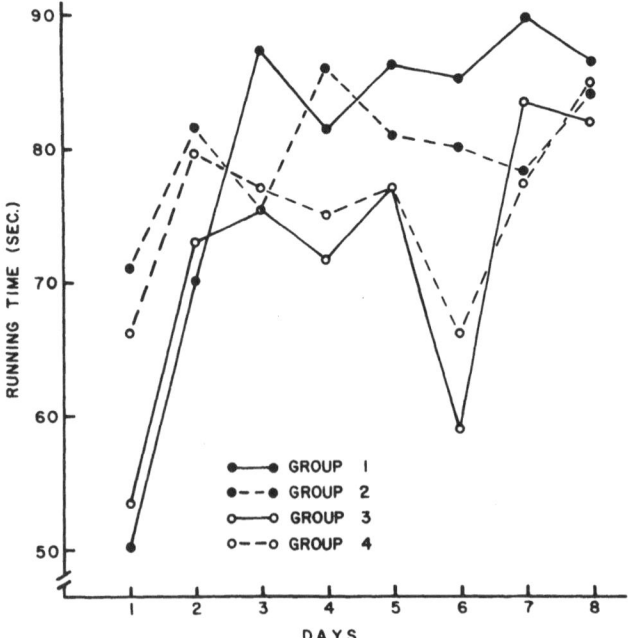

Fig. 1. Runway performance of *Dugesia dorotocephala* to food reinforcement.

goal box on the days not run. Group 3 ran in the runway the same as Group 1 but without reinforcement in the goalbox. Their feeding was given in the home tank on the day they were not being run. Group 4 was run in the runway without food reinforcement but was fed in the goal area on days not being run. Feeding consisted of 30-seconds access to the food in each case. Groups 1 and 3 were equated for runway experience and feeding. Groups 2 and 4 were equated for handling and runway experience. An analysis of variance of the experimental groups (1 and 2) and the control groups (3 and 4) indicates a significant difference at the 5% level. The experimental groups again have shown the poorest performance. The experiment is open to a number of criticisms and interpretations, but it is obvious that we are not manipulating our variables appropriately. In addition to other things, it would appear that increasing the intertrial interval and the length of the runway are called for. The habituation demonstrated by all groups is not due to satiation alone, however, since the same effects are shown with 20 trials and 96-hour intertrial intervals, and, in addition, the worms eat when placed in the goal area with food present and when fed in the home tank. The same effect has been demonstrated in our laboratory with another invertebrate, the starfish *Lucida clathrata*. Similar behavior in the planarian has been reported by Best and Rubenstein (1962b) and brought under experimental control. I am certain that under proper conditions runway acquisition in planarians can be demonstrated.

Another type of instrumental behavior, maze learning, has more consistently yielded positive results. Ernhart and Sherrick used a T maze in their regeneration study in 1959. Other studies by Best (1963 and 1965), Best and Rubenstein (1962a and b), Corning (1964 and 1966), Humphries (1961), T. Lee (1964), and Shinkman and Hertzler (1964) have reported positive findings. Pickett, Jennings, and Wells (1964) got negative results investigating the effect of RNA and cannibalism in maze learning. Rice and Lawless (1957) were not able to observe response alternation in a choice situation. Experiments by T. Lee (1964) and Shinkman and Hertzler (1964) have subsequently obtained positive results, however, for response alternation.

Avoidance conditioning has been reported for *D. dorotocephala* by Ragland and Ragland (1965). The authors also claim that *D. tigrina* shows little, if any, instrumental conditioning. R. M. Lee (1963) has shown that *Cura foremanii* can be trained to make an operant escape response, and Best (1965) has extended these findings. Since an operant response for worms does not involve the use of shock and response reliability is improved (see Cordaro and Ison, 1963; and Shinkman and Kornblith, 1965), we have been working recently with operant procedures. Carl Skeen and I obtained the data shown in Fig. 2. The data are for 16 light-escape contingent *C. foremanii* and their yoked control pairs. In our apparatus each experimental worm had to break a beam of light that was brighter than the test-chamber illumination provided by a 200-watt lamp located 5 in. above the apparatus.

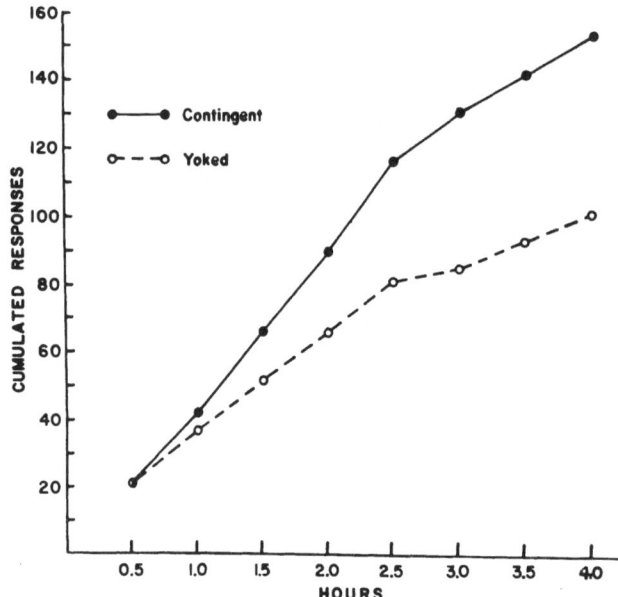

Fig. 2. Comparison of light-escape contingency *Cura formanii*
with their yoked control pairs for each half-hour of training.

The background illumination was 2.2 log ft-c, and the beam of light which
the worms had to break was 3.2 log ft-c. Additional differences from
Lee's (1963) procedure were that the overhead light was turned off for
only 60 seconds, and the S were run for only 4 hours. The cumulated
responses for each half-hour period are shown in Fig. 2. The experimental
group's terminal performance is significant at the 5% level for a one-
tailed test. The data reflect more the limitations of the apparatus and
procedure than the capability of *C. foremanii* to make the operant response.
An improved version of the apparatus and response system is being constructed.

Following the original demonstration of classical conditioning in
worms by Thompson and McConnell (1955), there were a number of
suggestions that the behavior observed was in fact due to sensitization to
light or shock, or to both, or that the results could be attributed to pseudoconditioning (see Bennett and Calvin, 1964; H. M. Brown, 1964; Halas
et al., 1961; Halas *et al.*, 1962; James and Halas, 1964b; and Jensen, 1965).
There have been a number of studies outside of McConnell's laboratory
which have had controls for light or shock, or both, which have reported
learning (e.g., Crawford, King, and Siebert, 1965; Crawford, Livingston,
and King, 1966; and Vattano and Hullett, 1964). A number of studies by
Kimmel and his students (Baxter and Kimmel, 1963; and Kimmel and
Harrell, 1964 and 1966) should also be included, although in his earlier

studies he did not rule out sensitization, alpha conditioning, or some other process. The recent paper by Kimmel and Yaremko (1966) is interpreted by Kimmel as good evidence for conditioning.

Conditioning in worms has been reported in a variety of other general circumstances. Griffard (1963) has used water flow as a conditioned stimulus and reported conditioning, although he also did not rule out sensitization. Fantl and Nevin (1965), Griffard and Peirce (1964), Jacobson (in this book), and Kimmel and Harrell (1964 and 1966) have reported differential conditioning to visual and vibratory stimuli. In an experiment using amino acid analysis of planarians, Crawford, King, and Siebert (1965) reported partial reinforcement effects during extinction, although not during acquisition. Kimmel and Yaremko (1966) used a larger number of subjects and obtained a clearer demonstration of partial reinforcement effects in that effects for both acquisition and extinction were observed.

The differential effectiveness of distribution of conditioning trials in worms received early attention by Cornwell (1960a and b). By appropriately distributing trials, a high percentage rate of responding has been reported (e.g., McConnell, Jacobson, and Kimble, 1959). A comparison of the effects of distribution of practice in the classical conditioning of planarians has been described by Crawford, Livingston, and King (1966), who report a significantly superior level of acquisition, with greater resistance to extinction, among subjects given distributed acquisition trials. The demonstration of the effectiveness of distribution of practice, together with the studies previously mentioned which show response alternation behavior, would lead one to believe that he should get spontaneous recovery of classically conditioned responses in worms. Such a study has been recently reported by Larry King and myself (1966). Thirty *C. foremanii* were given light-shock pairings 20 times a day for 5 days and were then divided into two matched groups. The performance of these groups is shown in Fig. 3. One group of worms was then given 40 extinction trials to light alone on the sixth day, while the second group was given 20 trials on day 6 and 20 trials on day 7. The performance of the two groups during extinction is shown in Fig. 4 in blocks of 10 trials. It can be seen in Fig. 4 that the third block of 10 trials for the second group, shown by the dashed line, is considerably greater than the comparable point for the first group receiving all their extinction trials on the single day. This difference is significant at less than the 1% level. The results correspond to effects observed under similar circumstances in higher organisms frequently referred to as spontaneous recovery.

Vattano and Hullett (1964), in their study on learning in planarians as a function of interstimulus interval, were apparently the first to examine backward conditioning in planarians. They compared two forward, two backward, one simultaneous and one pseudoconditioning group. After 25 light-alone adaptation trials they gave 250 acquisition trials to 12 groups of worms, 6 at a time, with a 1-second overlap of light and shock in all groups except the pseudoconditioning group. The two forward groups

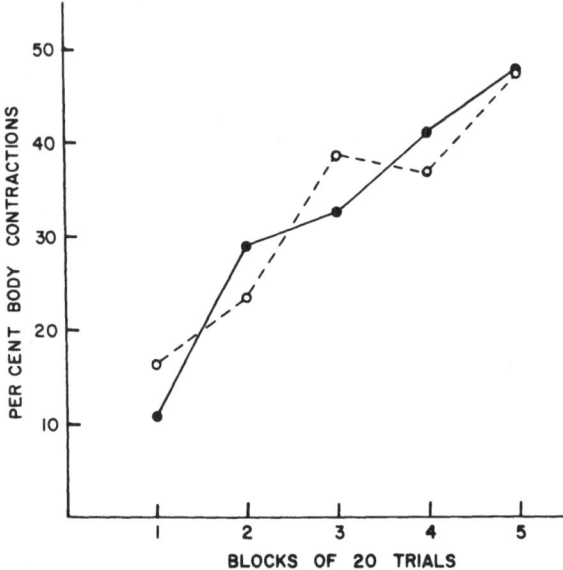

Fig. 3. Performance of the two matched groups during training and prior to the extinction trials testing for spontaneous
recovery.

received 3 seconds of light and 1 second of shock, and 2 seconds of light
and 1 second of shock. The two backward groups received 2 seconds of
shock and 1 second of light, and 3 seconds of shock and 1 second of light.
The simultaneous group received 125 unpaired 2-second presentations of
light and shock each. All groups were then given 25 extinction trials, and
the number of body contractions and cephalic turns were recorded.
Cephalic turns did not turn out to be as meaningful a measure as body
contractions. For the latter measure they found that the two forward-
conditioning groups were significantly different from all the other groups,
the backward-conditioning groups were the poorest, and the simultaneous
and pseudoconditioning groups were essentially the same. The results of
the experiment are not as straightforward as they appear upon first examination, however, since the backward groups had only half of the shock and
half of the light presentations given the other groups, and furthermore,
the duration of light and shock presentations in the forward and backward groups were not the same. Jacobson has reported in a previous
article in this book an experiment which did not have these difficulties but
which yielded similar findings.

On the basis of the preceding review of studies I believe that most
people would be inclined to conclude that worms are capable of a certain

amount of learning. The next study I am going to describe also investigates backward conditioning. The results are perhaps more controversial and exemplify the problem of learning versus sensitization. The study was initiated because I wanted to make a biochemical comparison between groups of worms receiving different experimental treatments with recognized effects.

Everyone "knows" that backward-conditioning procedures produce negligible results. As the result of preliminary studies I observed that the backward procedure employing what appears to be almost standard conditioning stimuli—the ubiquitous 3-second and 1-second combination of light and overlapping shock—resulted in strong body contractions which interfered with observation of the conditional response. In the present study the duration of shock was reduced, therefore, to 0.25 second, and, in addition, for convenience of conditioned-response observation, the onset of the unconditioned stimulus was made coincident with the offset of the conditioned stimulus. Three groups of 11 *C. foremanii* were given 3-second light and 0.25-second shock for 25 trials a day for 4 days. The

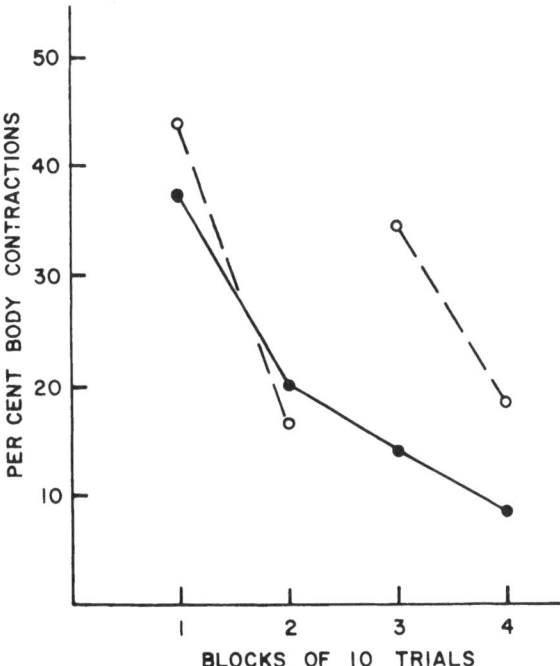

Fig. 4. Performance of the group receiving 40 extinction trials on a single day compared with that of the group having a 24-hour interval between the first and last 20 extinction trials.

groups were all given equal durations of light and shock but in a forward, backward, and pseudoconditioning manner. The pseudoconditioning group received light and shock presentations which were randomized by Gellermann series. In the forward and backward groups, there was an approximate 60-second intertrial interval. In the pseudoconditioning group, there was a minimum of 30 seconds between stimulus presentations. The shock source was 30 volts of alternating current with 25 K ohm series resistance.

The results of the 100 trials of acquisition are shown in Fig. 5 where it may be seen that the forward-conditioning procedure produced a significantly greater performance ($p < 0.002$) than either the backward or pseudoconditioning groups, which are essentially similar to each other. It may be seen that the terminal response levels are approximately 22% and 42%. On the fifth day all S were given 25 3-second light presentations. Figure 6 shows the 25 extinction trials in blocks of 5 trials. It may be seen that all three groups show a "reminiscence" effect in that the first 10 extinction trials all have a higher percentage of body contractions than the terminal acquisition percentages. In the over-all 25-trial comparisons there was an increase of 5.1%, 18.5%, and 3.7% for the forward, backward, and pseudoconditioning groups, respectively. The real surprise, of course, is the performance of the backward-conditioning group, whose frequency of body contractions on the 25 extinction trials is not significantly different from that of the forward group. The forward-conditioning group

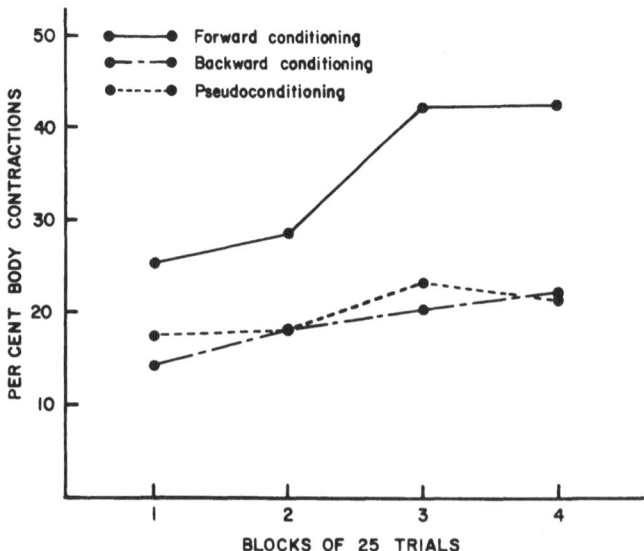

Fig. 5. Performance of forward, backward, and pseudoconditioning groups during acquisition training.

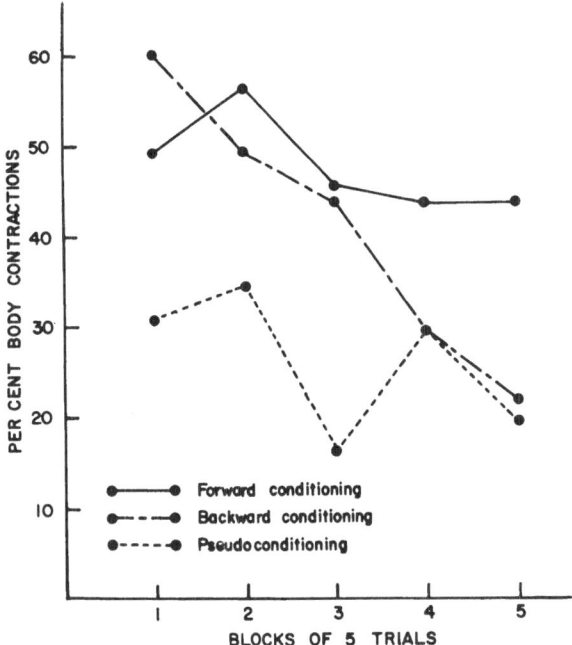

Fig. 6. Performance of forward, backward, and pseudoconditioning groups during extinction.

differed significantly from the pseudoconditioning group ($p < 0.002$),
and the backward and pseudoconditioning groups approach being significantly different from each other ($p < 0.10$).

There was 93% agreement between the experimenter and two judges
on a sample of the extinction trials. There was less agreement on the
observations of body contractions in the experimental (forward) group
than in either of the other two. The body contractions of the backward
group appeared to be qualitatively different from those of the forward
group in that there were very infrequent combined body turns and longitudinal contractions. This was the type of response which maintained the
response rate of the forward-conditioning group. The backward group,
on the other hand, showed almost entirely a longitudinal contraction in
which the worm would ball up completely. By "ball up" I mean a sharp
anterior contraction coupled with a less definite posterior contraction.
This resulted in a more defined response and a clearer pattern of extinction
in the backward-conditioning group.

The question whether or not the experiment is replicable is of interest,
of course. I have not repeated the experiment, but I agree that it needs
replication. If the results are valid they represent, on the one hand, evidence

that learning occurs in worms. On the other hand, they are readily inter-
pretable as being due to sensitization. The results are consistent with
Cornwell's (1960a) finding that conditioned responses may be observed
in the absence of overlapping conditioned and unconditioned stimuli.
They are also consistent with Brown's (H. M. Brown, 1964) report of
conditioned responses with a 5- to 15-second delay between conditioned
and unconditioned stimuli. It should not be too surprising to have someone fail to replicate the experiment, however, since a number of studies
using extinction measures have reported widely divergent results. H. M.
Brown (1964), Jacobson (1963), and McConnell (1964 and 1965a), e.g.,
have discussed this problem. Differences in resistance to extinction may be
seen in experiments by Baxter and Kimmel (1963), H. M. Brown (1964),
Corning and John (1961), Crawford, King, and Siebert (1965), Crawford,
Livingston, and King (1966), James and Halas (1964b), Kimmel and
Yaremko (1966), and Vattano and Hullett (1964), as well as in the experiments on spontaneous recovery and backward conditioning just described.
Obviously, we do not have enough information about the parameters
determining resistance to extinction in worms, or in rats either.

In my opinion the equivocal interpretations that are available for
this experiment and many others point out the bankruptcy of the term
learning for scientific usage. I am not suggesting that it is not a useful term
anymore than education is not a useful term. In our society, however,
"learning" and "garbage" are similar in that they are defined by exclusion.
This may be seen by examining any of our standard texts on the topic,
e.g., Hall (1966), Hilgard and Bower (1966), and Kimble (1961). Parsimony
alone demands that we restrict ourselves in the use of the term learning.
In my opinion it is rather clear that we can modify the behavior of worms,
but I am not convinced that they, or any other organisms, may be properly
described as learning. The assumption that worms learn does serve the
purpose of polarizing null hypotheses, thereby promoting the production
of dissertations and journal publications, and giving rise to controversies
and symposia. The assumption of learning in planarians also carries with
it the implication that all learning laws are held in common by all organisms.
This is a convenient belief, but it is more supported by fiction than fact.
Bitterman (1965a and b) has recently revived the notion of phyletic comparisons of learning in terms of quantitative and qualitative differences.
In spite of his stimulating research and provocative ideas, in my opinion
we just do not have enough data on a sufficient enough number of organisms
to support such a unified theory. I am much more inclined to believe that
an organism's capabilities are determined by morphological and biochemical factors which are not necessarily restricted to the development
of the nervous system. Rather than speculate that worms and other
organisms learn, develop habit formation, or demonstrate mental states,
we should restrict ourselves to performance measures of amplitude,
frequency, duration, and analogous measures. We should be better able
to systemize our data by restricting ourselves to such measures in defining

stimulus and response events and in varying our contingencies and con-
tiguities in the study of the behavior of organisms.

It would be convenient if we could establish a biochemical basis for
learning so that either a direct or correlated measure could be used as
an index. McConnell (1965a) has recently discussed the implication of
RNA and learning, and Gaito (in this volume) has summarized some of the
various theoretical points of view. A somewhat different approach has been
attempted in a recent study by Crawford, King, and Siebert (1965). Worms
were examined for differential amino acid groups, and Pearson product-
moment correlations between amino acid densities and performance were
determined. A follow-up study (King, Crawford, and Klingaman, 1965)
employed regenerated pairs of worms in an attempt to provide improved
biochemical and behavioral control. Although the correlations were in the
right direction and there was a trend toward increasing correlation with
increased training, the statistically significant correlations of the first
study were not found. It was also not clear whether the lack of positive
results was due to the effects of regeneration or differences in experimental
procedure or both. In another, unpublished, study by Crawford, Magos,
King, and Magos, classically conditioned worms were compared with a
light-only control group. In this experiment, as in all the experiments from
our laboratories in which we have worked with classical conditioning,
we have used apparatus essentially like that described by McConnell,

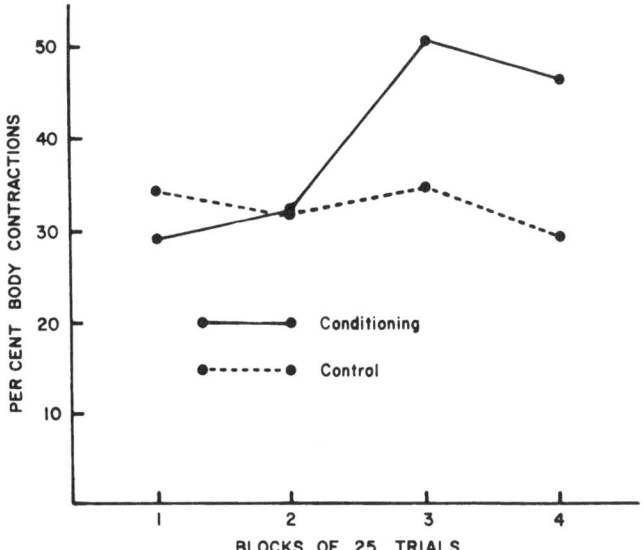

Fig. 7. Classically conditioned worms compared with a light-only
control group.

Cornwell, and Clay (1960). The behavioral data are shown in Fig. 7 where it can be seen that once again it looks as though the worms have learned. Amino acid analyses have also been run on untrained worms, and upon pseudoworms which have been given light-shock pairing and light alone and also have been left untrained. Our pseudoworms are portions of beef liver the approximate size of planarians. The worms from the study on operant behavior, spontaneous recovery, and backward conditioning previously described have also been given an amino acid analysis. The amino acid analyses are done by paper chromotography using intact worms. The solvent is a 4:1:1 solution of butanol, water, and glacial acetic acid. A 0.2% ninhydrin in saturated butanol solution is used as the developer. The chromatograms are placed in the heat chamber and then read for optical density when dried. The amino acids are usually grouped in four to seven spots, most often six, and we have found that spots 3 and 4 appear to be the most related to behavior. Spot 3 includes L-aspartic acid; glutamine; taurine; citrulline; DL-methionine; 3,4-dihydroxyphenylalanine; and serine. Spot 4 includes glutamic acid, L-proline, L-tyrosine, and L-threonine. Typically, spot 3 is positively correlated and spot 4 negatively correlated with behavior. Unfortunately the correlational analysis of the data is incomplete at this time. Preliminary examination of the Crawford, Magos, King, and Magos data indicates that the correlations between terminal behavior of all subjects and amino acids are perhaps meaningful but not always statistically significant. Higher correlations are observed when taken between amino acid spots and the experimental and control groups separately. The per cent distribution of spots for worms and pseudoworms also appears to be different. Low correlations indicating a consistent directional trend are also seen in the study on backward conditioning. Sample correlations from the operant and spontaneous recovery study are essentially similar, being 0.30 to 0.40 and 0.60 to 0.70, respectively. A tendency toward correlations in the latter two studies may perhaps be due to better response definition in the operant study and the larger number of worms employed in the spontaneous recovery study. Limiting factors in obtaining greater correlations are the inexact nature of paper chromatography, the variability of behavior, and the small number of worms employed. Another consideration is the smaller number of training trials employed in our later studies. We are also unable to say whether better results are obtained on acquisition data or on extinction data. Our first study (Crawford, King, and Siebert, 1965) obtained positive results following extinction. Jacobson, Fried, and Horowitz (1966) have recently failed to observe differential effects for extinction in their study with RNA. It is too early to evaluate these studies properly, but in the case of the amino acid experiments the best that may be hoped for is an index of performance, similar to an IQ or test score, which is related to the performance not of any individual subject but only of the group. We are a long way from taking a biochemical sample from an individual and using it to evaluate learning, performance, personality, emotion, or other psychological states.

REFERENCES

Austad, E., A preliminary attempt at food-reward conditioning in planarians. *Worm Runner's Digest*, 1965, **7**(2), 41.

Barnes, C. D., and B. G. Katzung, Stimulus polarity and conditioning in planaria. *Science*, 1963, **141**, 728.

Baxter, R., and H. D. Kimmel, Conditioning and extinction in the planarian. *Amer. J. Psychol.*, 1963, **76**, 665.

Bennett, E. L., and M. Calvin, Failure to train planarians reliably. *Neurosci. Res. Progress Bull.*, 1964, **2**(4), 3.

Best, J. B., Diurnal cycles and cannibalism in planaria. *Science*, 1960, **131**, 1884.

Best, J. B., Protopsychology. *Sci. Amer.*, 1963, **208**(2), 55.

Best, J. B., Behavior of planaria in instrumental learning paradigms. *Animal Behav.*, Suppl. 1, 1965, **13**, 69.

Best, J. B., and E. Elshtain, Unconditioned response to electric shock: mechanism in planarians. *Science*, 1966, **151**, 707.

Best, J. B., and I. Rubenstein, Environmental familiarity and feeding in a planarian. *Science*, 1962a, **135**, 916.

Best, J. B., and I. Rubenstein, Maze learning and associated behavior in planaria. *J. Comp. Physiol. Psychol.*, 1962b, **55**, 560.

Bitterman, M. E., Phyletic differences in learning. *Amer. Psychol.*, 1965a, **20**, 396.

Bitterman, M. E., The evolution of intelligence. *Sci. Amer.*, 1965b, **212**(1), 92.

Brown, F. A., Jr., Response of the planarian, *Dugesia*, to very weak horizontal electrostatic fields. *Biol. Bull.*, 1962a, **123**, 282.

Brown, F. A., Jr., Responses of the planarian, *Dugesia*, and the protozoan, *Paramecium*, to very weak horizontal magnetic fields. *Biol. Bull.*, 1962b, **123**, 264.

Brown, F. A., Jr., An orientational response to weak gamma radiation. *Biol. Bull.*, 1963, **125**, 206.

Brown, F. A., Jr., and Y. H. Park, Seasonal variations in sign and strength of gamma-taxis in planarians. *Nature*, 1964, **202**, 469.

Brown, H. M., Experimental procedure and state of nucleic acids as factors contributing to "learning" phenomena in planaria. *Unpublished doctoral dissertation*, University of Utah, 1964.

Brown, H. M., Jr., and E. C. Beck, Does learning in planaria survive regeneration? *Fed. Proc., Abstr.*, 1964, **23**, 254.

Cordaro, L., and J. R. Ison, Psychology of the scientist: X. Observer bias in classical conditioning of the planarian. *Psychol. Rep.*, 1963, **13**, 787.

Corning, W. C., Evidence of right-left discrimination in planarians. *J. Psychol.*, 1964, **58**, 131.

Corning, W. C., Retention of a position discrimination after regeneration in planarians. *Psychon. Sci.*, 1966, **5**, 17.

Corning, W. C., and E. R. John, Effect of ribonuclease on retention of conditioned response in regenerated planarians. *Science*, 1961, **134**, 1363.

Cornwell, P., Classical conditioning with massed trials in the planarian. *Worm Runner's Digest*, 1960a, **2**, 34.

Cornwell, P., A preliminary study of pseudo-conditioning under conditions of massed stimulation in the planarian. *Worm Runner's Digest*, 1960b, **2**, 101.

Crawford, F. T., F. J. King, and L. E. Siebert, Amino acid analysis of planarians following conditioning. *Psychon. Sci.*, 1965, **2**, 49.

Crawford, F. T., and L. W. King, Spontaneous recovery of a classically conditioned response in the planarian. *Psychon. Sci.*, 1966, **6**, 427.

Crawford, F. T., P. A. Livingston, and F. J. King, Distribution of practice in the classical conditioning of planarians. *Psychon. Sci.*, 1966, **4**, 29.

Cummings, S. B., and C. C. Moreland, Sensitization vs. conditioning in Planaria: some methodological considerations. *Amer. Psychol., Abstr.*, 1959, **14**, 410.

DeBold, R. C., W. R. Thompson, and C. Landraitis, Differences in responses to light between two species of planaria: *Dugesia tigrina* and *D. dorotocephala*. *Psychon. Sci.*, 1965, **2**, 79.

Ernhart, E. N., An informal report on two heads being better than one. *Worm Runner's Digest*, 1960, **2**, 92.

Ernhart, E. N., Conditionability of two-headed planaria. Paper read at Midwestern Psychological Association, Chicago, May, 1961 (*described by* Jacobson, 1963).

Ernhart, E. N., and C. Sherrick, Retention of a maze habit following regeneration in planaria (*D. maculata*). Paper read at Midwestern Psychological Association, St. Louis, May, 1959 (*described by* Jacobson, 1963).

Fantl, S., and J. A. Nevin, Classical discrimination in planarians. *Worm Runner's Digest*, 1965, **7**(2), 32.

Gaito, J., The possible role of RNA in learning and memory events, this volume, p. 23.

Griffard, C. D., Classical conditioning of the planarian *Phagocata gracilis* to water flow. *J. Comp. Physiol. Psychol.*, 1963, **56**, 597.

Griffard, C. D., and J. T. Peirce, Conditioned discrimination in the planarian. *Science*, 1964, **144**, 1472.

Halas, E. S., R. L. James, and C. S. Knutson, An attempt at classical conditioning in the planarian. *J. Comp. Physiol. Psychol.*, 1962, **55**, 969.

Halas, E. S., R. L. James, and L. A. Stone, Types of responses elicited in planaria by light. *J. Comp. Physiol. Psychol.*, 1961, **54**, 302.

Halas, E. S., R. C. Mulry, and M. DeBoer, Some problems involved in conditioning planaria: electrical polarity. *Psychol. Rep.*, 1962, **11**, 395.

Hall, J. F., *The psychology of learning*. Philadelphia: J. B. Lippincott Co., 1966.

Hartry, A. L., P. Keith-Lee, and W. D. Morton, Planaria: memory transfer through cannibalism reexamined. *Science*, 1964, **146**, 274.

Haynes, S. E., L. B. Jennings, and P. H. Wells, Planarian learning: nontransfer and nonfacilitation in a van Oye maze. *Amer. Zool., Abstr.*, 1965, **5**, 713.

Hilgard, E. R., and G. H. Bower, *Theories of learning*, 3rd ed., New York: Appleton-Century-Crofts, Inc., 1966.

Humphries, B., Maze learning in planaria. *Worm Runner's Digest*, 1961, **3**, 114.

Jacobson, A. L., Learning in flatworms and annelids. *Psychol. Bull.*, 1963, **60**, 74.

Jacobson, A. L., Learning in planarians: current status. *Animal Behav., Suppl.* 1, 1965, **13**, 76.

Jacobson, A. L., Classical conditioning and the planarian, this volume, p. 195.

Jacobson, A. L., C. Fried, and S. D. Horowitz, Planarians and memory. *Nature*, 1966, **209**, 599.

James, R. L., and E. S. Halas, A reply to McConnell. *Psychol. Rec.*, 1964a, **14**, 21.

James, R. L., and E. S. Halas, No difference in extinction behavior in planaria following various types and amounts of training. *Psychol. Rec.*, 1964b, **14**, 1.

Jensen, D. D., Paramecia, planaria, and pseudo-learning. *Animal Behav., Suppl.* 1, 1965, **13**, 9.

Kimble, G. A., *Hilgard and Marquis' conditioning and learning*, 2nd ed., New York: Appleton-Century-Crofts, Inc., 1961.

Kimmel, H. D., and V. L. Harrell, Differential conditioning in the planarian. *Psychon. Sci.*, 1964, **1**, 227.

Kimmel, H. D., and V. L. Harrell, Further study of differential conditioning in the planarian. *Psychon. Sci.*, 1966, **5**, 285.

Kimmel, H. D., and R. M. Yaremko, Effect of partial reinforcement on acquisition and extinction of classical conditioning in the planarian. *J. Comp. Physiol. Psychol.*, 1966, **61**, 299.

King, F. J., F. T. Crawford, and R. L. Klingaman, A further study of amino acid analysis and conditioning of planarians. *Psychon. Sci.*, 1965, **3**, 189.

Lee, R. M., Conditioning of a free operant response in planaria. *Science*, 1963, **139**, 1048.

Lee, T., An investigation of alternation behavior under conditions of free and forced choice. *Worm Runner's Digest*, 1964, **6**, 42.

McConnell, J. V., Memory transfer through cannibalism in planarians. *J. Neuropsychiat.*, *Suppl.* 1, 1962, **3**, S42.

McConnell, J. V., On the turning of worms: a reply to James and Halas. *Psychol. Rec.*, 1964, **14**, 13.

McConnell, J. V., Cannibals, chemicals, and contiguity. *Animal Behav.*, *Suppl.* 1, 1965a, **61**.

McConnell, J. V., ed., *A manual of psychological experimentation on planarians.* Ann Arbor: *Worm Runner's Digest*, 1965b.

McConnell, J. V., Comparative physiology: learning in invertebrates. *Ann. Rev. Physiol.*, 1966a, **28**, 107.

McConnell, J. V., Specific factors influencing planarian behavior, this volume, p. 217.

McConnell, J. V., P. R. Cornwell, and M. Clay, An apparatus for conditioning planaria. *Amer. J. Psychol.*, 1960, **73**, 618.

McConnell, J. V., A. L. Jacobson, and D. P. Kimble, The effects of regeneration upon retention of a conditioned response in the planarian. *J. Comp. Physiol. Psychol.*, 1959, **52**, 1.

McConnell, J. V., R. Jacobson, and B. M. Humphries, The effects of ingestion of conditioned planaria on the response level of naive planaria: *a pilot study* (or: "you are what you eat??"). *Worm Runner's Digest*, 1961, **3**, 41.

McConnell, J. V., R. Jacobson, and D. M. Maynard, Apparent retention of a conditioned response following total regeneration in the planarian. *Amer. Psychol.* (*Abstr.*), 1959, **14**, 410.

McConnell, J. V., and G. J. Mpitsos, Effects of the presence or absence of slime on classical conditioning in planarians. *Amer. Zool.* (*Abstr.*), 1965, **5**, 653.

Oye, P. van, Over het geheugen bij de Platwormen en andere Biologische waarnemingen bij deze dieren. *Natuurw. Tijdschr.*, 1920, **2**, 1.

Parker, G. H., and F. L. Burnett, The reactions of planarians, with and without eyes, to light. *Amer. J. Physiol.*, 1900, **4**, 373.

Pearl, R., The movements and reactions of fresh-water Planarians; a study in animal behavior. *Quart. J. Micr. Sci.*, 1903, **46**, 509.

Pickett, J. B. E., III, L. B. Jennings, and P. H. Wells, Influence of RNA and victim training on maze learning by cannibal planarians. *Amer. Zool.* (*Abstr.*), 1964, **4**, 411.

Ragland, R. S., and J. B. Ragland, Planaria: interspecific transfer of a conditionability factor through cannibalism. *Psychon. Sci.*, 1965, **3**, 117.

Reynierse, J. H., D. D. Jensen, and O. F. Schroeder, Shock duration and adhesion in the planaria, *Phagocata gracilis*. *Psychon. Sci.*, 1966, **4**, 335.

Rice, G. E., Jr., and R. H. Lawless, Behavior variability and reactive inhibition in the maze behavior of *Planaria dorotocephala*. *J. Comp. Physiol. Psychol.*, 1957, **50**, 105.

Shafer, J. N., and C. D. Corman, Response of planaria to shock. *J. Comp. Physiol. Psychol.*, 1963, **56**, 601.

Shinkman, P. G., and D. R. Hertzler, Maze alternation in the planarian, *Dugesia tigrina*. *Psychon. Sci.*, 1964, **1**, 407.

Shinkman, P. G., and C. L. Kornblith, Comment on observer bias in classical conditioning of the planarian. *Psychol. Rep.*, 1965, **16**, 56.

Taliaferro, W. H., Reactions to light in *Planaria maculata*, with special reference to the function and structure of the eyes. *J. Exp. Zool.*, 1920, **31**, 59.

Thompson, R., and J. McConnell, Classical conditioning in the planarian, *Dugesia dorotocephala*. *J. Comp. Physiol. Psychol.*, 1955, **48**, 65.

VanDeventer, J. M., and S. Ratner, Variables affecting the frequency of response of planaria to light. *J. Comp. Physiol. Psychol.*, 1964, **57**, 407.

Vattano, F. J., and J. W. Hullett, Learning in planarians as a function of interstimulus interval. *Psychon. Sci.*, 1964, **1**, 331.

Walker, D. R., Memory transfer in planarians: an artifact of the experimental variables. *Psychon. Sci.*, 1966, **5**, 357.

Walker, D. R., and G. A. Milton, Memory transfer vs. sensitization in cannibal plan-
    arians. *Psychon. Sci.*, 1966, **5**, 293.
Walter, H. E., The reactions of Planarians to light. *J. Exp. Zool.*, 1908, **5**, 35.
Wells, P. H., L. B. Jennings, and M. Davis, Conditioning planarian worms in a van Oye
    type maze. *Amer. Zool. (Abstr.)*, 1966, **6**, 295.
Westerman, R. A., Somatic inheritance of habituation of responses to light in planarians.
    Science, 1963, **140**, 676.

Chapter 16

Training Flatworms in a van Oye Maze

Patrick H. Wells

Department of Biology
Occidental College
Los Angeles, California

Classical conditioning by light-shock pairing and two-choice (left-right or light-dark) maze situations, the methods most widely used to train planarian worms, always introduce ambiguities and confounding factors in experimental design. The neutrality of light as a conditional stimulus is only relative (unconditional response to light by naive worms may resemble the conditional response sought in the experiment). It is difficult, therefore, to define conditioning operationally in such a way as to exclude sensitization rigidly from the interpretation of results. It is apparent that electric shock also has disadvantages as a stimulus for use with planarian worms. Shock, when applied to worms, is "whole body" shock which must exert its effect by simultaneous depolarization of a great many cells. The physiological consequences of this have not been defined. It is not at all certain that this experimental design is comparable to similar experiments on vertebrates in which shock, as the unconditional stimulus, is applied to an appendage and the response of the animal is mediated through neuronal channels.

Studies in plastic mazes are nonideal in that free movement of the animals is restricted and prodding may be necessary to initiate locomotion. Choice points do not necessarily offer stimulus alternatives to which the animal is highly adapted to respond, and results may be confounded by the presence or absence of slime trails. Rewards (such as return to "home bowl") and punishments (such as a rerun of the maze) are of uncertain significance in shaping the behavior of worms.

In both the above learning situations animals must be manipulated, handled, or moved prior to measurement of the response, and measurement of the response itself often is subjective. Although most experiments are run blind, decision as to what constitutes a response is highly subjective, and hence variable from laboratory to laboratory. Finally, these experimental situations do not necessarily exploit the repertoire of behavioral patterns adaptive to planarians in their natural environments. In view of their built-in limitations it is not surprising that, in addition to much useful information, light-shock conditioning and two-choice maze learning situations have produced replication difficulties and differences of interpretation, some of which are discussed elsewhere in this volume (see articles by Jacobson and McConnell).

One technique which at least partially avoids the difficulties outlined above is the type of flatworm maze first used by P. van Oye almost 50 years ago (van Oye, 1920). In a van Oye type of experiment worms are housed in a container of water which, upon presentation of an appropriate stimulus, also is the maze. Thus, no handling, prodding, or transferring of the experimental animals precedes any training trial or critical test. The goal consists of a bit of food suspended by a rod or wire so that it is midway between the bottom and the surface of the water in the container. In order to reach the food, worms must move (from whatever initial location) to the surface, across the surface, and down the rod to the goal. Naive worms usually do not reach the goal, but performance improves with experience in an appropriate training regimen.

In recent experiments we have used the van Oye maze to investigate aspects of learning in planarian worms (Haynes, Jennings, and Wells, 1965, and Wells, Jennings, and Davis, 1966). In all of our studies locally collected *Dugesia dorotocephala* were housed in 250-ml beakers of well water,[1] usually 10 worms per container. Glass rods, 3.5 mm in diameter and of suitable lengths, were drawn into bait hooks at one end and attached to other pieces of rod in the form of a T. The arms of the T rest on the sides of the beaker and support the hook when the goal is placed in the water. Rods were constructed so as to place the goal at the surface, 8 mm below the surface, 16 mm below the surface, 24 mm below the surface, or 32 mm below the surface.

A training regimen found to be satisfactory consists of five training trials at each of the five levels, one trial each day for a total of 25 days. During each training trial the goals were baited with pieces of beef liver about 1 cm³ in size. Four trials at each level on alternate days, for a total of twenty trials, also gave significant results (Haynes, Jennings, and Wells, 1965). This is a mass conditioning situation, performance not being closely monitored during training trials. Baited glass rods which reached the bottom of the containers were placed in control-group beakers during training trials, and, in some cases, starved controls were run. During tests of learning, the goal was set at 32 mm below the surface for all groups. Worms reaching the goal during a 1-hour test period (sometimes 1.5 hours) were counted and temporarily removed to other beakers so as to avoid recounts. In tests of this kind which we have performed, worms from trained groups reached the goal in significantly larger numbers than did worms from the control groups.

The above experiments are ambiguous, however, in that they do not clearly distinguish between the roles of conditioning and heightened

[1]Well water supplied by Sparkletts Water Co., 4500 York Blvd., Los Angeles, California.

response to stimulus gradients in worm performance. Chemical gradients radiate from food at the goal, and it is conceivable that worms are being trained to follow these, rather than to solve the maze. To test this possibility, worms were trained in the usual way, then tested with no food in the mazes. Clean, empty goals were placed in the mazes and 0.5 ml of the supernatant from homogenized, centrifuged liver was stirred into each container. In this way the chemical stimulus to which the worms were conditioned was presented in a gradientless fashion. Performance curves for our first experiment of this type are presented as Fig. 1. In a 1-hour test trial, 23 of 60 experimentals and 3 of 60 controls reached the goal (sig. by x^2, $p < 0.01$). Replications of this experiment gave 17 of 50 experimentals and 3 of 50 controls (sig., $p < 0.01$) and 15 of 50 experimentals versus 3 of 50 controls (sig., $p < 0.01$). These results indicated that trained worms can be directed to an empty goal by introduction of a gradientless chemical stimulus.

The van Oye technique also offers the opportunity for stimulus pairing. In two sets of experiments worms were kept in continuous darkness but exposed to light during the 1-hour training period each day for 25 days. In performance tests at the end of the training period, 12 of 50 experimentals versus 3 of 50 control worms reached a clean unbaited goal in 90 min (sig., $p < 0.05$), and in a repetition using more animals, 17 of 80 experimentals

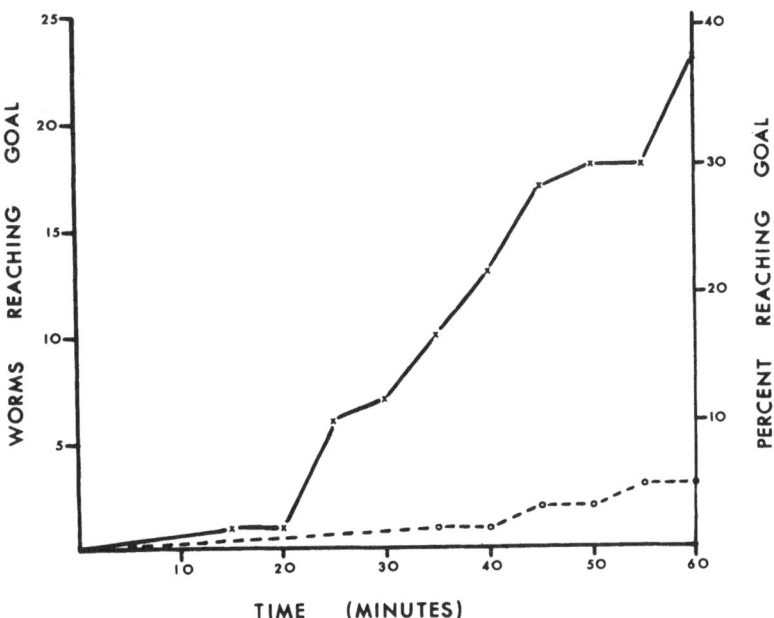

Fig. 1. Performance of trained (solid line) and untrained (broken line) planarian worms in a van Oye type of maze ($n = 60$ in each group). Curves (cumulative at 5-minute intervals) show the number of worms reaching an unbaited goal after exposure to a gradientless chemical stimulus.

and none of 80 controls reached the goal (sig., $p < 0.01$). In the van Oye maze situation, light is an unambiguously neutral stimulus. Presentation of light to naive worms does not result in measurable migration to an unbaited hook suspended 3 cm below the water surface. After its use as conditional stimulus in a light-food paired stimulus training regimen, exposure to light can elicit the conditional response of migration to the empty goal of the van Oye maze. Experiments with the van Oye maze provide new evidence that planarian worms can learn, at least at the level of simple conditioning. The van Oye maze has advantages over more widely used training devices which may make it a useful tool for additional studies of physiology and biochemistry of learning.

ACKNOWLEDGMENTS

I am grateful to the Research and Development Fund of Occidental College for support of the experiments discussed above and to my associates, Dr. Luther B. Jennings and Mr. Michael Davis, for permission to include our previously unpublished data on stimulus pairing.

REFERENCES

Haynes, Susan E., L. B. Jennings, and P. H. Wells, Planarian learning: nontransfer and nonfacilitation in a van Oye maze. *Am. Zoologist*, 1965, **5**, 713.
Jacobson, A. L., Classical conditioning and the planarian, this volume, p. 195.
McConnell, J. V., Specific factors influencing planarian behavior, this volume, p. 217.
Oye, P. van, Over het geheugen bij de Platwormen en endere Biologische waarnemingen bij deze dieren. *Natuurw. Tijdschr.*, 1920, **2**, 1–9.
Wells, P. H., L. B. Jennings, and M. Davis, Conditioning planarian worms in a van Oye type maze. *Am. Zool.*, 1966, **6**, 295.

Chapter 17

Major Factors in Classical Conditioning of Planarians: Stimulus Waveform and Neural Geometry[1]

Jay Boyd Best

Department of Physiology and Biophysics
Colorado State University
Fort Collins, Colorado

The classical conditioning paradigm for planarians of Thompson and McConnell (1955) has been used, with several modifications, for a number of interesting and provocative experiments on the mechanisms of memory storage (McConnell, Jacobson, and Kimble, 1959; Baxter and Kimmel, 1963; Corning and John, 1961; McConnell, 1964; Barnes and Katzung, 1963; and Hartry et al., 1964). In this paradigm the planarian is usually placed in a narrow plastic trough equipped with an electrode at each end and partially filled with water. The water forms a continuous conducting path between the two end electrodes. A trial consists of several seconds of brilliant illumination, constituting the conditioned stimulus, followed by approximately a second of electric shock, the unconditioned stimulus, administered through the water via the end electrodes. The response of the planarian during the lighted period preceding the shock is scored as a conditioned response. A response during the period of shock is scored as an unconditioned response.

The interpretation of the results of these various experiments has been equivocal, however, because of apparent discrepancies in the results of various laboratories (Bennett and Calvin, 1964; Halas et al., 1962; Van-Deventer and Ratner, 1964; Barnes and Katzung, 1963; and James and Halas, 1964).

It must be strongly emphasized that the discrepancies involve more than a disagreement between the McConnell group and those critical of their results. There are also marked discrepancies among the results of the critics. Thus James and Halas (1964b), VanDeventer and Ratner (1964), and Brown et al. (1966) report on acquisition of the conditioned response but question whether it is learning, whereas Bennett and Calvin (1964) report a failure in obtaining the acquisition.

[1] A portion of the research described was supported by Public Health Service grant No. MH07603 and National Aeronautics and Space Administration grant No. NSG-625.

Such fundamental variations in results suggest that one or more critical variables of the experimental situation have not been adequately described and reproduced. Unfortunately some of the studies which have been critical of the McConnell experiments have themselves done little to clarify the situation. In certain instances these sins of omission are obvious, such as the failure to mention the species used in the Hartry, Keith-Lee, and Morton (1964) study, the cavalier interchange of species in the VanDeventer and Ratner (1964) study, and the conclusion of these latter authors that water composition was unimportant because interchange of "creek water" of unspecified composition with "aged tap water" of unspecified composition produced no behavioral difference. There are others such as not pre-sliming the internal apparatus surfaces, refusal to use the same light pre-adaptation conditions, etc.

Intimation that the kind of electric current pulse used as the unconditioned stimulus may be one such variable is to be found in the study of Barnes and Katzung (1963) in which a monopolar square-wave train was employed instead of the inductorium output used in the Thompson and McConnell (1955), McConnell, Jacobson, and Kimble (1959), and Corning and John (1961) studies. An even earlier suggestion of this is to be found in the study of Halas, Mulry, and Deboer (1962). But while the Barnes and Katzung study indicated relevancy of the electric current wave form of the stimulus, it did not lead to reproducible results in other laboratories. In the Barnes and Katzung experiment, planarians, shocked with a $100 \, \text{sec}^{-1}$ monopolar square wave with a 5-msec pulse duration, exhibited a rapid monotonically increasing acquisition for anodally oriented ones. Bennett and Calvin (1964) did not obtain these results, nor did we in our laboratory.

A curious aspect of these studies has been the fact that the very ill defined "dirty" wave form of the Harvard inductorium has led to the most repeatable results. Thus, even though Bennett and Calvin found their results disappointing, it is clear from an examination of their data that, in those experiments in which the inductorium was used to administer the unconditioned stimulus, they actually did obtain an acquisition of the conditioned response to light. We had found very much the same kind of acquisition with the Harvard inductorium provided we used an oscilloscope to check the wave form and readjusted the vibrator so as to maintain the same kind of wave form.

In earlier papers (Best and Elshtain, 1966a and b) I had, from an examination of the biophysics of nerve excitation and histological information, derived a mathematical model which seemed to provide a rational basis for considering the effect of electrical stimulus wave form. In this paper I shall briefly recapitulate this basic formulation and then extend it to account quantitatively for further experiments in which an electrical unconditioned stimulus is employed in a Thompson and McConnell trough. Such experiments can be used to evaluate the constants of the model, and these values can be compared with those obtained by other workers in direct transmembrane measurements of large cells.

When an organism, such as a planarian, is immersed in a volume conductor, such as water, on which a temporally varying electric field is impressed, there is no reason to expect that the neurons of the central nervous system will be immune to direct excitation by the field. The situation is therefore not strictly analogous to the use of shock as an unconditioned stimulus for some animals, such as a sheep or a rat, in which only peripheral neural processes are subjected to direct excitation by the electric stimulus, for in the latter two cases the neural signal counterpart of the unconditioned stimulus is conducted to the brain via the normal sensory fiber tracts. Physiologically, a more direct analog would be the use of intracranial stimulation, for in these experiments the brain of the sheep or rat (as the brain of the planarian) is subjected to direct excitation by the unconditioned stimulus.

In considering some of these aspects of direct excitation, it is useful to look on a neuron as a sealed-ended tube oriented along the direction of the imposed electric field. This sealed tube is filled with ions which are free to migrate inside the tube but to which the walls and ends are relatively impermeable. The concentration of ions in the tube at position x, measured along the axis of the tube, at time t shall be denoted as $C(x, t)$. The origin $x = 0$ is arbitrarily but conveniently taken to be at the midpoint of the tube. The half length of the tube is L, so the ends of the tube are at $x = +L$ and $x = -L$. The electric field intensity shall be denoted as E, the diffusion coefficient of the ions as D, the gas constant as R, the absolute temperature as T, and the charge per mole of the ions as q.

After a sufficiently long time in the absence of an applied field, i.e., the resting state of the neuron, it is assumed that the ionic concentration becomes uniformly distributed along the neuron. Thus, initially, prior to application of a field,

$$dC/dx = 0 \quad \text{at } t \leq 0 \tag{1}$$

$$C(x, t) = C(0, 0) \tag{2}$$

The flux J of ions along the tube, i.e., the neuron, in the presence of an applied field of strength E will be given by the well-known expression

$$J = -D\frac{dC}{dx} + \frac{DqE}{RT}C \tag{3}$$

If these ions relevant to the excitation process are neither being consumed nor produced to an appreciable extent in the tube during the period in question, then the rate of change of the number of ions in one-half of the tube will be minus the rate of change of the number of ions in the other half. This rate of change will also be the cross-sectional area of the tube multiplied by the flux at its midpoint. The problem is simplified considerably by introducing the approximation

$$\frac{dC}{dx} \cong \frac{C(L) - C(0)}{L} \tag{4}$$

which will effectively imply that the distribution is linear and antisymmetric about the origin, i.e., the midpoint. Thus, within the accuracy of this approximation one can assert that

$$C(0, t) = C(0, 0) \qquad (5)$$

i.e., the concentration at the midpoint will remain constant at its initial value.

Now actually it is not necessary to examine the distribution at all x but only at $x = \pm L$, for clearly these will be the greatest and least values of C. If one supposes that the neuron will excite if the perturbation in C at any point of the bounding surface of the neuron exceeds some threshold value, then it is clear that this value will be exceeded at one of the ends if it is exceeded anywhere. All one needs to consider therefore is the behavior of C at $X = L$.

Even this can be simplified. It is probably only the relative perturbation that is important, i.e., the value of the dimensionless variable

$$y = \frac{C(L) - C(0)}{C(0)} \qquad (6)$$

In keeping with the discussion above, the differential equation describing the behavior of y would be

$$\frac{dy}{dt} + \frac{2D}{L^2} y = \frac{2Dq}{RTL} E \qquad (7)$$

This simple first-order linear differential equation can be readily solved for y. Thus, in the case of a square pulse of amplitude E and duration t, the solution would be

$$y = y(0)e^{(-2D/L^2)t} + \frac{ELq}{RT}\left(1 - e^{-2Dt/L^2}\right) \qquad (8)$$

The value of y would be zero at the beginning of a single pulse applied to a neuron initially at rest. The maximum value of the perturbation would occur at the end of the pulse. Thus, the maximum value of y occurring from a single pulse of duration t applied to a neuron of semilength L would be

$$y = \frac{ELq}{RT}\left(1 - e^{-2Dt/L^2}\right) \qquad (9)$$

The neuron will fire if y exceeds some critical threshold y^*. Hence, the condition of excitation is that

$$y \le y^* \qquad (10)$$

Using relations (9) and (10), one could therefore write the condition of excitation as

$$1 \le \frac{ELq}{RTy^*}[1 - e^{(-2D/L^2)t}] \qquad (11)$$

A very useful simplification in the form of relation can be achieved by defining a parameter λ, which shall be called the *rheobasic potential*. The reason for this designation will become apparent shortly. Thus, by defining

$$\lambda \equiv RTy^*/q \qquad (12)$$

one can write relation (11) in the form

$$1 \leq \frac{EL}{\lambda}(1 - e^{-2Dt/L^2}) \qquad (13)$$

From the definition of λ and the assumption that y^* is the same for all the neurons of the ensemble comprising the nervous system of the flatworm, it follows that λ will be the same for all the neurons of the ensemble. This assumption is subject to experimental test.

Consider the situation when, instead of a single pulse, a train of square pulses is used as the wave form of the electric stimulus field. Let t' designate the separation between pulses, and t, the width of one of the pulses. The amplitude of the pulses is E. If the number of pulses in the pulse train is sufficiently large, then the condition that a neuron of semilength L will be excited by the train is, according to our model, that

$$1 \leq \frac{EL}{\lambda} \frac{1 - e^{-2Dt/L^2}}{1 - e^{-2D(t+t')/L^2}} \qquad (14)$$

Thus relations (13) and (14) are comparable. Relation (13) described the condition of excitation for a neuron of semilength L by a single pulse, while relation (14) describes it for a train of such pulses.

It is worth discussing the nature of relations (13) and (14). The right-hand sides of both these relations can be identified with y/y^*. The essential features are most simply evident from relation (13), for which the behavior of y/y^* is shown in Fig. 1 for small and large values of L. It is interesting that the quantity $L^2/2D$ behaves as a time constant for the neuron.

An approximate value for this time constant can be calculated from available physical chemical data and the geometry of the neuron. At 25°C and 0.01 M concentration,

$$D \cong 10^{-5} \, cm^2/sec$$

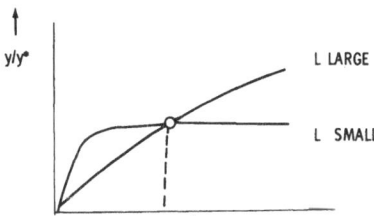

Fig. 1. Predicted buildup of excitatory state for large and small neurons following application of constant electric field at zero time.

for most of the ions, e.g., Na^+, K^+, Cl^-, Ca^{++}, and HCO_3^-, which might plausibly be involved in this process of direct electric excitation. One can, therefore, compute the time constant to be anticipated for various lengths of neurons. Thus the time constants $L^2/2D$ for neurons of lengths 2, 6, and 10 μ would be 0.5, 4.5, and 15 msec, respectively.

Notice in Fig. 1 that the value of y/y^* climbs more sharply as a function of t for small neurons than large ones, but plateaus at a lower level. The two curves cross at t^*. For values of t less than t^*, values of E can be chosen which will excite small neurons in preference to large ones. For values of t greater than t^*, values of E can be chosen which will excite long neurons in preference to short ones.

The quantity E has the units of field intensity (volts per centimeter), while L has the units of length (centimeters). The product EL, which occurs in both (13) and (14), will therefore have the units of potential (volts). It is clear from this and the forms of (13) and (14) that λ also has the units of potential, i.e., volts or millivolts.

An approximate expression can be given for the duration t^* of the stimulus pulse below which excitation of short neurons is favored and above which excitation of long neurons is favored. If L_1 and L_2 are used to denote the semilengths of classes of short and long neurons, respectively, then

$$t^* = \frac{L_1 L_2}{2D} \tag{15}$$

is the critical duration partitioning L_1 and L_2.

It is interesting to reexamine the Barnes and Katzung (1963) experiment in the light of relation (14). These authors used a monophasic square-wave pulse train in which the separation between pulses was equal to the pulse duration. Thus, under their conditions of experiment,

$$t = t' \tag{16}$$

so that (14) could be written as

$$1 = \frac{EL}{\lambda} \frac{(1 - e^{-2Dt/L^2})}{(1 - e^{-4Dt/L^2})}$$

The threshold value of E which would just produce excitation in a neuron of semilength L with a pulse train of this would be

$$\text{Thresh. } E = \frac{\lambda}{L} \frac{(1 - e^{-4Dt/L^2})}{(1 - e^{-2Dt/L^2})} \tag{17}$$

In the event that

$$t < \frac{L^2}{2D} \tag{18}$$

one can reduce Eq. (17) to a simpler form. Remembering that

$$e^{-4Dt/L^2} = 1 - \frac{4Dt}{L^2} + \text{neglected terms}$$

$$e^{-2Dt/L^2} = 1 - \frac{2Dt}{L^2} + \text{neglected terms}$$

one can write Eq. (17) as

$$\text{Thresh. } E = \frac{2\lambda}{L} \tag{19}$$

Let us now ask the circumstances under which Eq. (19) could be expected to apply and the kind of result which it predicts.

First, from my other paper in this volume and previous work (Morita and Best, 1965 and 1966), we know that the neurons of the brain involve ionic compartments about 2μ long, while those of the ventral nerve cords are at least 20μ long. This means that $L^2/2D$ is at least 50 msec for the ventral nerve cords and probably much longer. The value of t^* dividing those pulse durations which would preferentially excite the brain from those which would preferentially excite ventral nerve cords is about 5 to 10 msec. This means that, for a range of frequencies of about 10 to 40 sec^{-1} of a stimulating pulse train in which $t = t'$, one could anticipate that the ventral nerve cords would be the neurons preferentially subjected to primary excitation with Eq. (19) being essentially correct. A second feature is relevant. One would not expect the neurons of the brain neuropil to vary appreciably with the length of the planarian, but the neurons of the ventral nerve cords could be expected to be more or less proportional to the length of the planarian. One would thus anticipate, as a consequence of Eq. (19), that the threshold E for the planarian, denoted as E_p, would follow the relation

$$E_p \propto \frac{1}{L_p} \tag{20}$$

where L_p refers to the length of the planarian. Thus, one would expect, on the basis of relation (20), that the threshold field-intensity amplitude, for a stimulus train in which $t = t'$, would be independent of frequency (provided the frequency were kept in the range of about 10 to 40 sec^{-1}) and inversely proportional to the length of the planarian.

An experiment was conducted to test the prediction of relation (20). Planarians which had been maintained in the laboratory for a year on raw beef liver (twice weekly feedings) at a temperature of 69 to 72°F were used as subjects. A Grass Model S4 stimulator was used as the stimulus source. Testing for threshold was done by a modified "up-down" method. Pulse trains were 1 sec long. An experimental subject was administered a set of 10 such 1-sec test shocks in one session. The test shocks were administered 1.5 to 2 min apart. An illumination level of about 1 ft-c was used to observe the subjects during the test session. The test trough was preslimed by placing nontest worms in the troughs with water for 8 hr prior to its use in a session, removing the "slimer" worms, then rinsing the trough lightly with the water in which the test worms had been maintained.

The pulse trains employed as a stimulus were monophasic square waves in which $t = t'$, i.e., pulse separation equal to pulse duration. Subjects ranging from 0.9 to 1.6 cm in length when extended were tested using frequencies of 10, 20, 30, and 40 pulses/sec. The subjects were administered the shock only when they were extended, moving, and parallel to the axis of the trough. To avoid electrical "skin" effects, they were tested only when crawling on the bottom of the trough and not on the air-water interface. Although polarity orientation of the S was recorded in each test, there did not in this frequency range appear to be any significant difference in threshold between head-cathodal and head-anodal orientations although the responses elicited in the two cases differed somewhat.

The results are shown in Fig. 2. As predicted by relation (20), the threshold is not dependent upon the frequency (for the domain of frequencies in question), but does depend upon the length of the subject being less for long subjects than for short ones. In Fig. 3 the threshold field intensity E_p is plotted as a function of the reciprocal of the worm length, $1/L_p$. Relation (20) predicts a linear relationship between these with the regression line passing through the origin. As can be seen from Fig. 3, the theoretically predicted line accords well with the experimental points.

Encouraged by this success the analysis was extended further. For some purposes it is interesting to consider the results to be expected from a

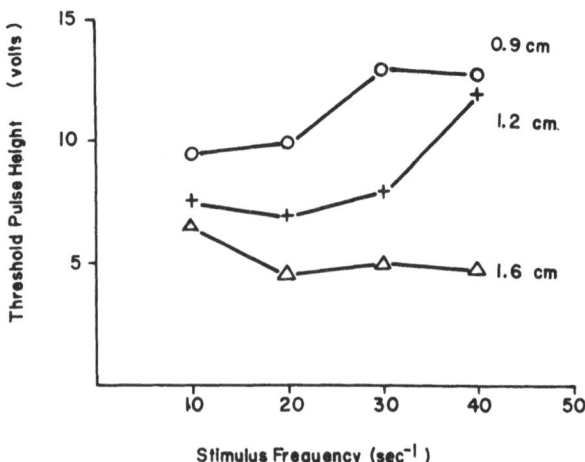

Fig. 2. Threshold pulse height of stimulus signal impressed on trough electrodes as a function of length of planarian and pulse frequency. In all cases, the stimulus was a 1-sec, monophasic, square-wave pulse train in which pulse separation equaled pulse duration.

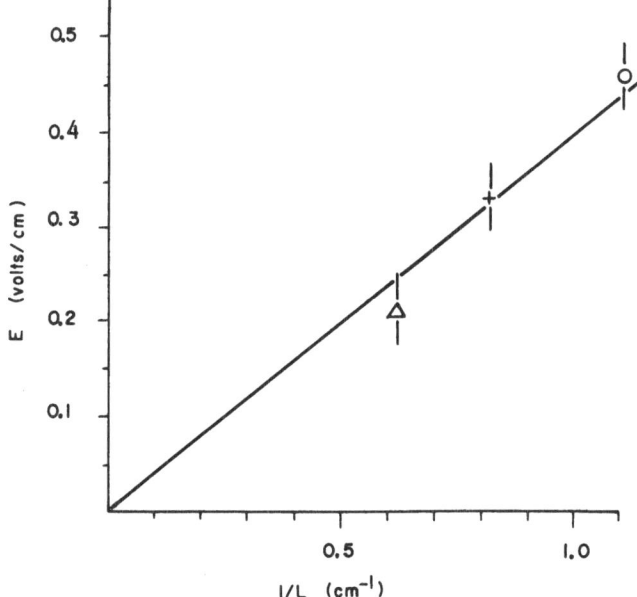

Fig. 3. Threshold field intensity as a function of the reciprocal of the length of the planarian. The theoretical curve is that predicted by relation (20). Experimental points were computed from data of Fig. 2.

single square stimulus pulse. Equation (13) gives the condition for primary excitation of a neuron of semilength L by such a pulse. The threshold field-intensity amplitude would be

$$E_t = \frac{\lambda}{L(1 - e^{-2Dt/L^2})} \tag{21}$$

for a neuron of length $2L$. Suppose that one applied a single square stimulus pulse of duration t to an ensemble of neurons having the same λ but a wide assortment of lengths and adjusted the amplitude to be just large enough to evoke excitation in the most easily excitable neurons of the ensemble. The most easily excited neurons will obviously be those for which E_t is a minimum.

Notice that Eq. (21) can be written in the form

$$E_t = f(t, L)\lambda \tag{22}$$

with

$$f(t, L) \equiv \frac{1}{L(1 - e^{-2Dt/L^2})} \tag{23}$$

Since λ is assumed to be the same for all the neurons of the ensemble, it is clear that E_t will be a minimum for those neurons for which $f(t,L)$ is a minimum. Thus, with the pulse duration t given, those neurons will be most easily excited which are of such a length that $f(t,L)$ is a minimum. Designate the minimum value of $f(t,L)$ with respect to L as $F_m(t)$. It is evident that $F_m(t)$ is a function of t but not of L. The threshold field intensity of the ensemble of neurons will be

$$E_p = \lambda F_m(t) \tag{24}$$

provided that one assumes that the values of L yielding the minimum value of $f(t,L)$ are, in fact, present in the ensemble.

What is the prediction made by Eq. (24)? Equation (24) predicts the pulse amplitude required to just evoke some kind of response from the worm for various values of pulse duration. The only arbitrary parameter which can be manipulated to fit the theoretical curve to the experimental points is λ, which enters into Equation (24) as a multiplicative constant, i.e., a scaling factor.

Figure 4 shows the curves $f(t,L)$. Each of the curves graphs $f(t,L)$ as a function of L for a given value of t. Note that each of these curves has a single minimum point. If each of these minimum values of $f(t,L)$ is plotted for each t, the result is the curve of $F_m(t)$. This is shown in Fig. 5.

For a given set of experimental data, the best estimate of λ can be calculated in the following way: Suppose that S_j is the standard deviation of the experimental value E_{pj} of the threshold pulse height for duration t_j. The estimate of λ will be given by

$$\frac{\sum E_{pj} F_m(t_j)/S_j^2}{\sum_j F_m^2(t_j)/S_j^2} = \lambda$$

Data of this kind was obtained in experiments on three different size classes of the planarian *Dugesia dorotocephala*. These experiments are described in detail elsewhere (Best, Elshtain, and Wilson, 1967). These are shown in Fig. 6.

We found first that the curve described by Eq. (24) fitted the data very well but that the measured value of λ was different for the different size classes. This value for λ was largest for the shortest size class of planarians, and smallest for the largest size class. Thus λ was found to be about 0.43 mv for the 0.7-cm planarians and 0.28 mv for the 1.6-cm ones. The precise reason for this variation is not clear, although I shall suggest one presently. The magnitude of λ is worthy of comment. According to the rationale used to derive Eq. (24), it would seem that 2λ should be interpreted as the amount the membrane must be depolarized for excitation to occur. Values determined by direct transmembrane voltage-clamp measurements on giant axons of squid or crustacea have yielded about 10 to 15 mv as the amount of depolarization requisite for excitation. Even correcting for the

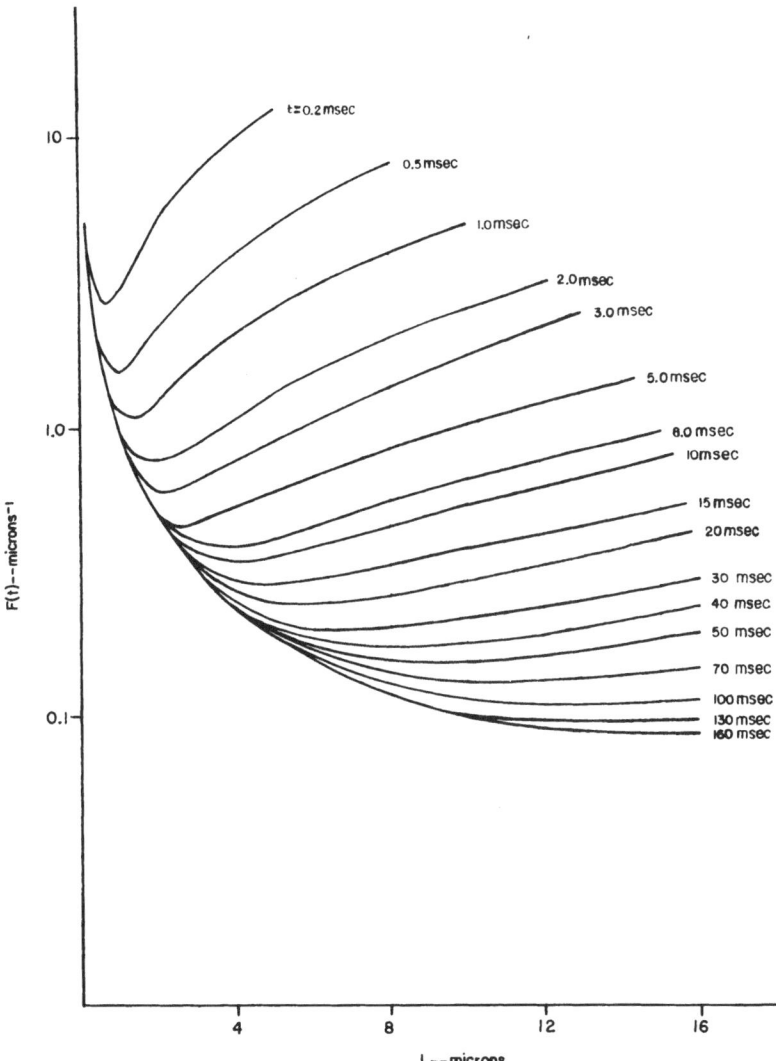

Fig. 4. The function $f(t, L)$ in equation (23) computed as a function of L for various values of pulse duration t.

different criterion of threshold used in these giant-axon experiments and in our own yields a difference of about 10 mv to be accounted for. Thus, on the basis of our model and experiments, the requisite excitatory membrane depolarization would appear to be about 1 mv for our planarians, compared with the 10 to 15 mv found for the giant axons of squid or crustacea.

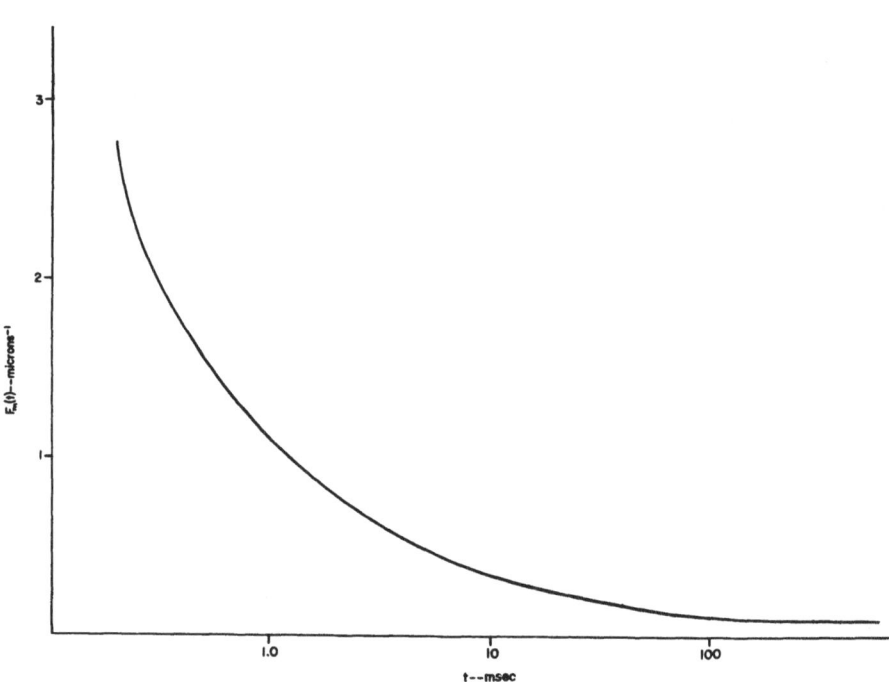

Fig. 5. The function $F_m(t)$ computed as a function of t from the series of minima shown in Fig. 4.

This discrepancy was somewhat disconcerting to us at first because one of the generalizable findings of electrophysiology has been the similarity of electrical properties of cell membranes across a wide range of cells. Thus, it has been found that the resting transmembrane potential tends to run about 75 to 80 mv whether the cell studied is the giant axon of the squid *Loligo*, the axon of the sartorius muscle of the frog, or that of the water plant *Nitella*. Assuming that our excitation model is essentially correct, as it seems to be from other tests applied to it, one would not expect to be an order of magnitude off in the value anticipated for the excitatory membrane depolarization.

A very strong hint as to the mechanism underlying this discrepancy is provided by another experiment of Frankenhauser and Hodgkin (1957) in which it was found that a fivefold decrease in the calcium ion concentration was sufficient to yield about a 10-v decrease in the amount the membrane must be depolarized to produce excitation. Thus, the lower value which we found for planarian neurons might be accounted for by a lower calcium concentration in the tissues of the planarian. That this may indeed be the case is indicated by the fact that the ionic composition of the medium found optimal for tissue culture of planarian cells is much

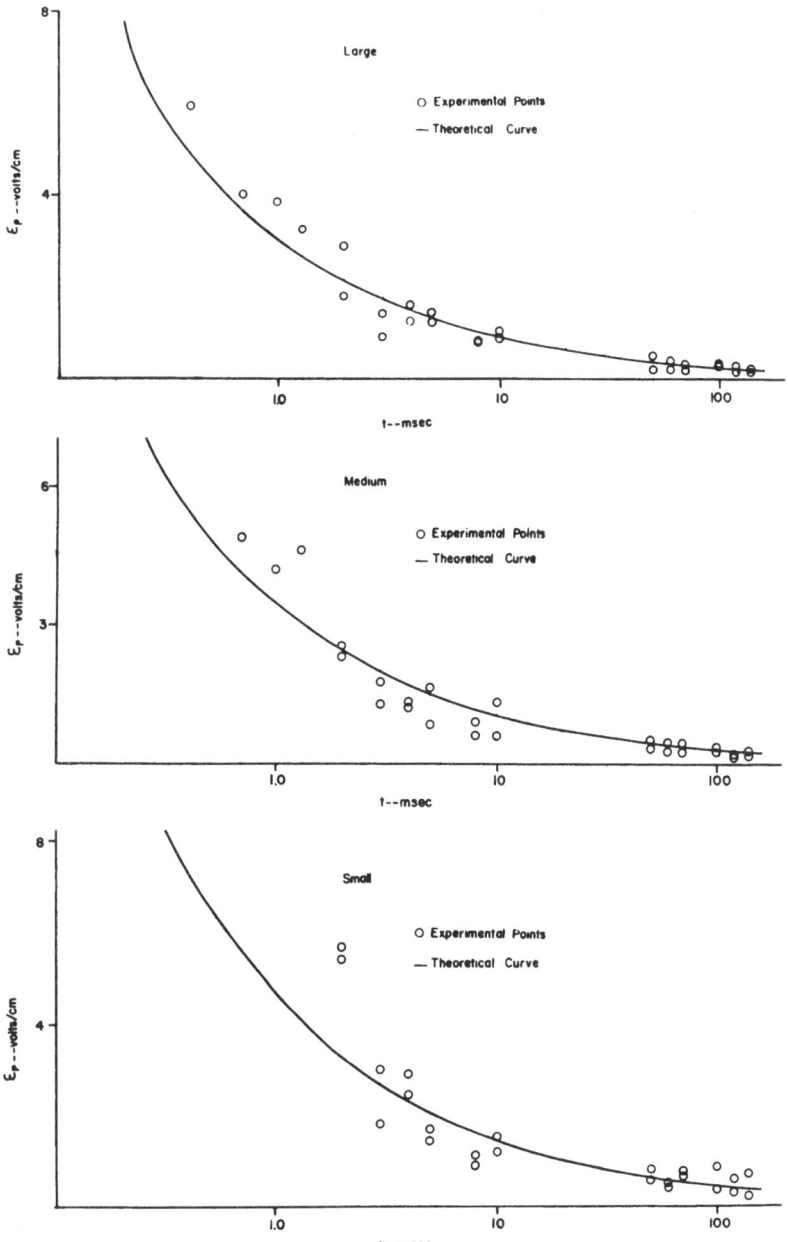

Fig. 6. The threshold field intensity E_p as a function of the pulse duration t for elicitation of a response from small, medium, and large *D. dorotocephala*. This field intensity is the pulse height of a single, monophasic, square-wave pulse. Experimental values are shown as circles. The theoretical curves were obtained from the function $\lambda F_m(t)$ with λ adjusted to minimize the sum of squares of the errors. Values of t are plotted on a logarithmic scale to accommodate the large range of values employed.

more dilute than either sea water or Ringer's solution. Also some preliminary determinations which we have made of the calcium content of planarian tissues gave a calcium content about one-fifth to one-tenth of that normal for the external medium of squid nerve or mammalian serum. We are currently pursuing this problem to ascertain:

1. How the tissue calcium content varies with the concentration in the external medium.
2. How λ varies with tissue calcium content.
3. Whether the difference in λ found for small, medium, and large planarians can be accounted for by a variation in tissue calcium concentration between these different sizes.

Other aspects of the electric-shock stimulation of planarians can be profitably discussed using this biophysical model of the excitation process. The lengths of the neurons which will be preferentially excited can be determined from the model for any specified waveform of the electrical stimulus. From this information and a knowledge of the histological distribution of neural lengths in various regions of the nervous system of the planarian, one can assert which regions of the nervous system will be subjected to primary excitation. By making the reasonable assumption that a contraction will be evoked in those portions of the worm containing the region of nervous system preferentially excited, one can use the model in conjunction with histological information to predict the kind of response which will be evoked by various waveforms. This dependence of the kind of behavioral response upon the stimulus waveform lies very close to the crux of the matter insofar as conditioning experiments are concerned. Just as topographical placement of the implanted electrode is critical for experiments involving intracranial stimulation of mammals or birds, so is this waveform interacting with neural size distribution critical to the planarians' conditioning experiments.

One can make some interesting predictions with regard to the stability of the response evoked by different waveforms. Using a single monophasic square-wave pulse, the threshold of field intensity for evoking some kind of response, irrespective of type, is given by Eq. (24). It will be recalled from the argument used to derive this from Eq. (22) that $F_m(t)$ was the minimum of $f(t,L)$ with respect to L for a given value of t. Corresponding to this minimum, there will be for any given t a unique value of L, which I shall denote as $L_m(t)$. When the stimulus pulse height is just at the threshold E_p for obtaining some kind of response from the planarian, it is the neurons of semilength $L_m(t)$ that will be excited. As the pulse width t is varied, $L_m(t)$ will also vary, and, as it does, one might anticipate a variation in the kind of response evoked by the stimulus pulse. We did an experiment of this kind (Best, Elshtain, and Wilson, 1967) in which we not only scored whether the worm responded or not, but also the kind of response. The stimulus was a single monophasic square-wave pulse with a range of duration from 0.7 to 140 msec. One does not in fact, under these circumstances, observe much variation in the kind of response evoked as the pulse duration is altered. This experimental consequence agrees rather well with the theoretical predictions of the model, since $L_m(t)$ can be computed to be 1.0 μ for the shortest pulse of 0.7 msec and 14.0 μ for the longest pulse of 140 msec. Thus, although the pulse duration was varied over a 166-fold range, the size of the neuron excited varied over only a 14-fold range.

The situation is quite different when pulse trains are used as the stimulus. Under these circumstances the anticipated range for L_m, and consequently the variation of the distribution of response type, becomes much more sensitive to the variation of the pulse width and frequency. Frequency of course would play no role in the single-pulse stimulus situation. It is worth inquiring into the theoretical reasons for this.

Figure 7 shows the behavior of Eq. (14) as a function of L for a succession of increasing pulse widths but with a fixed ratio of pulse width to pulse period. Since we are dealing in this instance with the reciprocal of the threshold pulse height, the higher the curve is, the smaller the pulse height required for excitation. Notice that, as L increases, all the curves start out approximating the line of slope 1, then bend over, and, for large L, approach the line of slope $t/(t+t')$. By slight changes in t and t', one can shift the lengths of neurons preferentially excited over a wide range. Thus, with such pulse trains the type of response evoked will be much more sensitive to alterations in the stimulus characteristics than was the case with the single pulse. Since the ventral cord neurons will vary with the length of the planarian, it is evident from this that the relative amount of primary excitation of ventral cord to head ganglion neurons will, for a given

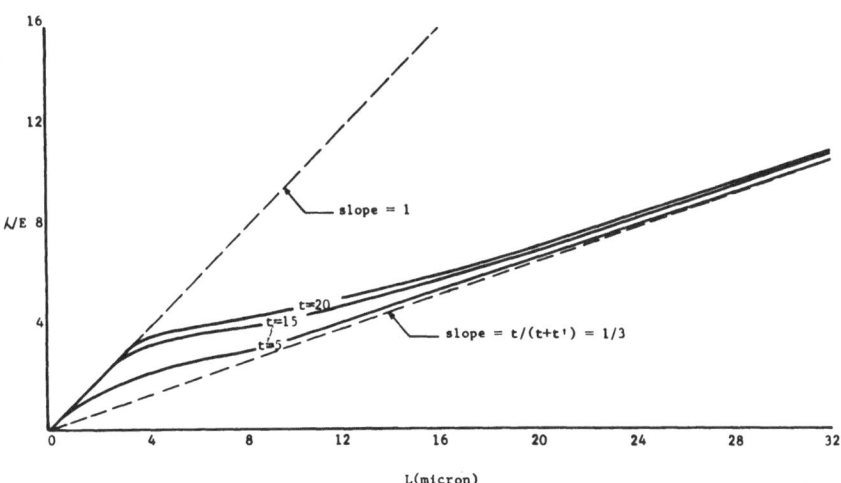

Fig. 7. Behavior of equation (14) as a function of L for increasing pulse widths with a fixed ratio of pulse width to pulse separation in stimulus pulse train.

waveform, depend upon the length of the planarian. By means of the theory, however, one can predict the way in which the waveform must be altered in order to preserve the same primary excitation relations between different portions of the planarians' nervous system.

With the clarification of at least one major source of experimental discrepancies and the resultant insight into the primary excitation mechanisms, there appears to be reason for optimism that the planarians will constitute a well-defined model system in which to study the processes of memory acquisition and storage.

REFERENCES

Barnes, C. D., and B. G. Katzung, Stimulus polarity and conditioning in planaria. *Science*, 1963, **141**, 728.

Baxter, R., and H. D. Kimmel, Conditioning and extinction in the planarian. *Am. J. Psych.*, 1963, **76**, 665.

Bennett, E. L., and M. Calvin, Failure to train planarians reliably. *Neurosci. Res. Program Bull.* (M.I.T.), 1964, **2**, 3.

Best, J. B., and E. Elshtain, Biophysics of unconditioned response elicitation in planarians by electric shock. *Worm Runner's Digest*, 1966a, **8**(1), 8.

Best, J. B., and E. Elshtain, Unconditioned response to electric shock: mechanism in planarians. *Science*, 1966b, **151**, 7r7.

Best, J. B., E. Elshtain, and D. Wilson, Single monophasic square wave electric pulse excitation of the planarian *Dugesia dorotocephala. J. Comp. Physiol. Psychol.*, 1967, **63**, 198.

Brown, H. M., R. E. Dustman, and E. C. Beck, Sensitization in planaria. *Physiology and Behavior*, 1966, **1**, 305.

Corning, W. C., and E. R. John, Effect of ribonuclease on retention of response in regenerated planarians. *Science*, 1961, **134**, 1363.

Frankenhauser, B., and A. L. Hodgkin, The action of calcium on the electrical properties of squid axons. *J. Physiol., London*, 1957, **137**, 218.

Halas, E. S., R. C. Mulry, and M. Deboer, Some problems involved in conditioning planaria: electrical polarity. *Psych. Reps.*, 1962, **11**, 395.

Hartry, A. L., P. Keith-Lee, and W. D. Morton, Planaria: memory transfer through cannibalism reexamined. *Science*, 1964, **146**, 274.

James, R. L., and E. S. Halas, No difference in extinction behavior in planaria following various types and amounts of training. *Psych. Rec.*, 1964a, **14**, 1.

James, R. L., and E. S. Halas, A reply to McConnell. *Psych. Rec.*, 1964b, **14**, 21.

McConnell, J. V., Cannibalism and memory in flatworms. *New Sci.*, 1964, **21**, 465.

McConnell, J. V., A. L. Jacobson, and D. P. Kimble, The effects of regeneration upon retention of a conditioned response in the planarian. *J. Comp. Physiol. Psychol.*, 1959, **52**, 1.

Morita, M., and J. B. Best, Electron microscopic studies on planaria, II: Fine structure of the neurosecretory system in the planarian *Dugesia dorotocephala. J. Ultrastructure Res.*, 1965, **13**, 396.

Morita, M., and J. B. Best, Some observations on the fine structure of the planarian nervous tissue. *J. Exp. Zool.*, 1966, **161**, 391.

Thompson, R., and J. V. McConnell, Classical conditioning in the planarian, *Dugesia dorotocephala. J. Comp. Physiol. Psychol.*, 1955, **48**, 65.

VanDeventer, J. M., and S. C. Ratner, Variables affecting the frequency of response of planaria to light. *J. Comp. Physiol. Psychol.*, 1964, **57**, 407.

Discussion

E. Halas: I was very interested in Dr. Jacobson's comments on acquisition and extinction being two different measurements of learning, and I would agree that they are two different measurements of learning. The reason that we have used extinction is that it is one method of trying to differentiate between this nasty problem of conditioning and sensitization, conditioning and pseudoconditioning, and so on. It is, for instance, a way of distinguishing between sensory adaptation and habituation.

In the work reported here, his data look very good. They differ from our data, however, in that we have never observed the low response rates in our pseudoconditioning control groups. When healthy animals have been trained, given a rest period, and then tested under extinction conditions, we find higher response rates. This is the major difference between our studies: Jacobson has been able to observe differences between experimental and control groups, while we have not. The reasons for these opposite findings are unknown.

Now one reason we have used extinction is that we have observed, from time to time in our control and experimental groups, a response enhancement, and the question becomes: Why this increase in responsivity? Is it due to conditioning, to sensitization, to temperature changes, etc.? We know from Ratner's experiments that temperature can affect responses to light. Factors such as these might be avoided by giving the animals a 24-hour rest period and then testing them. This will better differentiate between the groups on whether or not conditioning has occurred. Our own experience has been that these animals are highly variable; as an example, changing their water can make them sensitive.

A. Jacobson: As far as getting into the particular details of experiments in various laboratories is concerned, these have been covered in the exchange between Dr. Halas and Dr. McConnell (James and Halas, 1964a and b, and McConnell, 1964). There are a couple of points I should like to make here. It is really irrelevant whether I personally like acquisition as a measure of learning or whether Halas prefers extinction. My contention is that either measure, under appropriate conditions, can give you differences between experimental and control groups. A second point concerns those studies of Ratner and others which demonstrate that certain variables such as temperature can affect response tendencies; but, if you stop there and imply that these factors account for the differences between experiments, then you are not doing justice to the true situation. For one thing, the temperature variations that produced the effects were much

271

more extreme than the levels maintained in our laboratory and in the laboratories of others. Secondly, our control groups and the experimental groups are kept under the same conditions; we run them concurrently so that it would be difficult to attribute differences between groups to something like temperature. We do not test the experimental group one day then change conditions and test the controls the next day. The testing is carried out on all groups at the same time.

E. Bennett: Both Dr. Jacobson and Dr. McConnell have incorrectly referred to our conclusions (Bennett and Calvin, 1964), which state, ". . . at the present time the relevant factors necessary for reproducible and reliable training of planarians have not been described with sufficient precision to permit their use in the studies of the biochemical bases of learning." As researching men, they hurt themselves by taking out part of the conclusion in abstract form, by eliminating the word *reliable* in Jacobson's case, and by failing to indicate that we were really stating the need for a more exact definition of how to train, in the case of McConnell.

In addition, Jacobson refers to some of our data in which we did achieve some apparent training. I should like to point out that in these two experiments there was no control for observer bias. In the remaining experiments that were done, there was less evidence of conditioning; the data Jacobson cites represent our more successful attempts.

The thing we were trying to emphasize is that there are many factors which are involved in planarian training, if indeed they can be trained. It would be useful if these factors could be consistently described, especially for the biochemist who doesn't want to spend his life training. Some variables are known, but there are some inconsistencies. Jacobson reports a 60% level of response with training, yet, in studies in which people have not been able to repeat the regeneration and cannibalism findings, we see the criticism that they had not trained their animals to a high enough level. At first we were told to look for contractions, turnings, and what-have-you. Now, we are told that only contractions can be considered. Dr. Corning tells us that mass conditioning is fine, yet McConnell criticizes negative studies on the basis that they have used mass conditioning. In some reports we see that light increases responses; in others we see that there is habituation. We are told by some to train in the morning, others advocate the afternoon, and still another tells us to train according to whether the moon is out. I will not deny that any of these factors are important, but how many of these reports represent experiments that have been replicated several times? We need evidence of interexperiment reliability. And one wonders why there are all these inconsistencies.

A. Jacobson: We would certainly agree on the desirability of better specifying the methods for training planarians successfully. We are at the beginning stage in understanding all the variables that are critical. I agree with Bennett that there appear to be all sorts of variables and little evidence as to which ones are critical.

Now with reference to the two successful experiments by Bennett and associates that I referred to, it should be pointed out that these studies utilized planarians obtained from Dr. McConnell's cultures, whereas the less successful portions of the project used subjects obtained from a California supplier. Subject differences could be one reason for some of the variability. Also, in succeeding experiments by Bennett there were changes in the type of shock used, training schedules, etc. The changes in results could be explained by the changes in procedures. The main thing is to encourage scientists to look at the data for themselves. Too many have read the Bennett and Calvin conclusions and not their data.

S. Ratner: I should just like to point out that, after the initial excitement over the chemistry of learning and the realization that the Nobel Prize had vanished, there is now an effort to understand what is going on. Comments about factors emphasize that this is a noisy situation, and we must reduce some of this noise to get a clear picture of the thing we're after. I am reminded of the rat-learning literature where you do not have to go very far back to find absolutely no reference to an intertrial interval. We were unaware at that time of its relevance.

W. Corning: Several years ago, Dr. John and I became interested in using the planarian as a subject in behavioral and biochemical research. Since we had at that time the paper of Thompson and McConnell (1955) indicating that planarians could be conditioned, we expected our problems to lie in the biochemistry and not the behavior. This was not the case. Considerable time and effort was expended in trying to train flatworms by the techniques described by Thompson and McConnell. Finally, a desperation call to Dr. McConnell and some modifications in our techniques produced positive results. In other words, we made an attempt to learn more about the details of his procedures and, when they were combined with our own techniques, we were successful. I would suggest that part of the problem is in the replication of procedures; the studies of Halas and others were never a replication of any one study.

My last comment concerns the Bennett and Calvin paper and the qualifications that Dr. Bennett offers here. It is stated in the summary, "unsuccessful efforts to demonstrate learning in planaria by classical conditioning, habituation, operant conditioning and food reward methods are described." Now if Bennett and Calvin only meant to say that worms could not be trained *reliably*, this message did not get across. I, too, am frequently confronted with the statement, "Calvin says worms can't learn," and with a Nobel, people tend to believe this. However, I think that he is a better chemist than psychologist.

E. Halas: I was very interested in Dr. Corning's comments about procedure and that my procedures were not like the McConnell and Jacobson methods. Which procedures should I follow? They change their procedures every month, and they are still changing. If they had successful methods back in 1955 and 1959, why didn't they keep them?

W. Corning: I think the reply would have to be that, rather than beat a dead horse like the Thompson and McConnell paper, it would be better to improve methods, as McConnell has done and as we have done, and get results which was our experience at Rochester.

M. Cohen: With reference to Dr. Best's paper, I was wondering how he could distinguish between nerve and muscle stimulation. His effects could be due to direct muscle stimulation.

J. Best: There is no real assurance that we are not stimulating muscles. However, I can make predictions as to stimulus effects from the histology. Also, muscles have higher thresholds than neurons.

P. Cornwell: One point that needs to be emphasized is the use of latency as a measure of response. It might help to distinguish between an alpha response and the conditioned response. The latency measure could also help distinguish between two conditioning stimuli. Admittedly this measure could only be helpful if reliable latencies could be obtained.

E. Halas: McConnell has stated that 90% of the articles in scientific literature are difficult to replicate, and therefore it is not too surprising that his studies are difficult to replicate. I don't know where or how he obtains a percentage of 90%, but I agree that at least some experiments are difficult to repeat. But like it or not, as far as science is concerned, replication is "the name of the game," and I would not want to see the day when we accept data on the basis of faith rather than objective replication.

S. Ratner: I have one or two comments on the measure of latency in learning situations. For one, in the classical conditioning situation the amount of time in which one has to work is relatively short and the durations of these responses are often relatively long so that there are technical problems. Another problem alluded to by VanDeventer concerns response topography; when several responses are used as the criterion for the conditioned response, we find in a number of studies that there has been a systematic shift in what we call the "topography" of the response. There is also a shift in latency. Earlier and earlier portions of the response are all that is observed. These systematic shifts are independent of whether there is a behavior modification of the kind we have been talking about. (See Gardner, 1966.)

J. Best: We have some data on precipitation of fissioning by social isolation in planarians which I might include here. Generally, the finding is that, when a number of *D. dorotocephala* are placed in individual bowls, within 48 hours 80 to 85% of them will fission; with two per bowl, about 65% of them will fission; with 5 per bowl, 50% will fission; 10 per bowl, 35%; with 20 per bowl, about 18% to 20% will fission. The effect is not due to slime or some other inhibitory factor exuded by the worms because, if we set up the situation so that the water from a populated bowl is siphoned into a bowl with two worms, the fissioning is still found.

There are further complications. Smaller worms, around 0.8 to 0.9 cm long, show a lower fission rate. If longer worms are observed, there is a higher fission rate.

Now in our colony pans we observed that the population density tended to stay about constant even though we were removing worms. When a large number were removed, we found that the fissioning stopped. After a 6-month period, the effect returned.

What I think is going on here is a sort of feedback mechanism which serves to control population density. As far as we can tell, CO_2 and O_2 are not involved; it does not depend upon slime or any sort of metabolite. With the exclusion of factors such as these, we tend to think that the mechanism is contact; if they have not made contact with one another within a certain length of time, there is some sort of derepression of the fissioning.

Temperature does seem to be involved. If we do not get a temperature variation over a period of time, we can demonstrate the fissioning; after a variation in temperature, we do not get the effect for a period of time. What happens is that every animal that is on the verge of fissioning is caused to fission by the temperature change. This occurs within 48 hours so that if we check a week after the temperature shift, we find no fissioning. I'm not sure what this phenomenon has to do with conditioning experiments and their design, but there does seem to be a social isolation effect.

M. Jenkins: Are these animals asexual?

J. Best: When we first brought them out from Illinois, they were laying eggs, but that has dropped. We have not been able to determine anything from analyses of water composition.

D. Jensen: My comments concern the discrimination paradigm which McConnell was kind enough to refer to as his version of the "Cambridge-Jensen" experiment. I would like to point out that the experiments he mentions do not fit the criterion I mentioned in the 1964 Cambridge symposium (*Animal Behaviour, Suppl.* #1, 1965), that a stimulus other than shock should be used. We do not know what shock does to the animal. Observations in my own laboratory indicate that glands as well as muscles and nerves are stimulated. This is particularly relevant to the study of Griffard and Peirce (1964) because it has been shown that shock causes the animal to adhere to the substrate. This adhering is not always symmetrical—they may be turned to one side. The nature of the response recorded by Griffard and Peirce is also subject to question. Their experiment represents, I might add, one of those which is not described well enough in the literature to replicate. The situation works only if the animal is in the center of the Petri dish and stays there. There is evidence that the animal may stay there because the shock "glues" it in that position.

We need to understand the effect of shock, particularly with respect to the sense organs. Shock is a broad-band stimulus which might be affecting the sensitivity of the sense organs as well as producing a number of other effects. I would like to suggest that there is a confounding variable which can be thought of as the number of possible stimulus episodes in a particular stimulus modality. In the situation where the conditioned and unconditioned stimuli are paired (assuming that shock does affect the sense organs), there would be one stimulus episode per trial; in the unpaired situation, where the conditioned and unconditioned stimuli are separated, there would be two stimulus episodes. Now if adaptation occurs as a function of a number of stimulus episodes, it might explain the most remarkable things which are found.

H. Brown: I agree with the statement made by Dr. Jensen; it is very important to understand the effects of the stimulus parameters on the sense organs of the planaria so that the results of behavioral experiments utilizing light and electrical current can be intelligently interpreted. There is no question but that the polarity of an electrical field applied across the eye can alter the number of nerve impulses recorded from a single element of the optic nerve (H. K. Hartline, H. G. Wagner, and E. F. MacNichol, *Cold Spring Harbor Symposia on Quantitative Biology*, **17**, 138). I would like to add that the effects of current on photoreceptors are unlike the effects of light. Voltage–current curves obtained from photoreceptor cells (Fourtes, 1959) indicate that they are electrically inexcitable (H. Grundfest, *Cold Spring Harbor Symposia on Quantitative Biology*, **30**, 5), i.e., they act as ohmic resistors and changes of membrane conductance occur only when the membrane is activated by an appropriate stimulus, in this case, light.

J. McConnell: Even if you use something like touch as an unconditioned stimulus, you cannot avoid affecting the whole worm. The animal will contract, which causes the eyespots to be compressed or elongated, etc.

P. Applewhite: Regarding shock effects, we once took some of our smallest rotifers, which were about 90 μ in size, and subjected them to a large number of shocks administered every 2 seconds over a period of a few minutes. We cut them longitudinally and looked at them under the electron microscope. We found no evidence of structural damage. This doesn't mean that no damage occurred, but there wasn't anything obvious. We were using a-c shock.

D. Jensen: DC shock would probably have a greater effect than a-c. Dr. Miller and I have a publication showing that a-c shock delivered to the brains of rats has less effect on the threshold for central stimulation than d-c currents. There is little tissue damage resulting from a-c.

REFERENCES

Bennett, E. L., and M. Calvin, Failure to train planarians reliably. *Neurosci. Res. Program Bull.*, 1964, **2**, 3.
Fourtes, M. G. F., Initiation of impulses in usual cells of *Limulus. J. Physiol.*, 1959, **148**, 14.

Gardner, L. E., Habituation in the earthworm: retention and overhabituation. *Unpublished* doctoral dissertation, 1966, Michigan State University.

Griffard, C. D., and J. T. Peirce, Conditioned discrimination in the planarian. *Science*, 1964, **144**, 1472.

James, R. L., and E. S. Halas, No difference in extinction behavior of planaria following various types and mounts of training. *Psychol. Rec.*, 1964a, **14**, 1.

James, R. L., and E. S. Halas, A reply to McConnell. *Psychol. Rec.*, 1964b, **14**, 21.

McConnell, J. V., On the turning of worms: a reply to James and Halas. *Psychol. Rec.*, 1964, **14**, 13.

Thompson, R., and J. McConnell, Classical conditioning in the planarian, *Dugesia dorotocephala. J. Comp. Physiol. Psychol.*, 1955, **48**, 65.

Van Deventer, J. M., Habituation of responses to tactile stimulation in the planarian. *Unpublished* doctoral dissertation, 1966, Michigan State University.

Chapter 18

Regeneration and Retention of Acquired Information[1]

W. C. Corning

Department of Psychology
Fordham University
New York, New York

INTRODUCTION

In the present paper consideration will be given to the question of whether acquired habits can survive the process of tissue renewal in planarians. Regeneration involves retention in the sense that the tissue which is regrown is a replication of what was lost; cellular differentiation and specification proceeds until former structural and functional systems are restored. As to acquired information, we might ask whether this class of events is also established in systems which are perpetuated during regeneration. If so, then it is reasonable to assume that memory involves physiological substrates which are replicated and that the nature of these substrates might be delineated by effecting changes in specific components of the regenerative process.

PLANARIAN REGENERATION

In general, planarian regeneration proceeds through a series of well-documented stages. The initial tissue responses to a cut or wound serve as protective mechanisms: The surface of the cut is reduced as the surrounding mucles contract and an epithelial layer forms over the wound. Following this, the neoblasts (undifferentiated embryonic cells) migrate to the wound site and form a blastema. The neoblasts are totipotent and are the only source of the new tissue. As pointed out by Sengel in Chapter 8 and also by others (Brønsted, 1955, and Wolff, 1962), the differentiation of the brain appears to be a primary step toward regeneration when the anterior portion of the planarian has been removed. It is the first structure to develop, and its presence is necessary for the induction

[1] This paper was prepared with assistance from grant No. NSG 475, National Aeronautics and Space Administration.

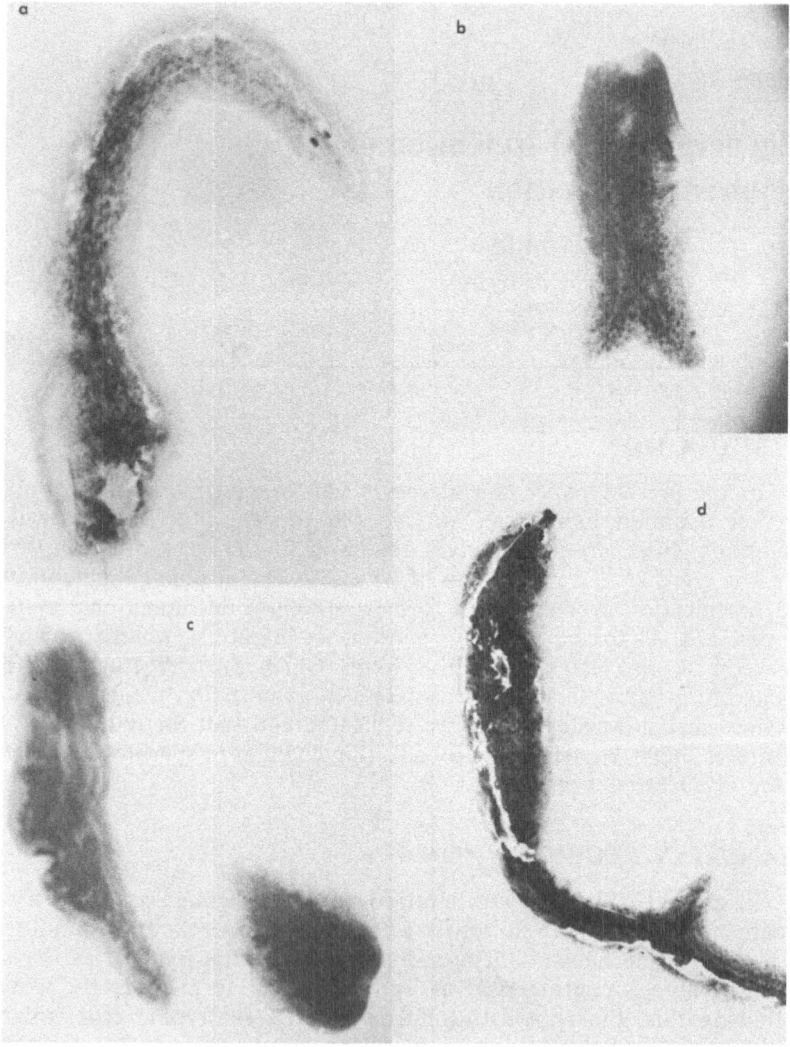

Fig. 1. (a) An animal in which budding has just started (note the bulges at the posterior end). (b) The posterior region of the original animal looked peculiar. As can be seen from the photograph, after the head was removed, the posterior portion looked as though it was going to divide longitudinally. (c) A bud that has just broken away from a posterior piece. The head had been removed one day earlier. (d) In this animal, the bud eventually developed eyespots.

of the eyespots. Neural connectivity between the cephalic ganglia and the eyes is unnecessary, as a brain homogenate will induce eye formation in animals raised without brains. After the establishment of the cephalic region, the more caudal areas differentiate. The completion of the head is followed by induction of a prepharyngeal region, which, in turn, induces pharyngeal development, etc., until the planarian is complete.

Mitotic activity may also be a necessary prerequisite for regeneration in some species. McWhinnie and Gleason (1957) have obtained evidence that mitosis precedes cell migration and differentiation in *Dugesia doroto-cephala*. Regeneration could be prevented by exposing sectioned planarians to colchicine, a substance which prevents cell division.

In addition to the sequential inductions observed during regeneration, there are also complex inhibitory processes at work. The brain, for instance, releases an inhibitory substance which diffuses throughout the planarian and acts to prevent the appearance of duplicate cephalic structures. Its potency decreases with distance along the worm; an anterior section grafted to a homologous region in another planarian will not differentiate any further, whereas, if it is grafted to a posterior area, development is completed. Occasionally, this inhibition is inadequate and freaks result. Jenkins (1963) attributes the appearance of bipolar planarians in her cultures to a lack of the inhibitory substance in the posterior parts of the planarians.

In our own cultures we have frequently observed growth abnormalities which have also been attributed to a breakdown in the potency of the anterior inhibition. The first indication of any abnormality was usually the appearance of a bud in the posterior regions. If the head of the planarian was removed, the bud would develop rapidly, a result due presumably to the complete lack of any anterior dominance. Some of the buds developed eyes, others developed into whole worms and separated from the original, while in other cases several more buds appeared. In all cases, the removal of the head stimulated development of the abnormal growth. Some examples of these abnormalities may be found in Fig. 1.

RETENTION AFTER TISSUE RENEWAL: POSITIVE EVIDENCE

In an attempt to localize the storage of acquired information in the planarian, McConnell and co-workers (1959) conditioned *D. dorotocephala*, sectioned them transversely, and, after regeneration of the two halves, measured the degree of retention by retraining them. One might have predicted that the anterior portion would display the greatest amount of savings since it is in this section of the planarian that the cephalic ganglia are located. Surprisingly enough, both halves displayed considerable retention, and furthermore, their retraining scores were approximately equal. The original subjects took an average of 134 trials to achieve a criterion of 23 responses to light in 25 trials; the regenerated head and tail sections required 40.0 and 43.2 trials, respectively, to reach the same criterion. Control groups demonstrated that regeneration did not sensitize planarians to light and that the level of retention was close to that observed in trained uncut animals left without treatment for the period required for regeneration to be completed in the other groups (4 weeks).

Further work established that the retention is manifested in second-generation animals (McConnell, Jacobson, and Maynard, 1959). Planarians were transected, and the anterior half was conditioned. When the anterior portion had regenerated the posterior region, the planarian was again divided and both halves were permitted to regenerate. Upon retraining, both halves demonstrated savings, although in this case the anterior derivative took significantly fewer trials than the posterior portion.

A major criticism of these two experiments centers around the use of light and shock as the conditioned and unconditioned stimuli. Planarians are photonegative to begin with, and the validity of using light as a conditioning stimulus has been questioned. Furthermore, there is evidence that, under certain conditions, shock can sensitize planarians to light. In reply to these criticisms, it should be pointed out that retention after regeneration is observed when the conditioned stimulus was the cessation of light rather than the onset of light (Cornwell, Cornwell, and Clay, 1961) and in studies (which are discussed below) where neither light nor shock are involved. Later work at Brookhaven National Laboratories indicated that the increment in light reactivity observed in trained subjects was not due to shock "sensitization"; animals that received unpaired light and shock treatments did not achieve criterion and had considerably lower response rates compared with groups receiving the paired light and shock training trials (Freed, 1966).

The basic findings of McConnell and his group were replicated in independent investigations conducted at the University of Rochester under the direction of E. R. John (Corning and John, 1961, and John, 1964). Our general procedures were as follows: (1) All subjects (*D. dorotocephala*) were first presented a series of habituation trials in which they only received the conditioned stimulus (light); these habituation trials served to reduce the naive response to light (see Fig. 2b). (2) They were then given training trials consisting of light-shock pairings as depicted in Fig. 2a. (3) When the subjects reached a criterion of 34 to 40 correct responses, they were sectioned transversely and allowed 14 to 18 days to regenerate. (4) Following regeneration, they were given extinction trials for 3 days and then retrained to criterion. The measures of retention were both the extinction data and the number of trials to criterion during retraining.

One major procedural modification in these studies was the utilization of a weaker light source. This modification was incorporated to minimize the possibility of pseudoconditioning and consequently to obtain a less equivocal assessment of experimental treatments.[2] Preliminary observations

---

[2] In later investigations carried out under similar conditions the strength of the light measured at the trough was 12 to 15 ft-c. Ambient illumination was less than 1 ft-c.

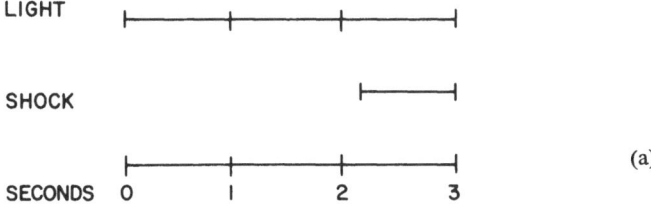

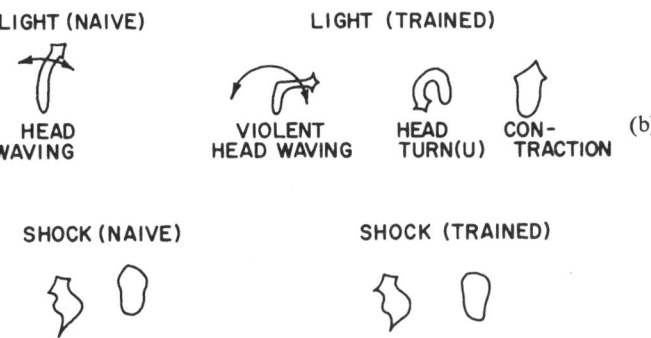

Fig. 2. (a) Relationship and duration of light and shock stimuli used to condition planarians. (b) Types of responses elicited by light and shock in naive and trained planarians.

had suggested that very intense light sources frequently elicited high response rates in naive subjects. The use of the weaker source required a greater number of trials to bring planarians up to criterion, but the differentiation between experimental and control groups was much sharper.

The results of these studies essentially confirmed what had been found by McConnell; the response scores of the regenerated head and tail segments were equal and much higher than naive subjects during extinction trials and during retraining trials. It took the regenerated subjects approximately half as many trials to reach criterion as it took the original animals. Evidence of retention was observed in animals tested 4 to 5 weeks later. It appeared that the physiological consequences of conditioning were locked in a system which was unaffected by the regenerative processes and that the posterior section was somehow able to impose the residual traces of acquired experience on the newly developing anterior tissue.

Westerman (1963) has been able to obtain evidence of retention after regeneration for a habituated response. When *D. dorotocephala* were given habituation trials consisting of 3 seconds of light, their reactivity to light diminished during each daily session of 25 trials. This drop in reactivity within each session could be attributed to sensory adaptation rather than habituation, particularly in the light of Brown's electrophysiological data. However, the responsivity of the subjects also decreased over successive days of habituation sessions, which suggests true habituation. The planarians were presented trials until they reached a criterion of 0 responses for 2 consecutive days, or a total of 50 trials without a response. Natural tail drops and surgically divided tails derived from these habituated animals demonstrated considerable retention when they were tested after regeneration; their initial level of responsivity to light was lower and they required fewer trials to achieve criterion.

While the Westerman study demonstrated retention for a different type of learning task, there remained some criticism over the use of light as a stimulus. It is often suggested that results would be much less controversial if a discrimination task were used. There are now several studies which pertain to this comment. Ernhart and Sherrick (1959) were the first to demonstrate that a position discrimination in a T maze is retained in regenerating planarians. Using a water-filled double-unit maze, they were able to train their subjects to a criterion of three consecutive errorless trials (six correct choices). The subjects were transected and after regeneration demonstrated savings of the previously acquired discrimination. The original subjects required an average of 58 trials to achieve criterion; the regenerated anterior and posterior portions took 29.5 and 28.6 trials to achieve the same criterion. Control subjects indicated that regeneration by itself did not predispose planarians to faster learning.

The Ernhart and Sherrick experiment used a darkened "goal-well" as reinforcement for correct negotiation of the maze. Humphries and McConnell, in a hexagonal maze where the planarian was required to make a black-white discrimination, used anodal shock as a punishment for an incorrect choice (1964). The subjects were able to achieve a 73 % correct response level after 150 trials. Tail segments (natural fissioning) which were dropped at the end of 30 trials displayed rather high correct response levels (75 %) when tested 2 to 3 weeks later. Curiously enough, tail drops that occurred after 60 trials exhibited a chance level of performance when tested after regeneration. It is possible that the negativity which is frequently observed in planarians subjected to long series of maze trials is also carried through regeneration. During the first stages of training planarians will display rapid learning, but with continued trials their performance often becomes erratic (Corning, 1964) and they may even show significant reversal of the trained preference (Best, 1962). The earlier tail drops of the Humphries and McConnell study most likely had not yet developed the negative reaction to the maze experience whereas the later tail drops had.

In our own T-maze studies we have been able to replicate the Ernhart and Sherrick finding and certain aspects of other investigations (Corning,

1964 and 1966). A position-discrimination task was used in which the subject (*D. dorotocephala*) had to select either the right or left arm of an elevated T maze in order to return to its home bowl. All subjects were first presented with 10 "free-choice" or preference-testing trials in which they could choose either arm to escape. Thereafter, trained subjects were forced to select the least preferred arm. When an incorrect choice was made, the animal received a poke at the anterior end with a small brush. This stimulation caused a sudden withdrawal of the animal. A correct response was recorded when the animal chose what was originally the least preferred arm and did not receive a poke for entering the wrong arm. Animals were trained until they achieved a criterion of 9 out of 10 correct trial runs. They were then sectioned transversely and allowed 18 to 26 days to regenerate. After regeneration was completed, they were given an additional 10 preference-testing trials and were then retrained.

The results of this study are summarized in Fig. 3. It is clear from these data that there is considerable retention of the discrimination in the

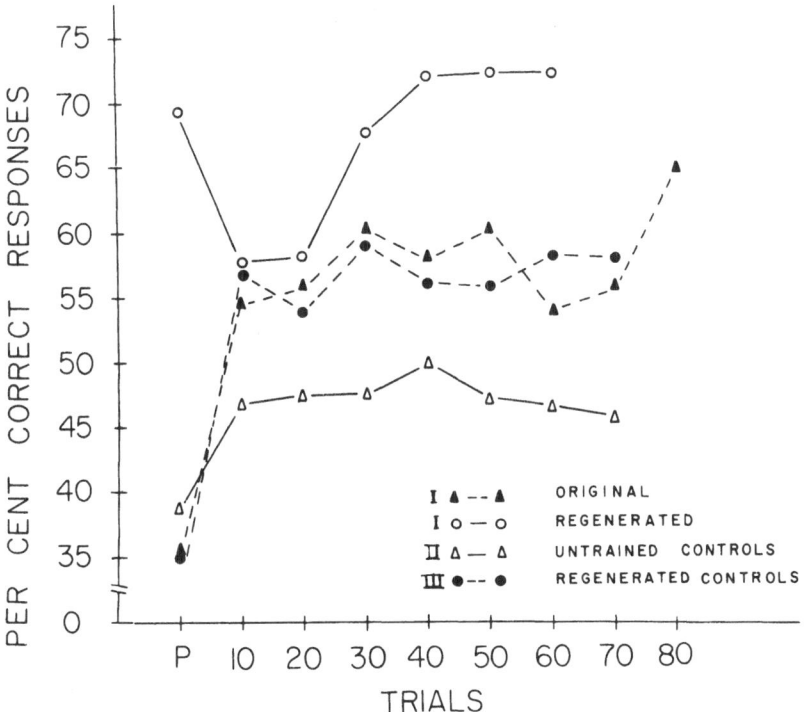

Fig. 3. Average per cent correct response performance of planarians in a T maze. "P" indicates preference tests.

regenerated portions of previously trained animals (group 1). Their performance during the preference trials given after regeneration demonstrated that the original preference had been reversed: the original subjects of group 1 chose the arm to which they were later trained an average of 35.3 %; after training to criterion, sectioning, and regeneration, the anterior and posterior regenerates displayed correct response levels of 68.6 % and 69.8 %. Upon initiation of retraining, the performance dropped for 20 trials and then rose to levels that were considerably higher than the original subjects displayed. Control groups rule out sensitization due to regeneration (group 3) and also demonstrate that untrained subjects permitted to select any arm to escape the maze (group 2) do not attain the levels displayed by the group 1 animals.

To summarize, there is now substantial evidence that several types of learning can be retained through regeneration; we have evidence for habituation, classical conditioning, and discrimination learning. There is also evidence that for each of these tasks, the posterior portion of the planarian is able to retain as much information as the anterior part. What mechanisms could effect such a continuation of information?

MECHANISMS FOR RETENTION

In earlier studies, our interest was focused upon the role of RNA in learning and memory by the theoretical paper of Hydén (1959) and by some preliminary unpublished findings of E. R. John and associates which dealt with ribonuclease effects upon discrimination performance in cats. We were interested in what effects ribonuclease might have upon the regenerative process in planarians and, more importantly, the effects of the enzyme upon retention of the acquired response. It was expected that, if ribonuclease could affect memory, its effect would be more pronounced on the posterior segment, since it would have to regenerate a dominant anterior portion in the presense of the enzyme, whereas the anterior segment would only have to grow a nondominant tail.

The literature concerned with ribonuclease effects in living systems is not at all clear on what the enzyme actually does inside the cell. There have been reports of mitotic abnormalities, inhibition of protein synthesis, permeability changes, a reduction of inductive capacity, etc. Little evidence is available concerning its effect upon planarians. Preliminary work at Rochester established that ribonuclease could produce abnormalities in regenerating *D. dorotocephala*. At concentrations of 0.1 mg/ml, regeneration was sometimes inhibited, but more often, animals would regenerate with structural anomalies. These anomalies included a lack of pigment in the head region, one-eyed animals, a lack of eyes, incomplete auricle development, and membrane invaginations. The effects were always observed in the regenerating *posterior sections*. Henderson and Eakin (1961) also studied the effects of ribonuclease and other agents on tissue differentiation in *D. dorotocephala*. A slice was made from an area posterior to the eyespots to the anterior tip of the animal, and the wounded planarians were placed in an isotonic solution to delay muscle contraction around the incision. Ribonuclease concentrations of 1 mg/ml evoked what was referred to as "disruptive" responses which were localized to the wound site; these responses were characterized by a disorganization of cells. The effect was reversible and was not observed with trypsin, chymotrypsin, lysozyme, etc.

To study the effects of the enzyme upon retention, a weaker (0.07 to 0.1 mg/ml) concentration was used. Conditioned and transected planarians were allowed to regenerate in pond water or in a pond-water solution containing ribonuclease. As mentioned previously, animals regenerating in pond water demonstrated the equal retention of the classically conditioned response. In the segments required to regenerate in the enzyme, we observed a differential effect on anterior and posterior derivatives. Whereas the response frequency of the anterior sections was comparable to that of animals regenerating in pond water, the posterior portions exhibited the response levels of a naive animal—there was no evidence that they had ever been trained. The results from this study suggested that ribonuclease had interfered with the processes by which the posterior section transmitted the experiential information to the developing anterior tissue. Various control groups demonstrated that the effect of ribonuclease upon conditioned-response retention was only observed in regenerating posterior segments and not in whole worms and also that neither regeneration nor the presence of the enzyme sensitized flatworms to light. More detailed considerations of these studies have been published elsewhere (Corning and John, 1961, and John, 1964).

Recently, research at the Academy of Sciences in Leningrad has confirmed certain aspects of the regeneration studies (Cherkashin and Sheiman, 1966). If the cerebral ganglia were removed in already conditioned *Ijimio tennis*, the subjects returned to a naive response level. Regeneration of the cerebral ganglia was accompanied by a return of the conditioned response. This study confirms the earlier work of McConnell: Memory appears to involve the entire central nervous system, and it is reestablished in the newly grown tissue. Further research by these investigators examined the effect of various enzymes and inhibitors upon conditioned-response retention. If whole worms were placed in a ribonuclease solution during conditioning, the animals did not learn; if already conditioned subjects were placed in ribonuclease, the conditioned response was inhibited but not lost; upon removal of the enzyme, it reappeared. Other agents such as deoxyribonuclease, 8-azaguanine, bromine uracil, and aurantine had no effect on retention. When already conditioned planarians were transected and placed in a solution of ribonuclease, there was a loss of the conditioned behavior. However, in contrast to the Corning and John study, no differential effect on anterior and posterior sections is reported.

Studies such as these which apply both macro- and micro-lesioning techniques in planarians should prove to be most valuable in attempts to understand the means by which the trace is propagated. The understanding of memory transfer in planarians may shed some light on the phenomenon of engram delocalization in higher brains. With respect to the planarian, there are several possibilities, some of which have been discussed previously (John, 1964). First, it is possible that there are photoreceptors in the epidermis of the planarian. The existence of light detectors along the length of the flatworm would allow conditioning to occur independent of the cephalic apparatus. Mitigating against this suggestion are some earlier data obtained with eyeless *D. dorotocephala*. With repeated light-shock pairings there was some increase in response to light, but the levels reached by these animals were not at all comparable to those of normal subjects. Another mode of transfer might occur via the neoblasts; transferal of information from nerve tissue to neoblasts would account for the reestablishment of memory traces in regenerating tissue. Related to this suggestion is the possibility that trace formation might take place directly in the neoblasts. A last possibility is that the trace is stipulated in the newly regenerated neural elements by the electrical activity of the original tissue or by the movement of intracellular macromolecules from the older tissue to the new tissue. The last suggestion, the stipulation of information directly from the original cells, would seem most likely, but how ribonuclease is able to interfere with this process remains to be determined. Clearly, much more research is needed to define the cellular events mediating information transfer during regeneration.

NEGATIVE EVIDENCE

In order to replicate the essential findings of the previously discussed studies, it is first necessary to demonstrate learning adequately and then determine that there is retention in the anterior and posterior segments after regeneration. Consideration of learning demonstrations may be found in Chapters 13 and 14 by Jacobson and McConnell. An extensive examination of conditioning and pseudoconditioning in a series of experiments spanning 7 years has convinced us that planarians are capable of learning providing a number of precautions are taken. In the use of light and shock in classical conditioning, it is necessary to determine that the increase in light responsiveness is due to learning and not to shock sensitization. As mentioned previously, we have satisfied ourselves that learning is the most reasonable explanation for the behavioral changes that are observed. Animals receiving random light and shock do not demonstrate comparable response levels. Furthermore, the finding that a nonlight-associated discrimination task is retained after regeneration provides what seems to us the most convincing evidence that planarians learn and that the nervous system is equipotential as far as the storage of the acquired information is concerned.

Equal retention between the regenerating segments of a planarian is not always observed. Agoston (1960) reports that anterior regenerates demonstrated greater retention of a classically conditioned response to light than the posterior sections and that larger pieces frequently retained more than smaller pieces. Roe (1963) used an ingenious hexagonal maze in which subjects (*D. dorotocephala*) were required to make a series of discriminations between white and black alleyways and found that only the anterior portions demonstrated any retention. It was suggested that a more complex form of learning such as the maze-discrimination task requires the cerebral ganglia for both acquisition and storage.

More recently, several investigations have been interested in variables that might offer alternative explanations for the learning reported by others. VanDeventer and Ratner (1964) report that worm size is inversely related to light sensitivity. Since transection and regeneration does result in smaller animals, it is suggested that this variable might account for the retention reported by others. A similar suggestion is made by Brown (1964), who in an extensive series of investigations has been generally unable to replicate much of the regeneration findings. In addition to finding that smaller worms are more reactive than larger ones, Brown also found that the origin of the worm along the axis is correlated with light sensitivity; anterior derivatives were more sensitive than middle or posterior derivatives. These differences persist up to a month, and the question is raised whether the regeneration time of other studies might have been too short. Furthermore, an examination of the relationship between light intensity and the origin of the regenerated animal indicated that at higher intensities (1000 ft-c) the anterior and posterior portions displayed equal response levels, whereas at lower intensities (500 ft-c) the anterior regenerates were more reactive to light than the posterior sections. Thus, according to Brown, the combination of size diminution due to regeneration and the intensity of the light sources could explain the equal retention reported by McConnell (McConnell, Jacobson, and Kimble, 1959) and the head and tail differences reported by Corning and John (1961) in their ribonuclease-treated groups.

However, Brown ignores the control groups that were used in these studies. Both the McConnell study and the Corning and John investigation incorporated controls which ruled out the alternative explanations offered by Brown and other critics. Regeneration did not sensitize the response of planarians to light, nor was there a differential effect of light upon naive anterior and posterior regenerates. In fact, at both weak and strong intensities the light sensitivities of head and tail portions have always been judged equal. The studies of Brown have yielded other results, mostly of a negative nature; no clear evidence of conditioning was obtained, and, in contrast to Westerman's study (1963), no evidence of habituation was found. A more detailed account of Brown's work may be found in the following chapter.

What I should like to consider at this point is the problem of replication. Certainly the failure of various investigators to repeat some of the basic planarian findings has been damaging to the general acceptance of this body of research. Part of the problem has been due to incomplete descriptions of experiments. In the published condensations of complex studies, important procedures are frequently omitted; in fact, the experimenter may be unaware at the time as to just how critical certain aspects of his techniques are (slime factors and shock parameters are good examples of this). Complicating matters is the fact that various positive studies have differed in techniques. Each investigator develops procedures and experimental designs which provide him with adequate control over the situation. The reasons for variation from one laboratory to another frequently remain obscure. In some cases, it is simply a matter of personal preference with respect to some criterion of response. For instance, McConnell's group has generally used more intense light sources, even though the base response rates of the planarians are higher. In our own studies we decided to use a minimal light intensity, one that elicited the lowest number of responses in naive animals and yet could be an effective conditioned stimulus when paired with shock. This creates a problem for the investigator who is attempting to replicate, and frequently he uses procedures derived from a number of different independent investigations. The Brown study is a good case in point. The intertrial interval was 30 seconds; this corresponds somewhat with Westerman's interval (30 to 60 seconds) but not with the interval reported by McConnell *et al.* (1959), which was, on the average, 84 seconds. Yet the findings are discussed with respect to both investigations. The shock source may also be critical; the variability of the Harvard inductorium may be most essential for conditioning. There are species differences to be considered. We have always found *D. dorotocephala* to be more readily conditionable than *D. tigrina*. They are also more active over long periods of time. Brown himself reports that *D. tigrina* "became quiescent 15 to 20 minutes after being introduced into the trough." Can we reasonably compare studies based upon two different species? There are a number of other variations which may also be critical. I am not trying to explain away Brown's study or any other negative study on the basis of procedural variation alone: What I am stressing here is that, if a replication of a certain finding is intended, then the procedures used to obtain that finding ought to be replicated also. One cannot invalidate a finding by implication (as the VanDeventer and Ratner study attempted to do with respect to temperature effects) or by using techniques based upon a number of different investigations. To do so leads only to further confusion and debate and to the necessity of holding conferences such as the present one.

SUMMARY

The evidence reported by various investigations would appear to warrant the conclusion that the mechanisms responsible for retention are distributed throughout the planarian. Retention after regeneration has been reported for habituation, classical conditioning, and discrimination learning. However, not all studies have found equal retention in the anterior and posterior regenerates, and, more recently, alternative explanations have been offered for the positive studies, i.e., the retention reported by others could be due to sensitization factors. It is felt that the positive studies provided adequate controls for these factors and that part of the failure to replicate results is due to the failure to replicate procedures.

REFERENCES

Agoston, E., Learning and regeneration in the planarian. *Worm Runner's Digest*, 1960, **2**, 53.

Best, J. B., and I. Rubinstein, Maze learning and associated behavior in planaria. *J. Comp. Physiol. Psychol.*, 1962, **55**, 560.

Brønsted, H. V., Planarian regeneration. *Biol. Rev.*, 1955, **30**, 65.

Brown, H. M., Experimental procedures and state of nucleic acids as factors contributing to "learning" phenomena in planaria. *Unpublished doctoral dissertation*, University of Utah, 1964.

Cherkashin, A. N., and I. M. Sheiman, The use of simple biological models in memory mechanisms. Paper presented at International Psychology Congress, Moscow, 1966. (See also *J. Biol. Psychol.*, 1967, **9**, 5.)

Corning, W. C., Evidence of right-left discrimination in planarians. *J. Psychol.*, 1964, **58**, 131.

Corning, W. C., Retention of a position discrimination after regeneration in planarians. *Psychonomic Sci.*, 1966, **5**, 17.

Corning, W. C., and E. R. John, Effect of ribonuclease on retention of conditioned response in regenerated planarians. *Science*, 1961, **134**, 1363.

Cornwell, G., P. Cornwell, and M. Clay, Retention of a conditioned response following regeneration in the planarian. *Worm Runner's Digest*, 1961, **3**(1).

Ernhart, E. N., and C. Sherrick, Retention of a maze habit following regeneration in planaria (*D. maculata*). Paper presented at Midwestern Psychology Association, May, 1959, St. Louis, Mo.

Freed, S., Endogenous biochemistry of planarians correlated with learning experiments. *Brookhaven Nat. Lab. Rep.*, No. 981(T-414), 1966.

Henderson, T. R., and R. E. Eakin, Irreversible alteration of differentiated tissues in planaria by purine analogues. *J. Exp. Zool.*, 1961, **146**, 153.

Humphries, B., and J. V. McConnell, Factors affecting maze learning in planarians. *Worm Runner's Digest*, 1964, **6**, 52.

Hydén, H., Biochemistry of the Central Nervous System, Vol. 3, *Proceedings of the 4th International Congress on Biochemistry*. New York: Pergamon Press, 1960.

Jenkins, M. M., Bipolar planarians in a stock culture. *Science*, 1963, **142**, 1187.

John, E. R., Studies on learning and retention in planaria, *in* M. A. B. Brazier, ed., *RNA and Brain Function; Learning and Memory*. Berkeley, Calif.: University of California Press, 1964.

McConnell, J. V., A. L. Jacobson, and D. P. Kimble, The effects of regeneration upon retention of a conditioned response in the planarian. *J. Comp. Physiol. Psychol.*, 1959, **52**, 1.

McConnell, J. V., R. Jacobson, and D. M. Maynard, Apparent retention of a conditioned response following total regeneration in the planarian. *Amer. Psychol., Abstr.*, 1959, **14**, 410.

McWhinnie, M. A., and M. M. Gleason, Histological changes in regenerating pieces of *Dugesia dorotocephala* treated with colchicine. *Biol. Bull.*, 1957, **11**, 371.

Roe, K., In search of the locus of learning in planarians. *Worm Runner's Digest*, 1963, **5**, 24.

VanDeventer, J. M., and S. C. Ratner, Variables affecting the frequency of response of planaria to light. *J. Comp. Physiol. Psychol.*, 1964, **57**, 407.

Westerman, R. A., Somatic inheritance of habituation of responses to light in planarians. *Science*, 1963, **140**, 676.

Wolff, E., Recent researches on the regeneration of planarians, *in* D. Rudnick, ed., *Regeneration*. New York: The Ronald Press Company, 1962.

Chapter 19

Effects of Ultraviolet and Photorestorative Light on the Phototaxic Behavior of Planaria[1]

H. Mack Brown

VA Hospital and Department of Neurology
University of Utah College of Medicine
Salt Lake City, Utah

INTRODUCTION

The negative phototaxic behavior of planaria (turns and contractions) is augmented when light is paired with an electrical field. It has been suggested that this phenomenon represents a learning process (Thompson and McConnell, 1955), i.e., that planaria can be conditioned to associate light and electrical stimuli. This conclusion is based on the tacit assumption that planaria are not sensitized by the electrical field used in conditioning procedures. However, it has been shown that this assumption is incorrect; planarian negative phototaxic behavior is augmented by light and shock that are temporally unrelated (Brown *et al.*, 1966b). This finding suggests that modified planarian behavior, widely attributed to "learning," actually represents sensitization. It has been suggested that the augmented phototaxis (conditioned response) of planaria is transferable to other worms by ingestion (McConnell, 1962), and that it is evident in both head and tail sections regenerated from conditioned, bisected worms, because both types of regenerate display similar exaggerated behavior in response to light stimuli (McConnell *et al.*, 1959). However, when head and tail sections were regenerated in the presence of ribonuclease, tail sections were less light responsive (showed "less resistance to extinction") than head sections following regeneration. This was attributed to the abolition of "memory" in the tail regenerates by ribonuclease (Corning and John, 1961). The validity of these conclusions is based on the assumption that regenerated head and tail sections are behaviorally similar to each other and to the animals from which they originated. However, recent studies show that short planaria are more responsive to light than long planaria (Brown, 1964a; Brown *et al.*, 1966a; and VanDeventer and Ratner, 1964).

[1] This paper is based on part of a thesis submitted to the faculty of the University of Utah in partial fulfillment of the requirements for the Ph.D. degree.

295

Consequently, regenerated (shorter) segments react more frequently to intense light stimuli than the animals from which they originated (Brown *et al.*, 1966a). This augmented behavior could not be attributed to procedures that occurred before the animals were sectioned. Moreover, head and tail segments, allowed to regenerate for 4 weeks or less, are not behavioral replicas of one another when tested with moderate (less than 500 ft-c) light stimuli. Under these conditions, tail regenerates are inherently less responsive than head regenerates (Brown *et al.*, 1966a). Thus, regenerated tail sections would be expected to show reduced phototaxis, or "less resistance to extinction," than regenerated head sections. Since a short regeneration period (2 weeks) and low light intensity (less than 100 ft-c) were conditions employed in the ribonuclease study, one may question the conclusion that the reduced responsivity of regenerated tail sections results from the effects of ribonuclease on retention.

It is the purpose of this paper to clarify the extent that conditioning procedures, duration of regeneration, and treatments that affect nucleic acids contribute to modifications of planarian negative phototaxic behavior.

Learning terminology (conditioned stimulus, conditioned response, conditioning, extinction, savings, retention, etc.) will be avoided in this paper, because of the confusion surrounding these terms and the logical inconsistency of applying them to planarian behavior, i.e., if the augmented phototaxis of planaria is due to sensitization and is not a conditioned response, it is illogical to use the term *extinction* to describe the decay of augmentation. However, because learning terms have been used extensively in recent literature describing planarian behavior, the application of measures used in this study to previous work will be indicated where appropriate.

METHOD

Apparatus

Receptacles for the worms consisted of clear acrylic round-bottomed troughs, 11 in. long, with inside cross-sectional areas of 0.5 cm^2. The light stimulus L was a 3-second pulse of uniform flux and minimal heat, obtained from a tungsten ribbon-filament lamp mounted beneath the trough. The lamps were operated at a color temperature of 2870°K. Illuminance was measured with an SEI photometer (Salford Electrical Instruments, Sussex, England). The illuminance at the receptacle was 500 ft-c. Ambient illumination was 1 ft-c. The shock stimulus S was a d-c pulse, 1 second in duration, from a full wave rectifier power supply. The resistivity of the spring water in which the animals were trained varied from time to time; therefore, constant current stimuli were used. It was found that 20 μamp were sufficient to elicit consistent contractions from the worms. To ensure reliable stimulus presentation and control mechanical vibration, RC networks were used to time the durations of L and S.

Trial Procedure

Dugesia tigrina (supplied by Ward's of California, Monterey, California) were used in the present study. The animals were kept in an aerated, ecologically balanced aquarium. Upon assignment to an experiment each animal was placed in a 90 by 15-mm petri dish half filled with spring water, so that positive identification of each animal could be made and a record of the animal's condition could be kept. The temperature of the water in the petri dish was maintained at the same temperature as the water in the training receptacles (21°C). The animals were observed individually. They were allowed to adapt to the trough and ambient light for 5 minutes before trials were presented. A light *L* trial was a 3-second light presentation. A light-shock *LS* trial consisted of a 3-second *L* and a 1-second *S* coincident with the final second of *L*. Because planaria are more responsive to light if mechanically irritated (Sgonina, 1939), it was necessary to refrain from agitating the worm after it was placed in the receptacle. This was accomplished by presenting daily trials in two consecutive blocks of 15 trials each and allowing 20 to 25 minutes between blocks. The minimum interval between trials was 30 seconds. The trough water was changed and the receptacle cleaned following each block of trials.

Behavioral Criteria

Trials were administered only when the animal was gliding smoothly along the trough (ciliary movement). Responses were judged by visual observation of the animal's behavior during exposure to *L*. The criterion for a response was either a longitudinal contraction or a head turning movement of at least 45° that occurred during the initial 2 seconds of *L*. Head-turning movements were approximately ten times more frequent than contractions. A contraction was the dominant response to *S* presented without *L*.

It is conventional practice to show that conditioning leads to an enduring behavioral modification (conditioned response) by demonstrating that the animal can be reconditioned in fewer trials (savings), or that the response gradually decays to a preexisting level when the conditioned stimulus is presented without the unconditioned stimulus (extinction). Savings and extinction are terms that imply retention of a learned response and should be avoided since the responsivity of planaria is modified by many factors not associated with learning. For these reasons the terms will not be used. The experimental procedures will be referred to simply as tests of negative phototaxis in planaria.

The tests of planarian negative phototaxis used in this study were based on two empirically derived criteria. The *L*-criterion (light only) was obtained to set the baseline level of planarian responsivity to *L* alone. For this purpose 12 planaria were given 30 *L* trials daily on 2 consecutive days. The animals responded 115 times in the 720 trials (16%); therefore a daily response of 16% (5 responses in 30 trials) was set as the baseline.

This is similar to values obtained in other studies (Bennett and Calvin, 1964; Corning and John, 1961; Thompson and McConnell, 1955). The percentage of responses of planaria to L when L and S were paired (LS-criterion) was established similarly. Four planaria were given 30 LS trials daily for 8 days. Each animal responded at least 10 times in a given block of 15 trials during the 8-day period; therefore, 10 responses in a block of 15 trials was set as the LS-criterion.

Three measures of phototaxis were based upon these criteria. (1) The number of trials required for responsivity to decay to the L-criterion level after responsivity had been raised to the LS-criterion by LS trials (extinction in conventional learning terminology). (2) The percentage of responses elicited during L-criterion trials after responsivity was raised to LS-criterion by LS trials (responses divided by trials). (3) The number of trials required to reattain LS-criterion compared with the number of trials required to reach the same criterion initially (savings in conventional learning terminology).

To investigate the effect of alteration of nucleic acids on phototaxis, planaria were exposed to 260-mμ ultraviolet radiation. The following procedures were employed.

Ultraviolet Sources

A hydrogen lamp and monochromator and a low-pressure mercury arc (G.E. 15T8) were used as sources of monochromatic 260-mμ light. The mercury arc was used predominantly because its high energy output at 260 mμ reduced the time required to obtain an effective exposure.

Ultraviolet Dosimetry

The total amount of energy to which the animals were exposed was determined by uranyl oxalate actinometry (Leighton and Forbes, 1930). The actinometer cell was placed in the light path used to irradiate the planaria. The method is based on the conversion of oxalic acid, sensitized by uranyl sulfate, to nonacidic photoproducts by ultraviolet photons. Following irradiation of a standardized sample of uranyl oxalate, the number of converted molecules was determined by titrating the reaction mixture with a standard solution of potassium permanganate; the number of incident photons and total energy were calculated.

Ultraviolet Treatment

To facilitate irradiation of the animals, they were immobilized by cooling or placed in a rectangular quartz cuvette and held in the light path mechanically.

Photorestoration Treatment

Alteration of nucleic acids induced by ultraviolet can be partially reversed by exposure to white light (photorestoration). This procedure was carried out on a group of planaria as a control for behavioral changes induced by ultraviolet alone. Immediately following ultraviolet treatment, the animal to be photorestored was placed in a 90 by 15-mm polystyrene petri dish half filled with spring water. Two fluorescent tubes (G.E. F15T8.D and F15T8.WWX) 6 in. beneath the dish supplied the desired wavelengths of photorestorative light (Dulbecco, 1955). The dorsal aspect of the animals was exposed by light reflected from a mirror placed over the top of the dish. Exposure time was 6 hours and the illuminance was 140 ft-c. Tests of phototaxis were begun 24 hours after the ultraviolet or combined ultraviolet-photorestoration treatments.

Behavioral Criterion of Ultraviolet Exposure

A pilot study was conducted with 33 planaria to determine the optimal ultraviolet exposure energy and the mode of exposure (dorsal, ventral, or both dorsal and ventral) that would have an effect on their performance to light stimuli without significantly influencing their appearance and general behavior. The planaria were given LS trials until LS-criterion was attained. They were then exposed to different total energies of ultraviolet light by varying the exposure time. It was found that a split exposure, half to the dorsal side and half to the ventral side, with a total energy of 5.2×10^7 ergs significantly altered their phototaxic behavior as measured by the percentage of responses elicited during L-criterion trials and the number of trials required to L-criterion. This effect seemed to be partially reversible by photorestorative light, i.e., photorestored worms were more responsive to L trials than worms irradiated with ultraviolet light. Lower energies had no effect on their behavior under light stimuli, and the effects at higher energies, especially if administered to a single side of the animal, were not photoreversible. In addition, the high-energy exposure caused undesirable effects such as impaired mobility and mottling of the epidermis.

RESULTS

Dimerization of Planarian Nucleic Acids

It is well established that nucleic acids strongly absorb 260-mμ light (ultraviolet); recent evidence indicates that the light absorption is followed by disruption of the hydrogen bonding in the nucleic acids with the production of intra- and interchain dimers (Beukers and Berends, 1961, and Wulff and Rupert, 1962). In animal cells these changes are reversible to some extent by a broad range of wavelengths in the visible region of the spectrum (Blum, 1959, and Jagger, 1958); white light is photorestorative.

The modification of planarian behavior by ultraviolet irradiation described in this paper probably results from dimerization of nucleic acids. Evidence that ultraviolet light causes dimerization of planarian nucleic acids was obtained from the following experiment. Twenty-five planaria with a total weight of 397 mg were cooled, and both aspects of the animals were exposed to 5.2×10^7 ergs of ultraviolet light. The total energy (10^8 ergs) was twice that used in the behavioral experiments, so that a greater yield of photoproducts could be obtained. Following irradiation, the animals were placed in a test tube and homogenized with a ground glass pestle. The nucleic acids were extracted by the method shown on the flow sheet in Fig. 1. This is a modification of the Schmidt-Thannhauser technique (Hutchison and Munro, 1961) that minimizes the use of heat, a necessary requirement for the detection of cytosine photodimers (Shugar, 1960). The absorption spectra of the RNA and DNA fractions, determined with an ultraviolet spectrophotometer (Beckman DU), were compared with absorption spectra of unaltered nucleic acids and photoproducts of irradiated nucleic acids (Fig. 2).

The absorption spectrum of the DNA fraction indicated that DNA dimers were present, as is shown by the fit of the dimer absorption curve (solid line) and the absorption curve of the DNA fraction (broken line). The absorption spectrum of unaltered nucleic acids is also shown (dotted line). There was little evidence for the presence of dimers in the RNA fraction. This was expected because cytosine photodimers are only obtained

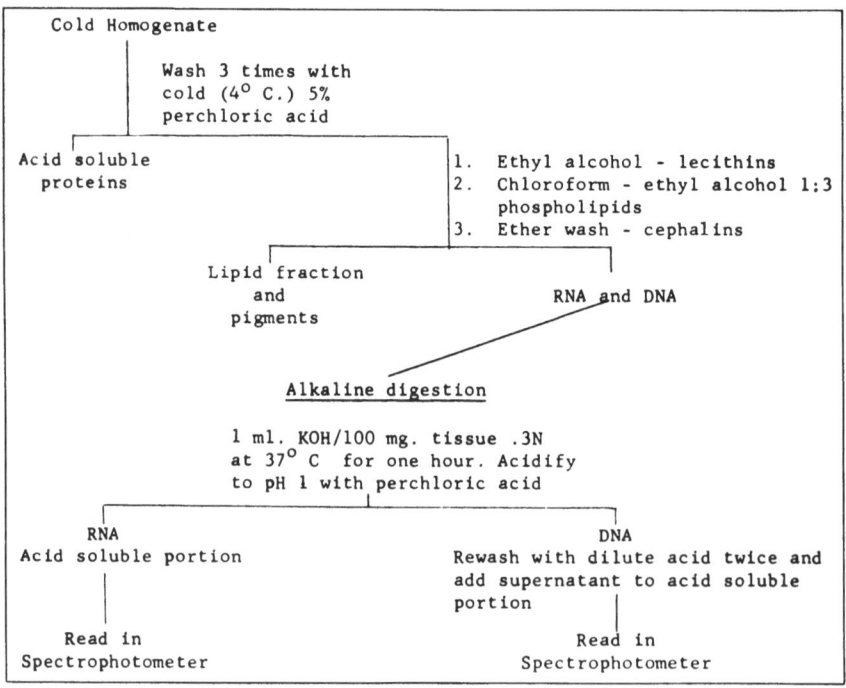

Fig. 1. Flow diagram of the procedure employed for the isolation of nucleic acids from planaria.

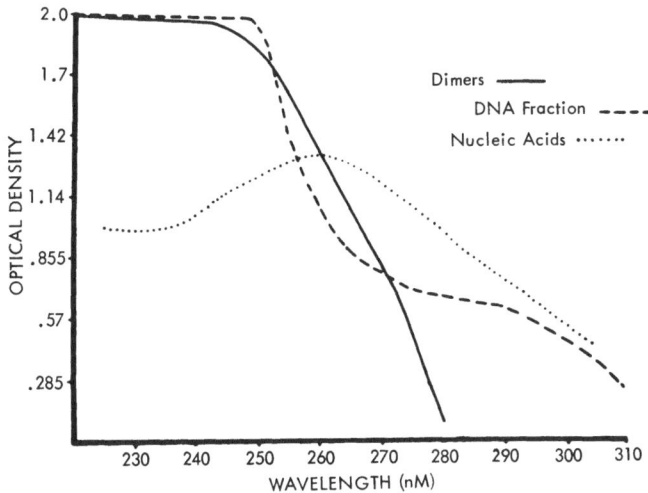

Fig. 2. Absorption spectra of nucleic acids showing evidence of dimerization in DNA fraction.

in small yields (Shugar, 1960) and the small amount obtained may have been converted to the monomer form by the heat used in alkaline digestion.

Effects of Ultraviolet and Photorestorative Light on Planaria Regenerated from Experienced and Naive Animals

The evidence cited above suggests that ultraviolet light causes chemical changes in the nucleic acids of planarian tissues and that the changes may be associated with altered phototaxis. In the following experiments the L-criterion and LS-criterion performance of regenerated planaria were evaluated to determine whether the effects of ultraviolet treatment are dependent upon prior experience of planaria with LS trials and whether the effect is detectable in planaria regenerated from both head and tail segments. In the first experiment the L-criterion performance of 79 animals in 6 experimental groups was evaluated, with the results shown in Table 1.

L-criterion performance. Three groups consisted of naive N planaria that received no LS trials prior to being irradiated and bisected. These animals were housed in their dishes for the period of time (8 to 12 days) usually required for animals receiving LS trials to attain criterion. One group received the ultraviolet treatment NUV, one group the photo-restoration treatment NPR, and animals in the third group were sham-irradiated NC. Following treatments, they were cut into two equal lengths, a head portion H and a tail portion T, and allowed to regenerate for 2 weeks. The number of planaria in each group was: $NUV = 8, NPR = 9, NC = 25$. Planaria in the other three groups were given experience E with LS trials until they attained the LS-criterion. They were then given radiation treatments and bisected. The *EUV* group had 14 animals; *EPR*, 10 animals; and *EC*, 13 animals. Following regeneration, all the animals were given *L* trials until they met the *L*-criterion. The number of trials required to attain criterion and the percentage of responses to criterion for each of the groups were compared statistically with *t-tests.* The relevant comparisons were *naive* versus *experienced, ultraviolet* versus *photorestorative* versus *control* treatments, and *heads* versus *tails* that received the same treatment. The total number of relevant comparisons was 72, 19 of which were statistically significant (Table 1).

The *t-test* analyses indicated that experience with *LS* trials prior to treatment and bisection had no effect on the phototaxis of regenerated planaria (trials to *L*-criterion and the percentage of responses elicited during *L*-criterion training). In 10 of 12 cases there were no differences between the measures of phototaxis of naive and experienced animals, and in one case (Table 1, line 1), the effects of prior experience appeared to have a negative effect. Naive, sham-irradiated heads *NC-H* required more trials until *L* criterion was met than experienced, sham-irradiated

TABLE 1

Results of *t-tests* comparing the *L*-criterion performance* (trials to *L*-criterion and percentage of responses until *L*-criterion was attained) of regenerated head *H* and tail *T* segments from experienced *E* and naive *N* planaria. The segments were given ultraviolet *UV*, photorestoration *PR*, or control *C* treatments prior to regeneration.

		Trials to *L*-criterion					Per cent response				
		Group	vs. Group	*t*	*P*	*df*	Group	vs. Group	*t*	*P*	*df*
	1.	*NC-H*	*EC-H*	2.31	0.05	12					
	2.	*NC-H*	*EUV-H*	4.06	0.01	11					
Heads	3.	*NC-H*	*NUV-H*	4.42	0.01	9	*NC-H*	*NUV-H*	4.62	0.01	9
vs.	4.	*NPR-H*	*EUV-H*	2.58	0.05	6					
Treat.	5.	*EC-H*	*EUV-H*	4.57	0.01	9					
	6.	*EC-H*	*NUV-H*	5.85	0.001	7	*EC-H*	*NUV-H*	5.20	0.01	7
	7.						*EUV-H*	*NUV-H*	2.89	0.05	6
	8.						*EPR-H*	*NUV-H*	6.89	0.001	5
Tails	9.	*NC-T*	*EUV-T*	3.23	0.01	24	*NC-T*	*EUV-T*	2.13	0.05	24
vs.	10.	*NC-T*	*NUV-T*	2.82	0.02	20	*NC-T*	*NUV-T*	2.18	0.05	20
Treat.	11.	*EC-T*	*EUV-T*	2.32	0.05	14					
Heads	12.	*NC-H*	*NC-T*	2.76	0.02	23	*NC-H*	*NC-T*	2.67	0.02	23
vs.	13.	*EC-H*	*EC-T*	2.41	0.05	11					
Tails	14.						*EUV-H*	*EUV-T*	2.31	0.05	12

* The group with the largest mean is given first (in the left-hand column) in comparisons. All results that were statistically significant are shown. See Brown (1964) for all other comparisons.

heads *EC-H*. There was one isolated case that indicated experience had a subsequent effect on phototaxis. Experienced heads irradiated with ultraviolet light *EUV-H* gave a higher per cent response than naive ultraviolet-irradiated heads *NUV-H*; however the two groups did not differ significantly in the number of trials to *L*-criterion (line 7).

Effects of treatments occurred independently of experience. Naive, sham-irradiated head regenerates *NC-H* were more responsive than both experienced *EUV-H* and naive *NUV-H* head regenerates irradiated with ultraviolet light (lines 2 and 3). Similar results were found for regenerated tail segments (lines 9 and 10). Sham-irradiated control animals from head and tail regions were also more reactive during *L*-criterion trials than equivalent animals treated with ultraviolet light (lines 5 and 11). In two cases photorestored head regenerates were more reactive than ultraviolet-treated head regenerates (lines 4 and 8). An *analysis of variance* of the effects of treatments on planaria, free of developmental effects, showed that both control and photorestored regenerates were more responsive than regenerates that had been irradiated with ultraviolet light (trials to *L*-criterion of heads and tails that received the same treatment were pooled and the differences among treatment means were compared statistically, $P < 0.05$).

The phototaxic behavior of regenerated planaria was dependent upon the developmental state of the regenerated segments as well as the radiation treatments. Under the conditions of these experiments all reliable differences showed that regenerated head segments were more responsive than regenerated tail segments regardless of previous experience or treatment (lines 12, 13 and 14).

The *LS*-criterion of phototaxis was also used to test the effects of experience and the differential development in head and tail sections regenerated for 2 weeks. It will be recalled that this measure of phototaxis relies on planaria's response to *L* when *L* and *S* are presented together. This procedure allows the comparison of the responsivity of regenerated planaria with that of the animals from which they originated and the effects of prior training on the retraining performance of uncut planaria.

LS-criterion performance. Thirty-six planaria divided among four experimental groups were used. Each of the animals in the groups was given *LS* trials until responsivity was raised to the *LS*-criterion. The planaria in three groups were bisected following ultraviolet ($n = 11$), ultraviolet + photorestorative ($n = 5$), and sham ($n = 11$) treatments. Animals in the fourth group ($n = 9$) were not cut or irradiated but remained in their dishes for the same period that animals in the other groups were allowed to regenerate (2 weeks). After the 2-week period each of the worms was once again given *LS* trials until the *LS*-criterion was met. Analyses (*t-tests*) were done to determine whether the mean number of trials to *LS*-criterion of regenerated planaria was reliably different from that of the parent animals (whole ones), and whether head and tail regenerates given the same treatments were reliably different from each other.

In addition, the two *LS*-criterion performances of uncut planaria were compared to determine whether experience with *LS* trials has an influence on the subsequent *LS*-criterion performance of planaria.

Prior experience with *LS* trials had no influence on the subsequent performance of uncut planaria or regenerated head and tail sections. The first and second *LS*-criterion performance of uncut, untreated planaria did not differ reliably, i.e., the worms required as many trials to criterion the second time as they did the first time (first, 121; second, 95; $t = 1.57$; $P > 0.05$). Similarly, the performance of head segments did not differ reliably from the performance of the original whole worms ($P > 0.05$) under the control, photorestorative or ultraviolet treatment, as shown in the second column of Table 2. Heads *tended* to be more responsive than whole worms. (The mean number of trials to *LS*-criterion of head regenerates in the control group was 81 versus 145 for whole animals; photorestored heads, 97 trials versus 135 for whole animals.)

Tail sections required at least twice as many *LS* trials to criterion as the head sections or whole animals; this was the case no matter which treatment the animals received. In the control group, regenerated tails required more trials to criterion than regenerated head segments or whole worms ($P < 0.001$), and tail regenerates in the ultraviolet-treatment group required more trials to *LS*-criterion than whole planaria ($P < 0.01$). Thus by all three tests of planarian phototaxis used in these experiments, the

TABLE 2

Results of *t-tests* comparing the *LS*-criterion performance of regenerated head segments, tail segments, and the whole planaria from which the regenerated animals originated.

				Trials to *LS*-criterion				
		Segment		Head			Tail	
	Means		*t*	*P*	*df*	*t*	*P*	*df*
	145	Whole	1.84	> 0.05	8	5.63	< 0.001	9
C	81	Head				5.28	< 0.001	9
	317	Tail						
	135	Whole	1.00	> 0.05	2	0.74	> 0.05	3
PR	97	Head				1.03	> 0.05	3
	235	Tail						
	116	Whole	0.64	> 0.05	6	3.39	< 0.01	9
UV	154	Head				1.78	> 0.05	9
	294	Tail						

responsivity of tails was consistently less than that of heads. When a moderately intense source (500 ft-c or less) is used for the light stimulus, head and tail sections cannot be regarded as behaviorally identical. This supports the conclusion of Brown et al. (1966a) that the phototaxis of head regenerates is greater than tail regenerates, providing a short regeneration period (less than 4 weeks) and a light source with an intensity below 500 ft-c is used.

DISCUSSION

To be useful, the concept of a conditioned response (an acquired response that is elicited by a neutral stimulus and occurs initially with volition and purpose) must be differentiated from innate, reflex behavior. If this distinction is not made, the terms *learning* and *conditioned response* become scientifically meaningless (Efron, 1966), because virtually any induced alteration in behavior can be construed as learning. It is unfortunate that this distinction has not been applied more rigorously in planarian learning studies. Most studies of planarian behavior have relied upon a reaction pattern that is elicited naturally from planaria by light. Thus light is not a neutral stimulus for planaria, and the response (negative phototaxis) is not acquired; in a strict sense, experimental augmentation of planaria's negative phototaxis should not be considered a conditioned response.

Aside from the above considerations, the evidence of the present study shows that planarian phototaxis can be modified by factors not associated with a learning process. Also, experimental augmentation of phototaxis by paired light and shock (*LS* trials) had no effect on the subsequent behavior of the animals. Thus there was no evidence of retention of augmented phototaxis. Similar findings have been reported previously (Baxter and Kimmel, 1963; Halas et al., 1961; and James and Halas, 1964), but McConnell has argued that extinction is not as efficacious a measure of retention in planarian learning studies as savings (McConnell, 1964). In the present study, there was no evidence of retention when either retention measure was used. Light trials, with or without shock trials, following radiation treatments, were simply a means of quantifying the functional integrity of the pathways that mediate the negative phototaxic behavior of planaria. They were not relevant as tests of retention.

Consistent differences between the response levels of regenerated head and tail segments found in this study have also been described previously (Brown, 1964a and b, and Brown et al., 1966a and b). This warrants comment because the finding is relevant to the conclusions of other investigators (Corning and John, 1961; McConnell et al., 1959; and Westerman, 1963). McConnell et al. (1959) suggested that planaria, regenerated from rostral and caudal sections of conditioned planaria, are behaviorally similar. It has since been shown that the intrinsic differences in responsivity of head and tail segments can be masked by light of the same intensity as that used by McConnell and co-workers (Brown *et al.*, 1966a). Thus it may be misleading to evaluate a variable by the performance of head and tail regenerates. This was clearly demonstrated in an experiment described by Brown (1964a). He examined the effects of ultraviolet and ultraviolet followed by photorestorative light on animals regenerated from head, middle, and tail segments. In this study planaria were given *LS* trials until they attained *LS*-criterion. Following this, the animals were trisected; the head portions received the ultraviolet treatment, the middle sections the photorestorative treatment, and tail sections were sham irradiated. After 2 weeks' regeneration, each of the animals was given *L* trials until *L* criterion was attained. When the data were analyzed there were no differences among the *L*-criterion performances of the three groups. It was shown in the present and other studies (Brown *et al.*, 1966a) that planaria regenerated from rostral regions are more responsive than regenerates from caudal regions and that this responsivity is graded along the axis of the parent animal (Brown and Beck, 1964). The fact that regeneration of a head ganglion takes longer in planaria grown from caudal regions (Wolff and Lender, 1950) is probably due to the time-graded regeneration field that exists along the major axis of planaria (Brønstead, 1955). Inasmuch as eye formation in regenerating planaria is dependent upon a head ganglion, segments taken from rostral regions develop a visual apparatus more rapidly than segments from more caudal regions. Thus head regenerates, following a short regeneration period, are more responsive than regenerated middle segments, which are in turn more responsive than tail regenerates. The present study showed that ultraviolet light had the nonspecific effect of reducing light responsivity in planaria, an effect that was somewhat mitigated if ultraviolet light was followed by photorestorative light. In the study described by Brown (1964a), the effect of the different developmental levels of the three regenerated segments on light responsivity interacted with and suppressed the effects of ultraviolet and ultraviolet plus photorestorative treatments.

Under the conditions of the experiments reported by Corning and John (1961), planaria regenerated from head and tail segments probably were not behaviorally equivalent. Thus their conclusion that ribonuclease abolished retention in regenerated tail sections is open to question. Also, it has been suggested that their ribonuclease may not have been active under the conditions of their experiments (John, 1964) which leads to the possible conclusion that the head-tail differences they described in their experimental groups may not have been due to the effects of ribonuclease on retention but to developmental differences in the regenerated animals. The conclusion that "habituation training survives regeneration" in planaria (Westerman, 1963) is also subject to the criticism that differences in performance of planaria regenerated from head and tail segments were not considered.

It should be pointed out that, despite the similarity in the conditions of this study and the ribonuclease study, there is a disparity in control results. Head-tail differences were only detected by Corning and John (1961) in groups treated with ribonuclease whereas, in the present study, head-tail differences were evident in the untreated control animals. An explanation of this discrepancy may be found in the number of extinction observations analyzed in the two studies. Corning and John limited their post-treatment extinction analysis to no more than 3 days (40 trials/day); they did not "extinguish" their animals to the baseline performance level. In the present study a minimum period of 8 days (30 trials/day) was used, and planarian behavior was evaluated until the baseline criterion was attained. It is suggested that results similar to those in the present study might have been obtained from the ribonuclease study had planarian behavior been evaluated over a longer period of time and had the statistical analysis not been limited to data from the first day that significant differences were obtained. This last point is pertinent because of the considerable day to day variability in performance of planaria to light stimuli (Bennett and Calvin, 1964, and Brown, 1964a).

SUMMARY AND CONCLUSION

1. Ultraviolet light caused a reduction in planarian phototaxic behavior, as determined by the number of times planaria turned or contracted to repeated presentations of a 3-second light stimulus of moderate intensity (500 ft-c). This effect was associated with dimerization of nucleic acids in planarian tissues.

2. The reduction in phototaxic behavior by ultraviolet light was reversible to some extent by white light (photorestoration).

3. The effects of ultraviolet and photorestoration were not altered by previous training procedures.

4. Training procedures did not cause modifications in planarian behavior that could be attributed to learning.

5. Planaria regenerated from head segments of bisected animals were more responsive to light stimuli of moderate intensity than were regenerated tail segments. This occurred independently of training and radiation procedures and was shown to confuse the evaluation of other experimental variables.

It is concluded that the negative phototaxic behavior of planaria can be utilized to gauge the effects of ultraviolet and photorestorative light on the general functional level of the animals. Otherwise, there is no evidence from this study to indicate that planaria possess any special attribute for the study of biological processes that underlie learning and retention.

ACKNOWLEDGMENTS

The author wishes to express his appreciation to Dr. E. C. Beck in whose laboratory this work was conducted, and to Drs. T. E. Ogden and E. S. Halas for critical readings of the manuscript.

REFERENCES

Baxter, R., and H. D. Kimmel, Conditioning and extinction in the planarian. *Amer. J. Psychol.*, 1963, **76**, 665.
Bennett, E. L., and M. Calvin, Failure to train planarians reliably. *NRF Bull.*, 1964, July-Aug., 3.
Beukers, R., and W. Berends, The effects of U.V.-irradiation on nucleic acids and their components. *Biochim. Biophys. Acta* , 1961, **49**, 181.
Blum, H. F., *Carcinogenesis by Ultraviolet Light*. Princeton, N.J.: Princeton University Press, 1959, pp. 75–87.
Brønstead, H. V., Planarian regeneration. *Biol. Rev.*, 1955, **30**, 65.
Brown, H. M., Experimental procedures and state of nucleic acids as factors contributing to "learning" phenomena in planaria. *Unpublished doctoral dissertation*, University of Utah, 1964a.
Brown, H. M., Photoreversal of the light-response decrement in Planaria due to ultraviolet light. *Amer. Psychol.*, 1964b, **19**, 484.
Brown, H. M., and E. C. Beck, Does learning in Planaria survive regeneration? *Fed. Proceed.*, 1964, **23**, 254.
Brown, H. M., R. E. Dustman, and E. C. Beck, Experimental procedures that modify light response frequency of regenerated Planaria. *Physiol. Behav.*, 1966a, **1**, 245.
Brown, H. M., R. E. Dustman, and E. C. Beck, Sensitization in Planaria. *Physiol. Behav.*, 1966b, **1**, 305.
Corning, W. C., and E. R. John, Effect of ribonuclease on retention of conditioned response in regenerated planarians. *Science*, 1961, **134**, 1363.
Dulbecco, R., Photoreactivation, *in* A. Hollaender, ed., *Radiation Biology*. New York: McGraw-Hill Book Company, 1955, p. 455.
Efron, R., The conditioned reflex: a meaningless concept. *Pers. Biol. Med.*, 1966, **9**, 488.
Halas, E. S., R. L. James, and L. A. Stone, Types of responses elicited in Planaria by light. *J. Comp. Physiol. Psychol.*, 1961, **54**, 302.
Hutchison, W. C., and H. N. Munro, The determination of nucleic acids in biological materials: a review. *Analyst*, 1961, **86**, 774.
Jagger, J., Photoreactivation. *Bact. Rev.*, 1958, **22**, 99.
James, R. L., and E. S. Halas, No difference in extinction behavior in Planaria following various types and amounts of training. *Psych. Rec.*, 1964a, **14**, 1.
James, R. L., and E. S. Halas, A reply to McConnell. *Psych. Rec.*, 1964b, **14**, 21.
John, E. R., Studies on learning and retention in planaria, *in* M. A. B. Brazier, ed., *Brain Function, II: RNA and Brain Function; Memory and Learning*. Los Angeles, Calif.: University of California Press, 1964, p. 168.
Leighton, W. G., and G. S. Forbes, Precision actinometry with uranyloxalate. *J. Amer. Chem. Soc.*, 1930, **52**, 31.
McConnell, J. V., Memory transfer through cannibalism in planarians. *J. Neuropsychiat.*, *Suppl.* 1, 1962, **3**, 42.
McConnell, J. V., On the turning worms: a reply to James and Halas. *Psych. Rec.*, 1964, **14**, 13.
McConnell, J. V., A. L. Jacobson, and D. P. Kimble, The effects of regeneration upon retention of a conditioned response in the planarian. *J. Comp. Physiol. Psychol.*, 1959, **52**, 1.
Sgonina, K., Vergleichende Untersuchungen uber die Sensibilisierung und den bedingten Reflex. *A. Tier. Psychol.*, 1939, **3**, 224.

Shugar, D., Photochemistry of nucleic acids and their constituents, *in* E. Chargaff and J. N. Davidson, eds., *The Nucleic Acids.* New York: Academic Press, Inc., 1960, p. 39.

Thompson, R., and J. McConnell, Classical conditioning in the planarian *Dugesia dorotocephala. J. Comp. Physiol. Psychol.*, 1955, **48**, 65.

VanDeventer, J. M., and S. C. Ratner, Variables affecting the frequency of response of Planaria to light. *J. Comp. Physiol. Psych.*, 1964, **57**, 407.

Westerman, R. A., Somatic inheritance of habituation of responses to light in planarians. *Science*, 1963, **140**, 676.

Wolff, E., and T. Lender, Sur le déterminisme de la régénération des yeux chez une planaire d'eau douce, *Polycelis nigra. C. R. Soc. Biol.*, 1950, **144**, 1213.

Wulff, D. L., and C. S. Rupert, Disappearance of thymine photodimer in ultraviolet-irradiated DNA upon treatment with a photoreactivating enzyme from baker's yeast. *Biochem. Biophys. Res. Comm.*, 1962, **7**, 237.

Chapter 20

The Biochemistry of Memory

James V. McConnell

Mental Health Research Institute
The University of Michigan
Ann Arbor, Michigan

It has now been some thirteen years since I first started working with planarians. Sometimes when I stop and look back at this worm's dozen of years, I wonder if I should not have told Bob Thompson to go have his head examined when he suggested back in 1953 that we see if we could not condition a flatworm. But one never has the choice of living one's life over again in a different fashion, and I am not really sure I would change the basic pattern anyhow. For I have learned a great deal of psychology from working with the worms; indeed I have probably learned more about the behavior of that strange laboratory species called *Homo scientificus* than I had any desire to learn. If I could change things, I would not give up the worms, and I would not give up any of the experiments I have done (although many could have been better performed), but I do believe that those of us early in the field made some classic mistakes in the fine art of public relations. We were, in our green and callow days, pretty naive about the sort of reactions we should elicit from our colleagues. Instead of glossing over the possible significance of our findings, we stupidly said out loud, for everyone to hear, that we had found some pretty intriguing things in the laboratory. Telling the truth about such findings is probably a basic mistake under the circumstances; had we glossed over the possible importance of the regeneration and cannibalism studies, had we neglected to state our own interpretations of these results, we should not have got into as many public battles with noted scientists and we should surely have had less trouble getting research funds. But then, had we done so, the field of biopsychology probably would not have been as exciting as it is now is; so it is obvious that you cannot win them all. And probably the battles we have got into have kept us on our intellectual toes as well as kept us lean and hungry.

As I am sure you all know by now, the first experiment that suggested that memories might be stored chemically in planarians was performed in the middle 1950's by Allan Jacobson, Dan Kimble, and me (1959). We classically conditioned planarians of the species *Dugesia dorotocephala* using light as a conditioned stimulus and shock from an inductorium as an unconditioned stimulus. When the animals had reached the criterion of 23 responses out of 25 consecutive trials, they were cut in half and allowed to regenerate for a month. A second group of animals, the "time-control" worms as we called them, were trained to criterion, then set aside to regenerate without being cut in half. A third group, the "regeneration-control" planarians, were first cut in half and allowed a month's regeneration and then were conditioned for the first time. The time-control animals, which were never cut, took about 180 trials to reach criterion the first time around; a month later, they retrained in but 40 trials. This savings was, of course, highly significant. The regenerated heads reached criterion also in 40 trials; the tails, in 43 trials. Subsequent studies, incidentally, showed that the tails often retained prior training significantly better than the heads. So far, so good. But what of the regeneration-control animals? Perhaps, as some critics have implied, planarians simply are sensitized by the regeneration. Well, the regeneration-control animals, far from showing any sensitization, showed exactly the opposite, they took significantly *longer* to learn than had the original animals. But, again of interest, the regenerated tails learned somewhat faster than the regenerated heads.

Now, if Jacobson, Kimble, and I had been smart, we would have scrapped that study immediately and performed another quite different one in which we showed that sensitization, or perhaps even habituation, could be retained in regenerating pieces of planarians. That finding would have been exciting enough, and perhaps people would have believed us since nobody cares much about sensitization or habituation anyhow. Instead, we had the audacity to say that we had data suggesting that memories could be stored throughout the planarian's body, and the battle was on. One noted zoologist told me, rocking back and forth in her chair as she spoke, that she could believe that planarians could learn, but she could not believe that they could remember for more than 5 minutes. When I asked her what data led her to this belief, she merely snorted at me and said that she needed no experiments to tell her obvious truths like that. Another zoologist told me our data could not possibly be true because, if the tail of a planarian could remember, a zoologist would surely have discovered the phenomenon years ago. And yet a third informed me that he would not believe our data even if he repeated the experiment in his own laboratory and got the same results as we did. Now, this third zoologist was an exceptionally truthful man; for, a year or so later, he did in fact repeat the regeneration work, he got positive results, and he still refuses to believe the findings.

The next regeneration study performed in our laboratory went a step further (Jacobson *et al.*, 1959). Large *D. dorotocephala* planarians were cut in half, and the tails were discarded. The heads were then immediately conditioned to the usual criterion and then set aside to regenerate. When they had regrown tails, they were again cut in half and now both pieces were allowed to reform. Both regenerates were then trained to the original criterion, and both groups showed significant savings. It should be pointed out, however, that in this study the head regenerates were superior to the tails, perhaps because the tails were completely reformed animals, while the heads retained the original brains and much of the structure of the original animals. It is interesting to note, by the way, that a group of biologists in Russia has recently repeated this work, carrying the animals through additional cuttings and regenerations; these biologists report equal and highly significant retention by both heads and tails (Cherkashin and Sheiman, 1966).

Not long after the first regeneration paper was published in 1959, Ernhart and Sherrick (1959) announced similar results for planarians trained to turn right and then left in a two-unit T maze. Both head and tail regenerates showed significant retention of the original maze learning, and the tails did slightly but not significantly better than the heads. In subsequent years, several other investigators repeated these various studies, sometimes with positive, sometimes with negative results (Corning, 1964; Jacobson, 1962; and Roe, 1963). An analysis of these various studies suggests that the transfer occurred whenever the maze habit was firmly established in the donor animals. Additionally, Westerman (1963a and b) reported that habituation to the onset of a bright light survived cutting and regeneration. Now, Westerman's study was particularly well controlled, and his experiment would seem to pose a knotty problem for those critics who appeal to sensitization as an explanation of the results of the earlier regeneration studies utilizing the light-shock conditioning paradigm. You see, Westerman found that the heads and tails of habituated animals were significantly *less* sensitive to light after they had regenerated, while his control regenerates did not show this decreased sensitivity. Indeed, the maze studies by Ernhart and Sherrick (1959) and by several others of us, as well as Westerman's experiment, make it difficult indeed to raise the specter of simple sensitization as an explanation of the phenomenon of retention of prior training following regeneration in the planarian. Incidentally, it is odd to note that few of the critics of the regeneration studies have ever bothered to cite, much less discuss, any of these later studies.

To continue with our story, however, it was in 1960 that we took the next so-called "giant leap forward." The regeneration experiments had suggested that memories might be stored throughout the planarian's body. If so, it seemed logical that one could transfer memories from one animal to another, just as one can transplant whole chunks of tissue from one planarian to another. After trying several techniques that failed to work, we hit upon the idea of feeding trained planarians to untrained cannibals. So we conditioned some worms to the usual criterion, then cut them in half, and fed them to hungry cannibals. We also chopped up untrained planarians and fed them to another group of cannibals. Then, a day or so later, all the cannibals were trained. We found that the cannibals that had eaten the trained worms were, from the first trials, significantly more responsive to the conditioned stimulus (light) than were the cannibals that had eaten control animals. Indeed, the control cannibals were scarcely different in their subsequent behavior than were planarians that had not been allowed to cannibalize at all. Oddly enough, the experimental cannibals did not always reach criterion sooner than the control cannibals, for the slope of the learning curve for the experimental cannibals was noticeably flatter than that for the control cannibals; but the experimental cannibals showed an initial response rate to the conditioned stimulus of some 50 %, while the controls responded less than 25 % of the time in the first 25 trials (McConnell et al., 1961; Humphries, and Jacobson, 1961; and McConnell, 1962). Shortly thereafter, Roy John and two of his students repeated this study; like our experiment, theirs was performed blind, and their results, including the slopes of the two curves, were almost identical to ours (John, 1964). They added one further touch, however. After several days of training, just as the curves for the experimental and the control cannibals were approaching each other, they gave both groups another feeding. Now, again, there was a highly significant increase in the response rate of the experimental cannibals but no change in response rate of the control animals.

Again, had we been wise, we would have scrapped our initial findings and performed a study in which we fed sensitized planarians to cannibals and showed that sensitization transferred cannibalistically. I do believe that most people would have believed us, for few people have bothered to think through just what sensitization is and how long-term sensitization could be mediated chemically in any organism. Once we had convinced people that sensitization could be passed along to another animal, we might have tried habituation, as Westerman did subsequently (1963a and b). Once we had shown that this too worked and after we had given the scientific world time to digest these findings, we might have been able to announce that learning could be transferred cannibalistically without raising nearly as much dust as we did by starting with the learning study first.

But we did as we did, and there is no undoing it now. The next step in the process was not ours, however, but that of Roy John and Bill Corning, for it was they who first came forth with evidence that one of the chemicals involved in memory storage in the planarian might be RNA. Ward Halstead and Holger Hydén had both suggested that one of the nucleic acids, probably RNA, might be the memory molecule, and Corning and John reasoned that they might test the hypothesis by letting bisected planarians regenerate in a solution of ribonuclease following training. As we all know, they found that the ribonuclease seemed to wipe out the memory in the regenerated tails, but not in the heads (Corning and John, 1961). It is important to note that Corning and John ran several additional groups to control for sensitivity of various kinds. They found retention in trained but uncut animals, in cut animals that regenerated in pond water, and in heads that regenerated in the ribonuclease solution. None of the control groups showed any significant sensitization. Several other groups of investigators, including the Russian biologists I mentioned earlier, subsequently showed that ribonuclease has the effect of wiping out memory in planarians, although the meaning of these data is not yet clear (Fried and Horowitz, 1964; Rieke and Shannon, 1964; and Schwartz and Sweet, 1966).

Building on this early work by Corning and John, my students and I undertook to test the RNA hypothesis in a slightly different way. We attempted to extract biologically active RNA from untrained planarians and inject it into untrained animals to see if we could get a transfer. As I have mentioned elsewhere, we made every mistake in the book; our only excuse for performing such a clumsy experiment is that it was the first of its kind. Nonetheless, we did get evidence that some kind of transfer mediated by our extract had taken place (Zelman et al., 1963). Luckily for us, other groups were able to replicate our early work using much more rigidly controlled conditions. The most noteworthy of these replications is that by Jacobson, Fried, and Horowitz who have recently shown that RNA extracted from the bodies of pseudoconditioned planarians does not cause a transfer when injected into recipient worms, nor does RNA extracted from untrained animals, while RNA taken from well-conditioned animals does cause a transfer (Jacobson et al., 1966a and b). Since the pseudoconditioned donors did not show any increased response rate to the conditioned stimulus, while the well-conditioned animals did, it would seem that no sensitivity to the light took place in the pseudoconditioned donors, hence one would hardly expect any transfer of sensitivity via RNA.

One might imagine that, with so much positive evidence in the literature suggesting that some form of acquired behavior tendencies could be transferred chemically from one planarian to another (Westerman, 1963a and b; McConnell et al., 1961; Humphries and Jacobson, 1961; McConnell, 1962; John, 1964; Rieke and Shannon, 1964; Zelman et al., 1963; Jacobson et al., 1966a and b; Fuchs et al., 1966; Kabat, 1964; Ragland and Ragland, 1965; Wells, 1963), the scientific world would have accepted the issue as closed. As most of you know, this is hardly the case. Indeed, skepticism has been expressed on all sides. If we ignore the critics who would not believe data no matter how carefully the experiments were performed, what leads people to be skeptical? The answer, I believe, is chiefly one study, that by Hartry, Keith-Lee, and Morton (1964), which I believe to be the most quoted but least understood experiment in the planarian literature. Therefore, I intend to examine this study in considerably greater detail than most of the other experiments mentioned so far.

Let us begin by looking at Table 1, which shows the results of this particular study. On day 1 of the experiment, the 24 animals in group I on the far right of the table were conditioned to the criterion of 23 responses out of any consecutive 25 trials. Half of these animals were then ground up and fed to the untrained cannibals in group C_1-ca, while the other 12 were labeled C_2-2 and set aside for 24 hours. During the training of the animals in group I, three other groups were given various experiences in the training troughs. One group was given shock, but no exposure to light, then was ground up and fed to the cannibals in group S-ca. A second group was given trials of light only, but no exposure to shock, then ground up and

TABLE 1

Performance during conditioning trials. Significance is indicated by underscoring. Any two means not underscored at any point by any one line are significantly different at the level indicated: $\cdots\cdots$, 0.01 level; ———, 0.05 level; and $------$, 0.10 level. L-ca, cannibals of group L (planarias exposed to light only); H-Ca, cannibals of group H (planarias which were handled only); C-Ca, cannibals of group C_1 (planarias which had been conditioned); C_2-2, second day performance of group C_2 (planarias which had been conditioned and not cannibalized); S-Ca, cannibals of group S (planarias exposed to shock only); F-Ca, cannibals of group F (naive, unstimulated planarias); N, naive, unstimulated, unfed planarias (group N); and I, performance on first day of planarias known to be naive and unfed (groups C_1 and C_2).

L-ca	H-ca	C_1-ca	C_2-2	S-ca	F-ca	N	I
			Mean trials to criterion				
58.0	60.8	67.5	69.9	88.3	90.0	153.9	157.3

L-ca	H-ca	C_1-ca	C_2-2	S-ca	F-ca	N	I
			Mean conditioned responses in first 25 trials				
16.9	14.7	13.9	12.7	11.6	10.2	7.6	7.1

fed to the cannibals in group L-ca. The third group was given nothing but handling, i.e., was poked or squirted just as often as the animals in group I but was never exposed to either light or shock, then was ground up and fed to the cannibals in group H-ca. The animals in group N were given no experience on day 1 and did not cannibalize at all. On the second day of the experiment, the animals in all the remaining groups were coded and run blind to the 23:25 criterion. As you can see, the animals that cannibalized the light-control and the handling-control animals were at least as responsive as the cannibals that ate the conditioned planarians. The cannibals that ate the shock-control worms were somewhat inferior, as were the cannibals that ate the untrained tissue, while all of these groups were superior to the untrained animals that had not cannibalized at all.

Now, what shall we make of these interesting data? At first blush, it would seem that, since most of the cannibals that ate control worms did as well as the cannibals that ate conditioned worms, there is no evidence that classical conditioning was transferred cannibalistically in this study, a conclusion with which I agree. But it is equally clear that *something* did transfer, for the groups on the left are significantly superior to the groups on the right of the table. Hartry, Keith-Lee, and Morton believe that their results can best be explained as a transfer of sensitization, and again I agree with them, but I should like to point out that those critics who have invoked the concept of sensitization in the past have almost uniformly stated that it was the shock that was the sensitizing agent, and such seems *not* to be in the case in this study.

We are still left with an interesting problem though: Why did not the conditioning transfer? One possible reason could be that, instead of feeding their cannibals chunks of tissue, as was done in all other cannibalism studies, these authors ground up the experienced worms into very tiny pieces. Now, whenever one grinds up living tissue, one releases ribonuclease; hence, when one extracts RNA from tissue, one always makes sure that the tissue is extremely cold prior to homogenation, since cold helps neutralize ribonuclease. If RNA is indeed the transfer molecule, it might have been badly degraded before the cannibals got around to eating the finely ground up tissue provided them by Hartry, Keith-Lee, and Morton. I shall come back to this point later. But there is a much more obvious fault with this otherwise excellent study. Look at the scores for group C_2-2. These animals were conditioned on day 1, then were allowed to sit untouched for 24 hours, and then were retrained. It took these animals 70 trials to reach criterion the second time around. When I remind you that Hartry, Keith-Lee, and Morton used massed training techniques and when I tell you that they used clean troughs rather than slimed troughs, perhaps you will begin to suspect what my criticism is. Frankly, I do not think the animals in group I were well conditioned at all; rather, I think they were mostly sensitized, for one would expect the animals in this group to reach criterion the second day in 40 trials or even much less if true conditioning had taken place on day 1. If only a small amount of conditioning took place on day 1, but a lot of sensitization took place, one would expect pretty much the results shown in this table. The fact that all the rest of us who obtained positive transfer results used spaced training and slimed troughs supplies some measure of support for my interpretation of the Hartry, Keith-Lee, and Morton results.

It was with this experiment in mind that Jessie Shelby and I last year began a study on the cannibalistic transfer of a maze habit in the hope that none of the objections raised to the conditioning studies would be applicable to this different type of experiment. We began by giving a large number of *D. dorotocephala* preference trials in the simple T maze I described in an earlier paper in this collection. We selected for the study only those animals that showed a weak preference for the light or the dark arm or no preference at all; animals with strong preferences were discarded. We then began training donor worms until they had reached the criterion of 9 correct responses out of 10 daily trials 2 days in a row. In all cases, throughout the experiment, animals were trained opposite to whatever initial bias they had shown during the preference tests. As soon as an animal had reached criterion, it was set aside to rest for a period of 2 or more weeks. Then all the trained animals were given additional trials until they again reached criterion; as you might suspect, they showed highly significant savings. They were then chopped in fairly large pieces and fed to cannibals, and the cannibals were then trained.

Now, suppose we take one of the victim worms as an example. Let us assume that, in the 20 preference trials, it chose the light arm 11 times, the dark arm but 9 times. It would then have been trained to go to the dark arm until it reached criterion. The cannibal to which it was fed would have been preference-tested too, and it too would be trained opposite to its initial bias. But to what cannibal should we feed a dark-trained victim worm? If the cannibal showed an initial preference for the light arm, it would be trained to go to the dark arm no matter what it ate. So, if we fed it a dark-trained worm, obviously the cannibal would be trained the same way as was the worm it ate. In such a case, we might speak of giving the cannibal "positive instructions," and we should expect it to learn rapidly if a transfer occurred. However, we could just as easily have fed this same cannibal a worm trained to go to the light arm, in which case, since the cannibal would be trained to go to the opposite arm than was the worm it ate, we could say we had given the cannibal "negative instructions," and we should expect it to learn more slowly. Likewise, we could feed the cannibal an untrained worm, in which case we might speak of "neutral instructions." And there is one more set of instructions that we could use that might not have occurred to you. What about feeding a cannibal part of a light-trained victim and part of a dark-trained victim, all at the same time? In this case, we could speak of giving the cannibal "conflicting instructions." As some of you may recognize, this is, in fact, a better-controlled replication of a study reported informally by a Tallahassee high-school student named Andy Moorer (1963). So far, we have run four animals in each of the groups in two complete replications, giving us a total of eight animals in each of the four groups. In both replications, the positive-instructions animals were significantly superior to all others. Surprisingly enough, the negative-instructions animals were second best, being significantly superior to the animals in the two other groups. In one replication, the conflicting-instructions animals were significantly inferior to neutral-instructions animals; in the second replication, the difference between these two groups was not significant. The conflicting instructions animals did, however, show more vascillation and indecision at the choice point, more turns and false starts than did the animals in the other groups. Needless to say, the whole experiment was performed blind, the mazes were carefully slimed, and the animals were given widely spaced trials.

Now, as best I can see, it is going to be rather difficult for anyone to explain these results as sensitization of some kind. And here, so it seems, we have preliminary evidence that the transfer effect is stimulus specific.

I am sure that this one study will not settle the issue, but perhaps it will help matters somewhat.

What will help settle the issue, I think, is for these writers who appeal to sensitization or pseudoconditioning as an explanation of the chemical studies to sit down and think through precisely what it is they are talking about. Those of us who have got positive results with attempts to transfer acquired behavioral tendencies via cannibalism or RNA injection have stated our position rather clearly. We believe that, when the animal is trained, some kind of chemical change takes place within the animal's body, probably within the nervous system, and that RNA is somehow involved in mediating this change. Hence RNA is likewise somehow involved in the storage of engrams or memories. (In my opinion, it is likely that RNA is not the only chemical involved and that different types of experiences are mediated by quite different molecules, but that is an un-proven assumption on my part.) We assume that, whatever the chemicals are, if they are taken from the body of a trained animal and incorporated in the body of an untrained animal, they will so affect the neurons of the untrained animal that the subsequent behavior of these neurons is more like the behavior of the donor animal's neurons than was the case prior to the incorporation. Much of this explanation is highly theoretical and subject to change without prior notice. But I submit that the phenomenon of transfer itself is on rather good grounds. We may continue to argue about what it is that gets transferred and what the mediator is, but by now we all should be willing to admit that something is passing from one animal to another chemically that affects the recipient's behavior. When you stop and think about it, that is a rather remarkable accomplishment in and of itself, a phenomenon that was not even dreamed about some ten years ago. Even in the event that it turns out that learning per se cannot be transfer-red, the cannibalism and RNA injection studies in planarians may still have made an interesting contribution to our understanding of animal behavior.

And, now, at last, let us look at sensitization.

It seems to me that, if someone states that a given behavioral change is due to sensitization rather than to learning, he should have a clear-cut idea what learning is and what sensitization is. Anyone who has ever tried to define learning in an exact fashion knows the difficulties involved; probably the same difficulties obtain in any definition of sensitization. Nor is it always easy to separate the two, since the behavioral changes we typically ascribe to such a form of learning as classical conditioning obviously are due in part to sensitization. But let us for the moment assume that sensitiza-tion is a generalized increase in responsivity to all incoming stimuli, not due to any pairing of the conditioning and the unconditioned stimuli, while learning is an increase in responsivity that is due to the pairing of the conditioning and the unconditioned stimuli, and, further, let us assume that use of the proper controls allows us to differentiate between the two. Now, what can we say about sensitization? I mean, what is our *theory of sensitivity*? We all have theories of learning, and sometimes they help us predict what the animal is going to do in certain circumstances. We expect learning, e.g., to be reflected in what is typically called a *learning curve*, i.e., a negatively accelerated rise in performance level that eventually reaches asymptote. But what of sensitization? Does it increase monotonically and with a negative acceleration? Jensen appears to suggest that this is the case, but there are few data that support this position. And how long does sensitization last? In the case of learning, one expects some forgetting over time, but considerable savings should occur too, and the forgetting that does occur could be due to interference from other habit patterns. Would one expect sensitization to stay around a long time, and, if there is a time-dependent decrement in sensitization, would resensitization later on take fewer trials than did the original sensitization? Would any observed decrement be due to interference from other habits or other sensitizations? And what about extinction? If one omits the unconditioned stimulus in a conditioning paradigm, one typically gets very rapid extinction. Does sensitization extinguish too, and, if so, how? Is the extinction due to the omission of the unconditioned stimulus? Does the sensitization disappear entirely, or would one expect very rapid resensitization after extinction had taken place? Would electroconvulsive shock given immediately after sensitization trials knock out the sensitization in the same fashion as it knocks out learning? After extinction, does one get spontaneous recovery of sensitization or reminiscence or even latent sensitization? And what chemicals are involved in the mediation of the sensitization response? RNA? Proteins? Polypeptides? Large molecules or small?

If one wishes to explain all of the conditioning studies as showing mere sensitization due to noxious stimulation rather than showing true learning, then one must have in mind a type of sensitization that builds up slowly from one day to the next and that remains in rather full force for weeks or even months when the animal is not being trained, yet disappears in a matter of a few trials when the conditioned stimulus is presented without the unconditioned, and yet reappears very rapidly indeed when retraining is instituted. I ask you, what kind of "mere sensitization" is that? If one wishes to invoke sensitization to explain the Griffard and Peirce study (1964) or any of the other experiments showing conditioned discrimination, then one must realize that one is advocating a type of sensitization that is clearly stimulus specific. Davis (1963) has shown that giving planarians electroconvulsive shock immediately after training knocks out the memory of the conditioning; yet would not one expect convulsive shock to increase rather than to decrease sensitization?

It seems to me that, unless one attempts to answer some of these questions, one should not appeal to sensitization as an explanatory concept. To say, as Hartry, Keith-Lee, and Morton do, that their cannibalism results may be due to "changes in nutritional or metabolic factors or resulting changes in the degree of activation or sensitization" without specifying what other behavioral modifications one would expect if this were the case seems to be begging the issue. For instance, if the sensitization that Hartry, Keith-Lee, and Morton speak of is due to the build-up of some evanescent chemical in the animal's body, one would expect it to dissipate if one waited a sufficient length of time. Thus, if one waited 24 hours after training before feeding the experienced animals to the cannibals, one would not expect much transfer to take place. Yet, when Roy John and his students did precisely this, they found excellent transfer (John, 1964). The sensitization change, therefore, cannot be very evanescent. Also, there are several studies whose results suggest that the application of ribonuclease destroys the transfer effect. Does this mean that sensitization is mediated by RNA?

In summary, I think it is about time that the theorists who hold to what Walker and Milton (1966) call "the sensitization hypothesis" either put up or shut up. They must tell us exactly what their hypothesis predicts in the various situations I have asked about, or they should admit that their sensitization hypothesis is fuzzy to the point of uselessness.

Of course, anyone who thinks the problem through carefully will probably conclude that the battle is largely based on semantics. The behavioral changes we typically refer to as being due to learning in a classical conditioning situation, e.g., are actually the result of the combined effects of a great many different but related factors inside the animal (McConnell, 1966). Conditioning almost always includes behavioral modifications due to sensitization to the conditioned and the unconditioned stimuli, to the experimental procedure, and to various forms of external but adventitious stimulation in the environment; conditioning also includes large amounts of habituation. An animal in a maze does not merely learn to turn left or right; rather it learns thousands of things about the experimental procedure, and various types of sensitization and habituation must occur in maze training as well as in classical conditioning. It would be rather frightening if the *only* factors that transferred chemically were those due to the pairing of the conditioned and unconditioned stimuli or, on the other hand, if *none* of the factors that transfer were in any way influenced by the pairing of these stimuli.

For it seems to me that, when the theorists who hold to the sensitization hypothesis are forced to account for all the data, they will find they have swallowed the camel and are now straining at a worm. The fact that, in their terms, long-lasting sensitization occurs, that it is chemically mediated, that it shows many of the characteristics of so-called "true learning," and, most of all, that it can successfully be transferred from one organism to another via cannibalistic ingestion or RNA injections or both—these facts are so incredible that to argue whether this sensitization is or is not true learning seems to be emphasizing semantic trivialities rather than looking at data logically.

Let me then make a suggestion. If we stop arguing about whether worms can learn or whether their behavioral modifications are due to mere sensitization, we could start all over again and avoid bringing into comparative psychology all of the tedious, trivial arguments that have plagued experimental psychology since the time that Hull and Tolman took up their rats and began battling. We could begin by gathering correlations between environmental changes and behavioral changes in planarians, not bothering too much about what labels we assign to the processes involved. And, since we have an intriguing new tool, the transfer phenomenon, let us put it to work for us as cleverly as possible. Suppose we find, as I suspect will be the case, that some components of an acquired behavioral change will transfer, while others will not. Likewise, suppose we find that the various components of the acquired behavioral change are mediated by quite different chemicals. What an interesting lever such findings would give us in an attempt to pry out the details of what really goes on inside an organism when its behavior changes! Perhaps we shall eventually come up with new and better definitions of learning and sensitization. More likely, we may eventually coin new terms to cover our better understanding of the biological bases of behavior, terms that hopefully will lack the surplus meanings and emotional connotations of the words we have used too long in the past. I do believe that the current studies on the chemistry of memory could play a significant part in such an undertaking. It would be a real pity if we chose instead to continue the semantic battles and word play that so many of us have engaged in recently.

REFERENCES

Cherkashin, A. N., and I. M. Sheiman (Sci. Center Biol. Res., Acad. Sci., Puschino, U.S.S.R.), The use of simple biological models in memory research. Paper presented at the *Int. Congr. Psychol., Moscow, Aug.* 6, 1966, *Symp., Biological Bases of Memory Traces.*

Corning, W. C., Evidence of right-left discrimination in planarians. *J. Psych.*, 1964, **58**, 131.

Corning, W. C., and E. R. John, Effect of ribonuclease on retention of conditioned response in regenerated planarians. *Science*, 1961, **134**, 1363.

Davis, G., Memory loss through retrograde amnesia in the planarian. *Worm Runner's Digest*, 1963, **5**(2), 77.

Ernhart, E. N., and C. Sherrick, Jr., Retention of a maze habit following regeneration in Planaria (*D. maculata*). Paper read at *Midwestern Psychol. Assoc., St. Louis,* May, 1959.

Fried, C., and S. Horowitz, Contraction—a leaRNAble response. *Worm Runner's Digest*, 1964, **6**(2), 3.

Fuchs, A., R. Harrington, R. Lariviere, and T. Robinson, Degree of learning and degree of memory transfer in planarians. *Worm Runner's Digest*, 1966, **8**(2), 28.

Griffard, C. D., and J. T. Peirce, Conditioned discrimination in the planarian. *Science*, 1964, **144**, 1472.

Hartry, A. L., P. Keith-Lee, and W. D. Morton, Planaria: memory transfer through cannibalism re-examined. *Science*, 1964, **146**, 274.

Humphries, B., and R. Jacobson, The effect of ingestion of conditioned planaria on the response level of naive planaria, II. *Worm Runner's Digest*, 1961, **3**, 165.

Jacobson, A. L., An attempt to demonstrate transfer of a maze habit by ingestion in Planaria. University of Michigan *dissertation*, 1962.

Jacobson, A. L., C. Fried, and S. D. Horowitz, Planarians and memory, I: Transfer of learning by injection of ribonucleic acid. *Nature*, 1966a, **209**, 599.

Jacobson, A. L., C. Fried, and S. D. Horowitz, Planarians and memory, II: The influence of prior extinction on the RNA transfer effect. *Nature*, 1966b, **209**, 601.

Jacobson, Reeva, J. V. McConnell, and D. M. Maynard, Apparent retention of a conditioned response following total regeneration in the planarian. *Amer. Psychol., Abstr.*, 1959, **14**, 410.

John, E. R., Studies on learning and retention in planaria, *in* M. A. Brazier, ed., *Brain Function, Vol. II*. Los Angeles, Calif.: University of California Press, 1964, 161–182.

Kabat, L., Transfer of training through ingestion of conditioned planarians by unconditioned planarians. *Worm Runner's Digest*, 1964, **6**(1), 23.

McConnell, J. V., Memory transfer through cannibalism in planarians. *J. Neuropsychiat.* (*Suppl.* 1), 1962, **3**, S42.

McConnell, J. V., Comparative physiology: learning in invertebrates. *Ann. Rev. Physiol.*, 1966, **28**, 107.

McConnell, J. V., A. L. Jacobson, and D. P. Kimble, The effects of regeneration upon retention of a conditioned response in the planarian. *J. Comp. Physiol. Psychol.*, 1959, **52**(1), 1.

McConnell, J. V., R. Jacobson, and B. Humphries, The effects of ingestion of conditioned planaria in the response level of naive planaria. *Worm Runner's Digest*, 1961, **3**, 41.

Moorer, A., Letter to the Editor, *Worm Runner's Digest*, 1963, **5**(2), 108.

Ragland, Rae S., and J. B. Ragland, Planaria: interspecific transfer of a conditionability factor through cannibalism. *Psychon. Sci.*, 1965, **3**, 117.

Rieke, J., and L. Shannon, The effects of deoxyribonuclease and ribonuclease on the transfer of learning by cannibalism in planarians. *Worm Runner's Digest*, 1964, **6**, 7.

Roe, Kiki, In search of the locus of learning in planarians. *Worm Runner's Digest*, 1963, **5**(2), 16.

Schwartz, M. C., and S. Sweet, An investigation of the relation between RNA and learning in planarians. *Worm Runner's Digest*, 1966, **8**(2), 25.

Walker, D. R., and G. A. Milton, Memory transfer vs. sensitization in cannibal planarians. *Psychonom. Sci.*, 1966, **5**, 293.

Wells, P. H., Experiments on conditions of learning in planarian flatworms. *Worm Runner's Digest*, 1963, **5**(1), 58.

Westerman, R. A., Somatic inheritance of habituation of responses to light in planarians. *Science*, 1963a, **140**, 676.

Westerman, R. A., A study of habituation of responses to light in the planarian *Dugesia dorotocephala*. *Worm Runner's Digest*, 1963b, **5**(2), 6.

Zelman, A., L. Kabat, R. Jacobson, and J. V. McConnell, Transfer of training through injection of "conditioned" RNA into untrained planarians. *Worm Runner's Digest*, 1963, **5**, 14.

Discussion

R. Kenk: As far as I know, the experiments concerning the transmission of training by regeneration have been made chiefly on *Dugesia dorotocephala*. If the posterior or postpharyngeal part of a trained *D. dorotocephala* or *D. tigrina* is allowed to regenerate a new head and anterior portion, then the regenerated part is claimed to consist of untrained, naive tissue formed from the blastema which has grown out of the posterior or trained half. However, regeneration in planarians is a combination of two processes: the formation of new tissues from neoblasts or undifferentiated cells which have emigrated from the old tissue (epimorphosis); and a rearrangement of the old, differentiated tissues which retain their functional characteristics and only change their position to conform with the new situation (morphallaxis). Both processes occur simultaneously, but the proportion between epimorphosis and morphallaxis varies in the different species. Thus *D. dorotocephala* and *D. tigrina* regenerate mainly by morphallaxis, and *Cura foremanii,* by epimorphosis. In *Dugesia* the blastema forming the new anterior end is very small and apparently develops only into the new head: the pharynx arises in the old tissue, developing from accumulations of neoblasts. In *Cura* the blastema is extensive, appears as a definite outgrowth of the old tissue, and develops into the entire prepharyngeal part of the body including the new pharynx. We illustrate this diagrammatically in Fig. A, which indicates the old (trained) tissues by stippling. The

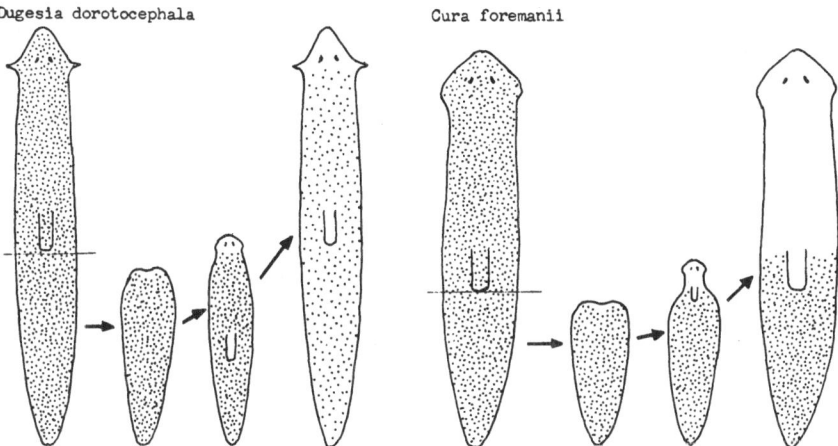

Fig. A. Regeneration differences in two species of planarians.

323

striking difference between individual species has been pointed out already by the earlier investigators of planarian regeneration, e.g., C. M. Child (1906); Child's "regeneration" is epimorphosis, his "redifferentiation" is morphallaxis; *Planaria simplicissima* is *C. foremanii*, and *Planaria maculata* is a synonym of *D. tigrina*. Thus it is not possible to consider the regenerated half of *D. dorotocephala* (or *D. tigrina*) to consist of naive tissue. *Cura* would be more appropriate as an object for regeneration experiments.

J. Best: In our sections we have noticed how little difference there is between the neoblasts and the nerve cell bodies. It looks to me that the cell bodies of neoblasts meet the specifications of neurons except that they haven't sent out any roots yet. I wonder just how much modification a neoblast has to undergo before it becomes a nerve cell.

W. Corning: I wonder if Dr. Brown would specify the location of his light source. In the thesis write-up, I note in the methodology section that the light was below the trough, and then later on it appears to be above.

H. Brown: The light source was located above the trough in the studies that I have just finished reporting (Brown, Dustman, and Beck, 1966a and b). The location of the source was below the troughs in the ultraviolet studies (Brown, H. M., Chapter 19 of this volume). The bottom mounting allowed better access for observing the behavior of the animals. There was no difference found in the baseline response levels of the animals under either condition as long as the light intensity was the same. Also, head-tail differences in the frequency of response to light stimuli was the same under both conditions.

P. Wells: Dr. McConnell has touched upon a number of things in his paper, and there is insufficient space here to comment upon all of them. I agree that it is difficult to differentiate operationally between the words learning and sensitization as they have been used by many writers. But, rather than argue about the meanings of words, we have devoted our attention to designing experiments which would test the validity of the hypothesis of memory molecules. Our main concern was that most of the transfer findings involved experiments in which light-shock pairings were used. Our reasoning was that, if there were an encoding of memories in a molecule, it ought to be possible to demonstrate the transfer of other kinds of learning.

In an early experiment we did obtain data to support the finding that, by allowing naive worms to cannibalize light-shock trained subjects, there was a more rapid conditioning of the naive worms. However, when this sort of experiment was repeated with animals trained in T mazes or the Van Oye type of maze, we did not observe any kind of transfer. It appears that the transfer of training is not a general phenomenon. Now Dr. McConnell has also presented some new T-maze data that may force us to reexamine the question, but, based upon our own findings, we would have to reject the general hypothesis of a molecular coding of memory (Pickett *et al.*, 1964 and 1965).

Coupled with these experiments, we did something else to examine the role of RNA in memory. If RNA is involved in memory, it should require a specific RNA to enhance performance of a specific task. However, exogenous RNA (placed in the medium) was able to increase the efficiency of maze learning; this was yeast RNA, and under these circumstances there was no difference between animals that cannibalized trained or control subjects (Pickett *et al.*, 1965). With these results we have moved closer to a contemporary Gaito position than to a memory molecule position.

E. Bennett: Dr. McConnell mentioned that he was not thinking very far ahead when he published his revolutionary results and that he should have hidden them under a barrel and made them out as something else. I would like to suggest that, if the key experiments had been replicated sufficiently and with a large enough population of subjects to begin with, we would all have been satisfied. The McConnell, Jacobson, and Kimble (1959) experiment used an *n* of about 5 in each group. The RNA transfer experiments that have been reported are poor, and, in the case of Fried and Horowitz (1964), the authors themselves admit this. If studies such as these look as if they might get us somewhere, why haven't they been exploited? Again, there are few repetitions. The Westerman (1963) experiment, e.g., is another key one that needs repeating. I don't know what has happened to Westerman, but surely he can see the potential of his finding. Why hasn't he exploited it?

A. Jacobson: I think the statement Dr. Bennett made should be qualified somewhat. Take, e.g., the recent work we reported in *Nature* (Jacobson *et al.*, 1966): Fried, Horowitz, and I followed up their initial work using 25 subjects in each group. Our original cannibalism work was repeated five times, and in that series of experiments we had 27 subjects per group.

REFERENCES

Brown, H. M., R. E. Dustman, and E. C. Beck, Experimental procedures that modify light response frequency of regenerated Planaria. *Physiol. Behav.*, 1966a, **1**, 245.

Brown, H. M., R. E. Dustman, and E. C. Beck, Sensitization in Planaria. *Physiol. Behav.*, 1966b, **1**, 305.

Child, C. M., *Archiv. fur Entwicklungsmechanik*, 1966, **20**, 411.

Fried, C., and S. Horowitz, Contraction: a leaRNAble response? *Worm Runner's Digest*, 1964, **6**, 3.

Jacobson, A. L., C. Fried, and S. D. Horowitz, Planarians and memory. *Nature*, 1966, **209**, 599.

McConnell, J. V., A. Jacobson, and D. P. Kimble, The effects of regeneration upon retention of a conditioned response in the planarian. *J. Comp. Physiol. Psychol.*, 1959, **52**, 1.

Pickett, J. B. E., L. B. Jennings, and P. H. Wells, Influence of RNA and victim training on maze learning by cannibal planarians. *Amer. Zool.*, 1964, **4**, 158 (Abstr.).

Pickett, J. B. E., L. B. Jennings, and P. H. Wells, Influence of RNA and victim training on maze learning by cannibal planarians. *Worm Runner's Digest*, 1965, **7**, 31.

Ratner, S. C., and J. M. VanDeventer, Effects of water current on responses of planaria to light. *J. Comp. Physiol. Psychol.*, 1965, **60**, 138.

Westerman, R. A., Somatic inheritance of habituation of responses to light in planarians. *Science*, 1963, **140**, 676.

PART III

Other Invertebrate Preparations

Chapter 21

Memory and the Microinvertebrates

Philip B. Applewhite and
Harold J. Morowitz

Department of Molecular Biophysics
Yale University
New Haven, Connecticut

CHOICE OF THE EXPERIMENTAL ORGANISM

The choice of an experimental organism in biology often determines what experiments can be performed, and thus may set limits on the physical level at which information can be gained. If the human brain with its 10^{10} neurons were chosen as the experimental object, the specification of all the neuronal interconnections would be too difficult and information about it at this level could not be obtained. Electron microscopy of the rat's brain has been done, but an electron microscopic mapping of the entire rat's brain is not feasible. Also, it is not convenient to use the smallest metazoan (50 μ) to study the gross chemical composition of a brain, nor is it convenient to insert electrodes in the smallest protozoan (5 μ). This is to say the experimental level one is working at eliminates certain organisms because of size restraints.

Furthermore, what is found out about a brain is a function of how it is looked for. Specifically, "which pattern of connexions will be found depends on which set of inputs and outputs is used" (Ashby, 1958). Any number of learning experiments, which treat the whole animal as a black box, reveal little of the physical and chemical changes inside. Electroencephalogram studies, e.g., give clues primarily to surface electrical events and "until the nature and source of brain rhythms are clearly known, inferences as to the meaning of the changes introduced through conditioning must remain highly tentative" (Morrell, 1961). If we want information about memory at the fine structural and molecular level, an experimental organism must be chosen that allows this information to be found. There is no guarantee that the organism chosen will have a brain that functions like one even of a similar species, but there is inferential evidence that is encouraging. Namely, "the basic biochemical reactions upon which cell growth and division depend are the same, or very similar, in all cells, those of microorganisms as well as those of higher plants and animals"

TABLE 1
Some Common Micrometazoa

Freshwater organism	Approximate size range, mm	Average life span	Reproduction	Digestive system	Circulatory and respiratory systems	Vision
Catenulida Macrostomida Neorhabdocoels	0.3–11.0	Indefinitely	Hermaphroditic Transverse fission Sexual	Pharynx Enteron Protonephridia opening to surface	None	a. Pair of light-refractive bodies b. One or more pigment cells and one or more retinal cells
Nematodes	0.2–10.0	Days to years	Most dioecious Some hermaphroditic	Buccal tube Pharynx Intestine Rectum	None	a. Dermal sensitivity b. One or more pigment cells and one or more retinal cells (may have cuticle lens)
Gastrotrichs	0.07–0.6	Few weeks	Hermaphroditic but male system degenerate Parthenogenesis	Pharynx Intestine Rectum Protonephridia opening to surface	None	Eye spots containing pigmented cells
Rotifers	0.05–2.0	Few weeks	Dioecious Some parthenogenesis	Buccal tube Mastax Esophagus Stomach Intestine Protonephridia emptying into	None	Simple ocelli containing few photo-receptor cells per ocellus

	Size (mm)	Lifespan	Reproduction	Digestive system	Circulatory	Sensory
Bryozoa	1.0	Few weeks–few months	Hermaphroditic Sexual and asexual	Pharynx Esophagus Stomach Intestine	None	Dermal sensitivity
Tardigrades	0.06–1.2	Several months	Dioecious Some parthenogenesis	Buccal tube Pharynx Esophagus Intestine Rectum	None	Pair eye spots, each containing single pigmented cell
Copepods	0.3–4.0	Few months	Dioecious	Esophagus Stomach Intestine Excretory organs	Simple heart or None	Simple ocelli containing few photoreceptor cells per ocellus
Ostracods	0.35–7.0	Few months	Dioecious Some parthenogenesis	Esophagus Stomach Intestine Excretory organs	None	Idem (lens)
Cladocera	0.2–18.0	Few months	Dioecious Parthenogenesis common	Esophagus Stomach Intestine Rectum	Simple heart	Idem (or compound eye)
Water mites	0.4–3.0	Several months	Dioecious	Buccal tube Pharynx Esophagus Stomach Intestine	Tracheae with at least one pair of openings on body	Idem (may have cuticle lens)

(Watson, 1965). Therefore, we can "find out much about many fundamental metabolic processes by studying one organism" (Williams, 1963).

Finding the relationship of structure and function to memory changes seems particularly appropriate at this time. With the electron microscope morphological changes due to behavior may be detected, and with autoradiography function can be localized in the cellular structure to less than 0.1μ in favorable cases. The use of the electron microscope for these purposes imposes a physical restriction on the size of the organisms used, particularly since it would be profitable to deal with the entire nervous system. With this approach the largest dimension of a whole brain that could be conveniently dealt with would be about 100μ. If magnified only 10,000 times (allowing a 100-Å resolution to be observed visually), the brain would be 1 m long.

For these reasons, experimental organisms of less than 1 mm in adult length have been chosen. Organisms this small are called the micro-invertebrates and include the so-called micrometazoa and virtually all the protozoa. The protozoa would be particularly good organisms in which to study the basis of memory because of their simplicity. Choosing the simplest experimental organism that exhibits the phenomenon to be studied eliminates a wide variety of irrelevent phenomena that may mask it. While protozoa lack the neurons of the metazoa, they do possess a fibrillar network that presumably may carry out the same functions (Kudo, 1966). Electron microscopy has been done on many of them (Pitelka, 1963), and they exhibit a wide variety of behaviors (Jennings, 1962, and Warden *et al.*, 1940) that could be used to elicit learning. The only disadvantage is the difficulty of working with these small organisms one at a time, or in training many at once.

Saltwater microinvertebrates have not been used by us because of the relative ease of obtaining and culturing the freshwater species. The particular micrometazoa we are using and a brief description of them are presented in Table 1; the species of interest are those under 1 mm in length. All of them have some sort of photosensitivity. Only the cladocera and some copepods have any sort of circulatory system, while only the water mites have a respiratory system. Thus, these animals consist essentially only of reproductive, digestive, and sensory systems. The nematodes, gastrotrichs and rotifers have a near (Birky and Field, 1966) constant (Hyman, 1940) number of cell nuclei in each species, reaching an approximate total of less than 1,000 nuclei. This becomes important when it is desirable to keep track of each cell. The brain sizes of some of these micrometazoa are indicated in Table 2. It is interesting to note that each of them would fit within a neuron of the cat cerebellum, measuring about 60 by 50 by 40 μ. More detailed accounts of these invertebrate nervous systems can be found elsewhere (Bullock and Horridge, 1965).

The choice of the proper organism may be crucial in advancing the chemistry of learning, much as viruses and bacteria advanced molecular genetics. "The virus . . . furnished us with the key to molecular genetics.

TABLE 2

Brain Sizes of Some Common Micrometazoa

Organism	Member of	Average length, mm	Approximate brain volume, μ^3 *
(*Taxonomic representation*)			
Stenostomum, genus	Turbellaria	0.6	15,000
Macrostomum, genus	Turbellaria	0.8	15,000
Nematode, class	Aschelminthes	1.0	5,000
Lepidodermella, genus	Gastrotricha	0.15	10,000
Philodina, genus	Rotifera	0.5	20,000
Monostyla, genus	Rotifera	0.15	2,000
Colurella, genus	Rotifera	0.05–0.15	As small as 300
Plumatella, genus	Bryozoa	1.0	15,000
Alonella, genus	Cladocera	0.5	25,000
Ostracod, order	Crustacea	1.0	100,000
Cyclopoida, suborder	Copepoda	1.0	100,000
Tardigrade, class	Arthropoda	0.7	100,000

*Within a factor of 2, or so.

Unfortunately, no such simple experimental object has been made available in the case of neurology" (Eigen and De Maeyer, 1965). Perhaps, a member of the microinvertebrates will qualify.

LEARNING EXPERIMENTS

Before the microinvertebrates can be used as model systems for studying the chemistry and physics of memory, it must be demonstrated that they are capable of learning. There have been many learning experiments with protozoa (McConnell, 1966, and Thorpe, 1963), but true learning in them has been questioned (McConnell, 1966). Of the learning experiments done with freshwater metazoa (Carthy, 1958 and 1965; Hyman, 1940; Jennings, 1962; McConnell, 1966; Thorpe, 1963; Warden et al., 1940; and Schone, 1961), only one has been reported for a micrometazoa (Soest, 1937). The animal was *Stenostomum*, but the results from the only control experiment do not clearly indicate learning was achieved.

We have been able to habituate 10 different species of microinvertebrates to two different intensities of mechanical shock. Specific details of the method and the results have been described (Applewhite and Morowitz, 1966), but the results are summarized in Figs. 1 to 3. Figures 1 and 2 compare the number of stimuli used to habituate the organisms as a function of the intensity of the stimulus. These figures also indicate on the abscissa how the organisms respond to the stimulus over time. Figure 3 offers a different measure of habituation: the time the animal's shell is closed in response to the stimulus. The conclusions, supported by statistical

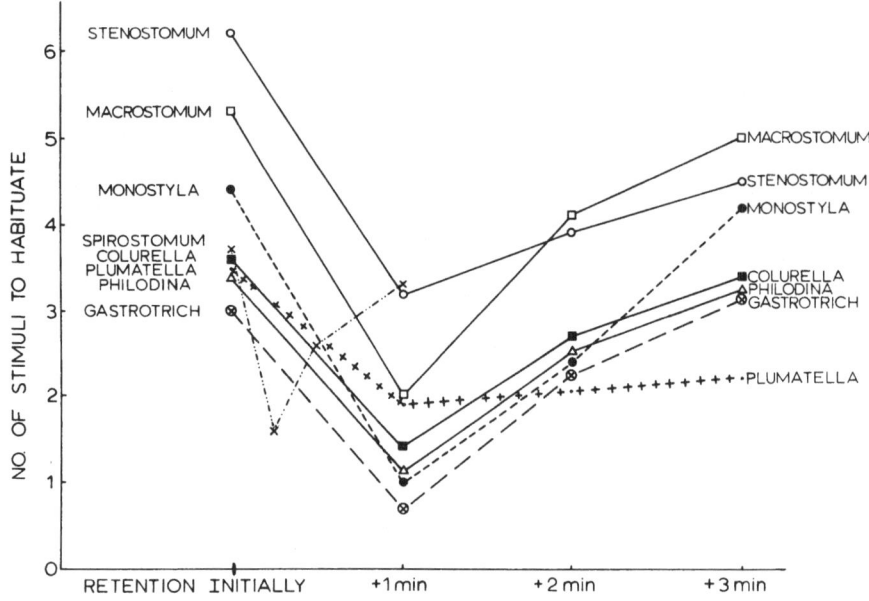

Fig. 1. Habituation to mechanical shock, intensity 1, $n = 20$ for each species.

tests, are: the more intense the stimulus, the longer it takes to habituate the organism and the longer the habituation lasts. An exact explanation of why this occurs may only come from detailed anatomical and electrophysiological studies of the organisms.

We have also reported on maze training of ostracods and copepods and classical conditioning of ostracods (Applewhite and Morowitz, 1966). Briefly, the maze was linear and consisted of a series of chambers separated by clear plastic partitions, each with a hole 1 mm wide in the corner. The chamber each animal was in was light, and the chamber ahead was dark. The hole was closed off after the animal went through it, and the animal remained in the dark chamber for 30 seconds. Then, this chamber was lighted, the one ahead was darkened, and the hole to it opened. Initially it took 46 seconds on the average for 25 copepods to go through the hole, and an average of 66 seconds for 30 ostracods. By the time they had gone through the seventh and last hole, the times were 8 seconds for the copepods and 12 seconds for the ostracods. Therefore, they learned the position of the hole, and, when retested after periods of from 1 to 7 minutes, they went into the next chamber in less time than it took them the first time. The maze was cleaned after each animal to eliminate any trail, and, while the holes were consistently on one side of the chamber, half of the animals had them on one side, half on the other side. When the holes were alternated, first on one side, then on another, the animals failed to learn their position.

Also their activity was constant throughout the experiment, and they were presumably adapted before the experiment began to the light-dark changes. In the classical conditioning of the ostracod, it was found that, for 20 animals, 13.3 light-shock pairings (5 seconds of light plus shock given every 115 seconds) were necessary before the light alone would cause them to close their shells. After an average of 4.6 presentations of the light alone every 60 seconds, they no longer closed their shells in response to it. Control groups of different combinations of stimulus orders produced no response to light.

Most recently, we have found classical conditioning in *Stenostomum* and avoidance conditioning in the copepod *Paracyclops fimbriatus poppei*. For classical conditioning, each *Stenostomum* was placed in its culture medium in a plastic training chamber measuring 1 by 1 by 1 mm, with platinum strip electrodes on two sides opposite each other. The animals were kept in a 14-hour-dark, 10-hour-light cycle and used at random times. Each was placed in the chamber for 15 minutes before the experiment began with the only light being a 15-watt substage illuminator (with a blue ground-glass filter) placed 1 ft away from the microscope mirror.

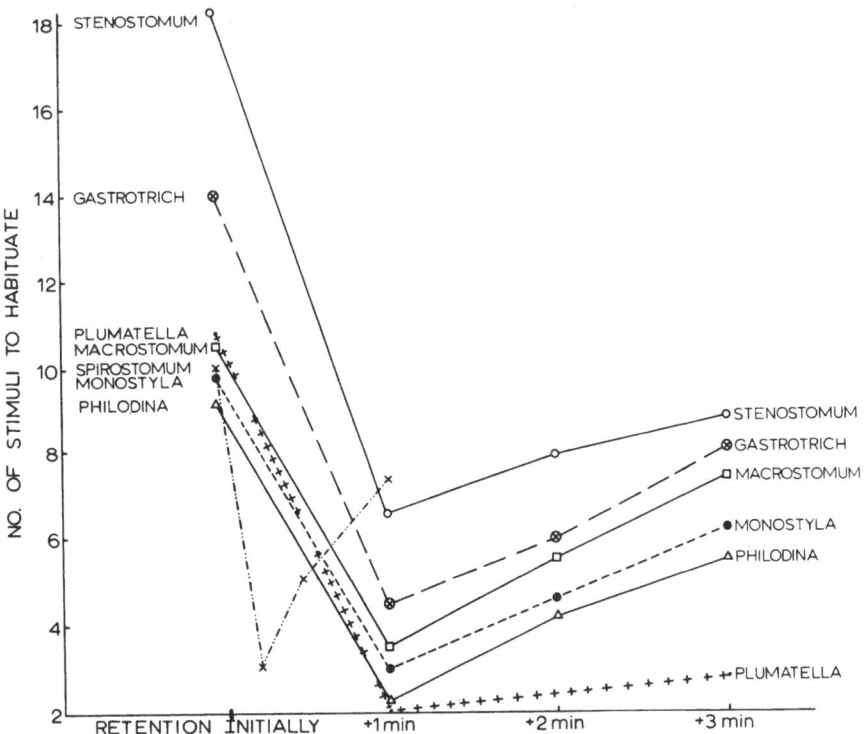

Fig. 2. Habituation to mechanical shock, intensity 3, $n = 20$ for each species.

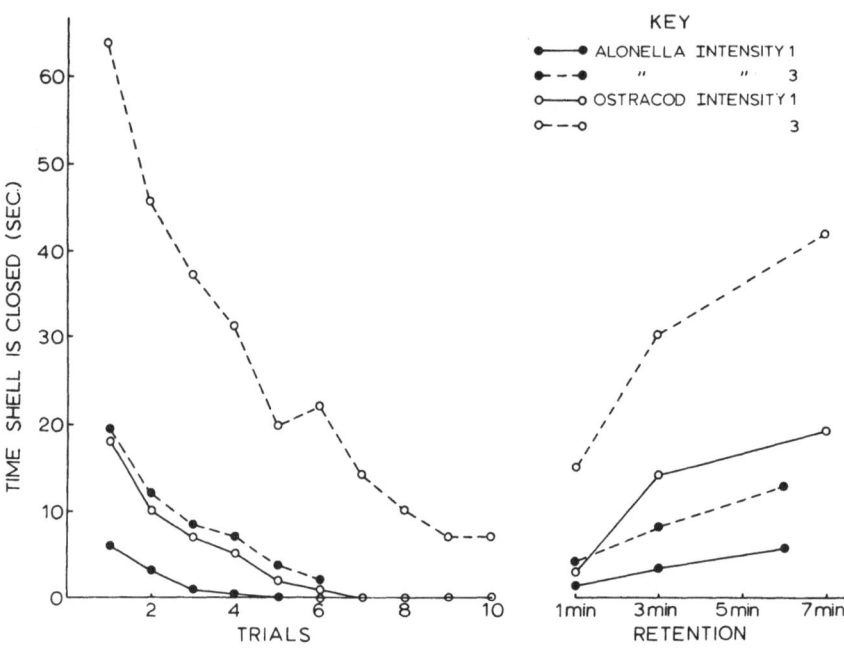

Fig. 3. Habituation to mechanical shock, Alonella and Ostracoda, $n = 20$ for each species.

The culture medium and the ambient temperature varied from 20 to 21°C during the course of the experiment. The conditioning stimulus was the light from a 150-watt bulb in a reflector placed 6 in. from the chamber. The light was turned on for 5 seconds every 30 seconds and gave an increase in illumination of about 500 times over that provided by the substage illuminator. After 4 seconds of the light, an electric shock of 5 volts and 100 msec duration in the biphasic mode was administered by a Grass stimulator; this caused *Stenostomum* to stop moving. The light went off after the 5 seconds, then after 25 seconds it went on again and the procedure was repeated. After each experiment, the chamber was cleaned and rotated 90° before reuse. Fifteen animals were used, and each was considered conditioned when it stopped moving four times in a row every time the light was flashed, separated by an interval of 25 seconds. It took a mean of 27.3 stimuli (SD 12.3) to condition them this way. This response was extinguished with a mean of 4.2 (SD 1.6) stimuli by presenting the 150-watt light for 5 seconds every 30 seconds until the animal no longer responded to it. One point of interest is that animals with more than one zooid could not be classically conditioned with these few stimuli, presumably because *both* the individual organisms would have to be conditioned and respond at exactly the same time. Three control groups of 12 animals per group were run. In the first group, each animal was left in the chamber with only the ambient light on for 14 minutes, the mean time for conditioning to occur (27.3 stimuli multiplied by 0.5 minute). Then the 150-watt light was presented for 5 seconds every 30 seconds for a total of 14 minutes. For the second group, a shock of 5 volts for 100 msec duration was given every 30 seconds for 14 minutes. Then the 150-watt light was presented for 5 seconds every 30 seconds over a period of 14 minutes. The third group was given only the light every 30 seconds for 5 seconds over a period of 14 minutes. None of these groups showed any response to the light.

In the avoidance conditioning experiment, each copepod was placed in its culture media in a plastic training chamber measuring 3 by 1.5 by 2 mm deep, with platinum strip electrodes opposite each other on the 3-mm sides to equalize any electrolysis effects. The bottom and end of half of the chamber were painted black and this half was covered over with a black top. The only illumination came from a Thermolyne lamp 6 in. from the microscope. The animals were kept in a 12-hour-dark, 12-hour-light cycle and run at random times. The chamber was cleaned after each use and rotated 90°. Each copepod was placed in the chamber for 10 minutes before the experiment began. The temperature remained constant in the chamber during the course of each experiment. Four groups of animals were used, with 20 in each group. For all the groups, the time each animal spent in the dark side over a period of 2 minutes spent in the chamber was recorded. Then, for the experimental group, a shock of 2.5 volts for 100 msec duration in the biphasic mode was given by a Grass stimulator every time the animal went in the dark during the next 3 minutes. No more than two shocks were given to the animal when it was in the dark, and an average of 30 shocks were given over the 3-minute period. Then, the time spent in the dark by each animal for the next 2 minutes was recorded. After this period, 1 minute elapsed before measurements were taken for the next 2 minutes. For control group 1, no shocks were given during the shock period. For control group 2, thirty shocks were given randomly during the shock period. For control group 3, the shocks were given when each animal was in the light part of the maze. The results are presented in Table 3. For the experimental group, the animals spend less time in the dark side of the chamber after they are shocked for going into that side. The figures for both the "time in the dark during 2 minutes after shock" and the "time in the dark during 2 minutes after 1 minute rest" are significantly different ($p < 0.01$) from those for the "time in the dark during 2 minutes before shock," as judged by the Friedman two-way analysis of variance test. In control groups 1 and 2 there are no significant differences. The results from control group 3 indicate the copepod cannot be trained to avoid the light part of the chamber by this method. It cannot be argued that the electric shock produces some sort of chemical or physical reaction with the dark part of the chamber that is unpleasant to the copepod. If this were the case, it should also occur in control group 2 when the shock is given randomly, since the electrodes span both the dark and the light chamber areas and unpleasant effects should also be present here.

TABLE 3

Avoidance Conditioning Copepods

$n = 20$ For Each Group

	\multicolumn{9}{c}{Shock 2.5 v, 100 msec given in dark over 3 min (avg. 30 shocks)}								
	Time, sec, in dark during 2 min before shock			Time in dark during 2 min after shock			Time in dark during 2 min after 1 min rest		
	1st min	2nd min	Total	1st min	2nd min	Total	1st min	2nd min	Total
Experimental Group									
Mean	29.1	29.8	58.9	20.3	25.7	46.0	26.4	26.6	53.0
SD	3.6	2.9	4.6	3.1	3.0	4.3	3.3	4.0	5.2
Control 1, No Shock						Rest			
Mean	29.7	30.4	60.1	30.6	30.2	60.8 1	30.4	31.6	62.0
SD	3.8	3.5	5.2	3.0	3.9	4.9 min	3.4	4.7	5.8
Control 2, Random Shock									
Mean	30.6	28.9	59.5	29.9	31.2	61.1	28.7	30.7	59.4
SD	4.1	3.8	5.6	3.6	3.9	5.3	4.1	4.0	5.7
Control 3, Shock Given in Light									
Mean	30.8	31.1	61.9	32.0	30.2	62.2	30.9	30.5	61.4
SD	4.6	4.1	6.2	3.9	3.5	5.2	3.4	3.7	5.0

DIRECTION OF FUTURE RESEARCH

It seems likely that any structural changes associated with the memory code will be subtle; otherwise they might interfere with other biological processes. If there are such changes, the electron microscope will probably be necessary to resolve them. Where any such changes in the nervous system would take place is not known, so the entire nervous system should be monitored. In the case of a protozoan, monitoring the whole animal is not so difficult a task because of its much smaller size. However, until learning has been adequately demonstrated in it, we have concentrated on the micrometazoan fine structure, in particular, a rotifer and gastrotrich about 100 μ in length. These two species can be habituated to mechanical shock, and we are investigating higher forms of learning in them.

As previously indicated, the rotifer and gastrotrich possess cell nuclei constancy, and this coupled with their small size offers an ideal object for study. A three-dimensional reconstruction at the fine structural level of their nervous system would give information about how an entire brain and peripheral ganglia were constructed and interrelated. Furthermore, a comparison of the synaptic connections and cell locations of two members of the same species would indicate whether they were genetically coded, were random, or were a result of a different environment.[1] This would necessitate, for a start, taking two eggs from the same parent, maintaining the environment constant for them, and then comparing their cell locations and synaptic interconnections by using serial sections. At present, electron microscopy has been done only on a few micrometazoa (Fahrenbach, 1963; Eakin and Westfall, 1965; Koehler, 1965; Mattern and Daniel, 1966a and b; and Lansing and Lamy, 1961) but not yet on a complete nervous system.

An advantage of using autoradiography with the electron microscope on these small animals is that the location of the radioactive compounds can be detected anywhere in the animal. Thus, if a particular metabolic inhibitor also inhibits learning but is localized outside the nervous system, no false conclusions would be drawn about its true relationship to learning. Conversely, if learning is not affected but the compound is localized in the nervous system, a reasonable conclusion would be that what it inhibited is not related to the memory code. With a brain of less than 200 cells, *each* cell can be checked for localization.

There are many drugs that have an effect upon learning and memory (McGaugh and Petrinovich, 1965), and, in addition to doing autoradiography with metabolic inhibitors and precursors, it would be desirable to do it with some of these drugs. From the dosage rates given for these drugs, the number of molecules acting on each cell can be computed. The fewer the molecules acting, the more specific are the action sites. Autoradiography with these drugs would give very specific information about localization. Those drugs that affected learning and memory without causing any fine structure damage[2] and were localized in the nervous system would be of particular interest. Further work would be required to show changes in these sites accompanying learning and memory changes over time.

REFERENCES

Applewhite, P. B., and H. J. Morowitz, The micrometazoa as model systems for studying the physiology of memory. *Yale J. Biol. Med.*, 1966, **39**, 90.

Ashby, W. R., *An Introduction to Cybernetics*. London: Chapman and Hall, Ltd., 1958.

Birky, C. W., Jr., and Bonnie Field, Nuclear number in the rotifer *Asplanchna*: intraclonal variation and environmental control. *Science*, 1966, **151**, 585.

Boloukhere-Presburg, M., Effet de l'actinomycine D sur l'ultrastructure des chloroplastes et du noyau d'*Acetabularia méditerranéa. J. Microscopie*, 1965, **4**, 363.

Bullock, T. H., and G. A. Horridge, *Structure and Function in the Nervous Systems of Invertebrates*. San Francisco, Calif.: W. H. Freeman, 1965.

---

[1]It could be argued that all the synaptic interconnections of the human brain could not be genetically coded. If there are 10^{10} neurons with an average of 10 interconnections each, the DNA would have to code 10^{11} of them. If only one base pair of 3.4 Å in length is needed to code each interconnection, the DNA has to be 34 m long, whereas that of man is only about 85 cm. This calculation assumes, of course, that the DNA is identical in each cell.

[2]There is evidence, e.g., that actinomycin D reduces nuclear size and causes nucleolar fragmentation (Boloukhere-Presburg, 1965).

Carthy, J. D., *An Introduction to the Behavior of Invertebrates*. London: George Allen and Unwin, Ltd., 1958.

Carthy, J. D., *The Behavior of Arthropods*. San Francisco, Calif.: W. H. Freeman, 1965.

Eakin, R., and J. Westfall, Ultrastructure of the eye of the rotifer *Asplanchna brightwelli*. *J. Ultrastructure Res.*, 1965, **12**, 46.

Eigen, M., and L. C. M. De Maeyer, Information storage and processing in biomolecular systems. *Neurosci. Res. Program Bull.*, 1965, **3**(3).

Fahrenbach, W. H., The sarcoplasmic reticulum of striated muscle of a cyclopoid copepod. *J. Cell Biol.*, 1963, **17**, 629.

Hyman, L. S., *The Invertebrates*, Vols. 1–5. New York: McGraw-Hill Book Company, 1940–1959.

Jennings, H. S., *Behavior of the Lower Organisms*. Bloomington, Ind.: Indiana University Press, 1962 (reissue).

Koehler, J., A fine structure study of the rotifer integument. *J. Ultrastructure Res.*, 1965, **12**, 113.

Kudo, R. R., *Protozoology*. Springfield, Ill.: Charles C. Thomas, Publisher, 1966.

Lansing, A., and F. Lamy, Fine structure of the cilia of rotifers. *J. Cell Biol.*, 1961, **9**, 799.

Mattern, C. F. T., and W. A. Daniel, The flame cell of rotifer. *J. Cell. Biol.*, 1966a, **29**, 552.

Mattern, C. F. T., and W. A. Daniel, The stomach cell of rotifer. *J. Cell Biol.*, 1966b, **29**, 547.

McConnell, J. V., Learning in invertebrates. *Ann. Rev. Physiol.*, 1966, **28**, 107.

McGaugh, J. L., and L. F. Petrinovich, Effects of drugs on learning and memory, *in* C. C. Pfeiffer and J. R. Smythies, eds., *International Review of Neurobiology*, 1965, **8**, 139.

Morrell, F., Electrophysiological contributions to the neural basis of learning. *Physiolog. Rev.*, 1961, **41**, 443.

Pitelka, D. R., *Electron Microscopic Structure of Protozoa*, New York: The Macmillan Company, 1963.

Schone, H., Complex behavior, *in* T. H. Waterman, ed., *The Physiology of Crustacea*, Vol. 2. New York: Academic Press, Inc., 1961.

Soest, H., Dressurversuche mit rhabdocoelen turbellarien. *Z. Vergleichende Physiol.*, 1937, **24**, 720.

Thorpe, W. H., *Learning and Instinct in Animals*. London: Methuen & Co., 1963.

Warden, C. J., T. N. Jenkins, and L. H. Warner, *Comparative Psychology*, Vol. 2. New York: The Ronald Press Company, 1940.

Watson, J. D., *Molecular Biology of the Gene*. New York: W. A. Benjamin, Inc., 1965.

Williams, R. J., *Biochemical Individuality*. New York: John Wiley and Sons, Inc., 1963.

Chapter 22

Some Ionic, Chemical, and Endogenous Factors Affecting Behavior of Hydra

Howard M. Lenhoff

Laboratory for Quantitative Biology
Department of Biology
University of Miami
Coral Gables, Florida

INTRODUCTION AND REVIEW

There are many reasons why research with hydra may interest psychologists who work primarily with Planaria. Historically, the first experiments in which the feeding of a cut-up planaria to another organism led to a major scientific advance were carried out using hydra over 200 years ago by Abraham Trembley (1744). Besides being responsible for many great advances in biology, such as making the first animal grafts using hydra, Trembley also noted that hydra, fed on pieces of planaria, retained the black color of their prey. Although Trembley of course did not attempt to show that the hydra learned from ingestion of the planaria, through these feeding experiments, he is credited with having made the first vital stain (Baker, 1952).

In the present paper I give examples from experiments with hydra showing (a) how both the environment and physiological state of an organism affect its behavior, (b) how the environment and endogenous factors interact to give altered behavioral responses, and (c) how through control of these factors behavioral responses can be regulated and predicted with accuracy. Such information obtained with hydra may also be of value to the study of Planaria and of other lower invertebrates whose behavior also may be greatly affected by similar physical, chemical, and endogenous factors.

Other than emphasizing the importance of being aware of and controlling such factors, this paper does not purport to add any knowledge to the chemistry of learning per se. I do not claim that hydra is the best research animal with which to study learning. In fact, I have yet to be convinced that hydra can learn at all. Instead, it is becoming increasingly apparent that this animal has evolved many receptor-effector systems that take care of its primary behavioral needs, most of which revolve around food

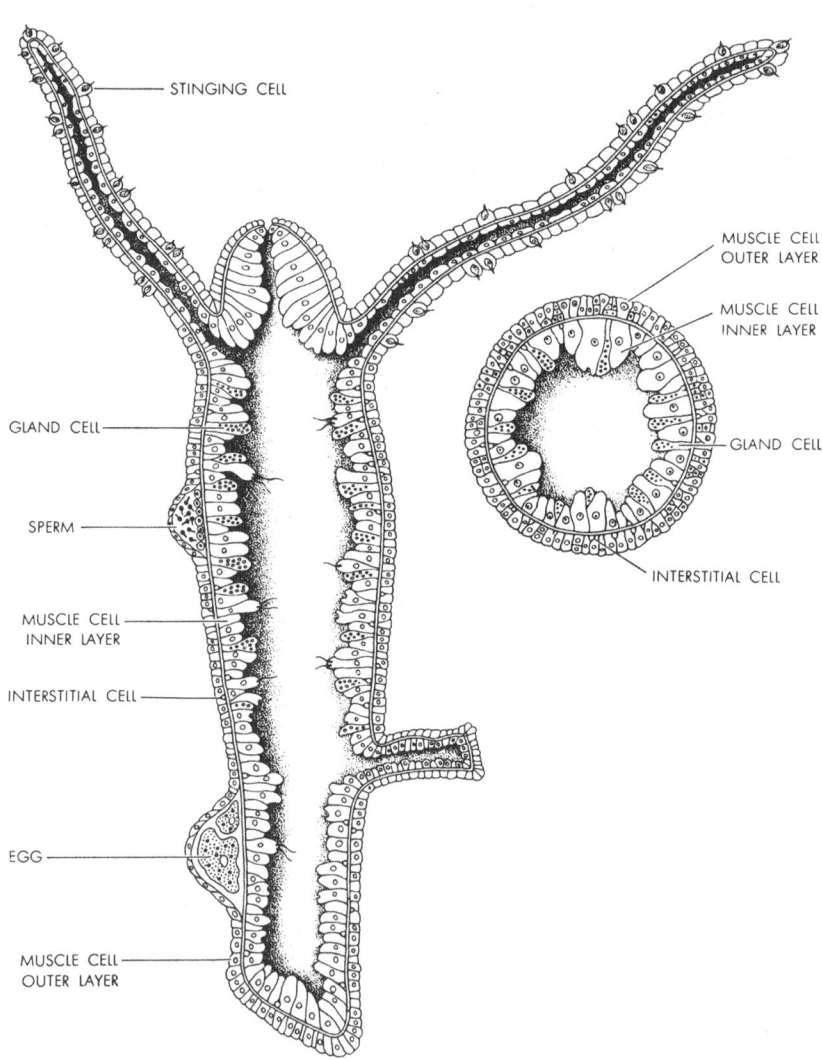

Fig. 1. Representative illustrations of sagittal and cross section of a hydra. From "The Sex Gas of Hydra" by W. F. Loomis. Copyright © 1959 by Scientific American, Inc. All rights reserved.

and defense. I welcome, nevertheless, suggestions for experiments which might show learning in hydra.

In my presentation I shall (a) review briefly the structure of hydra, (b) illustrate how much of the animal's behavior is markedly influenced by a single chemical compound—reduced glutathione, (c) describe some of the various effects that ionic and endogenous factors have on the gluta-thione-activated feeding response, and (d) describe for the first time a new type of behavioral response stimulated in part by glutathione.

Structure of Hydra

Like Dr. Loomis, I was first attracted to the use of hydra as an experi-mental animal because it is a relatively simple metazoan. As can be seen in Fig. 1, a hydra, which is about 0.25 in. in length, is constructed like a two-ply hollow tube made up of an outer epithelial layer (ectoderm) and an inner gastrodermal layer (endoderm). At the posterior end of the tube is the basal disc with which the hydra usually attaches to a surface, and at the anterior end is a mouth surrounded by a ring of tentacles. (Dr. Jensen views a hydra as a "planaria pharynx with tentacles.")

Hydra have about seven basic types of cells. Conclusive evidence by electron microscopy for the existence of nerve and sensory cells has not yet been firmly established (Lenhoff and Loomis, 1961) although there have been many papers published on the light-microscopy evidence for such cells. The recent electron micrographs of Lentz and Barnett (1965) give indication that there are nerve cells in hydra. There are no muscle cells, but rather the major ectodermal and endodermal cells have contractile fibers; the ectodermal fibers are arranged to contract longitudinally, whereas the endodermal fibers contract circularly.

Of special interest to the behavioral biologist is the nematocyte, a cell which houses the nematocyst. A nematocyst is a highly complex intracellular secretion product consisting of a capsule containing an eversible thread continuous with the capsule at one end and stoppered there by an operculum (Fig. 2). When stimulated to discharge, the thread everts and, depending upon the type of nematocyst, either attaches to an adjacent surface or penetrates the prey, simultaneously injecting a toxin. The nematocyst has been considered an independent effector, although this view has been questioned (see Picken and Skaer, 1966).

Effects of Reduced Glutathione on the Behavior of Hydra

In nature, most of the behavioral responses that I shall discuss start with the eversion of the nematocyst thread. Such eversion takes place when a moving prey accidentally contacts the nematocyst-laden tentacles of a hydra. The precise stimulus for nematocyst discharge is still unknown, although it is thought to be of a mechanical-chemical nature. On discharge of one particular type of nematocyst, the stenotele (Fig. 2), the prey is wounded by the harpoonlike action of the everted thread. Present among

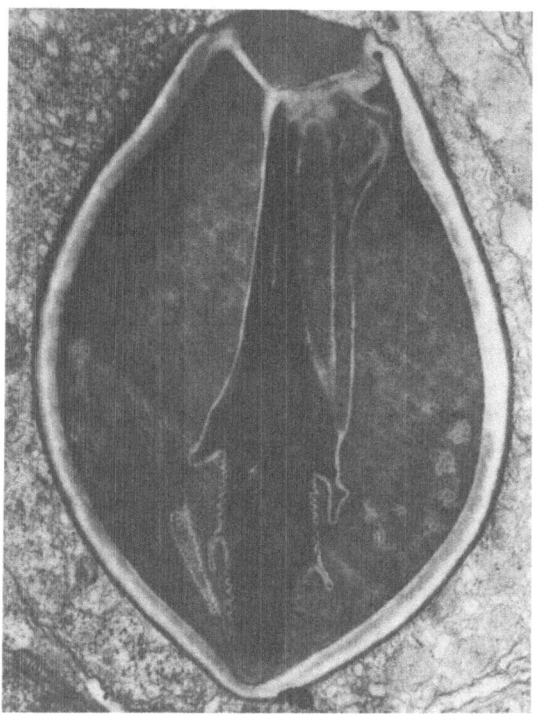

Fig. 2. Electron micro-
graph of a sagittal section
of nearly mature stenotele.
Reprinted by courtesy of
Dr. George Chapman and
the Loomis Institute for
Scientific Research, Inc.

the many chemicals emitted from the prey's wounds is the ubiquitous
tripeptide glutathione (Fig. 3). This chemical, in trace amounts, specifically
activates the feeding behavior of *Hydra littoralis* (Loomis, 1955).[1]

The feeding behavior of hydra consists of many steps. In essence,
the glutathione causes the tentacles to sweep inward toward the longitudinal
axis of the animal (Fig. 4b), to writhe and contract (Fig. 4c), and to bend
toward the mouth (Fig. 4d). As the captured prey, impaled on the tentacle
nematocyst, is drawn to the mouth, glutathione released from the prey
stimulates the mouth to open. On contact with the mouth, the prey is
ingested; the actual process of ingestion, however, is not glutathione
controlled.

To obtain quantitative data on the feeding response, we used two
measurements: the times elapsed between the moment the hydra was first
exposed to glutathione and the instant the mouth initially opened (*initial*

[1] Glutathione also activates the feeding response of some other species of hydra,
of the Portuguese man-of-war (*Physalia physalis*), and of the colonial hydroid *Cam-
panularia* (Lenhoff and Schneiderman, 1959). The feeding response of *Cordylophora*,
on the other hand, is activated by proline (Fulton, 1963), not glutathione. The specific
feeding activators for other coelenterates have not been determined.

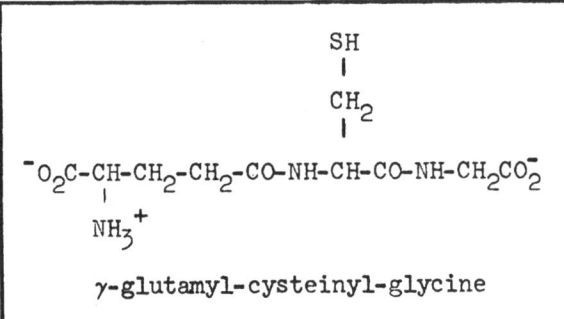

Fig. 3. Reduced glutathione.

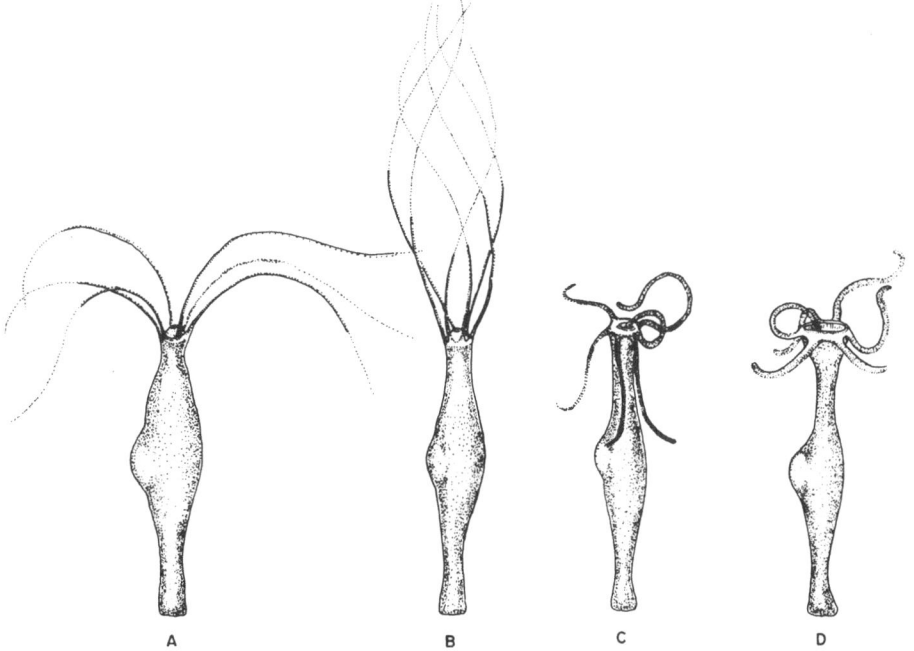

A   B   C   D

Fig. 4. Stages of the feeding response to reduced glutathione. (a) A hydra in the absence of the glutathione is shown with the mouth closed and the tentacles outstretched and relatively motionless. (b) After the addition of glutathione, the tentacles first begin to writhe and sweep inwards toward the central vertical axis of the animal. (c) Next, the tentacles bend toward the mouth, and the mouth opens; shown in this composite drawing are the various positions that a tentacle takes before contracting. These movements, culminating in mouth opening, usually all take place within half a minute. (d) This shows how a hydra looks during the greater portion of the feeding reflex, its mouth open wide and the tentacles in various phases of contraction. Frequently, the tips of the tentacles are observed within the hydra's mouth, as shown in (c) and (d). Reprinted from the *Journal of General Physiology* by courtesy of Rockefeller Press.

time, or t_i) and finally closed (*final time*, or t_f). The length of time that a mouth was open therefore equaled $t_f - t_i$ (Lenhoff, 1961).

Groups of five animals were used for each measurement of $t_f - t_i$. The standard deviation from the mean within each group was small (Lenhoff, 1961). In fact, most animals employed responded to glutathione nearly in synchrony and thus provided a basis for a quantitative and reliable feeding response assay. In order to get the animals to respond in this synchronous manner, we used hydra that were genetically alike, were in the same stage of development, and had been grown in a rigorously defined and controlled environment (Lenhoff, 1961a, and Loomis and Lenhoff, 1956).

Using this simple behavioral assay, it was possible to show that hydra had a unique specificity for the glutathione molecule: (a) The thiol group is not necessary, (b) the intact and unaltered tripeptide backbone is required, (c) there is a high affinity of the receptor for the glutamyl part of the tripeptide, and (d) the α-amino group of glutathione is probably required for the association of glutathione with the receptor (Fig. 5). These facts are the results of a series of studies using various glutathione analogs and amino

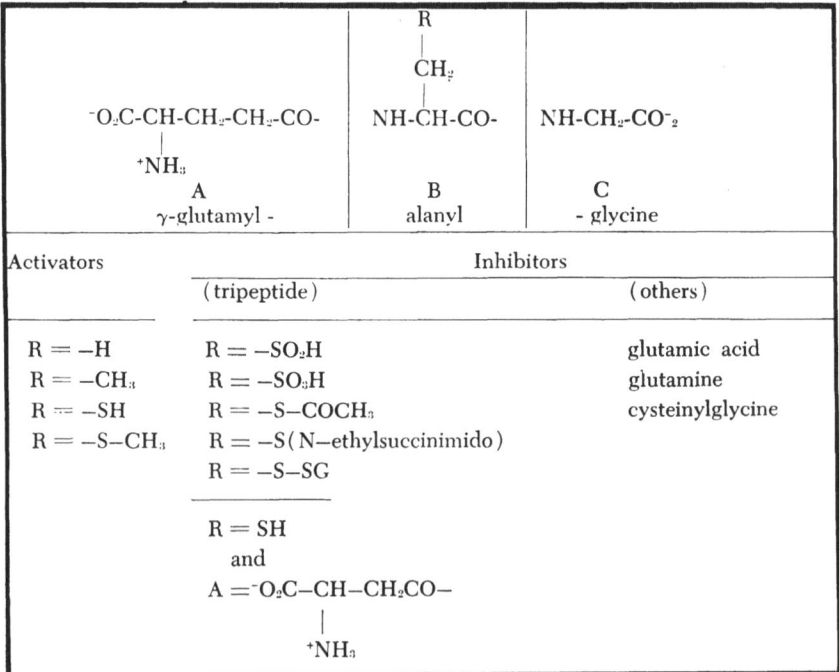

Fig. 5. Activators and inhibitors of the feeding response. Reprinted by courtesy of the Loomis Institute for Scientific Research, Inc.

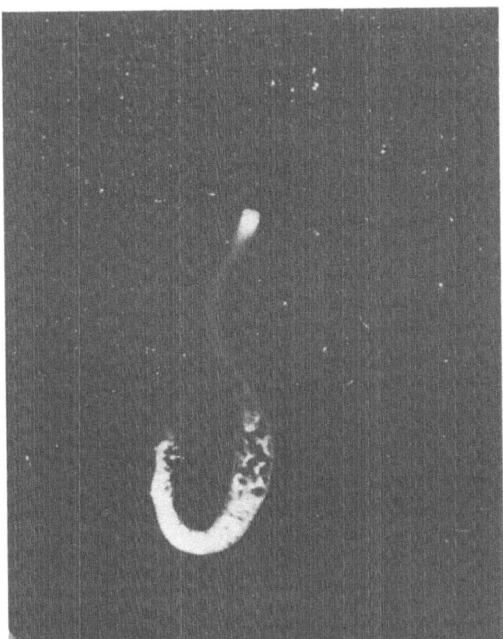

Fig. 6. Isolated *Physalia* gastrozooid. The mouth is the uppermost part at the end of the narrow cylindrical neck. Reprinted with the permission of *Biological Bulletin*.

acids either as competitive inhibitors or substitutes for glutathione (Cliffe and Waley, 1958; Lenhoff and Bovaird, 1961; and Loomis, 1955).

Hence, glutathione, in activating the feeding response of hydra, behaves differently than it does in all other known biological systems requiring glutathione for their functioning. The hydra system is the only one dependent not upon the thiol of the glutathione molecule but rather on the intact structure of the peptide backbone of the molecule.

Our studies thus far cannot demonstrate that glutathione is oxidized, destroyed, or altered in any way during its action on hydra. If such changes in the glutathione molecule do occur, they are at a level undetectable by our means, which have included radioautography, chromatography, and sulfhydryl analysis (Lenhoff, 1961b). Furthermore, it would appear that glutathione has to be constantly present at the receptor site in order for the feeding response to be effected. That is, hydra placed in a solution of glutathione will give a feeding response; however, within 1 minute after the hydra are removed from the glutathione solution and put into a solution free of glutathione, they cease to respond and close their mouths.

But, if the same animals are placed back in a solution of reduced glutathione, they reopen their mouths; this procedure can be repeated many
times during an hour and will give the same result (Lenhoff, 1961a).
Apparently the equilibrium between glutathione and the receptor is rapidly
attained.

Before continuing this discussion of the glutathione-activated feeding
response in hydra, it is worth noting that in this case the line between
physiology and behavior becomes very thin, if one exists at all. That is to
say, addition of the tripeptide glutathione to hydra leads to all sorts of
contractions, tentacle writhings, bendings, and mouth opening. In the same
manner, addition of the octapeptide hormone oxytocin to a female mammal
leads to contractions of the uterus. Yet the study of oxytocin activation of
the uterus is considered to be in the realm of physiology, not behavior.
Actually, at the basic level, these two peptide-stimulated phenomena may
be similar. In each of these instances a specific peptide supposedly combines
with its respective receptor to trigger a biological event which involves the
shortening of contractile fibers. If we dared extend the analogy, we could
suggest that perhaps hormone receptors evolved from "behavioral chemoreceptors."

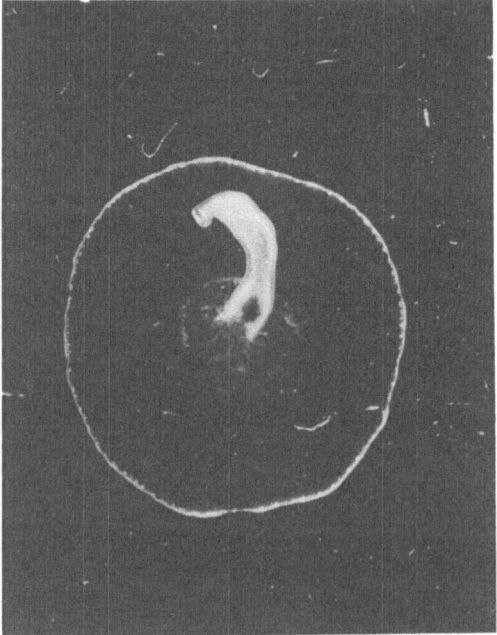

Fig. 7. A gastrozooid induced to spread by reduced
glutathione. Reprinted with the permission of
Biological Bulletin.

Rather than debate the merits of this subject, consider for this discussion on invertebrate behavior that I use glutathione in a way analogous to that of a psychologist using electric shock, i.e., I use glutathione to initiate a response that can be called behavioral. The difference between glutathione and shock is that shock, as Dr. Jensen has pointed out, may be deleterious to the animal, whereas glutathione is a natural stimulator of a complex series of contractions and relaxations known as *the feeding behavior*.

In order to give you an example of how glutathione can affect the behavior (or physiology) of a coelenterate, I should like to use as a vivid illustration a relative of hydra, the marine hydrozoan, the Portuguese man-of-war. The man-of-war (*Physalia physalis*) is a colonial coelenterate which has numerous specialized zooids attached to a float. The function of capturing food is carried out by one type of zooid, and the function of digesting the food is carried out in a cooperative fashion by the feeding polyps, called gastrozooids. Figure 6 shows a single gastrozooid of the hundreds present in a large man-of-war. This single gastrozooid will lie in the dish rather quietly until the juice from a fish or a weak solution of reduced glutathione is added (Lenhoff and Schneiderman, 1959). As soon as the glutathione is added, the gastrozooid writhes, twists, and turns until

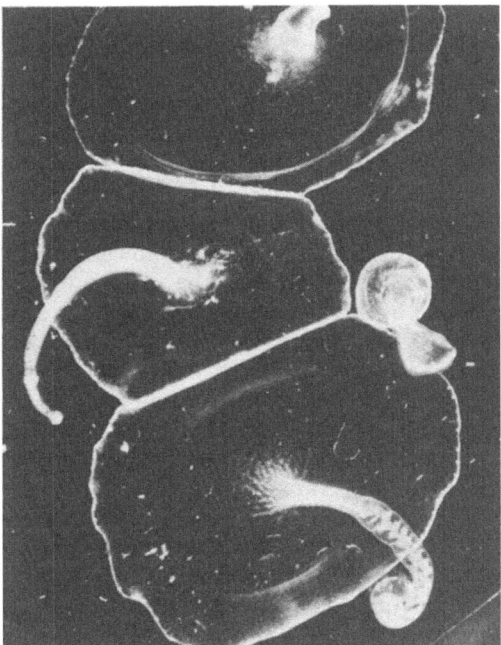

Fig. 8. Several gastrozooids that were induced to spread by reduced glutathione. Reprinted with the permission of *Biological Bulletin*.

its lips attach to the glass container and then it begins to spread as though it were attempting to engulf the dish. Figure 7 shows the same gastrozooid that has been in reduced glutathione for a half hour; the 1-mm diameter gastrozooid tube is now a disc of more than 20 mm. Two or three of these gastrozooids close to each other form a mosaic pattern (Fig. 8), and, in a similar manner, hundreds of them pressing against the surfaces of a fish can envelop that fish, forming a complete "stomach" around it.

THE EFFECTS OF ENVIRONMENTAL IONS ON THE EXTENT OF THE FEEDING BEHAVIOR

Hydra exposed to glutathione would not respond unless calcium ions were in the environment (Lenhoff and Bovaird, 1959). This observation led us to study the effects of major ions that would normally be present in a pond upon the feeding behavior of hydra. We felt that, in order to quantify the effects of glutathione and to study the mechanism of its action on hydra, we had to be aware of how the feeding response is affected by the various ions that are present in the solutions in which we keep the animals or by the ions in reagents with which we treat hydra—or even by ions that may leak out of the hydra themselves.

Calcium

Hydra placed in a solution free of added calcium gradually gave a lessened response to added glutathione (Lenhoff and Bovaird, 1959). The ability of the animals to respond to glutathione was restored by the addition of any of a number of calcium salts (Lenhoff and Bovaird, 1959). The rate at which hydra lost their responsivity to glutathione during the period immediately following the removal of environmental calcium was pH dependent; polyps at pH 6 were able to respond to glutathione for longer periods than hydra at a higher or lower pH (H. M. Lenhoff, *unpublished observations*). Regardless of the pH, however, after about an hour in a medium free of added calcium, none of the animals responded and all had to have calcium added back at levels around 10^{-4} M in order for full activity to be restored. Addition to the water of ethylenediamine-tetraacetic acid led to inhibition of the feeding response within minutes, presumably through chelation of calcium.

The requirement for calcium ions could not be replaced by any metallic ion except strontium. Apparently strontium (which has a molecular configuration· similar to that of calcium) is able to substitute partially in the reaction(s) of the feeding response that requires calcium ions. Although strontium allowed a weak feeding response to take place, the presence of this ion in place of calcium eventually led to the death of the animals.

Magnesium

Environmental magnesium ions were not required for the feeding behavior but rather, when in high concentrations (around 10^{-3} M), would inhibit it. This inhibition was completely overcome by calcium ions. Hence, during this so-called "magnesium anesthesia," the magnesium probably acts by competing with calcium for some site in the glutathione receptor-effector system (Lenhoff, 1961b, and Lenhoff and Bovaird, 1959).

Sodium

Sodium ions are not required in the environment for the feeding response to occur; instead, they too compete with calcium and thus inhibit (Lenhoff, 1961b and *unpublished observations*). Since sodium is monovalent, it is less effective than magnesium in competing with calcium. Nonetheless, we routinely include some sodium in our experimental test solution in order to lower the duration of the feeding response to glutathione so that an average response is around thirty minutes. Also, the presence of added sodium (10^{-4} M) provides a "buffer" to minimize the inhibitory action of trace amounts of extraneous sodium that may leak from the experimental animal or leach from the reagent bottles.

Potassium

Potassium ions in relatively low concentrations inhibit the glutathione-activated feeding response even in the presence of 10^{-3} M $CaCl_2$ (Lenhoff, 1965 and *unpublished observations*). The inhibitory action of potassium ions was discovered when we began to grow our stock cultures of *H. littoralis* in a culture medium containing potassium ions (Lenhoff, 1966). (As a rule, we grow our hydra in a potassium-free medium because potassium is not an absolute requirement for hydra growth.) Animals grown in the potassium medium gave responses to glutathione that were weaker than those normally elicited (Lenhoff, 1965 and *unpublished observations*).

To test the role of potassium ions on the responses activated by glutathione, the following experiments were carried out: Sparse cultures of hydra (about five hundred animals per fifteen hundred milliliters of culture medium) were grown in a potassium-free solution (Lenhoff, 1961a) for 1 week prior to testing. Then they were placed in a fresh aliquot of the same medium which contained a known concentration of KCl; the animals were removed at specified intervals and tested for their ability to respond to added glutathione. With increased time of exposure to potassium ions the hydra's response to glutathione decreased (Lenhoff, 1965). Furthermore, the greater the concentration of added potassium, the greater was the inhibition of the response activated by glutathione (Table 1). Sodium ions had no such effect under these conditions. When the potassium-inhibited hydra were placed back in a potassium-free medium, they regained their ability to respond fully to glutathione within 2 to 5 hours.

Potassium does not seem to be acting by competing with calcium ions, because the presence of calcium has no effect on overcoming the potassium inhibition. It may be possible that the potassium ion effect may be related to the electrophysiological changes in potential which recently have been

TABLE 1

Inhibition of Feeding Response by K$^+$

KCl concentration, molarity	Half-time for inhibition of response activated by glutathione, minutes
9×10^{-3}	12
7×10^{-3}	32
3×10^{-3}	50
6×10^{-4}	126

The animals were grown in a standard medium of 10^{-3} M $CaCl_2$ and 10^{-4} M $NaHCO_3$, pH 6.9, for 5 days. Before the experiment, the animals were fasted 2 days. The animals were then placed for periods of time varying from 1 minute to 3 hours in a solution of the standard medium containing KCl, after which they were removed, washed, and tested in the standard medium for the ability to give a feeding response to glutathione.

shown to occur in hydra exposed to reduced glutathione (see Rushforth's article in this volume). Hydra, like most organisms, have a high concentration of intracellular potassium ions (Lenhoff and Bovaird, 1960a), which possibly are exuded when they carry out an electrophysiological response. Perhaps the high concentration of potassium ions outside the animal may affect the potassium flow and therefore affect the degree of feeding response.

The role of environmental ions in affecting hydra behavior is not restricted to the feeding response. *Hydra littoralis* exposed to cesium or rubidium ions while the environmental sodium ions are low becomes extremely sensitive to mechanical agitation. This hypersensitivity was particularly noticeable in animals fed daily and maintained in a solution of 10^{-3} M $CaCl_2$, 10^{-3} M tris-(hydroxy)-methylaminomethane, and 10^{-5} M CsCl for 7 days, then transferred to a solution containing added NaCl at 10^{-3} M but otherwise of the same composition (Lenhoff, 1966).

This pronounced irritability of cesium-treated *H. littoralis* is not characteristic of members of this species and resembles the hypersensitive response to agitation of *Chlorohydra viridissima*. Perhaps the cesium treatment may be useful in the elucidation of transmission pathways or the effector system involved in the contraction response, or both. These results may have particular bearing upon Dr. Rushforth's experiments in which the quantitative control of the contraction response of hydra to agitation has been studied intensively and which are described more fully in his paper that immediately follows in this volume (Chapter 23).

Another Important Environmental Ion, the Hydrogen Ion

Through our study of the effects of the hydrogen ion, we have learned much about the properties of the glutathione receptor and about its combination with glutathione. For this reason our pH studies have been rather extensive. In addition, through knowledge of the subtle effects of pH, we have devised experiments that should eliminate some of the confusion that has prevailed in recent years around controversial points related to the feeding response. Therefore, in order (a) to provide the necessary background for understanding how the hydrogen ion can have major effects on the behavior of hydra, and (b) to illustrate how an invertebrate can be investigated with precision and accuracy, I shall discuss the subject of effects of pH on hydra in some detail.

First we must consider an analysis of the union of glutathione with the receptor, using concepts borrowed from enzymology, because it was from such considerations that the subtle effects of pH were learned. Accordingly, a plot of the activity of the receptor-effector system (i.e., the duration of the feeding response) versus glutathione concentration (Fig. 9) was interpreted much as an enzymologist interprets a curve showing the saturation of an enzyme by its substrate, or coenzyme (Lenhoff, 1961a and b). Each curve illustrates that a maximum response takes place at

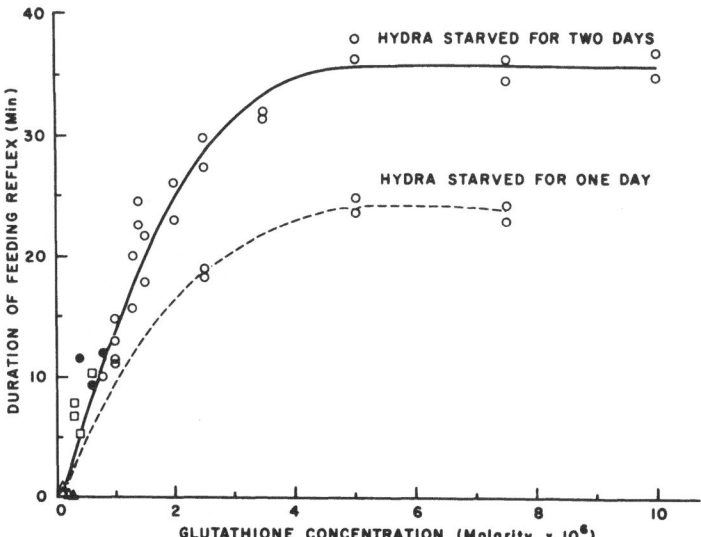

Fig. 9. Effect of glutathione concentration on the duration of the feeding response. Each point represents the mean for five hydra; the type of symbol indicates the number of hydra in the group of five responding to glutathione; i.e., O, five; ●, four; □, three, △, two; and ▲, one. Reprinted from the *Journal of General Physiology* with the permission of Rockefeller Press.

glutathione concentrations of 5×10^{-6} M and greater. A maximum response is considered here to be analogous to the maximum velocity of an enzyme; i.e., both occur during saturation of the specific site involved.

Such glutathione-saturation curves could not be used, however, to determine accurately the dissociation constant between glutathione and its receptor in a way analogous to that by which the Michaelis constant is determined, i.e., for hydra, by determining the concentration of glutathione eliciting a half-maximum response. Although Fig. 9 shows that the duration of the response decreased with the lowering of glutathione concentration, closer inspection shows that the curves do not intersect the origin. This deviation from linearity occurred at glutathione concentrations presumed to have been below "threshold." Thus, to use the saturation data, we had to find a repeatable, unequivocal, and objective means of determining the dissociation constant.

Equation for Determining the Dissociation Constant[2]

Consider in general terms that the combination of an activator A with its receptor R gives the complex AR which initiates a series of events leading eventually to the observed biological effect $\mathscr{E}$, such as a feeding response.

$$A + R \rightleftharpoons (AR) \rightarrow \mathscr{E} \tag{1}$$

The dissociation constant for the complex is K_A.

$$K_A = \frac{(A)R}{AR} \tag{2}$$

Assume that the total number of receptors R_T equals $R + AR$; that the total amount of activator A_T is $A + AR$, or simply A because AR is negligible compared to A; that the effect is proportional to the amount of receptor complexed with activator, i.e., $\mathscr{E} = k(AR)$; and that the maximum effect $\mathscr{E}_M$ is proportional to the total number of receptors, because they would all be tied up with the activator, i.e., $\mathscr{E}_M = k(R_T)$.

By a series of rearrangements and substitutions, followed by taking the reciprocals and multiplying through by (A), we derive

$$\frac{(A)}{\mathscr{E}} = \frac{1}{\mathscr{E}}(A) + \frac{K_A}{\mathscr{E}_M} \tag{3}$$

A plot of this equation, which represents a straight line, deemphasizes data at activator concentrations below threshold and hence allows us to derive K_A. The equation is analogous to the second form of the Lineweaver–Burk (Lineweaver and Burk, 1934) equation for the combination of enzyme and substrate and like the equation developed by Beidler (1954) for mammalian taste chemoreceptors.

[2] Lenhoff, 1963, 1965, and 1967.

As shown in Fig. 10, a plot of $(A)/\mathscr{E}$ versus (A) gives a straight line at most glutathione concentrations. The slope of the line is $1/\mathscr{E}_M$, and the extrapolated intercept is $K_A/\mathscr{E}_M$. If the resultant line is further extrapolated through the abscissa, it intersects at $-K_A$. Unlike data published in the Lineweaver–Burk (1934) and Beidler (1954) plots, at low (subthreshold) concentrations of glutathione the curve swings asymptotically upward. This deviation is always present, is repeatable, and gives an indication of being a useful quantitative index of threshold. At higher glutathione concentrations the line is straight and can be used to determine accurately the dissociation constant.

Using this type of plot, we have been able to determine the dissociation constant of glutathione and the receptor of *H. littoralis* under a variety of conditions. Typical values range between 10^{-7} and 10^{-5} M. This constant is significant in at least four ways. First, its smallness indicates that the receptor has a high affinity for glutathione. Second, 10^{-7} to 10^{-5} M is well within the range to be expected under natural conditions of feeding. Third, this number provides a means of quantitatively characterizing the receptor; i.e., the glutathione receptor of *H. littoralis* can be said to have a dissociation constant of, say, 5×10^{-7} M under the given conditions. This constant is a characteristic of the receptor and is nearly the same no matter what the nutritional state of the hydra. Fourth, the changes in the dissociation constant that take place as the pH is changed can be used to determine the pK's of ionizable groups at the receptor site (see below).

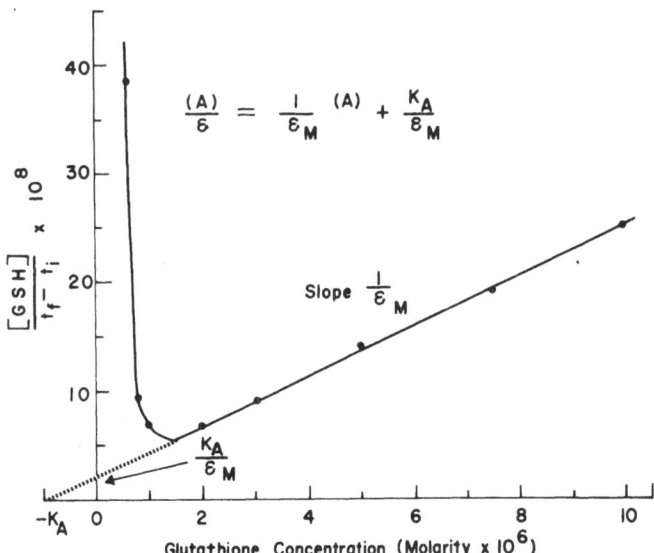

Fig. 10. Plot for determining constants of the combination of gluta-thione with its receptor. Reprinted with the permission of *American Zoologist*.

The other parameter of the equilibrium equation which is useful in analyzing the feeding behavior of hydra is the maximum effect $\mathscr{E}_M$. This number, expressed as the maximum time that a mouth can remain open to glutathione under a given set of conditions, provides a useful measure of the relative number of complete receptor-effector systems. A completed receptor-effector system is defined as one composed of all the components necessary for it to function in combination with glutathione.

Effect of pH on K_A

By measuring changes in K_A caused by pH, we were able to determine the pK's of ionizable groups at the receptor site that were involved in the combination with glutathione (Lenhoff, 1965). These analyses also were analogous to those used by enzymologists to determine pK's of ionizable groups at the active site of enzymes, For our purposes we needed an equilibrium equation, like Dixon's for enzymes (Dixon, 1953 and Dixon and Webb, 1958), which took into account the influence of pH on the dissociation constant. This modified equation assumed that, if the activator, receptor site, or activator-receptor complex ionized, then in the expression for equilibrium [Eq. (2)] each component (A, R, AR) equals its respective concentration multiplied by a term which is a function of pH. For example, if the activator ionized, then the total concentration of free activator, A_t, would be A times its respective pH function, f_a(pH). The logarithmic form of the equation is

$$pK_A = pK_A^0 + \log f_{ar}(\text{pH}) - \log f_r(\text{pH}) - \log f_a(\text{pH}) \qquad (4)$$

Here pK_A refers to the negative logarithm of the dissociation constant of A from AR, while pK_A^0 is the same constant if none of the components has ionic groups; if none ionizes, then these constants are equal. [The derivation of Eq. (4) is explained in Lenhoff, 1967.]

Equation (4), which indicates that a plot of pK_A against pH will consist of a series of straight lines joined by short curved parts, holds true for the glutathione-hydra system (Fig. 11). The results follow almost exactly the predictions from the modified Dixon equations. The following interpretations were made:

1. Ionizable groups at the receptor site participated in binding glutathione because significant variations in pK_A occurred with pH change.

2. The concave downward inflections at pH 4.6, 4.8, 6.5, and 7.6 represented pK's of ionizable groups at the receptor site. These pK's most likely do not represent ionizable groups at glutathione (Wieland, 1954), which are either below pH 4 (2.1 and 3.5) or above pH 8 (8.7 and 9.6). If the receptor site is protein, then the determined pK's may represent two β-carboxyls of peptide aspartic acid (or γ-carboxyls of peptide glutamic acid), an imidazole group, and a terminal α-amino group, respectively.

3. The horizontal lines indicated pH ranges which do not affect the combination of glutathione with the receptor site.

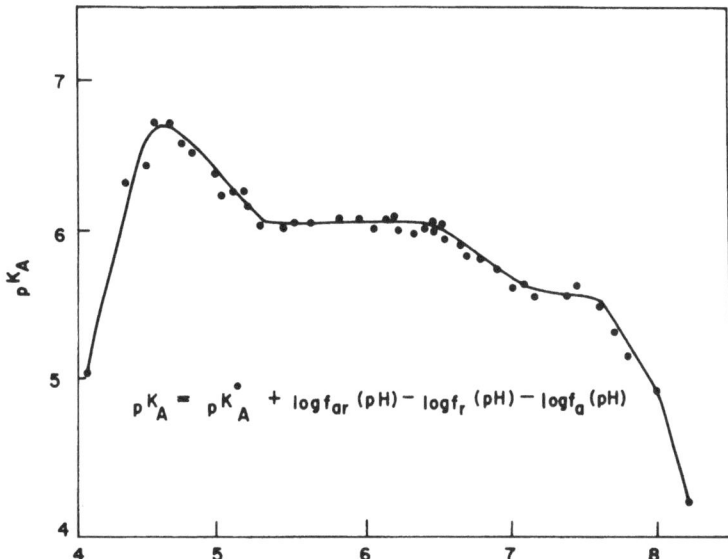

Fig. 11. Effect of pH on the dissociation constant K_A between glutathione and its receptor. Reprinted with the permission of *American Zoologist*.

4. The quenching of the charges (see Dixon and Webb, 1958) at around pH 4 and 8 indicated that groups at pH 4.6 and 7.6 may be associated with complementary charged groups on glutathione.

A Possible Mechanism for the Binding of Glutathione to the Receptor Site

It is likely that the charged groups at the receptor site bind complementary charged groups on glutathione. The positively charged α-amino group of glutathione may neutralize a negatively charged carboxyl of the receptor, while the terminal carboxyl of the glycyl moiety of glutathione may bind to a positively charged group of the receptor's terminal α-amino group. Similarly, the groups represented by pK's at 4.8 and 6.5 may be involved in the binding or may be sufficiently close to the receptor site to be displaced somewhat during the binding process. These displacements would be represented by the pK values indicated by the concave upward bends at pH 5.2 and 7.0.

The suggested binding mechanism points out the rigid specificity of the receptor for glutathione, but does not tell us what happens after the combination occurs. Since we cannot detect any chemical alteration in glutathione during the activation and since glutathione has to be constantly present at the receptor for activation to take place (Lenhoff, 1961a and b), we conjecture that activation occurs in the following manner: Glutathione may activate the receptor by causing a reversible modification of the tertiary structure of the receptor molecule, perhaps much in the way that an allosteric protein is activated by its effector.

Use of pH Data to Resolve Some Seemingly Inconsistent Results

The pH experiments, in addition to demonstrating the presence of ionizable groups at the receptor site that are affected by combination with glutathione, also provide the basis for experiments that help to resolve some seemingly inconsistent observations on the feeding behavior (see Lenhoff, 1965). For example, the experiments illustrated in Fig. 11 point up how slight alteration of the hydrogen ion concentration in the macro-environment around the receptor site can have major effects on the dissociation constant between glutathione and the receptor. These changes, therefore, must affect the number of functional receptor sites, which in turn may change the number of functional receptor-effector systems. Such a conclusion is borne out by measuring the maximum response $\mathscr{E}_M$ to glutathione over the same pH range (Fig. 12), because the $\mathscr{E}_M$ is probably proportional to the number of functional receptor-effector systems. As might be expected, a plot of $\mathscr{E}_M$ against pH gives a pattern related to that given by the plot pK_A against pH.

Analysis of $\mathscr{E}_M$ measurements taken at different pH values helps to clarify some otherwise confusing observations made independently by myself (Lenhoff, 1961b) and Burnett et al. (1963) on hydra and by Fulton (*personal communication*) in analogous experiments on *Cordylophora*. For example, hydra exposed to glutathione at pH 7.6 until they no longer respond will begin to exhibit a feeding response in less than a minute when offered live *Artemia* or *Artemia* extracts. Burnett et al. (1963) interpreted

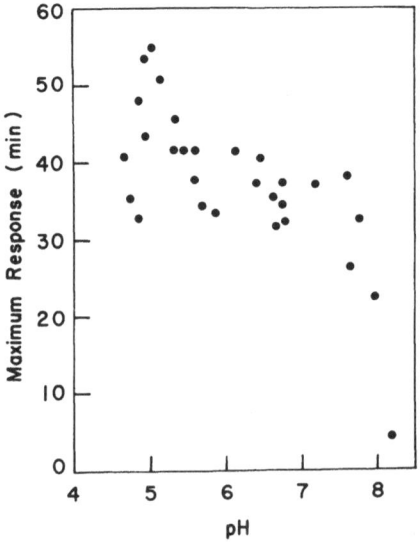

Fig. 12. Effect of pH on the maximum feeding response $\mathscr{E}_M$ in *H. littoralis*. Reprinted with the permission of *American Zoologist*.

this observation to indicate that a feeding activator other than glutathione is released and induces the hydra to ingest the brine shrimp. An alternative explanation would now appear more plausible: Since *Artemia* extracts have a pH of about six and since hydra culture solution (pH 7.6) has a weak buffer capacity, the release of shrimp fluids during food capture could reduce the area immediately around the receptor by as much as two pH units, which would result in a greater number of functional receptor-effector systems (Fig. 12 and Lenhoff, 1965); in the presence of already saturating glutathione concentrations, the feeding response is able to begin again. Such activation by increased hydrogen ions, however, will occur at saturating concentrations of glutathione and not in the nonspecific manner proposed by Balke and Steiner (1959).

This alternative explanation is supported by the following experiments: Groups of five *H. littoralis* were exposed to solutions of saturating glutathione (5×10^{-5} M) at set pH values until the animals stopped exhibiting the feeding response. Next, they were washed and placed in a solution buffered at a different pH but also containing 5×10^{-5} M glutathione. As predicted by Fig. 12, animals transferred from pH 6.0 to 7.6 and from pH 4.8 and 4.3 did not exhibit a supplementary feeding response. On the other hand, animals transferred from pH 4.3 and 4.8 responded again for an average of 12 minutes, and those transferred from pH 7.6 to 6.0 gave a supplementary response averaging 9.6 minutes.

From these experiments it can be expected that, when hydra are exposed to saturating amounts of glutathione at pH 4.8 until they no longer respond, then these animals will not ingest newly captured shrimp. This proved always to be the case (Lenhoff, 1965).

The glutathione response of *H. pirardi* was also shown to be pH sensitive, animals responding for 85 minutes at pH 7.25 and 130 minutes at pH 6.75. In related experiments on the contraction responses of hydra, Rushforth (1965) has shown that this measure of responsiveness to glutathione is also pH dependent.

Effect of Anions on Feeding Response

Too little work has been done on the effect of anions on the feeding response to make any forthright statements about their action. Their order of effectiveness in increasing the duration of the feeding response in hydra was $Cl^- > Br^- > I^- = NO_3^-$ (Lenhoff and Bovaird, 1959). Phosphate ions enhanced the feeding response of *Cordylophora* to proline, possibly by complexing the inhibitory magnesium ions (Fulton, 1963).

The effects of the anions are most obvious in measurements of the duration of the feeding response. For example, the maximum response of hydra in histidine buffer was nearly twice that obtained with hydra in tris-maleate buffer. On the other hand, the K_A's measured in each of these buffers at the same pH were nearly identical (Lenhoff, *unpublished observations*); this experiment provides further proof that the K_A is a property of the system.

ENDOGENOUS FACTORS AFFECTING THE FEEDING RESPONSE OF HYDRA

Until now, I have been discussing environmental factors affecting the degree of feeding response carried out by hydra in the presence of reduced glutathione. Of equal importance in affecting the degree of response are endogenous factors, i.e., those resulting from changes in the physiology of the animal. Therefore, I should like to discuss how the feeding behavior is influenced by: (1) the prior exposure of hydra to food or glutathione or both; (2) the prior exposure to substances released by hydra into the environment (which then act back on the hydra). Lastly, I shall discuss (3) the interactions of the glutathione-activated feeding response with another hitherto undescribed chemically activated behavioral response of hydra. In the paper following mine in this volume, Dr. Rushforth discusses still another aspect of endogenous interactions with the glutathione-activated feeding response, i.e., the interactions of the glutathione receptor-effector system with the receptor-effector systems involved in the contractions induced either by light or by mechanical agitation.

Effect of Fasting on the Feeding Response

Inspection of Fig. 9, which describes the effect of glutathione concentration on the duration of the feeding response, shows that hydra fasted for 2 days give a maximum response significantly greater than the response given by animals fasted for 1 day.[3] I interpret these data to indicate that hydra fasted for 2 days have more completed receptor-effector systems than hydra fasted for 1 day. That is, when all the receptor-effector systems are completed, the hydra is capable of carrying out a maximum response. Why the duration of the feeding response of hydra starved for 2 days is greater than that of hydra starved for 1 day is not clear. Possibly hydra fasted for 2 days have accumulated more substrate to supply energy for the contractions. This conjecture gains support from the next experiment.

Most of the experiments thus far described indicate that the hydra usually respond to the continuous presence of glutathione for a period averaging between 30 and 40 minutes, depending upon the particular set of experimental conditions. From such data I postulated that, as one result of glutathione activation, some substance in the receptor-effector system is consumed and that the concentration of that substance limits the duration of the feeding response to 30 to 40 minutes.

If this postulate were true, then one might expect that, after the hydra carried out a maximum response, there would be a period during which they would give no further response to a fresh solution of glutathione, followed by another period in which they would regain their ability to

---

[3] It is interesting to note that, although the hydra fasted for 2 days respond longer than hydra fasted for 1 day, approximately the same concentration of glutathione is required to elicit a half-maximal response on both groups of hydra. This finding provides further evidence that the K_A is a constant property of the animal.

respond maximally. This proved to be the case, as shown in Fig. 13. In the experiment shown there, large numbers of hydra were exposed to glutathione for 40 minutes. The animals, all of which completed a maximum feeding response, were then washed and placed in glutathione-free culture solution. At various intervals, groups of these animals were exposed to a fresh solution of glutathione. The results in Fig. 13 show that during the first hour the animals gave little or no response. By the tenth hour, however, the hydra had regained their ability to give a 15-minute response, and, after 1 day, they were able to respond maximally. Extending the interval between exposures to over 70 hours did not result in any further increase in the length of their response to fresh glutathione.

This lag and the gradual resumption of the ability of hydra to respond maximally to stimulation by glutathione are interpreted to signify the period for the resynthesis of some substance (called X) which we postulate to be limiting in the receptor-effector system.

In psychological terminology, this inability of hydra to respond to a new dose of glutathione immediately after they have given a maximal response to an initial solution of glutathione might be interpreted as an example of habituation. I prefer not to call it habituation, however, as that might imply a short-term learning, or retention. Instead, I interpret the results described in Fig. 13 as more akin to fatigue, during which the limiting substance is consumed. Later, after the substance is replenished, the animal can again respond maximally (Lenhoff, 1961b).

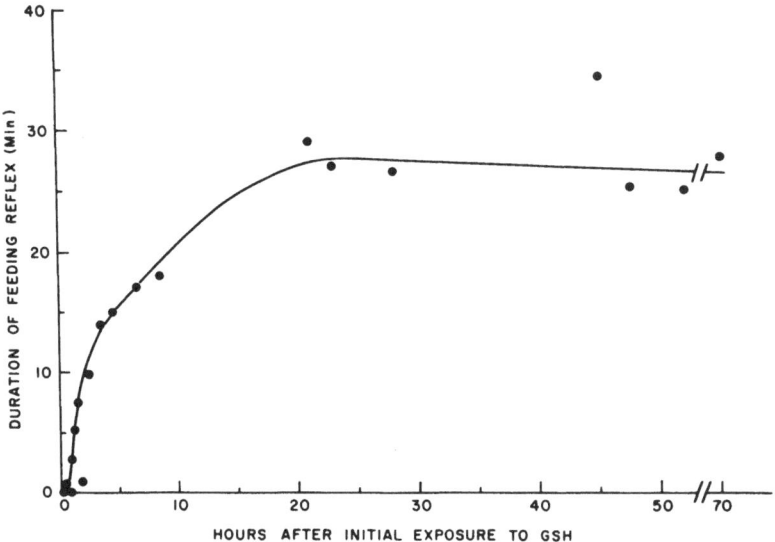

Fig. 13. Time for the recovery of the ability of *H. littoralis* to respond to glutathione. Reprinted by courtesy of Loomis Institute for Scientific Research, Inc.

We have evidence that a hydra's ability to regain its competency to respond maximally to a second dose of glutathione after having just given a maximal response to the previous dose can be inhibited by placing the animal in a weak solution of cyanide during the 24-hour recovery period (Brown and Lenhoff, *unpublished observations*). Hence, some cyanide-sensitive step, probably an oxidation-reduction reaction, is required in the resynthesis of the postulated substance X.

This inhibition is not due to a general poisoning of the animal. Within a day after the animals are removed from the cyanide solution, their ability to respond to glutathione is completely restored. Furthermore, cyanide does not seem to have any noticeable deleterious effect on hydra exposed to a 10^{-4} M solution for periods of a few days. For example, in 10^{-4} M cyanide, hydra can contract and respond to mechanical stimuli; in addition, hydra in cyanide can be made to capture brine shrimp but not to ingest them. The inability of hydra to ingest captured food in the presence of cyanide led to an experiment which demonstrated that fasted animals in cyanide are not able to respond to glutathione. Thus, the step involved in the immediate consumption of X or its conversion to some utilizable form is also cyanide sensitive.

To provide more evidence concerning the existence of the limiting substance X and to learn something about its role in the execution of the response, we carried out some temperature experiments. If the reactions of the feeding reflex (Fig. 14) are depicted as involving the conversion of the limiting substance X to Y, then one might expect two major results from lowered temperatures. First, a slight drop in temperature should lower the rate of all the thermochemical reactions. By slowing down the reaction that converts X to Y, we should therefore slow down the rate at which the supply of X is depleted and thus increase the length of time that the mouth remains opened, albeit opened not as widely as it would be at a higher temperature. This proved to be the case, as shown in Table 2; as the temperature approached 15°C, the hydra responded for nearly 100 minutes.

As a second effect of lowering the temperature, the limiting reaction may proceed so slowly that the optimum (threshold) conditions necessary for the feeding response are not maintained. Thus, when hydra are held below

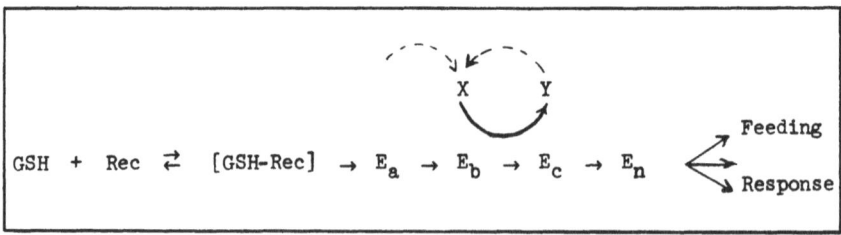

Fig. 14. Schematic outline of the glutathione receptor-effector system. Rec represents the receptor; E_a, E_b, E_c, and E_n, enzymes; X, the postulated limiting substance; and Y, its metabolic product.

TABLE 2

Effect of Temperature on the Duration of the Feeding Response

Temperature, °C	$t_f - t_i$,* minutes	Temperature, °C	$t_f - t_i$,* minutes
6.2	5.9	18.1	86.9
8.9	36.4	18.6	55.0
10.3	59.8	19.7	59.2
12.5	70.7	20.6	55.0
14.5	60.0	21.9	35.1
15.4	88.6	24.1	29.4
16.3	99.7	25.3	21.5
		27.7	19.7

* The terms t_f and t_i are defined on pages 344 and 346.

15°C, they are observed to open their mouths for a few minutes, then close, open, etc., until they finally stop responding. The total duration of the responses below 15°C becomes progressively less until the hydra barely respond (Table 2). In fact, when the temperature is lowered from 20 to 5°C, the mouth takes longer to open (Lenhoff, 1961b); such results are now interpreted to mean that, as the temperature is lowered, it takes longer for the completion of all the reactions (including the limiting one) leading to mouth opening.

Experiments such as those indicating the role of consumable substrate in affecting the degree of the feeding response point up the importance of: (1) being aware of the time of the last feeding in relation to the time that the behavioral experiment is carried out; (2) controlling the amount of food offered, the regularity of feeding, and possibly the kind of food; and (3) controlling the temperature. Such factors may be extremely important in hydra and planaria—animals that carry out most of their digestion intracellularly and that lack a defined circulatory system to distribute nutriments.

Substances Emitted by Hydra

Another endogenous factor that may influence the degree of the feeding response is the effect of substances that leach out of the hydra themselves and act back upon the animals. The first indication that material coming from the hydra affects the degree of the feeding response came from experiments using trypsin (Lenhoff and Bovaird, 1960b) as an artificial activator of feeding. [Trypsin acts in a nonspecific manner in activating feeding in *Hydra littoralis* and *Physalia* (Lenhoff and Bovaird, 1960b) and *Cordylophora* (Fulton, 1963).]

The degree of the feeding response activated by trypsin varied depending upon the locations of the animals in the culture dish; those that were apart from others responded best, while hydra from groups did not respond as well or at all. In order to illustrate quantitatively these crowding effects, we measured the trypsin activation of a response using unfed hydra that were kept at different densities for 2 days and washed either once or twice daily or not at all. The data in Fig. 15 show that the hydra at greater densities responded best when the water was changed twice a day; this effect was especially evident with hydra at a density of 10 per ml. We conclude that some substances diffusing from the hydra accumulate and, in an unknown manner, depress the feeding response activated by trypsin.

The crowding experiments indicate that material naturally leaching from hydra can affect the feeding response. It is even possible that hydra, while being stimulated with glutathione, may be giving off the inhibitory compound(s) into the environment at an increased rate. This possibility is suggested by experiments in which the same solution of glutathione was used to activate a maximum response in a series of three successive groups of hydra. Each group gave a good response to the glutathione (Lenhoff, 1961a and b); thus, these experiments showed that the hydra ended their feeding response in the presence of glutathione, not because the glutathione was destroyed but, rather, because some intrinsic change had taken place within the animal. We interpret this change to be the result of the consumption of compound X. However, review of various experiments of this nature indicates that, although each successive group of hydra gave a good response, there was often a significant decrease in the response of hydra exposed to the solution of glutathione that had been used on previous groups of animals. Such experiments suggest that either the glutathione concentration was in some manner slightly lowered after these transfers or some inhibitory factor gradually accumulated in the environment.

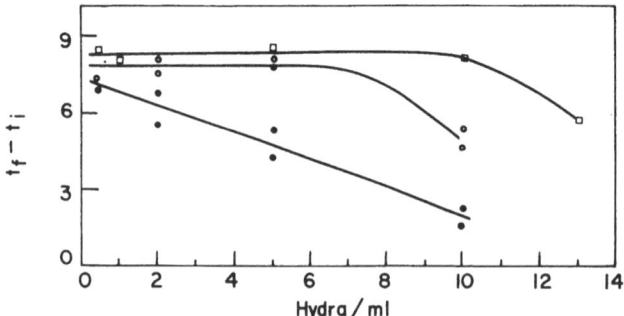

Fig. 15. Influence of crowding on feeding response activated by trypsin. The animals were not fed during the experiment. The culture solution was changed either twice daily □, once daily ○, or not at all ●. Reprinted with the permission of *American Zoologist*.

This postulated inhibitory factor may be the same factor as that released by hydra in crowded cultures. At the moment, we have no clue to the nature of the inhibitory substance, but it may be that potassium from the hydra accumulates in the environment until it reaches inhibitory concentrations; we have already shown that concentrations of potassium as low as 10^{-3} and 10^{-5} M inhibit the feeding response activated by either glutathione or trypsin, respectively (Table 1 and Lenhoff, 1965).

Regardless of the nature of the substance(s) from hydra that accumulate in the environment, their inhibitory effects can be minimized as follows: Maintain a low population density in the stock culture of animals to be experimented upon; change the bathing medium frequently, or occasionally agitate the culture in order to admix the postulated inhibitor(s) with the rest of the culture solution so that its concentration will be minimal.

Such experiments point out the importance of being aware of the effect on the feeding behavior of material leaching from the hydra. It does not seem unreasonable to suppose that such factors may also influence the extent of other behavioral responses of the animal.

Interactions of the Glutathione Receptor-Effector System with Another Chemically Mediated Receptor-Effector

I have been stressing the importance of controlling the environment and of being aware of the endogenous metabolism of the animal in order to get repeatable, quantitative, and unequivocal manifestations of the animal's behavior. I should now like to describe some experiments in which we discovered a new behavioral response by carrying out the glutathione activation of feeding under conditions in which the environment and metabolism could not be precisely controlled. That is, in all experiments heretofore mentioned we customarily used hydra fasted for 2 days to avoid the complications resulting from food and digestion.

In discovering the phenomenon which I shall now describe, called *neck formation*, we used hydra having their gastrovascular cavity full of food and fluids. Such hydra, presented with a live *Artemia* nauplius, *Artemia* extract, or a solution of reduced glutathione, formed a tight constriction in the region just below the hypostome and sometimes extending over the adjacent one-third of the body tube. The duration of the constriction roughly corresponded with the period of mouth opening, and the glutathione had to be continuously present for neck formation to persist over an extended time.

When the swollen hydra were presented with a live *Artemia* nauplius, the neck constriction formed, the mouth opened, and the nauplius was swallowed. During ingestion, the nauplius was carried down through the constriction, apparently by peristalsis, and into the fluids of the swollen gastrovascular cavity. Hence, it would appear that these neck constrictions allow hydra to retain in the gut previously ingested food while they are swallowing newly captured prey.

Dr. R. Blanquet and I have analyzed the factors responsible for this behavioral manifestation of a neck formation which allows the animal to ingest its food. Three factors are required to be present simultaneously in order to produce neck formation. First, the hydra must be swollen with fluid. Second, present in the fluid must be a substance released inside the gastrovascular cavity after food has been ingested—a substance we have recently identified as the amino acid tyrosine. And, third, reduced glutathione must be present outside the animal; during feeding, the glutathione comes from the injured newly captured prey. If any one of these three factors is not present, then no neck formation occurs and the hydra will maintain its inflated shape. Thus, in hydra we have one of the earliest examples of the interaction of different chemically activated receptor-effector systems leading to specific contractions associated with a behavioral response.

This example of neck formation has been presented here particularly to illustrate the value of studying the behavior of your animal under a variety of experimental conditions. Both "ideal" and "unideal," "natural" and "unnatural" conditions should be tried. The "natural" conditions help us to find behavioral phenomena, while the "unnatural" can help us to understand mechanism as well as to find other behavioral phenomena.

CONCLUSIONS

I have given some examples of how the quantitative aspects of hydra's feeding behavior are affected by the major ions in pond water, by pH, by metabolism, and even, indirectly, by the stimulus itself. I have also given examples of how, by not being aware of these subtle effects, the investigator may misinterpret the data, and of how, accordingly, slight changes in conditions may give different results from laboratory to laboratory. I have given an example, and Dr. Rushforth gives others, showing how, even within a simple animal, one behavioral response may affect the extent of other behavioral responses. Furthermore, in the case of neck formation, we have an example of how three different kinds of receptors must be activated simultaneously to elicit a response.

In conclusion, I suggest that in behavioral experiments using invertebrates—particularly those which I call "naked" (those having a significant number of cells exposed directly to the environment—in addition to the controls for pseudoconditioning and for sensitization, there should be suitable controls for the effect on the animal of environmental factors. Similarly, before proceeding too fast with a study of invertebrate behavior, perhaps now is an appropriate time to bolster our knowledge of the physiology and biochemistry of these lower forms.

In this paper I have given many examples of environmental and endogenous factors affecting the behavior of hydra. Perhaps my findings may have little bearing on experiments dealing with the behavior of Planaria—but possibly they might.

REFERENCES

Baker, J. R., *Abraham Trembley of Geneva*, London: Edward Arnold (Publishers), Ltd., 1952.

Balke, E., and G. Steiner, Über du chemische Nahrungswahl von *Pelmatohydra oligactis* Pallas. *Naturwissenschaften*, 1959, **46**, 22.

Beidler, L. M., A theory of taste stimulation. *J. Gen. Physiol.*, 1954, **38**, 133.

Burnett, A. L., R. Davidson, and P. Wiernick, On the presence of a feeding hormone in the nematocyst of *Hydra pirardi*. *Biol. Bull.*, 1963, **125**, 226.

Cliffe, E. E., and S. G. Waley, Effect of analogues of glutathione on the feeding reaction of hydra. *Nature*, 1958, **182**, 804.

Dixon, M., The effect of pH on the affinities of enzymes for substrates and inhibitors. *Biochem. J.*, 1953, **55**, 161.

Dixon, M., and E. C. Webb, *Enzymes*. New York: Academic Press, Inc., 1958.

Fulton, C., Proline control of the feeding reaction of *Cordylophora*. *J. Gen. Physiol.*, 1963, **46**, 823.

Lenhoff, H. M., Activation of the feeding reflex in *Hydra littoralis*, I: Role played by reduced glutathione, and quantitative assay of the feeding reflex. *J. Gen. Physiol.*, 1961a, **45**, 331.

Lenhoff, H. M., Activation of the feeding reflex in *Hydra littoralis*, in H. M. Lenhoff and W. F. Loomis, eds., *The Biology of Hydra and of Some Other Coelenterates*. Coral Gables, Fla.: University of Miami Press, 1961b, p. 203.

Lenhoff, H. M., On the mechanism of the glutathione-receptor of *Hydra littoralis*. *Int. Congr. Zool., Proc. XVI, Washington, D.C.*, 1963, **3**, 69.

Lenhoff, H. M., Some physicochemical aspects of the macro- and microenvironment surrounding hydra during activation of their feeding behavior. *Amer. Zool.*, 1965, **5**, 515.

Lenhoff, H. M., Influence of monovalent cations on the growth of *Hydra littoralis*. *J. exp. Zool.*, 1966, **163**, 151.

Lenhoff, H. M., Chemical perspectives on the feeding response, digestion and nutrition of selected coelenterates, *in* B. Scheer and M. Florkin, eds., *Chemical Zoology*, *Vol. II*. New York: Academic Press, Inc., 1967.

Lenhoff, H. M., and J. Bovaird, The requirement of trace amounts of environmental sodium for the growth and development of *Hydra*. *Exp. Cell. Res.*, 1960a, **20**, 384.

Lenhoff, H. M., and J. Bovaird, The enzymatic activation of a hormone-like response in *Hydra* by proteases. *Nature*, 1960b, **187**, 671.

Lenhoff, H. M., and J. Bovaird, Action of glutamic acid and glutathione analogues on the *Hydra* glutathione receptor. *Nature*, 1961, **189**, 486.

Lenhoff, H. M., and J. Bovaird, Requirement of bound calcium for the action of surface chemoreceptors. *Science*, 1959, **130**, 1474.

Lenhoff, H. M., and W. F. Loomis, eds., *The Biology of Hydra and of Some Other Coelenterates*: 1961. Coral Gables, Fla.: University of Miami Press, 1961, p. 69.

Lenhoff, H. M., and H. A. Schneiderman, The chemical control of feeding in the Portuguese man-of-war, *Physalia physalis*. L., and its bearing on the evolution of the Cnidaria. *Biol. Bull.*, 1959, **116**, 452.

Lentz, T., and R. Barnett, Fine structure of the nervous system of *Hydra*. *Amer. Zool.*, 1965, **5**, 341.

Lineweaver, H., and D. Burk, The determination of enzyme dissociation constants. *J. Am. Chem. Soc.*, 1934, **56**, 658.

Loomis, W. F., Glutathione control of the specific feeding reactions of hydra. *Ann. N.Y. Acad. Sci.*, 1955, **62**, 209.

Loomis, W. F., and H. M. Lenhoff, Growth and sexual differentiation of hydra in mass culture. *J. Exp. Zool.*, 1956, **132**, 555.

Picken, L. E. R., and R. J. Skaer, A review of researches on nematocysts, *in* W. J. Rees, ed., *The Cnidaria and Their Evolution*. New York: Academic Press, Inc., 1966, p. 19.

Rushforth, N., Inhibition of contraction responses of hydra. *Amer. Zool.*, 1965, **5**, 505.

Trembley, A., *Mémoires pour servir à l'histoire d'un genre de polypes d'eau douce à bras en forme des cornes.* Leyden, The Netherlands: J. and H. Verbeck, 1744.

Wieland, T., Chemistry and properties of glutathione, *in* S. Colowick, ed., *Glutathione.* New York: Academic Press, Inc., 1954.

Chapter 23

Chemical and Physical Factors Affecting Behavior in *Hydra*: Interactions Among Factors Affecting Behavior in *Hydra*

Norman B. Rushforth

Department of Biology
Western Reserve University
Cleveland, Ohio

INTRODUCTION

Coelenterates are animals of great interest to the behavioral physiologist, since they possess the most primitive nervous system found in the animal kingdom. This system has diffuse organization consisting of a two-dimensional nerve net. In sea anemones, e.g., the network functions as a single all-or-nothing conducting element. Stimulation at any point on the surface of the anemone generates impulses in the net resulting in the same evoked response (Pantin, 1935). Such impulses have the properties of threshold and refractory period, as do the nervous impulses of higher animals.

The structural simplicity of many coelenterates coupled with their superficially stereotyped behavior has intrigued workers studying mechanisms underlying behavioral control. Starting with Trembley's classical studies (1744) there have been numerous efforts to give "complete" analysis of the behavior patterns exhibited in this phylum. However, such apparent simplicity has proved deceptive, and many features of coelenterate behavior have as yet resisted analysis. Some attempts have been made to determine if coelenterates are capable of learning. In general, the results of these efforts have been inconclusive (for a review of these studies see Ross, 1965), but nevertheless the problem has considerable importance. Is learning solely an attribute of an organized central nervous system? Can modification of behavior occur in the absence of complex nervous structure? Such questions have added interest in view of the recent finding that some coordinating systems controlling behavior in coelenterates are of a non-nervous nature (Mackie, 1965).

The present investigations were undertaken to study behavioral modifications in the freshwater *Hydra*. They are extensions of earlier behavioral studies of the coelenterate *Hydra pirardi* Brien (Rushforth, 1965a). The results of such investigations illustrate the complexity of interactions among factors affecting behavior in this animal. They demonstrate the importance of analyzing physiological mechanisms rather than emphasizing operational definitions of stimulus-response relationships in studying behavioral modifications.

RESPONSE OF *HYDRA* TO MECHANICAL STIMULATION

The earliest recorded account of the contraction of hydras to mechanical agitation was given by Trembley (1744) in his observations of *Hydra viridis*. Later Wagner (1904) described the response of this species to intermittent mechanical stimulation. He observed that the animals contract when their culture dishes are tapped. If the animals are stimulated in this way at 1-second intervals, they first contract and then reexpand. Thus, he concluded that the hydra adapts to a nonlocalized mechanical stimulus recurring at frequent intervals. Wagner found, however, that, if the interstimulus interval is increased to enable the hydra to reexpand after each contraction, the animal does not adapt but continues to contract to each mechanical stimulus. He concluded that adaptation was dependent on the stimuli being rapidly repeated and maintained that the recovery from the acclimatizing effect must be of very short duration.

Our first experiments on the response of *Hydra* to mechanical stimulation were performed with *H. pirardi* (Rushforth, 1965a). In contrast to Wagner's results, we found that *H. pirardi* adapts to nonlocalized mechanical agitation even when the interval between the stimuli is long enough to allow the animal to expand completely. The process of adaptation to this stimulus was not considered to be one of muscular fatigue, since contractions were readily evoked in hydras habituated to mechanical stimulation by a different stimulus, that of light. Rather, the decrease in sensitivity of the hydras to mechanical agitation was thought to be an example of habituation.

Since there are marked differences in the contraction response to light among different species of *Hydra* (Burke and Rushforth, 1966), investigations were made of the response to mechanical stimulation of species other than *H. pirardi*. Most of the experiments reported here were performed using the green *H. viridis*.[1] These animals were cultured by modifying a method by Loomis and Lenhoff (1956) and raising the hydras in a solution containing 1.5×10^{-3} M $CaCl_2$, 1.2×10^{-3} M $NaHCO_3$, and 1.2×10^{-4} M Na_4EDTA at 21°C.

In order to quantify the contraction response of *H. viridis* to mechanical stimulation, groups of 10 hydras, which had not been fed for 24 hours, were placed in 50 ml of culture solution in a preparation dish on a rotatory shaker. After the animals became attached to the bottom of the dish, the dish was rotated at a fixed speed for a predetermined time period and the

---

[1] The strain of *H. viridis* used was a European Green Normal strain (EGN) obtained from cultures raised by Professor Paul Brien in Brussels.

numbers of animals contracting were recorded. The rotation speeds were variable and calibrated in revolutions per minute. The sensitivity of *H. viridis* to this mechanical stimulus may be represented by a response curve: a plot of the proportions of a group of 50 hydras contracting to pulses of agitation at various rotation speeds (Fig. 1a). It is seen from such a plot that the proportion of hydras contracting increases with the increasing strength of the stimulus. If hydras are exposed to intermittent stimulation at a fixed rotation speed (e.g., a 3-second period of agitation at 130 rpm every 30 seconds), they become less responsive to the stimulus with continued exposure. This reduced sensitivity is demonstrated by (1) a decrease in the proportion of hydras contracting with an increasing number of stimulus trials, the relationship being approximately exponential in form (habituation curve, Fig. 1b), and (2) a lowering of the response curve from the initial value after 6 hours of intermittent stimulation (Fig. 1c).

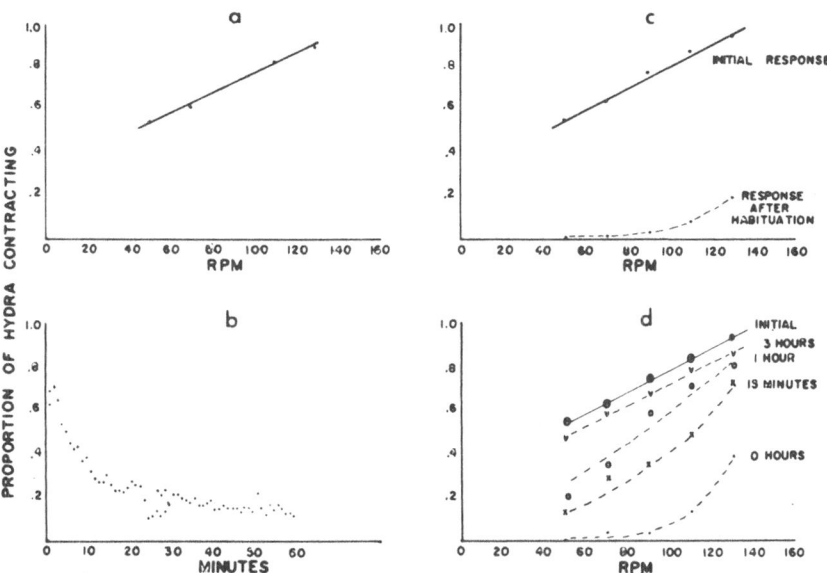

Fig. 1. (a) Sensitivity of *Hydra viridis* to mechanical stimulation: response curve. Each point represents the average proportion of hydras in five groups of 10 animals, that contracted within a 2-second period that followed a 2-second shaking period. (b) Effect of intermittent mechanical stimulation: habituation curve. Each point represents the average proportion of hydras that contracted within a 2-second period that followed a 3-second period of stimulation at 130 rpm every 30 seconds, based on three groups of 10 animals. (c) Effect of habituation on the sensitivity to mechanical stimulation: *solid line*, response curve of hydras prior to intermittent agitation; *broken line*, response curve of hydras after 6 hours of intermittent mechanical agitation. These curves are based on three groups of 10 animals. (d) Decay of sensitivity to mechanical stimulation following habituation: response curves of 30 hydras prior to intermittent agitation (*solid line*) and at various intervals (0 hours, 15 minutes, 1 hour, and 3 hours) following habituation during which the animals were not stimulated.

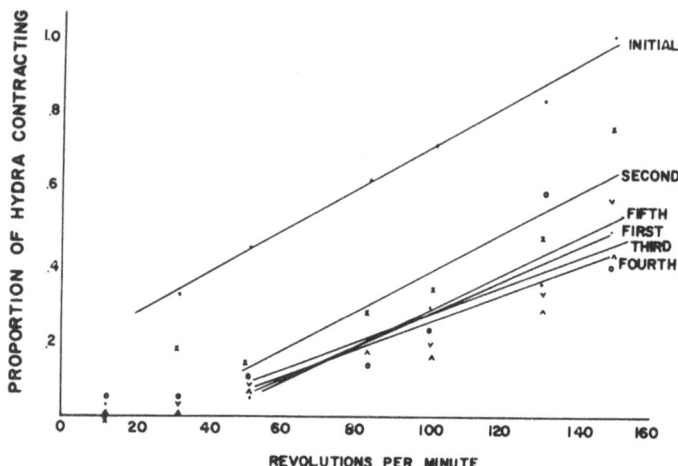

Fig. 2. Effect of successive habituations on the sensitivity of *Hydra viridis* to mechanical stimulation: response curves of 30 hydras prior to intermittent agitation (initial) and after 15 minutes following five successive habituation regimes consisting of 6 hours of intermittent stimulation (first, second, third, fourth, and fifth).

TABLE 1. Effects of Mechanical Stimulation on *Hydra viridis* Under Different Regimes

		Proportion of hydra contracting:	Initially and (after 1 hour)
	Interstimulus interval, min:	1	5
Stimulus strength, rpm	*Length of stimulus*, sec		
52	1	0.25 (0.04)	0.20 (0.18)
	3	0.27 (0.03)	0.26 (0.24)
	5	0.31 (0.06)	0.31 (0.36)
81	1	0.24 (0.06)	0.30 (0.26)
	3	0.29 (0.06)	0.33 (0.32)
	5	0.36 (0.06)	0.36 (0.34)
102	1	0.33 (0.05)	0.36 (0.27)
	3	0.35 (0.06)	0.36 (0.30)
	5	0.35 (0.06)	0.45 (0.37)
129	1	0.44 (0.08)	0.42 (0.36)
	3	0.40 (0.11)	0.44 (0.38)
	5	0.61 (0.17)	0.54 (0.67)
150	1	0.70 (0.07)	0.66 (0.69)
	3	0.72 (0.11)	0.72 (0.72)
	5	0.86 (0.10)	0.83 (0.81)

In previous studies of the sensitivity of *H. pirardi* to repeated mechanical agitation, downward shifts in the response curves following habituation and their restoration to prehabituation levels were observed. It was noted that following habituation the animal's sensitivity was fully restored after approximately four hours. In more recent experiments with *H. viridis* a similar time was found for the extinction of habituation. Animals were habituated for a period of 6 hours after determining the initial response curve. Response curves were then determined immediately (0 hours), at 15-minute intervals during the first hour, and then at hourly intervals following the cessation of intermittent stimulation. After 4 hours the response curve is not significantly different from the initial value ($p > 0.10$). It is seen from Fig. 1d, however, that sensitivity is considerably increased after a period of only 15 minutes.

The following experiment was undertaken to determine if successive exposures to mechanical stimulation would have a cumulative effect on the decrease in sensitivity of the hydras. The response curve of a group of 50 *H. viridis* was determined, and the response curves were determined both 15 minutes and 4 hours following cessation of the stimulation. This regime was repeated five times. Plots of the successive response curves taken 15 minutes after termination of intermittent stimulation, together with the initial prehabituation curve, are presented in Fig. 2. No systematic change in the responsiveness of the hydras is seen with the increasing numbers of habituations. In addition, it was found that, after 4-hour periods during which the animals were not stimulated, the response curves were not significantly different ($p > 0.10$) from prehabituation values for any of the five exposures. Thus, there is no cumulative effect on the animal's decrease in sensitivity to mechanical stimulation beyond that attained during the first habituation.

A series of experiments was undertaken to determine the role of various parameters on the decrease in sensitivity of *H. viridis* with repeated stimulation. Groups of 100 hydras were exposed to different regimes of mechanical agitation in order to ascertain the effects of changes in (1) the length of the stimulus interval (the shaking period in seconds), (2) the stimulus strength (the speed of rotation in revolutions per minute), and (3) the interstimulus interval (the time between successive shaking periods in minutes). The proportions of hydras contracting initially and after 1 hour of repeated stimulation under such regimes are given in Table 1. An analysis of these results indicates that the initial proportion of hydras responding increases with (1) increasing stimulus strength ($p < 0.001$) and (2) length of the stimulus ($p < 0.001$) but is independent of the interstimulus interval ($p > 0.10$). At the end of 1 hour, the proportions of animals contracting are significantly less ($p < 0.001$) for groups of hydras stimulated at a short interstimulus interval (1 minute) than those of groups agitated every 5 minutes. Such proportions are significantly ($p < 0.001$) greater, for hydra exposed to higher stimulus strengths and longer stimulus intervals. In Fig. 3 it is seen that hydras shaken for periods of 5 seconds every minute at both 150 and 52 rpm habituate, the rate of habituation being greater at the lower rotation speed. However, when the interstimulus interval is increased to 5 minutes, no habituation takes place at either rotation speed.

Recently Thompson and Spencer (1966) reviewed the literature on habituation. They propose nine parametric relations for stimulus variables to characterize habituation in intact organisms. We may summarize the results of studies of the response of *Hydra* to repeated mechanical stimulation, both those presented here for *H. viridis* and those reported earlier using *H. pirardi*, in terms of these nine criteria. For the two species of *Hydra* studied:

1. Repeated applications of the mechanical stimulus result in a decreased response (habituation). The decrease is approximately a negative exponential function of the number of stimulus presentations (Fig. 1b).

2. If the stimulus is withheld, the response recovers with time (Fig. 1d)

3. The greater the frequency of stimulation (i.e., the shorter the inter-stimulus interval), the more rapid and pronounced is habituation (Table 1 and Fig. 3).

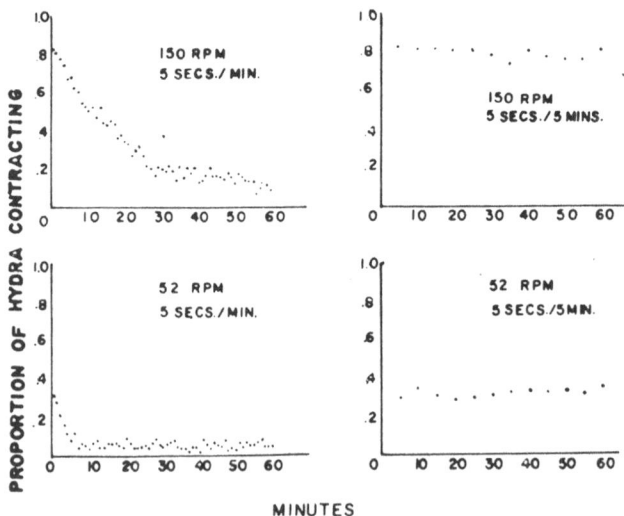

Fig. 3. Habituation curves of *Hydra viridis* exposed to different regimes of intermittent stimulation: *upper left*, hydra agitated at 150 rpm for 5 seconds every minute; *lower left*, hydra agitated at 52 rpm for 5 seconds every minute; *upper right*, hydra agitated at 150 rpm for 5 seconds every 5 minutes, and *lower right*, hydra agitated at 52 rpm for 5 seconds every 5 minutes. Each curve is based on 10 groups of 10 hydras.

4. The weaker the stimulus, the more rapid is habituation (Fig. 3). This result is the opposite of that expected if the process were one of fatigue.

Each of these results is in accordance with parametric relations considered to characterize the process of habituation. Other properties used to define habituation operationally have either not been observed or only been partially demonstrated for the response of *Hydra* to mechanical agitation. For example:

5. No stimulus generalization to other stimuli has been found for hydras habituated to mechanical agitation. Indeed, the contraction of hydras exposed to light when habituated to mechanical stimulation was considered as evidence that the habituated response did not result from muscular fatigue (Rushforth, 1965a). The concept of stimulus generalization for a habituated response, however, has not gained universal acceptance, and the specific nature of the habituating stimulus is often stressed (Bullock, 1966).

Thompson and Spencer point out that:

6. Habituation may be potentiated. If a repeated series of habituation experiences are administered, habituation becomes successively more rapid. *Hydra* habituates faster on exposure to intermittent stimulation a second time (Rushforth, 1965a), but successive administrations of the stimulus do not result in further changes in sensitivity. In the present studies no cumulative effect on the responsiveness of *H. viridis* occurred after several habituations (Fig. 2). Thus, only partial potentiation of habituation is found for hydras response to mechanical stimulation.

The remaining criteria suggested in defining habituation are currently under investigation for this system:

8. The effects of repeated stimulation beyond the asymptotic response level.

9. Dishabituation.

We may conclude that the response of *Hydra* to mechanical agitation conforms with several of the parametric criteria advocated to define habituation operationally. It must be stressed, however, that nothing is known of the physiological mechanisms involved in the response. This fact has added importance since later investigations have shown that a decrease in sensitivity to mechanical agitation is not unique for habituation in *Hydra*, but decreased sensitivity may be effected using stimuli of quite different modalities. Both photic and chemical stimuli may be used to reduce the sensitivity of *Hydra* to a mechanical stimulus. Experiments demonstrating such effects will now be considered.

EFFECTS OF LIGHT ON THE SENSITIVITY OF *HYDRA* TO MECHANICAL STIMULATION

The light sensitivity of *Hydra* was first noted by Trembley (1744), who described the animal's movements toward a light source. Wilson (1891) showed that such phototaxis was dependent on the wavelength of the light. Both *H. viridis* and *H. fusca* were found to be most sensitive to blue light. Although no specific light receptors in *Hydra* have been found, when animals are sectioned below the tentacles and exposed to unilateral light, the apical region migrates toward the light source while the body region does not (Feldman and Lenhoff, 1960). Other workers have noted that *Hydra* undergoes complex orientation movements in response to light. The animals are maximally sensitive to the blue region of the spectrum (Haug, 1933, and Passano and McCullough, 1963). Singer *et al.* (1963) investigated the contraction response of *H. pirardi* to light and showed that the reaction time was found to be shortest in blue light (400 to 450 mμ) and increased markedly above 500 mμ. Passano and McCullough (1962) noted that blue light induces frequency changes in rhythmical electrical potentials recorded from this animal. Such changes were not observed for wavelengths above 500 mμ. The spectral sensitivities of such responses suggest that carotenoid pigments are involved in photoreception in *Hydra*. Although no specific photoreceptor pigment has been isolated, recently Krinsky and Lenhoff (1965) have extracted carotenoids from *H. pirardi* and *H. littoralis* which absorb in the blue region. The absorption spectra of ether and digitonin extracts of green *H. viridis* also indicate the presence of carotenoid in this species, but there is an additional peak in the red at 670 mμ probably representing the chlorophyll component (Rushforth, *unpublished observations*).

Recent studies have shown inhibitory effects of light on components of behavior in *Hydra*. Passano and McCullough (1964) discovered that endogenous electrical potentials which precede contractions could be inhibited by light in *H. pirardi* and *H. littoralis*. Such effects were also observed with localized illumination of the subhypostomal region of the animal. They attributed this to suppression of the pacemaker system associated with contractions of the longitudinal muscles. Inhibition of contractions could also be produced indirectly by illuminating the basal region of a hydra (Passano and McCullough, 1965). Such inhibition was thought to result from effects on the rhythmic potential pacemaker systems in the lower stalk of the animal.

Light inhibits contractions induced by mechanical stimulation in dark-adapted *H. viridis* (Rushforth, 1965b). *H. viridis* placed for 24 hours in darkness has significantly lower response curves when tested in light (about 70 ft-c) than hydras exposed for 24 hours to light of this intensity (Fig. 4a). The following experiment was undertaken to investigate the responsiveness of dark-adapted hydras to mechanical stimulation when tested in the dark and to determine the extent of inhibition by light. Groups of 30 hydras were placed in the dark for 24 hours and their contractions to mechanical stimulation determined before exposure to light. Since previous studies indicated that hydras are insensitive to wavelengths above 500 mμ, the contractions were observed in light from a photographic safelight housing a Wratten OC filter. The number of hydras extended prior to shaking and

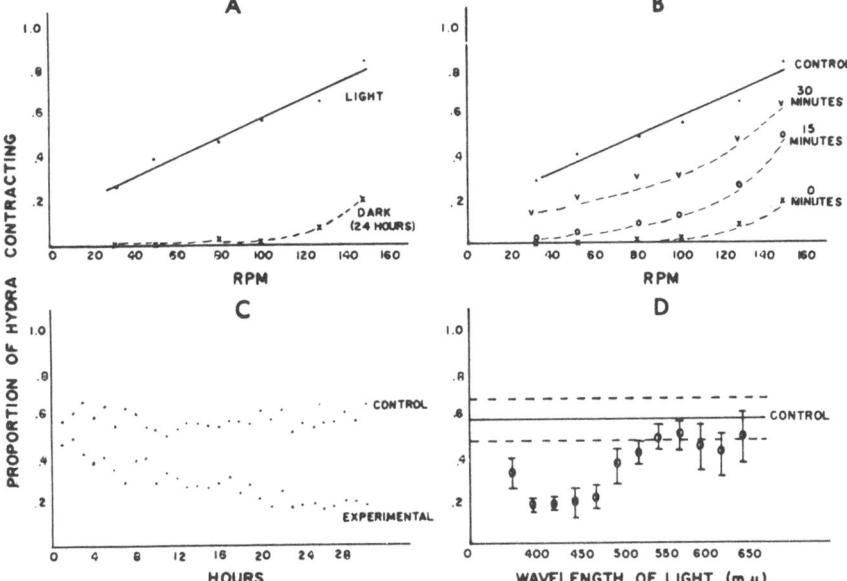

Fig. 4. Effect of light on the sensitivity of dark-adapted *Hydra viridis* to mechanical stimulation. (a) *Solid line* (light), response curve of 30 hydras tested in ambient light; *broken line* (dark 24 hours), response curve of 30 hydras dark-adapted for 24 hours and tested in ambient light. (b) Response curves of 150 hydras dark-adapted for 24 hours: *solid line* (control), tested in the dark; *broken line* (0 minutes), immediately on exposure to ambient light, and *broken lines*, after 15 and 30 minutes respectively in ambient light. (c) Effect of the length of dark adaptation on the sensitivity of hydras to mechanical stimulation. Each point represents the average proportion of a group of 60 hydras contracting to five shaking pulses at 130 rpm; the pulses were 1 minute apart and of 5-second duration. The experimental values were given by groups of hydras dark-adapted for different time periods and tested in ambient light, whereas the control values were given by groups of light-adapted hydras tested at the same times. (d) Sensitivity of dark-adapted hydras to mechanical stimulation when exposed to light of different wavelengths. Mean proportions and 95% confidence intervals are plotted for groups of 30 hydras dark-adapted for 24 hours, contracting to mechanical stimulation (five successive pulses of 5 seconds of shaking every minute at 130 rpm) when exposed to monochromatic light. The control values (*solid line*, the mean; *broken lines*, the upper and lower 95% confidence limits) were determined for a group of 30 hydras dark-adapted for 24 hours and tested in the dark.

the number contracted after a 5-second shaking pulse at each of six different rotation speeds were recorded. Pulses of agitation were given at 1-minute intervals, and the different rotation speeds were used in a random sequence. The results of five successive experiments were combined to give composite response curves for a group of 150 animals. The composite response curve for these 24-hour dark-adapted hydras tested in the dark (Fig. 4b, control) demonstrates that such animals are quite responsive to the shaking stimulus. The hydras were then retested 15 minutes later in ambient light (about

70 ft-c). It was observed that on exposure to light the hydras elongate to approximately one and a half times their dark-adapted lengths. In this extended state they are insensitive to mechanical stimulation. The composite response curve of hydras tested on exposure to light was significantly depressed (Fig. 4b, 0 minutes) and quite similar to the response curves of animals habituated to mechanical stimulation seen previously. When the hydras were retested in light at successive 15-minute intervals, it was found that the sensitivity to mechanical stimulation was restored. After 30 minutes the animals had regained much of their former sensitivity to the stimulus (Fig. 4b, 30 minutes), and after 2 hours the response curve was not significantly different ($p > 0.10$) from the control levels. Thus, although light inhibits the response to mechanical agitation, the hydras adapt to the light stimulus. This process demonstrates the complex interaction of two factors, light and mechanical stimulation, in the behavior of *Hydra*.

The amount of dark adaptation determines the extent of the inhibition by light of contractions to shaking, as seen in the following experiment. A group of 15 hydras, dark-adapted for a predetermined number of hours, and a control group of 15 light-adapted hydras were placed on the rotator. The animals were given five shaking pulses at 130 rpm; the pulses were 1 minute apart and of 5-second duration. The average proportions of hydras contracting in a total of 60 animals tested in both the control group and the dark-adapted experimental group are plotted in Fig. 4c. It is seen that the degree of inhibition increases with increasing numbers of hours of dark adaptation and levels off to a plateau value at about twenty-four hours. Thus, there is evidence of very slow dark adaptation in *Hydra*.

Previous reports have shown that behavioral responses to light, such as migration, orientation, contraction, and changes in electrical activity, are selectively induced by the blue region of the spectrum. It is therefore of interest to determine the spectral sensitivity of inhibition by light of mechanically induced contractions. Groups of 30 *H. viridis* were dark adapted for 24 hours. Light of specific wavelengths was directed onto the animals, and their contractions to five successive pulses of 5 seconds of shaking at 130 rpm were determined. The light source used was a Bausch and Lomb 250-mm grating monochromator housing a 150-watt Sylvania Tru-Flector DFA bulb. The relative intensities of the light at the various wavelengths were recorded by a Clairex 602 photoconductive cell and milliammeter connected to a 200 to 235-volt dc-regulated power supply. The proportions of hydras contracting to mechanical stimulation when exposed to monochromatic light, corrected for differences in the relative intensities at the different wavelengths, are given in Fig. 4d. The mean proportions and 95 % confidence limits are plotted for wavelengths at 25-mμ intervals over the range of 375 to 650 mμ. These values may be compared with the corresponding quantities for the control animals which were dark adapted for 24 hours and tested in the dark. It is evident that inhibition is greatest in the blue region (400 to 475 mμ) and decreases above 500 mμ.

The properties of light and dark adaptation together with spectral sensitivity are characteristic of photosensitive processes in general. They are not exclusively attributable to specialized photoreceptors but are also found in many animals with diffuse sensitivity to light. Such properties are thus fundamental to photosensitive processes. North and Pantin (1958) observed the phenomenon of light and dark adaptation in the light response of the sea anemone *Metridium senile*. They found that the entire body surface of *Metridium* is photosensitive. Unilateral illumination produces local contractions of the parietal musculature, which causes the animal to bend toward the source of light. Maximum sensitivity occurs in the blue-green region of the spectrum at about five hundred millimicrons, but there is considerable variability in the response of different animals. Light adaptation is more rapid than in *Hydra*, the photosensitive system being virtually adapted in about ten minutes. Sea anemones possess no complex sense organs, and there is no evidence of local concentrations of pigment. Isolated mesenteries contract with local illumination even when anesthetized with magnesium chloride, which abolishes neuromuscular action, and this suggests that the muscular system itself is light sensitive (North and Pantin, 1958). North and Pantin concluded that the photoreceptors in *Metridium* are contained in the epitheliomuscular cells but are of microscopic cell-organell size or of ultramicroscopic size as in enzyme systems. They do not implicate the sensory nerve cells of the ectoderm.

The rate of dark adaptation found here in studies of *H. viridis* is much slower than that observed in several other animals. Hecht (1918 and 1919) showed a progressive shortening of the reaction time with increasing length of time in the darkness for *Mya* and *Ciona*. The functional relationship was that of a rectangular hyperbola for both animals. The process was complete in 45 minutes in *Mya* but continued in *Ciona* for 200 minutes. Steven (1950) found that dark adaptation in a larval lamprey was complete in 20 to 30 minutes, while North and Pantin (1958) recorded an increase in light sensitivity in *Metridium* up to 90 minutes of dark adaptation. The parameters of dark adaptation in such systems appear to be compatible with the kinetics of a bimolecular chemical reaction by which a photosensitive product is produced from two precursors (Hecht, 1927). The increase in sensitivity in darkness is thought to result from the accumulation of this product at the receptor surfaces. Considerably more precise measurement must be made with *Hydra* to ascertain whether or not such a model is applicable.

Several studies have shown that many invertebrate nerve cells are light sensitive, although some of the neurons involved have other functions in addition to photoreceptor properties. Prosser (1934) showed that the sixth abdominal ganglion of the crayfish is light sensitive. A small number of central neurons, which are also second-order interneurons for tactile stimuli, mediate the light response (Kennedy, 1958a and b). Microelectrode recordings indicate that light reduces the depolarization of these cells. Spectral sensitivity measurements show a maximum response near 500 mμ (Bruno and Kennedy, 1962). Von Fritsch (1911) and Scharrer (1928), working on *Phoxinus*, and Young (1935a and b), studying lampreys, have discovered photochemical effects in nerve cells. Some molluscan neurons containing carotenoid-protein complexes have also been shown to respond to light. The best-known examples are the visceral ganglion cells of *Aplysia* (Arvanitaki and Chalazonitis, 1961). While such cells are probably not functional photoreceptors, since they are well shielded from illumination, nevertheless they show on-off responses to light. From intracellular recordings it is seen that both primary inhibition and excitation occur: Inhibition is associated with hyperpolarization, and excitation with depolarization. The discharge patterns resulting from such processes are related to the degree of pigmentation.

Diadema and other light-sensitive echinoids give a variety of responses which Millot and Yoshida (1960a and b) ascribe to complex nervous interaction. In *Diadema* they describe shadow reflexes resulting from a reduction of ambient light. This is due not to a primary effect of light on the receptors, but rather to adaptation and inhibition in the animal's nervous system. The distribution of photosensitivity corresponds almost exactly with the distribution of nerve elements in the superficial nerve layer. In *Hydra* one may speculate on the possibility of photosensitivity of the nerve cells. Histological studies of McConnell (1932) and more recently Burnett and Diehl (1964) show concentrations of both sensory and ganglion cells in the subhypostome and the bases of the tentacles. Removal of the tentacles abolishes the contraction response to mechanical agitation (Rushforth, 1965a). Surgical experiments and studies of localized illumination have implicated the subhypostome in the light-induced contractions. There is, however, insufficient evidence to determine the photoreceptive structures in *Hydra*. Sensory cells in the epithelium of *Hydra*, with apical structures similar to modified cilia, have been observed with the electron microscope (Lentz, 1965). Yet light sensitivity of the epitheliomuscular cells, as in *Metridium*, has not been eliminated. We must be aware of the possibility of a photoreceptor biochemically, but not histologically, differentiated. The present studies do demonstrate behavioral inhibition in *Hydra* linking the two stimuli, light and mechanical agitation. However, the receptor structures and underlying physiological process remain to be determined.

EFFECTS OF CHEMICAL STIMULATION ON THE SENSITIVITY OF *HYDRA* TO MECHANICAL AGITATION

The experiments just described demonstrate the inhibitory effects of light on the contraction response to mechanical stimulation. Strong light stimulation, however, may also excite contractions in *Hydra* (Rushforth *et al.*, 1963). Contractions induced both by light and mechanical agitation may themselves be inhibited by a third factor, chemical stimulation. These effects are further examples of complex interactions among factors influencing the behavior of the animal.

The contraction response of *Hydra* to intermittent light stimulation is suppressed when the animal feeds on *Artemia salina*. Such inhibition is also produced by extracts of *Artemia* or the tripeptide reduced glutathione (GSH). Inhibitory effects of GSH on the light response have been studied in some detail for *H. pirardi* and several strains of *H. viridis* (Rushforth, 1965b). Procedures used to quantify this effect show that the modes of action of GSH are similar to those discovered for the mouth-opening response of *H. littoralis* (Lenhoff, 1961a and b and 1965), which are discussed in this volume. The inhibitory activity is dependent on (1) the concentration of GSH, (2) the pH of the medium, (3) the nutritional state of the hydra, and (4) its previous exposure to GSH. The animals adapt to 10^{-5} M GSH so that after approximately one hour the frequencies of light-induced contractions are restored to control levels. The adaptation to GSH is not due to degradation of the glutathione molecule but to processes occurring in the animal. Removal of GSH removes inhibition; therefore GSH must be constantly present for inhibition to take place. The molecular configuration of GSH is quite specific in its inhibitory effects. However, the sulfhydryl group is not essential for inhibition since the S-methyl analog of GSH is active. Analogs with sterically large groups substituted for the sulfhydryl moiety, such as oxidized glutathione and S-acetyl glutathione, have no inhibitory effect. These compounds reduce the inhibitory effect of GSH, which indicates competition for the GSH receptor. Excision of the tentacles significantly reduces inhibition of light-induced contractions by 10^{-5} M GSH but has little effect on the light response in the absence of GSH. This result suggests that the receptors for GSH are predominantly located in the tentacles.

The following studies were undertaken to investigate the inhibitory effects of GSH on the contraction response of *Hydra* to mechanical stimulation. Earlier observations with *H. pirardi* show that contractions induced by shaking are suppressed when the animal feeds on *Artemia salina* or is exposed to extracts of *Artemia*. Inhibition is also observed when GSH is present in the environment of the animal (Rushforth *et al.*, 1964). In *H. viridis* marked inhibition of mechanically induced contractions occurs with 10^{-5} M GSH. As seen in Fig. 5a, the response curve of hydras exposed to GSH at this concentration is considerably lower than that of control animals. The hydras gradually adapt to GSH so that the sensitivity to mechanical stimulation is restored (Fig. 5b). Adaptation to 10^{-5} M GSH takes up to 7 hours and is much slower than adaptation to light. The degree of inhibition depends on the concentration of the GSH. Response curves for *H. viridis* are systematically lowered as the concentration is increased over the range of 10^{-7} to 10^{-5} M GSH (Rushforth, 1965b). However, the decrease in sensitivity levels off, which suggests saturation of the glutathione receptors at higher concentrations. Both the initial inhibition and the course of adaptation to this stimulus are dependent on the concentration of GSH, as seen from the following experiment. Groups of 30 hydras were given five pulses of agitation of 2-second duration at 130 rpm on exposure to various concentrations of GSH. The proportions

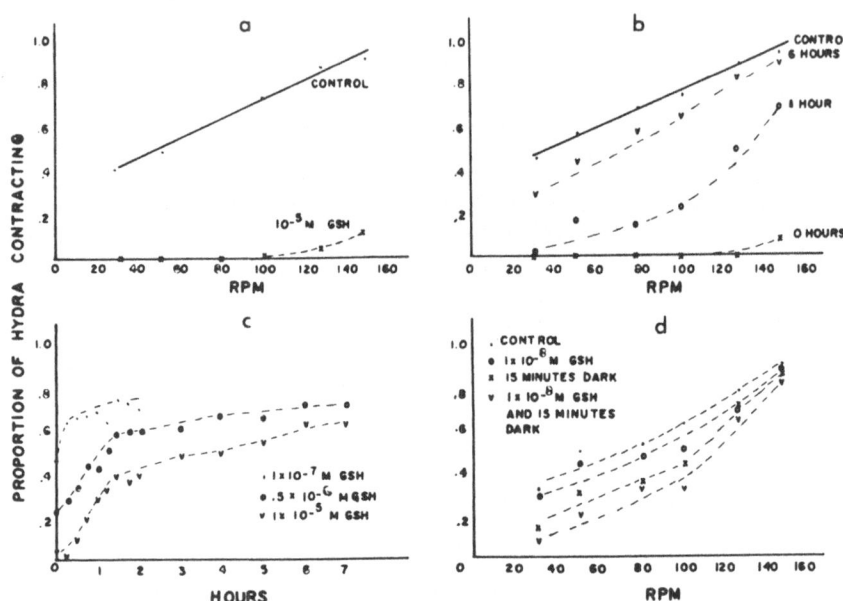

Fig. 5. Effect of reduced glutathione (GSH) on the sensitivity of *Hydra viridis* to mechanical stimulation. (a) *Broken line*, response curve for a group of 30 hydras exposed to 10^{-5} M GSH, and *solid line*, the response curve of a group of 30 control animals. (b) Response curves of a group of 30 hydras: *solid line* (control), before exposure to GSH; *broken lines*, immediately following exposure to 10^{-5} M GSH and after 1 and 6 hours' exposure. (c) Adaptation of hydras to various concentrations of GSH measured by changes in the sensitivity to mechanical stimulation. Each symbol represents the mean proportion of a group of 30 hydras contracting to five successive pulses of agitation of 2-second duration at 130 rpm when exposed to GSH at a given concentration. (d) Effect of joint stimulation by light and GSH on dark-adapted hydras: response curves for light-adapted hydras (control), light-adapted hydras exposed to GSH (1×10^{-8} M GSH), hydras dark-adapted for 15 minutes (15 minutes dark), and hydras dark-adapted for 15 minutes and exposed to GSH (1×10^{-8} M GSH and 15 minutes dark). All four groups were tested in ambient light and consisted of 30 animals.

of hydras contracting at various time periods following exposure to GSH are plotted in Fig. 5c. It is seen that, at 1×10^{-7} M GSH, restoration of sensitivity to mechanical agitation is complete within an hour, but adaptation to 1×10^{-5} M GSH is much longer, requiring 6 to 7 hours for completion.

The properties of this system are quite similar to those found for GSH inhibition of light-induced contractions. For example, experiments with analogs of GSH which have sterically large groups substituted for the sulfhydryl group show competitive effects like those found in studies of the light response. Both oxidized glutathione and S-acetyl glutathione reduce inhibition by GSH. The inhibitory effects of GSH are additive with those produced by light stimulation in dark-adapted hydra. For example, animals dark-adapted for only 15 minutes and exposed jointly to light and 1×10^{-8} M GSH have significantly ($p < 0.01$) reduced response levels to mechanical stimulation (Fig. 5d). The mean proportions of hydras contracting under joint stimulation are lower than those of dark-adapted animals subjected singly to 10^{-8} M GSH or light. Thus, there may be similarities in the underlying physiological processes involved in inhibition by light and GSH.

ELECTROPHYSIOLOGICAL STUDIES

The results of the behavioral investigations presented above illustrate responses in *Hydra* of considerable complexity. The animal responds to a chemical, mechanical, or photic stimulus. Such responses are modified, however, by joint exposure to stimuli of differing modalities. The response to intermittent mechanical stimulation satisfies several criteria used to characterize habituation operationally. However, reductions in the sensitivity to mechanical agitation of a comparable magnitude can be achieved using both light and GSH. In addition, for suitably chosen concentrations of GSH and lengths of time in the dark, the curves depicting adaptation to GSH and light, as measured by mechanically induced contractions, are equivalent to those describing extinction of the habituated response (Rushforth, *unpublished observations*). It is therefore necessary to study the physiological bases of such processes if distinctions are to be made between habituation and other forms of behavioral inhibition in *Hydra*. I should like to conclude by reporting some preliminary observations on electrophysiological correlates of some of these responses.

Passano and McCullough (1963) were the first to show that several distinct sets of electrical potentials could be recorded from intact, unstimulated *Hydra*. Such potentials are produced by spontaneously active coordinating systems, some of which are correlated with overt behavior of the animal. One pacemaker system produces relatively large electrical pulses which precede contractions of the longitudinal muscles of the ectoderm. These potentials were originally termed "contraction-burst" potentials (CB's), since Passano and McCullough (1964) reported that they usually occur in bursts. Using improved external electrodes (Josephson, 1967), long-term recordings have been made from essentially unrestrained animals of several species. Since many hydras give such potentials as single, widely spaced pulses rather than in bursts (Rushforth, 1966), we prefer to designate such potentials as *contraction pulses* (CP's).

Passano and McCullough (1965) propose that the conducting system for these potentials is the ectodermal nerve net. However, their large size, 30 mv recorded externally, suggests they are of epitheliomuscular origin rather than the product of the small and sparse nerve cells of this net. Indeed strong evidence for this is a recent discovery of a transepithelial potential in *Hydra* (Josephson and Macklin, 1967). Contraction pulses result from a change in this potential maintained across the body wall of the animal.

The most characteristic feature of the behavior of *Hydra* consists of

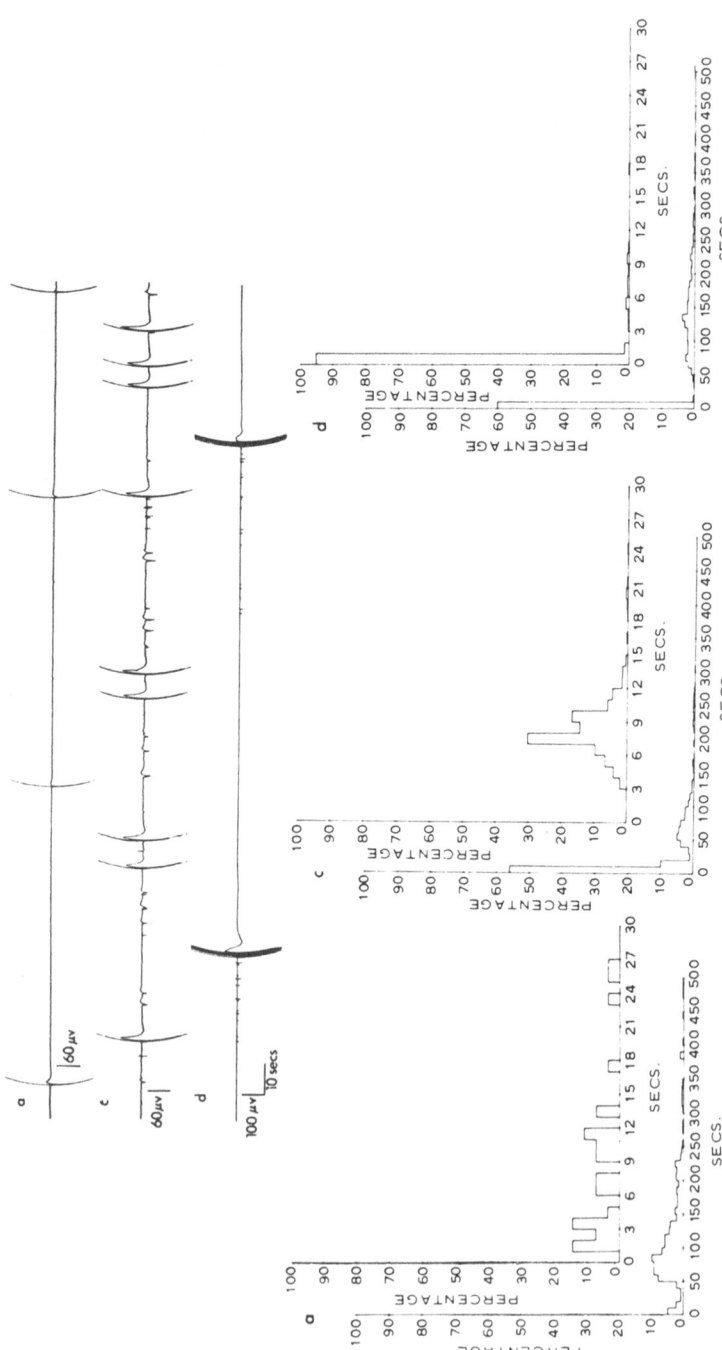

Fig. 6. Electrical records and interpulse histograms for contraction pulses of three species of *Hydra. Upper records*: (a) Contraction pulses from a *Hydra pseudoligactis*, (c) a *Hydra pirardi*, and (d) a *Hydra littoralis. Lower*: Percentage histograms of intervals between successive contraction pulses taken from 12-hour records of (a) a *Hydra pseudoligactis*, (c) a *Hydra pirardi*, and (d) a *Hydra littoralis*. The lower histograms of each pair depict percentage for all intervals in the 12-hour record; upper histograms depict percentage for intervals less than 30 seconds.

the periodic contractions of the body column. The contraction responses to light and mechanical stimuli are superimposed on this background of endogenous activity. Spontaneous contractions of unstimulated hydras of several species are currently under investigation. The temporal characteristics of such activity are being studied using 12-hour records of contraction pulses. Contraction pulse patterns for various species of *Hydra* are quite different. For example, differences are evident in the electrical records shown in Fig. 6. Most *Hydra pseudoligactis* give single contraction pulses similar to those in Fig. 6a. When the intervals between the successive pulses are plotted in the form of a percentage histogram with 10-second interval classes (see the lower histogram of Fig. 6a), we obtain a distribution with relatively few intervals less than 10 seconds. The percentage histogram with 1-second interval classes for intervals less than 30 seconds in length (upper histogram of Fig. 6a) shows no distinct modal class. In contrast, most *H. pirardi* give bursts of potentials. For the hydra whose record is shown in Fig. 6c, the bursts consist of two or three potentials, occasionally separated by single pulses. The percentage histogram for the intervals from a 12-hour record for this animal (lower c, Fig. 6) shows a large proportion of intervals in the class of 0 to 10 seconds. Such intervals constitute the short intervals occurring in bursts. There are relatively few intervals in the class of 20 to 30 seconds, but there is a secondary modal class from 40 to 50 seconds. Intervals falling in this last class are long intervals following or preceding bursts or intervals between single pulses. This bimodal histogram is quite characteristic of the patterned electrical activity of contractions in *H. pirardi*. The histogram with 1-second interval classes for intervals smaller than 30 seconds is unimodal, the modal class being 6 to 7 seconds. The bursts of potentials given by a *H. littoralis* (Fig. 6d) are quite different from those of *H. pirardi*. Bursts consisting of five or six pulses within 1 or 2 seconds occur at intervals sometimes several minutes in length. Not all *H. littoralis* show burst activity of this kind, however, since some animals of this species are quite similar to *H. pseudoligactis* in their nonburst character. Records taken from several *H. viridis* indicate that the contraction pulse pattern of this species corresponds quite closely to that of *H. pirardi*. We find, therefore, that the patterns of spontaneous contraction in *Hydra* are quite variable and are species specific.

Are the behavioral responses to external stimuli different for the various species? Earlier studies indicated that, while *H. pirardi* contract to mechanical and photic stimulation, *H. pseudoligactis* are relatively insensitive to such stimuli (Rushforth, 1965a). Recently we have discovered electrophysiological correlates of such differences in the light response of the two species. Figure 7 depicts part of an electrical recording taken from a *H. pseudoligactis*. The high gain used to show the small rhythmic potentials (about 30 to 40 μv) considerably distorts the larger contraction pulses (about 5 to 10 mv). The top record shows a single contraction pulse and a rhythmic potential within a time span of 4 minutes. When the animal is stimulated with light for 2 minutes (about 600 ft-c), followed by a 2-minute period of relative dark, the electrical activity of the hydra is considerably altered (Fig. 7b and c). In the 2-minute periods of strong light illumination there is a marked increase in the number of rhythmic potentials, while, in the 2 minutes of ambient light, bursts of contraction pulses are induced. After several intermittent exposures to 2 minutes of light the animal's contraction pacemaker system is excited (Fig. 7d) but is gradually restored to control levels (Fig. 7e). Thus, in *H. pseudoligactis* light inhibits contractions and their associated contraction pulses, while exciting rhythmic potentials. In contrast, in *H. pirardi* there is an increase in the number of contraction pulses during exposure to light. The normal burst pattern of this animal is changed from that seen in Fig. 8a and b to bursts containing more pulses, which occur exclusively during the 2-minute periods of light (Fig. 8c and d). Thus, in *Hydra* there are electrophysiological correlates of both excitatory and inhibitory effects of light.

Spontaneous contraction pulses and those induced by light or mechanical agitation in *H. pirardi* are inhibited when the animal feeds on *Artemia*

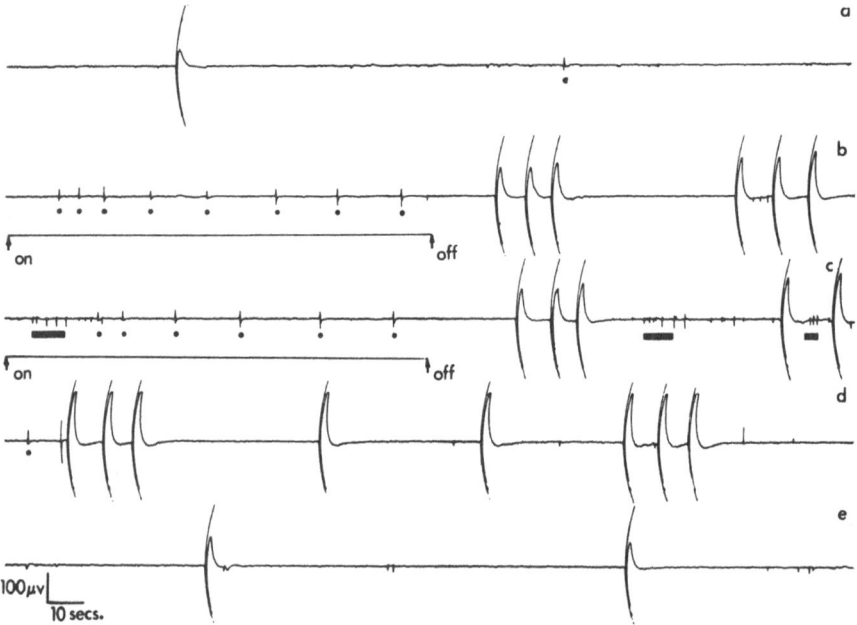

Fig. 7. Effect of light on the electrical activity of *Hydra pseudoligactis. Upper record*: Four-minute record of a hydra in ambient light; (b and c) consecutive 4-minute records of the animal exposed to 2 minutes of direct-light stimulation (on to off), followed by 2 minutes in ambient light; (d) record of the animal in ambient light following a regime of 10 successive light exposures of 2-minute duration each, followed by 2 minutes of ambient light (as in b and c); (e) 4-minute record of the hydra in ambient light taken 30 minutes after record d with no direct-light stimulation during this period. The closed circles represent rhythmic potentials, and the closed rectangles represent bursts of tentacle pulses.

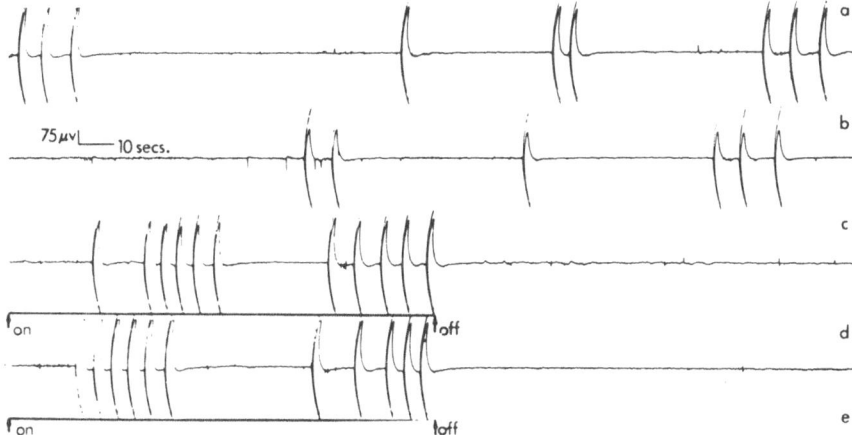

Fig. 8. Effect of light on the electrical activity of *Hydra pirardi*. (a and b) Records of a hydra in ambient light; (c and d) records of the animal exposed to periods of 2 minutes of direct-light stimulation (about 600 ft-c, on to off), followed by 2-minute periods of ambient light. The total record consists of consecutive periods of 4-minute duration each. Rhythmic potentials and tentacle pulses are not easily identified in this record.

or is exposed to extracts of *Artemia* or GSH. In Fig. 9 such inhibition of light-induced contraction pulses by 10^{-5} M GSH is shown. In the record shown in Fig. 9a we observe a burst of contraction pulses in a 2-minute period of light stimulation. When GSH is placed in the environment of the hydra, contraction pulses are inhibited but light-induced rhythmic potentials are not suppressed (Fig. 9b and c). Gradually the hydra adapts to the GSH, and the contraction pulses are restored, frequently as single pulses at first (Fig. 9d) and then as bursts (Fig. 9e). Thus, GSH inhibits electrical potentials associated with contractions in *Hydra*.

CONCLUSION

In this paper several examples of interaction among factors controlling the behavior of the freshwater *Hydra* have been given. The responses of the animal are superimposed on an endogenously active system so that contractions to external stimuli occur as modifications of spontaneously produced contraction patterns. *Hydra* contracts to photic and mechanical stimuli. Light also, however, has inhibitory effects on the animal and can be used to suppress contractions induced by mechanical agitation. Spontaneous contractions and those produced by light and mechanical agitation may themselves be inhibited by chemical stimulation. Such inhibition is quite specific to GSH and a single closely related analog, S-methyl glutathione. In addition, suppression of mechanically induced contractions is significantly greater when hydras are exposed jointly to light and GSH

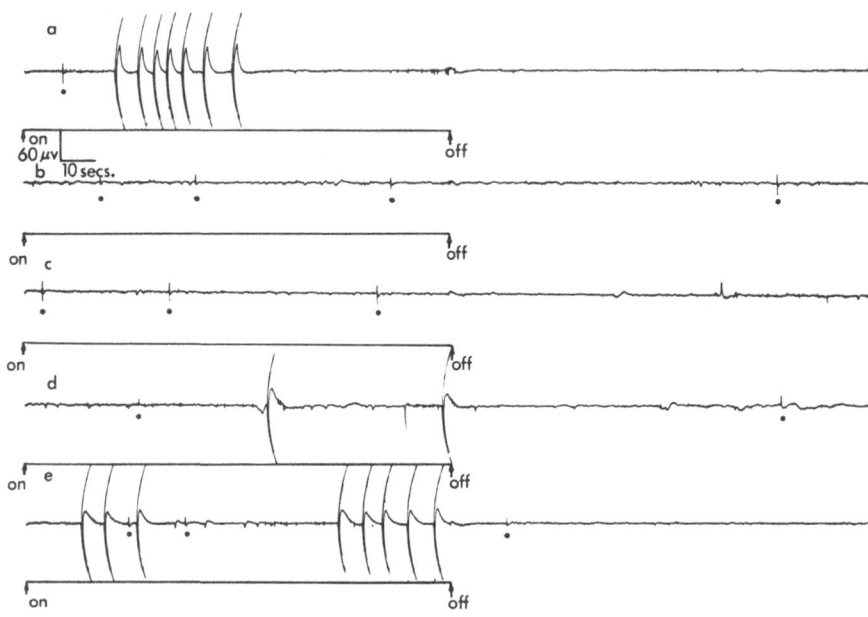

Fig. 9. Effect of light and GSH on the electrical activity of *Hydra pirardi*. (a) Record of a hydra exposed to 2 minutes of direct-light stimulation (about 600 ft-c, on to off), followed by 2 minutes of ambient light; (b and c) consecutive 4-minute records of the animal after exposure to 10^{-5} M GSH; (d) record of the animal after 40-minute exposure to 10^{-5} M GSH; (e) record taken 60 minutes after exposure of the animal to 10^{-5} M GSH. In b, c, d, and e the hydra was exposed to the same light regime as in a. The rhythmic pulses are designated by closed circles; tentacle pulses are not easily identified in this record.

than when either of these stimuli are administered alone. Such effects demonstrate the control of one receptor-effector system in *Hydra* by another. Although some of the properties of these systems have been elucidated, the physiological mechanisms involved appear complex and as yet largely unknown.

The response of *Hydra* to intermittent mechanical stimulation exhibits several characteristics compatible with a process of habituation. However, changes in the sensitivity of the animal to this stimulus similar to those resulting from repeated exposures can be induced both by light and chemical stimulation. It is clear that the physiological bases of such processes must be determined before distinctions are made between habituation and other forms of behavioral inhibitions in *Hydra*.

ACKNOWLEDGMENTS

Much of the research discussed in this paper was supported by grants from the National Institutes of Health. I wish to acknowledge with gratitude the assistance of Mr. Paul Bakunas, Mr. Otto Morgenstern, Miss Bonnie Edwards, and Miss Marilyn Wiley. I wish to thank Dr. R. K. Josephson and Dr. E. Robson for reading an earlier version of this paper and suggesting many improvements, some of which are incorporated in this published version.

REFERENCES

Arvanitaki, A., and N. Chalazonitis, Excitatory and inhibitory processes initiated by light and infra-red radiations in single identifiable nerve cells (giant ganglion cells of *Aplysia*), in E. Florey, ed., *Nervous Inhibition*. New York: Pergamon Press, 1961, pp. 194–231.

Bruno, M. S., and D. Kennedy, Spectral sensitivity of photoreceptor neurons in the sixth ganglion of the crayfish. *Comp. Biochem. Physiol.*, 1962, **6**, 41–46.

Bullock, T. H., ed., Simple systems for the study of learning mechanisms. *Neurosci. Res. Program Bull.*, 1966, **4**, 105–233.

Burke, D. S., and N. B. Rushforth, Inhibition of pacemaker activity in *Hydra*. *Amer. Zool.*, 1966, **6**, 525.

Burnett, A. L., and N. A. Diehl, The nervous system of *Hydra*, I: Types and distribution of nerve elements. *J. Exp. Zool.*, 1964, **157**, 217–226.

Feldman, M., and H. M. Lenhoff, Phototaxis in *Hydra littoralis*: Rate studies and localization of the "photoreceptor." *Anat. Rec.*, 1960, **137**, 354–355.

Fritsch, K. von, Beitrage zur physiologie der pigmentzellen in der Fischant. *Pflüng. Arch. ges. Physiol.*, 1911, **138**, 319–343.

Haug, G., Die Lichtreaktionen der Hydran (*Chlorohydra viridissima* und *Pelmatohydra oligactis*). *Z. vergleich. Physiol.*, 1933, **19**, 246–303.

Hecht, S., The photic sensitivity of *Ciona intestinalis*. *J. Gen. Physiol.*, 1918, **1**, 47–66.

Hecht, S., Sensory equilibrium and dark adaptation in *Mya arenaria*. *J. Gen. Physiol.*, 1919, **1**, 545–58.

Hecht, S., The kinetics of dark adaptation. *J. Gen. Physiol.*, 1927, **10**, 781–809.

Josephson, R. K., Conduction and contraction in the column of hydra. *J. Exp. Biol.*, 1967 (in press).

Josephson, R. K., and M. Macklin, Transepithelial potentials in hydra. *Science*, 1967, **156**, 1629–1631.

Kennedy, D., Responses from the crayfish caudal photoreceptor. *Amer. J. Ophthalmol.*, 1958a, **46**, 19–26.

Kennedy, D., Electrical activity in a "primitive" photoreceptor. *Ann. N.Y. Acad. Sci.*, 1958b, **74**, 329–336.

Krinsky, N. I., and H. M. Lenhoff, Some carotenoids in *Hydra*. *Comp. Biochem. Physiol.*, 1965, **16**, 189–198.

Lenhoff, H. M., Activation of the feeding reflex in *Hydra littoralis*, in H. M. Lenhoff and W. F. Loomis, eds., *The Biology of Hydra and of Some Other Coelenterates*. Coral Gables, Fla.: University of Miami Press, 1961a, pp. 203–232.

Lenhoff, H. M., Activation of the feeding reflex in *Hydra littoralis*, I: Role played by reduced glutathione, and quantitative assay of the feeding reflex. *J. Gen. Physiol.*, 1961b, **45**, 331–344.

Lenhoff, H. M., Some physiochemical aspects of the macro- and micro-environments surrounding *Hydra* during activation of their feeding behavior, *Amer. Zool.*, 1965, **5**, 515–524.

Lentz, T. L., and R. L. Barnett, Fine structure of the nervous system of *Hydra*. *Amer. Zool.*, 1965, **5**, 341–356.

Loomis, W. F., and H. M. Lenhoff, Growth and sexual differentiation of *Hydra* in mass culture. *J. Exp. Zool.*, 1956, **132**, 55–568.

Mackie, G. O., Conduction in the nerve-free epithelia of Siphonophores. *Amer. Zool.*, 1965, **5**, 439–53.

McConnell, C. H., The development of the ectodermal nerve net in the buds of *Hydra*. *Quart. J. Micr. Sci.*, 1932, **75**, 495–509.

Millott, N., and M. Yoshida, The shadow reaction of *Diadema antillarum Philippi*, 1: The spine response and its relation to the stimulus. *J. Exp. Biol.*, 1960a, **32**, 363–374.

Millott, N., and M. Yoshida, The shadow reaction of *Diadema antillarum Philippi*, 2: Inhibition by light. *J. Exp. Biol.*, 1960b, **32**, 376–389.

North, W. J., and C. F. A. Pantin, Sensitivity to light in the sea anemone *Metridium senile* (L): adaptation and action spectra. *Proc. Roy. Soc. B*, 1958, **148**, 385–396.

Pantin, C. F. A., The nerve-net of the Actinoza, I: Facilitation. *J. Exp. Biol.*, 1935, **12**, 119–138.

Passano, L. M., and C. B. McCullough, The light response and the rhythmic potentials of *Hydra. Proc. Nat. Acad. Sci.*, 1962, **48**, 376–382.

Passano, L. M., and C. B. McCullough, Pacemaker hierarchies controlling the behavior of hydras. *Nature*, 1963, **199**, 1174–1175.

Passano, L. M., and C. B. McCullough, Co-ordinating systems and behavior in *Hydra*, I: Pacemaker system of the periodic contractions. *J. Exp. Biol.*, 1964, **41**, 643–664.

Passano, L. M., and C. B. McCullough, Co-ordinating systems and behavior in *Hydra*, II: The rhythmic potential system. *J. Exp. Biol.*, 1965, **42**, 205–231.

Prosser, C. L., Action potentials in the nervous system of the crayfish. *J. Cell. Comp. Physiol.*, 1934, **4**, 363–377.

Ross, D. M., The behavior of sessile coelenterates in relation to some conditioning experiments. *Anim. Behav., Suppl.*, 1965, **1**, 43–55.

Rushforth, N. B., Behavioral studies of the coelenterate *Hydra pirardi* Brien. *Anim. Behav., Suppl.*, 1965a, **1**, 30–42.

Rushforth, N. B., Inhibition of contraction responses of *Hydra. Amer. Zool.*, 1965b, **5**, 505–513.

Rushforth, N. B., An analysis of spontaneous contraction pulse patterns in *Hydra. Amer. Zool.*, 1966, **6**, 524.

Rushforth, N. B., A. L. Burnett, and R. Maynard, Behavior in *Hydra*: contraction responses of *Hydra pirardi* to mechanical and light stimuli. *Science*, 1963, **139**, 760–761.

Rushforth, N. B., I. T. Krohn, and L. K. Brown, Behavior in *Hydra*: inhibition of the contraction responses of *Hydra pirardi. Science*, 1964, **145**, 602–604.

Scharrer, E., Die lichtempfindlichkeit blinder elritzen. *Z. vergleich. Physiol.*, 1928, **7**, 1–38.

Singer, R. H., N. B. Rushforth, and A. L. Burnett, The photodynamic action of light on hydra. *J. Exp. Zool.*, 1963, **154**, 169–174.

Steven, D. M., Some properties of the photoreceptors of the brook lamprey. *J. Exp. Biol.*, 1950, **27**, 350–364.

Thompson, R. F., and W. A. Spencer, Habituation: a model for the study of the neuronal substrates of behavior. *Psychol. Rev.*, 1966, **73**, 16–43.

Trembley, A., Mémoires pour servir à l'histoire d'un genre de polypes d'eau douce à bras en forme de cornes. Leyden, The Netherlands: J. and H. Verbeck, 1744.

Wagner, G., On some movements and reactions of *Hydra. Quart. J. Microscop. Sci.*, 1904, **48**, 585–622.

Wilson, E. B., The heliotropism of *Hydra. Amer. Naturalist*, 1891, **25**, 414–433.

Young, J. Z., The photoreceptors of lampreys, I: Light-sensitive fibers in the lateral line nerves. *J. Exp. Biol.*, 1935a, **12**, 229–238.

Young, J. Z., The photoreceptors of lampreys, II: The functions of the pineal complexes. *J. Exp. Biol.*, 1935b, **12**, 254–270.

Chapter 24

Annelids and Learning: A Critical Review[1]

Stanley C. Ratner[2]

Michigan State University
East Lansing, Michigan

The study of annelid learning appears to be moving from its second infancy. The brief flurry of work about fifty years ago seemed to establish clear evidence of annelid learning for a number of species. But the annelids and problems associated with interpretation of their learning were then retired into brief statements in general texts that alluded to "the facts." The move from the second infancy of the study of annelid learning began after a long delay when investigators began applying more stringent definitions of learning and requiring more precise specification of conditions associated with learning. Thus, the literature on annelid learning now contains more than 50 references and the repertoire of an infant probably contains more than 50 units. But the problem is the degree of coherence and development of these units.

In the present chapter I shall discuss some of the units in the second infancy of annelid learning. These will include: a description of some major reference works on the behavior and learning of annelids; a review and critical examination of recent work on maze and alley learning, classical conditioning, and habituation; and the presentation of some characteristics of habituation that are derived in a large measure from studies of habituation of annelids.

SOME SOURCE MATERIAL ON ANNELIDS AND LEARNING

This section is designed to give an overview of some of the general sources presently available for the student of behavior and learning in annelids. The section is not meant to be exhaustive. Rather it is organized into three parts with each including materials that are relevant for investigators interested in the study of annelids. The first section deals with general treatments of each class of annelids; the second deals with specific reviews of behavior and learning in annelids; the third section deals with

[1]Some of the research reported in this chapter was supported by a research grant GB-917 from the National Science Foundation.
[2]Present address: Beloit College, Beloit, Wisconsin.

sources that give general information on the handling, maintenance, and testing of the worms. Each section is organized chronologically.

General Reviews of Biology and Behavior

Background information on the structure and function of the nervous system of annelids was recently enriched by the publication of Bullock and Horridge's (1965) treatise on invertebrates. They critically review and summarize the literature in more than 120 pages and pay attention to comparative analyses both among and within the three major classes of annelids, polychaetes, oligochaetes, and leeches. Behavior is treated in detail in terms of movement patterns, although learning is only briefly discussed.

Laverack (1963), Dales (1963), and Mann (1962) have also published recent reviews. Laverack's book, *The Physiology of Earthworms*, is highly readable, somewhat comparative, and includes a chapter on behavior and sensory processes. The material on learning is confined to several pages in which the early T-maze studies of Yerkes and Heck provide the main focus. Dales (1963) treats annelids in general and is particularly attentive to classification and evolution. Dales' chapter on sensory process and behavior deals primarily with polychaetes and oligochaetes with a treatment of learning very similar to Laverack's (1963). Mann's book, *Leeches* (*Hirudinea*), published in 1962, fills the gap on biological aspects of this class of annelids. Mann also considers classification and general biological processes with one chapter on senses and behavior. Habituation, orientation, and effect of hunger are discussed briefly in this chapter.

Traité de Zoologie, edited by Grassé (1959), also contains an excellent section on biology and behavior of annelids. Special attention is paid to neural function and the ablation experiments that have been used to investigate behavior and neural function. But before the early 1960's and the material that is included by Grassé, the standard reference on oligochaetes was Stephenson's book published in 1930. Obviously, this is still an important and useful treatment of the biology and behavior of oligochaetes. Stephenson includes chapters and exhaustive references on sensory processes, behavior, and learning. Finally, Darwin's study (1881), an analysis of earthworms and their ecology, must be noted. A number of his observations still stand and are cited.

Reviews of Annelid Learning

The publication in 1965 of the report of a symposium on Learning and Associated Phenomena in Invertebrates contains three sections on annelid learning. Arbit (1965) discusses a number of studies of T-maze learning of annelids and reports attempts, that failed, to modify this learning by drugs that affect neural processes. Clark (1965) cites more than five studies of learning using several species of polychaetes and discusses a number of these studies and their results in detail. Studies of habituation and T-maze behavior are included. Ratner (1965) reviews 15 studies of classical conditioning. Particular attention is paid to critical analyses of how these studies conform to classical conditioning procedures and their congruence with findings on classical conditioning with other species.

Jacobson (1963) reports an exhaustive review of studies of annelid learning up to 1961. Approximately 30 studies are noted and classified by the learning process under study, i.e., habituation, classical conditioning, and instrumental learning. Thorpe (1956) does a similar but less detailed job with the literature available to him. He deals with annelid learning in several pages with a classification of habituation and association learning.

Warden, Jenkins, and Warner (1941) and Maier and Schneirla (1935) give detailed discussion of sensory processes, discriminations, and learning of annelids. Warden *et al.* (1941) include more than 40 references in their discussion of annelid behavior, including some that deal with earthworm communication. Maier and Schneirla (1935) give a similar but more abbreviated treatment of most of the same topics, including discussions of the two available, so-called, conditioning studies and the T-maze studies of Yerkes and Heck.

Handling, Maintenance, etc.

A compendium on culture methods for invertebrates, originally published in 1938, was recently reprinted and contains 20 pages on culture methods for various species of annelids (Lutz *et al.*, 1959). Barrett (1947) has written an entire book on the maintenance of earthworms and ecological factors associated with their survival. Barrett's book contains material on relevant soil conditions, acidity, etc. Ratner and Gardner (1967) have summarized a great deal of this information in a publication designed to aid the student of behavior in studies of the earthworm, *Lumbricus terrestris*. Methods are included for studies of behavior in semi-natural habitats and for laboratory studies of behavior modification, particularly habituation.

CRITICAL ANALYSIS OF ANNELID LEARNING

With few exceptions, reviews of annelid learning have been generous in conclusions about the adequacies of the demonstrations of such learning. I believe the reviewers have been concerned primarily with substantiating the assumption that such learning occurs. Thus, they have deemphasized examining the studies critically. I share the assumption that annelids learn, so that in the present section we shall inquire about the characteristics and adequacies of the demonstrations of the processes.

We shall examine adequacies in the *reliability* and *validity* of the learning demonstrations. Reliability can be roughly assessed concerning the consistency of the behavior patterns within subjects by comparing behavior in one session with that in another, and concerning the consistency between experiments by comparing the results of comparable procedures. Validity, as usual, is a more difficult assessment. We shall approach validity through *concept or construct validity*. That is, we shall ask if the procedures used are analogous to those used in other learning situations. Do the behavior changes conform to those found in other learning situations and do the variables that have been found to be relevant have comparable effects in annelid learning? In addition, validity will be assessed by the congruence of what is already known about annelids with what is concluded from the studies of their learning.

The largest and most complex body of literature deals with maze and alley learning of annelids. This will be treated first. Classical conditioning will be treated next, and the last part of this section will deal with habituation. The reviews will not be exhaustive; rather, representative studies and processes will be selected for analysis for each class of annelids and each learning procedure.

Maze and Alley Learning

Experimenters have used species of annelids in a number of different maze and alley learning situations. The situations are difficult to classify because it is often not clear what aspects of the situation were essential for learning, nor is it clear whether learning was demonstrated. In addition, failures to find learning may be important but are only rarely reported. In the context of this discussion through analogy with learning situations used with mammals, a maze is a labyrinth with several alternative pathways, and an alley is a straight tube with only one pathway.

In general, maze and alley procedures have involved observation of *approach*, or *forward-going*, *response* of annelids. Within this grouping, experimenters have used three types of procedures and they have met with varying amounts of success.

Evans' (1963b and 1966) studies with nereid polychaetes illustrate one procedure that in mammalian learning is called *passive-avoidance learning*. In this case, forward-going movements through an alley are inhibited by shocking worms at the end of the alley. Retention (remembering) of this inhibition follows a relatively regular course, with zero retention after 24 hours. The replication of the learning procedure with other species and the regularity of the retention function provide the main evidence for the reliability and validity of this procedure with polychaetes. However, a review of the results of control groups in alley studies by Ratner (1964) and Reynierse and Ratner (1964) shows comparable results. Forward-going responses are gradually inhibited in earthworms, *L, terrestris*, that move toward an open area at the end of the alley. Other groups moving to mossy areas, like polychaetes approaching an open-ended tube without shock, continue approaching these areas. Unlike the results for the polychaetes, the results for *L. terrestris* show retention of the inhibited response for at least 24 hours.

A second procedure using approach responses involves studies of approach to a goal area preceded by, or paired with, the elicitation of *defensive responses* to prodding, electric shock, light, or acid. This is the most frequently used procedure and comes directly from Yerkes' T-maze study (1912). The low level of reliability reported by Yerkes with species of oligochaetes has been repeatedly noted. Datta (1962, p. 533) in a carefully conducted study reports: "Probability of correct initial choice grows within sessions and falls to chance in the interval between sessions . . . spacing of trials to the extent of 25 minutes prevents the development of better-than-chance probability of correct choice . . . ." Datta notes systematic changes toward preference for one arm over another that were resistant to change in the location of the positive or negative stimuli. A recent study with *L. terrestris* reported by Kirk (1966) verifies the difficulty of studying the learning of approach when it is associated with defensive responses. In this case an alley was used, but a shock grid was placed at the starting portion of the alley. No approach learning was found and inhibition of approach was noted. Could this be another case of the learning of inhibition of approach?

Results of some studies suggest the possibility of approach learning associated with defensive responses in oligochaetes. Robinson (1953) and Yerkes (1912) report some increase in correct responding. But these investigators and others prodded their worms. It still seems appropriate to conclude that in these cases "learning the maze by worms may be accounted for as the learning by the experimenter to guide the movements of a flexible object, the worm" (Ratner, 1964, p. 31).

In addition, another line of information fails to articulate with observations of oligochaetes learning to go to one side of a maze (place responding). No evidence exists for the presence of receptors to indicate to the worm the direction of its turn. However, statocysts are reported for some polychaetes (Bullock and Horridge, 1965, p. 749), and ablation experiments suggest their function in the movement and orientation of these polychaetes.

Approach associated with defensive responses has been reliably

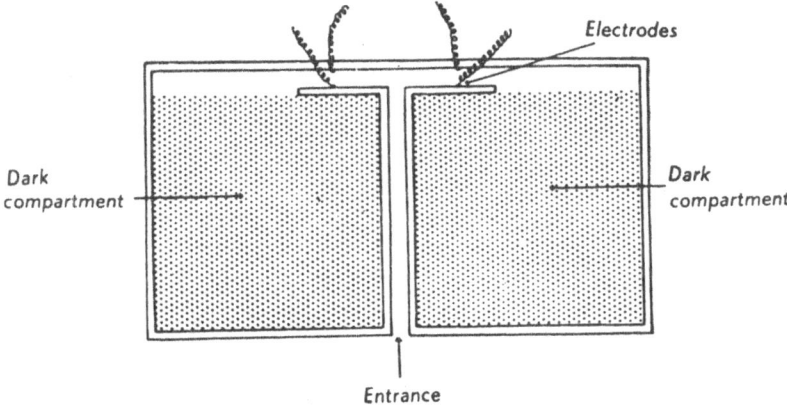

Fig. 1. A T maze for learning with associated defensive responses for polychaetes (from Evans, 1963b).

demonstrated with several species of nereid polychaetes (Evans, 1963a and b, and Flint, 1965). Figure 1 shows a diagram of the apparatus used in these studies. Evans (1963a and b) showed that approach to the correct arm must be followed by a several-minute period in a darkened chamber or else learning to approach does not occur reliably. Rather, inhibition appears, and we have another instance of learning to remain still. This procedure with polychaetes has general validity, but most of the variables known to affect learning have not yet been investigated in this situation. For example, observations of extinction, retention, and intertrial interval effects have not been reported.

The third procedure used to study maze and alley learning with annelids involves approach to a goal without eliciting defensive responses. Experimenters have used this procedure with alleys and T mazes for species of polychaetes and oligochaetes. Copeland (1930) studied one specimen of nereid polychaetes in a situation in which the worm approached the end of the tube and got a piece of clam. Copeland and Brown (1934) showed a similar procedure in a more complex situation. The reliability and validity of these procedures as demonstrations of alley learning are weak because of the small numbers of worms tested, the ambiguities of the measures, and the absence of important control procedures. But the procedures and results have validity in their congruence with things known about the behavior of polychaetes.

At least three studies of approach to a goal (without defensive responses) that have been reported used *L. terrestris* in alleys (Ratner, 1964; Reynierse and Ratner, 1964; and Kirk, 1966). Viewed together, these studies suggest low reliability and validity of this procedure with oligochaetes. In each of these studies the worms moved down an alley to approach a moss-filled goal box where they stayed for at least several minutes. The experimenters recorded the time required for the worm to traverse the tube and enter the box. Ratner (1964) and Reynierse and Ratner (1964) report decreasing response time for worms in this situation, but Kirk (1966) does not find this nor did Thompson (*personal communication*) in two attempts to replicate the Ratner alley procedure. In construct validity, the situation looks appropriate for *L. terrestris*, but studies by Ratner (1964) and Reynierse and Ratner (1964) of extinction of the learned alley response show unusually rapid extinction and unusually rapid relearning of the alley response by *L. terrestris*.

Datta (1962) studied approach to a goal (without defensive responses) using a T maze and found statistically significant changes in choices of the correct arm of the maze, but not all of the groups of worms, *L. terrestris*, showed these changes and for those that did the performance rose only to about 70% correct choice.

Summary of Maze and Alley Learning

Three types of maze and alley learning situations were identified and the adequacy of each as a preparation for study of annelid learning was evaluated. *Passive avoidance* when approach responses in an alley are inhibited by electric shock or light at the end of the alley has been demonstrated with polychaetes (Evans, 1963b) and oligochaetes (Ratner, 1964 and Reynierse and Ratner, 1964). The method appears to have greater validity and reliability with nereid polychaetes (Evans, 1963a and b) than it does with oligochaetes, although aspects of the validity of this situation with polychaetes require testing. *Approach learning associated with defensive responses*, such as the preparation used by Yerkes (1912) with the T maze, has been most frequently used with oligochaetes. But the results suggest that this is a weak preparation in both validity and reliability. A third procedure using mazes and alleys involves *approach responses to a goal area without associated defensive responses*. In this case, the worm moves through an alley or T maze to a goal area without shock, prodding, or other such stimulation. Again the data suggest this is a weak preparation for the oligochaetes, particularly *L. terrestris*. That is, as it has been used it is quite unreliable as a learning preparation.

Miscellaneous Studies of Approach Responses

Several other experimental procedures have been used in the study of approach and avoidance behavior of annelids, particularly oligochaetes. Krivanek (1956) attempted to study retention of a learned turning response with *L. terrestris*. He held a glass plate in front of the worm and forced it to turn to the right or to the left. The identification of the appropriate response was difficult and tests for retention were strongly confounded by extinction processes. Malek is reported (Jacobson, 1963, p. 90) to have used Darwin's test for leaf pulling with *Lumbricus herculeous*. The worms approached and pulled leaves of different shapes toward the burrow.

The weakness of mazes as a preparation for the study of learning by oligochaetes makes the several studies of latent learning using these species particularly difficult to interpret. In addition, these studies have met with indifferent successes (Bharucha-Reid, 1956, and James and Woodruff, 1965). No reports of approach learning by species of leeches have been found.

Classical Conditioning

Ratner (1965) has recently made a critical review of classical, Pavlovian conditioning studies that used annelids. Classical conditioning appears to be adequate in reliability and validity using oligochaetes in the Ratner-Miller conditioning ring. This procedure uses a bright light as the unconditioned stimulus and a low-level vibratory stimulus as the conditioned stimulus. The response of the annelid to the bright light is rising and withdrawing the anterior portions of its body, and a modified version of this response becomes elicited by the conditioned stimulus after 30 to 40 conditioning trials. Figure 2 shows a diagrammatic representation of this conditioning apparatus.

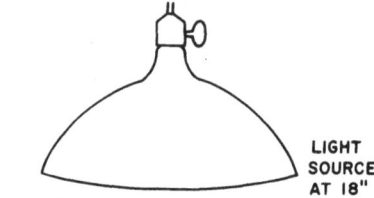

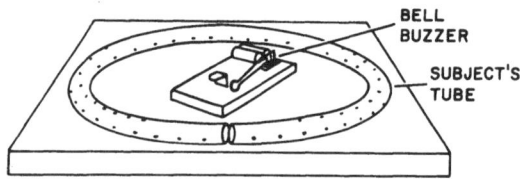

Fig. 2. Conditioning ring for oligochaetes (from Ratner
and Miller, 1959a).

A number of independent replications of the results of the conditioning
procedure have been published. In addition, a number of variables and
processes, including intertrial interval, trials, interstimulus interval,
extinction, and partial reinforcement effect, suggest the validity of classical
conditioning with annelids, particularly oligochaetes.

Some of the investigators who have studied conditioning with annelids
are Raabe (1939), Ratner and Miller (1959a and b), Wyers, Peeke, and
Herz (1964), and Herz, Peeke, and Wyers (1964). The Herz et al. (1964)
study replicated the findings of Ratner and Miller (1959a) with L. terrestris
and showed that the temperature of the worm, 23 to 25°C versus 10 to
12°C, significantly affected the level of conditioning. As expected, the
lower temperature was associated with a lower level of conditioning.

No studies that can be interpreted as studies of classical conditioning
have been found for species of leeches. One report of behavior change with
two specimens of the polychaetes, Hyroides dianthus, is given by Ada
Yerkes (1906). In this case, withdrawal responses to a touch were elicited by
a previously habituated stimulus, a shadow. This preparation can be regarded as a weak one in view of the low level of reliability and the indeter-
minate validity.

Ratner (1965) has noted that one of several important control procedures necessary to establish conditioning is the assessment of the base
rate occurrence of the response being conditioned. This is particularly acute
with studies of invertebrates because their response repertoire is limited
and unconditioned responses to an unidentified class of stimuli may imitate
conditioned responses. Thus, the reliability of the preparation is reduced.

TABLE 1

Medians of frequencies of forward and backward responses based on 120 trials for a base-rate control group and a light-stimulated group (Ratner and Gardner, in preparation).

Group	Freq. forward	Freq. backward
Base rate	28	26
Stimulated	17	70

Ratner and Gardner recently studied the movements of the earthworm, *L. terrestris*, which were stimulated by light pulses, the usual unconditioned stimulus. The movements of light-stimulated worms were compared with movements of base-rate control worms. Table 1 shows that both groups made forward and backward responses, but only the backward responses differentiated the movements of the light-stimulated group from those of the base-rate control group that had been observed "as if the light had gone on." These results further suggest the importance of the base-rate control group and suggest that the conditioned response in the Ratner–Miller ring probably consists only of the backward responses. Further evidence of the base rate of responding above zero is shown by Peeke, Herz, and Wyers (1965), who observed *L. terrestris* before, during, and after a period of stimulation and found that a small but consistent amount of responding occurred both before and for some period after periods of stimulation. Thus, experimental study of conditioning with *L. terrestris* requires careful specification of responses in order to maintain a reliable and valid preparation. The extent to which this "noise" may enter the conditioning of polychaetes has not yet been established.

Habituation

Experimenters have studied all three classes of annelids for the effects of repeated stimulation on their responses to this stimulation. These studies, studies of habituation, indicate that annelids are particularly useful for the analysis of habituation. In addition, habituation appears to be a particularly interesting and important form of behavior modification that can be considered a form of learning (Ratner and Denny, 1964, p. 231).

The classes of stimuli that experimenters have used in studies of habituation include photic, mechanical, thermal, electric, and chemical. Jacobson (1963), Mann (1962), Clark (1965), and Thorpe (1956) review some of the results of studies of habituation with species of annelids.

Clark's study (1960a) illustrates aspects of the procedures and findings of habituation using nereid polychaetes. *Nereis pelagica*, e.g., that live in glass tubes, respond to sudden stimulation by partial or complete withdrawal into the tubes in which they live. Repeated presentations of stimuli such as shock, moving shadow, and decrease in light intensity lead to a regular reduction in the frequency of withdrawal, until withdrawal no longer occurs when the stimulus is presented. A number of replications of these findings have been made with other species of polychaetes. However, Clark (1965, p. 90) identifies a class of stimuli, a touch at either end, whose presentation does not lead to habituation with carnivorous polychaetes. Clark suggests that in this case the responses are feeding responses and therefore do not habituate. These findings will be discussed below in the section on characteristics of habituation.

The validity of habituation of responses for species of polychaetes appears to be sound. Jacobson (1963, p. 84) and Clark (1960a and b) discuss findings that show that a number of variables, such as stimulus intensity and stimulus duration, have expected effects on the rate of habituation. In addition, the nature of the process is consistent with other things known about the species studied (Clark, 1960b).

Studies of habituation with oligochaetes have concentrated primarily on the earthworm, *L. terrestris*. Kuenzer (1958) reports detailed studies of habituation of the intact worm and studies of habituation and spread of habituation from single segments of the worm. He studied habituation of withdrawal responses with thermal, mechanical, and electrical stimuli. All of the studies of classical conditioning with *L. terrestris* (see above) include groups of worms that receive repeated stimulation using only one of the stimuli that are paired for the conditioned groups. Ratner and Stein (1965) also studied habituation to photic stimulation with *L. terrestris* and Gardner (1966) studied habituation, overhabituation, and retention of habituation of withdrawal responses to mechanical, vibratory stimulation.

The reliability and validity of habituation with oligochaetes appear very adequate, particularly when mechanical stimuli are used. Gardner (1966), using the apparatus shown in Fig. 3, found positive correlations of greater than 0.90 in repeated sessions of habituation of withdrawal responses of *L. terrestris* with mechanical stimulation. The consistency of findings from study to study also suggests the high degree of reliability of this preparation for the study of habituation.

The validity of this preparation is suggested by findings on the effect of variables such as intertrial interval (Ratner and Stein, 1965), stimulus intensity (Morgan, Ratner, and Denny, 1965), and retention intervals (Gardner, 1966). Short intertrial intervals reduce frequency of responses; intensity of a photic stimulus has a positive relation both to the initial level of responding and to the rate of habituation; and retention of a habituated response proceeds in the manner of the classic remembering curve found with other species.

One older and very extensive study (Gee, 1912) and one relatively recent study (Kaiser, 1954) have shown habituation in species of leeches. Gee (1912) observed that mechanical stimulation gradually failed to stop undulating movements of the leech *Dina microstoma*. He also studied the

Fig. 3. Habituation apparatus with vibratory stimulation for oligochaetes. Note the components of the withdrawal response in which the anterior portions are withdrawn and the posterior flattened. The hook of the anterior few segments is not visible. (From Gardner, 1966.)

effects of shadow on withdrawal responses and cessation of undulation of the leech and found that this stimulus led to more rapid habituation than mechanical stimulation. Responses to mechanical stimulation habituated in 9 to 12 trials for five specimens and habituation continued for 10 more presentations of the stimulus. Kaiser (1954) also studied habituation of the cessation of undulatory movements of *Haemopis* using shadow, a decrease in the level of illumination, as the stimulus. A very systematic relation was found between the number of presentations of the shadow and the duration of the cessation of undulation. In addition, the less distinct the shadow was; the more rapid, the habituation. These studies suggest that habituation is appropriately studied with leeches, and the small amount of study of their behavior modification further recommends them for work.

Characteristics of Habituation

Habituation has been studied with all the major classes of annelids and adequate preparations exist for such study with each class. In so far as habituation is a valid class of learning, and it seems to be, detailed study of habituation and possible chemical changes associated with it may be advisable.

However, the characteristics of habituation have not yet been fully specified, although Harris (1943), Thorpe (1956), and most recently Thompson and Spencer (1966) have dealt with the question. Review of the studies of habituation, particularly with annelids, and consideration of a number of results from studies in our laboratory suggest that habituation has a number of characteristics that are important for further analysis of the process. These characteristics are important both because of their direct relation to the process of habituation and because of their implications for control procedures in studies of habituation.

Changes in response topography. Observations of annelids during habituation trials clearly show that the form of the worm's response to the stimulus undergoes change with successive presentations of the stimulus. Gardner (1966) describes this clearly in the case of the response of *L. terrestris* to vibration. At first the worm makes a very noticeable response that involves both withdrawal of its anterior portions and hooking of the posterior portions. The withdrawal component habituates first but the hooking component remains, then the hooking component habituates. Corning (*personal communication*) has reported an equally systematic example of change in response topography in the crab, *Limulus*. In this case, early stimulation leads all legs to move and gradually fewer and fewer pairs of them respond to stimulation.

Gee (1912) and Kaiser (1954) report another form of change in response topography in habituation. In this case the duration of the cessation of undulation of the leech gradually shortens until zero seconds of cessation occur. Changes in response latencies during habituation also occur, judged from studies of responses to photic stimulation.

One of the implications of this characteristic of change in response topography during habituation concerns the measurement of responses. Since the correlation between the components of the response is less than perfect, measurement of all components, particularly the last one to habituate is important, if fine-grained analysis of habituation is to be made.

Interaction of response systems; effects of concurrent stimulation. One of the truisms of behavior analysis is that an organism's response to one thing is affected by his response to other things. While the implications of this are rarely noted, they are very important in studies of habituation. A search of the literature regarding responses of invertebrates to various classes of stimuli yields a number of observations that suggest that the amount of response, the frequency of response, or the form of the response, is affected by other concurrent sources of stimulation. So for annelids, e.g., an unpublished study by Ratner and Gardner tested responses of *L. terrestris* to light under three levels of contact with the environment, thigmotaxis. Table 2 shows the frequencies of response to light for earthworms in tubes of different diameters. Notice that a high degree of contact, the most narrow tube, is associated with a low frequency of response to light. Gee (1912) reports another type of interaction system for the leech, in which case, hunger, an empty crop, is associated with highly repeated responses

TABLE 2

Medians of the percentages of the time that worms move forward or backward during a 6-second period for a base-rate control group and a light-stimulated group with three levels of contact, i.e., tube size (Gardner and Ratner, in preparation).

Group	Response	Small tube, %	Middle tube, %	Large tube, %
Base rate	Forward	7	31	25
	Backward	2	1	1
Stimulated	Forward	10	28	30
	Backward	13	45	45

to stimulation, whereas a full crop is associated with positive thigmotaxis and negative response to light.

Differential habituation of orienting versus consummatory responses. The activities of organisms have been divided by Tinbergen (1951) into *orienting* and *consummatory* parts, and a number of investigators of behavior have used this classification. The report by Clark (1965, p. 90) includes the comment that a portion of the feeding response of polychaetes does not habituate. A number of studies in my laboratory with mammals (Askew, 1966, and Leibrecht, 1967) and studies of responses of annelids to light all suggest that the consummatory aspect of responses may only show *refractory phases* (brief decrements with repeated elicitation). But these consummatory components may not show longer-term decrements of the sort found when stimuli for orientation are repeatedly presented. A stimulus for orientation for the earthworm, e.g., is a low level of vibration, and responses to this stimulus habituate and remain habituated for more than 5 days.

While this characterization of habituation by differential habituation of orienting and consummatory responses is not yet fully satisfactory, the observation of differential rates and levels of habituation and differential degrees of retention of habituation appears to be accurate. This method for identification of the behavior systems or classification of the systems may be weak.

Individual differences in habituation. A final characteristic of habituation that is rarely mentioned but appears when data from individual worms are presented is the characteristic of large individual differences among the animals that are tested. These appear even when the experimenter has carefully done the same thing with each individual in the experiment. A number of experimental problems arise when individual differences are conspicuous. One relates to the stability and the generality of these differences. Findings by Gardner (1966) suggest that individual worms differ greatly from each other in rate of habituation; e.g., some habituate to vibratory stimulation in 24 trials and others in 80 trials. But these individual differences are stable; i.e., slow habituaters in one session are slow in another, etc. Presently, no findings are available regarding the generality of the individual differences across types of habituation procedures. On the other hand, if individual differences in a single test procedure are not stable, this suggests that the procedure or measurement are "noisy," i.e., unreliable.

Another question about individual differences relates to the sources of these variations which, if stable, suggest genetic or other biological factors. Thus, individual differences in habituation should not be a basis of irritation to the experimenter; rather they may be a basis for systematic study of sources of variation in a relatively reliable and valid measuring context that relates to behavior modification.

GENERAL SUMMARY

Annelid learning is critically reviewed in several ways. General source materials on annelid behavior and learning are noted and described. Then studies of annelid learning are analyzed for their adequacy by general evaluations of their reliability and validity. Maze and alley learning situations are viewed in three categories and each category is found to be adequate with polychaetes but less adequate with oligochaetes. Classical conditioning procedures are relatively adequate with oligochaetes but unanalyzable for polychaetes owing to the small number of classical conditioning studies with species in this class. Habituation is found clearly and adequately in a variety of annelids and test situations. The studies suggest that special attention should be given to several characteristics of the process. The characteristics are: changes in response topography during habituation; interaction of response systems, the effects of concurrent stimulation; differential habituation of orienting versus consummatory responses; and individual differences in habituation.

REFERENCES

Arbit, J., Learning in annelids and attempts at chemical modification of this behavior, *in* D. Davenport and W. H. Thorpe, eds., *Learning and Associated Phenomena in Invertebrates. Anim. Behav., Suppl.,* **1**, 1965, 83–87.

Askew, H., The effects of stimulus duration on the habituation of a head-shake response. M.A. Thesis, 1966, Michigan State University, E. Lansing, Mich.

Barrett, T. J., *Harnessing the Earthworm.* Boston: Bruce Humphries, 1947.

Bhatucha-Reid, Rodabé, Latent learning in earthworms. *Science,* 1956, **123**, 222.

Bullock, T. H., and G. A. Horridge, *Structure and Function in the Nervous Systems of Invertebrates, Vol. I.* San Francisco, Calif.: W. H. Freeman, 1965.

Clark, R. B., Habituation of the polychaete *Nereis* to sudden stimuli, I: General properties of the habituation process. *Anim. Behav.,* 1960a, **8**, 82.

Clark, R. B., Habituation of the polychaete *Nereis* to sudden stimulation, II: Biological significance of habituation. *Anim. Behav.,* 1960b, **8**, 92.

Clark, R. B., The learning abilities of nereid polychaetes and the role of the supraoesophageal ganglion, *in* D. Davenport and W. H. Thorpe, eds., *Learning and Associated Phenomena in Invertebrates. Anim. Behav., Suppl.* **1**, 1965, 89–100.

Copeland, M., An apparent conditioned response in *Nereis virens. J. Comp. Psychol.*, 1930, **10**, 339.

Copeland, M., and F. A. Brown, Jr., Modification of behavior in *Nereis virens. Biol. Bull.*, 1934, **67**, 356.

Dales, R. P., *Annelids*. London: Hutchinson & Co. (Publishers), Ltd., 1963.

Darwin, C., *The Formation of Vegetable Mould through the Action of Worms with Observations on Their Habits*. London: John Murray, Publishers, Ltd., 1881.

Datta, Lois-Ellen, Learning in the earthworm, *Lumbricus terrestris. Amer. J. Psychol.*, 1962, **75**, 531.

Evans, S. M., The effect of brain extirpation on learning and retention in nereid polychaetes. *Anim. Behav.*, 1963a, **11**, 172.

Evans, S. M., Behavior of the polychaete *Nereis* in T-mazes. *Anim. Behav.*, 1963b, **11**, 379.

Evans, S. M., Non-associative avoidance learning in nereid polychaetes. *Anim. Behav.*, 1966, **14**, 102.

Flint, P., The effect of sensory deprivation on the behaviour of the polychaete *Nereis* in T-mazes. *Anim. Behav.*, 1965, **13**, 187.

Gardner, L. E., *Habituation in the earthworm: retention and overhabituation*. Ph.D. Thesis, 1966, Michigan State University, E. Lansing, Mich.

Gee, W., The behavior of leeches with especial reference to its modifiability. *Univ. Calif. Pub. Zool.*, 1912, **2**, 197.

Grassé, P. P., *Traité de Zoologie, Tome V*. Paris: Masson et Cᵗᵉ Editeurs, 1959.

Harris, J. D., Habituatory response decrement in the intact organism. *Psychol. Bull.*, 1943, **40**, 385.

Herz, M. J., H. V. S. Peeke, and E. J. Wyers, Temperature and conditioning in the earthworm, *Lumbricus terrestris. Anim. Behav.*, 1964, **12**, 502.

Jacobson, A. L., Learning in flatworms and annelids. *Psychol. Bull.*, 1963, **60**, 74.

James, J. P., and A. B. Woodruff, Latent learning in earthworms. *Psychol. Rep.*, 1965, **16**, 406.

Kaiser, F., Beitrage zur Bewegung physiologie der Hirudineen. *Zool. Jahrb. (Allg. Zool.)*, 1954, **65**, 59.

Kirk, W. E., Effects of shock and light intensities and goal box conditions on runway performance of *Lumbricus terrestris*. M.A. Thesis, 1966, Ohio University, Athens, Ohio.

Krivanek, J. O., Habit formation in the earthworm, *Lumbricus terrestris. Physiol. Zool.*, 1956, **29**, 241.

Kuenzer, P. P., Verhaltenphysiologische Untersuchungen über das Zucken des Regenwurms. *Z. Tierpsychol.*, 1958, **15**, 31.

Laverack, M. S., *The Physiology of Earthworms*. New York: Pergamon Press, 1963.

Leibrecht, B., Effects of repeated sessions on habituation of a head-shake response. M.A. Thesis, 1967, Michigan State University, E. Lansing, Mich.

Lutz, F. E., P. S. Welch, P. S. Galtsoff, and J. G. Needham, *Culture Methods for Invertebrate Animals*. New York: Dover Publications, Inc., 1959.

Maier, N. R. F., and T. C. Schneirla, *Principles of Animal Psychology*. New York: McGraw-Hill Book Company, 1935.

Mann, K. H., *Leeches (Hirudinea)*. New York: Pergamon Press, 1962.

Morgan, R. F., S. C. Ratner, and M. R. Denny, Response of earthworms to light as measured by the GSR. *Psychon. Sci.*, 1965, **3**, 27.

Peeke, H. V. S., M. J. Herz, and E. J. Wyers, Ganglia removal and photically driven activity in the earthworm (*L. terrestris*). *Psychon. Sci.*, 1965, **3**, 187.

Raabe, S., Zur Analyze der Assoziationbildung bei *Lumbricus variegatus. Z. vergleich. Physiol.*, 1939, **26**, 611.

Ratner, S. C., Worms in a straight alley: acquisition and extinction or phototaxis? *Psychol. Rec.*, 1964, **14**, 31.

Ratner, S. C., Research and theory on conditioning of annelids, *in* D. Davenport and W. H. Thorpe, eds., *Learning and Associated Phenomena in Invertebrates. Anim. Behav., Suppl.*, **1**, 1965, 101–108.

Ratner, S. C., and M. R. Denny, *Comparative Psychology*. Homewood, Ill.: Dorsey Press, 1964.

Ratner, S. C., and L. E. Gardner, Studies of the behavior of earthworms, *in* A. Stokes, ed., *Laboratory Manual of Animal Behavior*. San Francisco, Calif.: W. H. Freeman, 1967.

Ratner, S. C., and K. R. Miller, Classical conditioning in earthworm *Lumbricus terrestris. J. Comp. Physiol. Psychol.*, 1959a, **52**, 102.

Ratner, S. C., and K. R. Miller, Effects of spacing of training and ganglia removal on conditioning in earthworms. *J. Comp. Physiol. Psychol.*, 1959b, **52**, 667.

Ratner, S. C., and D. G. Stein, Responses of worms to light as a function of intertrial interval and ganglion removal. *J. Comp. Physiol. Psychol.*, 1965, **59**, 301.

Reynierse, J. H., and S. C. Ratner, Acquisition and extinction in the earthworm, *Lumbricus terrestris. Psychol. Rec.*, 1964, **14**, 383.

Robinson, J. S., Stimulus substitution and response learning in the earthworm. *J. Comp. Physiol. Psychol.*, 1953, **46**, 262.

Stephenson, J., *The Oligochaeta*. Oxford, Eng.: Oxford University Press, 1930.

Thompson, R. F., and W. A. Spencer, Habituation: a model phenomenon for the study of neuronal substrates of behavior. *Psychol. Rev.*, 1966, **73**, 16.

Thorpe, W. H., *Learning and Instinct in Animals*. Cambridge, Mass.: Harvard University Press, 1956.

Tinbergen, N., *The Study of Instinct*. Oxford, Eng.: Clarendon Press, 1951.

Warden, C. J., T. N. Jenkins. and L. H. Warner, *Comparative Psychology: Plants and Invertebrates*. New York: The Ronald Press Company, 1941.

Wyers, E. J., H. V. S. Peeke, and M. J. Herz, Partial reinforcement and resistance to extinction in the earthworm. *J. Comp. Physiol. Psychol.*, 1964, **57**, 113.

Yerkes, Ada W., Modifiability of *Hydroides dianthus. J. Comp. Neurol.*, 1906, **16**, 441.

Yerkes, R. M., The intelligence of earthworms. *J. Anim. Behav.*, 1912, **2**, 332.

Chapter 25

Some Cellular Correlates of Behavior Controlled by an Insect Central Ganglion[1]

Melvin J. Cohen

Department of Biology
University of Oregon
Eugene, Oregon

A major problem in neurobiology is to relate the wealth of functional information about single neurons to the factors that determine the behavior of the whole animal. In considering this question one is immediately struck by two apparently paradoxical observations. The first concerns the large number of functional similarities found in neurons throughout the animal kingdom. Many of the physical and chemical parameters associated with propagation and transmission in neurons are common to nerve cells of an annelid worm, a lobster, a squid, or the cerebral cortex of man (Bullock and Horridge, 1965). In contrast to this commonality of certain properties in units of the nervous system, one is faced with the enormous variety of animal behavior. How does behavioral variability result from neural systems built with basically similar units?

Integrated behavior arises from the combined activity of a population of neurons. The pattern of connections between members of a neuron population may therefore be one major factor which determines the specific behavioral capability of a neuronal group. If one assumes that the pattern of connections between neurons determines the specific behavioral output of a nervous system, then there are three different questions one might ask about this parameter:

1. What are the properties of a specific connectivity pattern that critically determine the behavioral ability of the system; i.e., is there a specific wiring diagram in a system concerned with the perception of taste compared with one involved in acoustic responses?

2. What are the factors that control the formation of connections between one unit and another in an excitable system?

3. How are connections modified through use or disuse?

To examine these questions of cell connections and their relationship to behavior, a suitable preparation should satisfy the following conditions:

---

[1] This work was supported by PHS research grant 5 RO1 NB 01624.

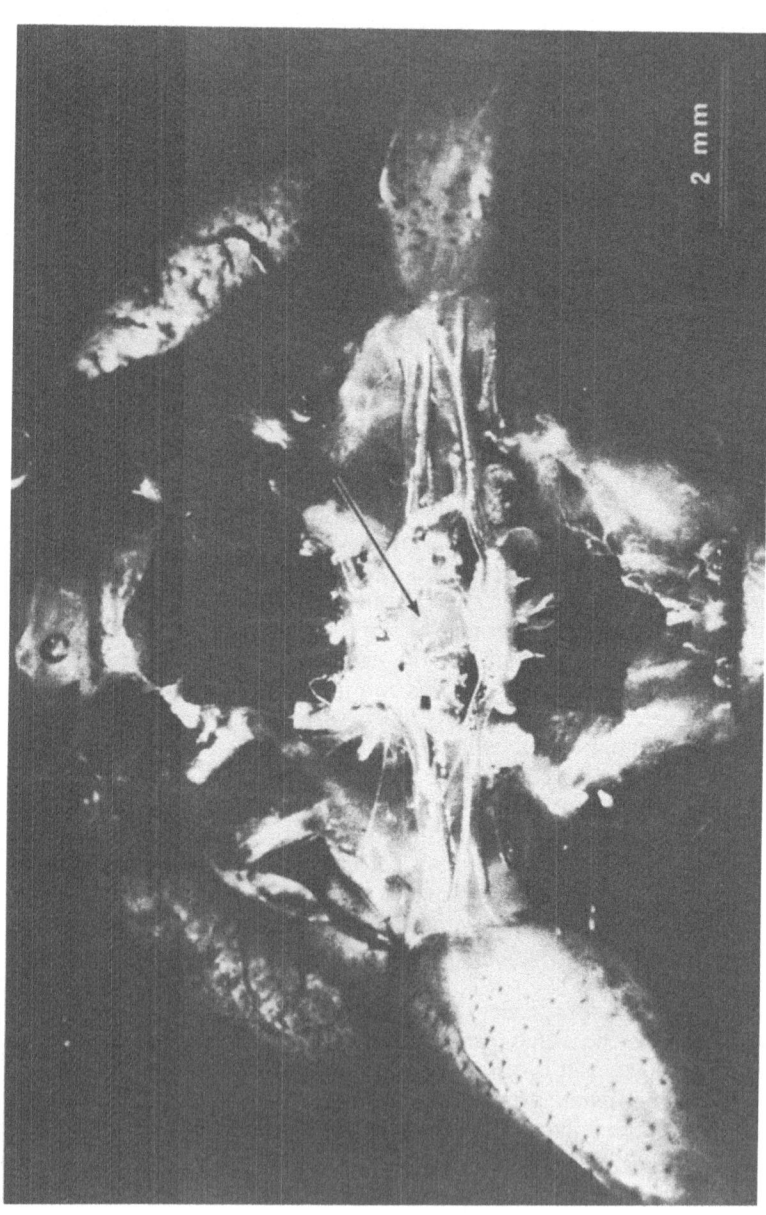

Fig. 1. Ventral view of an isolated prothoracic ganglion preparation used in the learning experiments. The head has been removed and the connectives between the pro- and mesothoracic ganglia have been cut. The arrow points to the ganglionic mass seen surrounded by the tubes of the tracheole system (modified from Eisenstein and Cohen, 1965).

The individual units within the system should be identified and recognizable from one preparation to another. The connections between units must be open to experimental modification. The system should give rise to a discrete bit of behavior so that changes at the cellular level can be related to particular alterations in the behavioral pattern. The thoracic ganglia of the cockroach *Periplaneta americana* satisfy most of these criteria. These ganglia are involved in a variety of integrative activity ranging from the control of locomotion (Pringle, 1940, and Wilson, 1965) to learning (Horridge, 1962, and Eisenstein and Cohen, 1965). In addition, the central and peripheral neurons of *Periplaneta* show an extraordinary capacity for regeneration in the adult individual (Bodenstein, 1957). This paper deals with a series of studies in the cockroach which attempt to relate the cellular events occurring in a small central ganglion to behavior evoked by this system.

LEARNING IN AN ISOLATED GANGLION

Our initial concern was to define the smallest bit of central nervous system in this animal that could still give rise to a bit of integrative behavior. We used a modification of the avoidance conditioning technique first developed by Horridge (1962) and determined that a single prothoracic ganglion isolated from the rest of the central nervous system could learn (Eisenstein and Cohen, 1965). The anterior and posterior connectives of the prothoracic ganglion were severed, as seen in Fig. 1. The peripheral nerves innervating the prothoracic legs were left intact. The isolated ganglion preparation was arranged so that an attached prothoracic leg received a shock by making contact with a saline solution when the leg assumed the extended position seen in Fig. 2(A). When the leg received the shock, it flexed as seen in Fig. 2(B). This broke contact with the saline and the leg no longer received shocks in the elevated portion. These animals were called *Positional* (*P*) animals after the terminology of Horridge (1962) because the shock was always received in a fixed relationship to the leg position. Each *P* animal was wired in series with another isolated ganglion preparation such that the prothoracic leg of the second animal received a shock each time the *P* animal was shocked. The second member of the pair is called the *Random* (*R*) animal because the relationship between its leg position and the shocks received is random. The *P* and *R* pairs were run for 45 minutes as a training period, and the number of shocks taken by the *P* animal were plotted against time (Fig. 3). Following the training period, the *P* and *R* members of a pair were disconnected and rewired so that there was no connection between the two. Each member could now avoid the shock by lifting its own leg out of the saline. These animals were now given a chance to avoid shock once again during a testing period and the results were plotted as shown in Fig. 3. In the test period, the *P* group very rapidly assumed an elevated leg position and took fewer shocks than during training. The *R* group at first took more shocks than even the naive *P* animals. After approximately 20 minutes, the *R* animals began

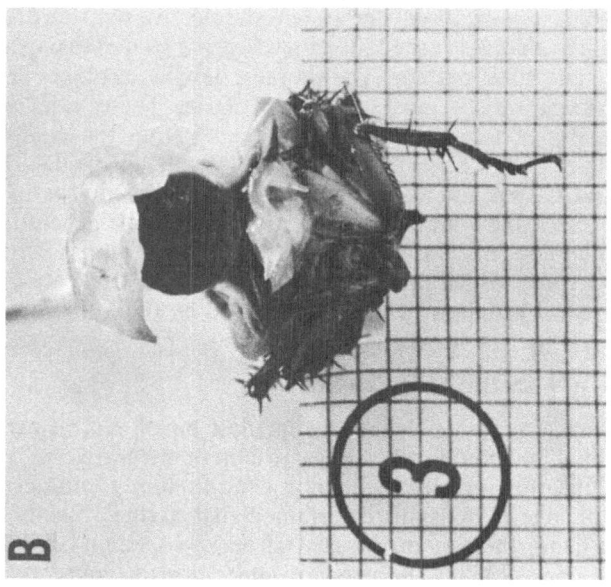

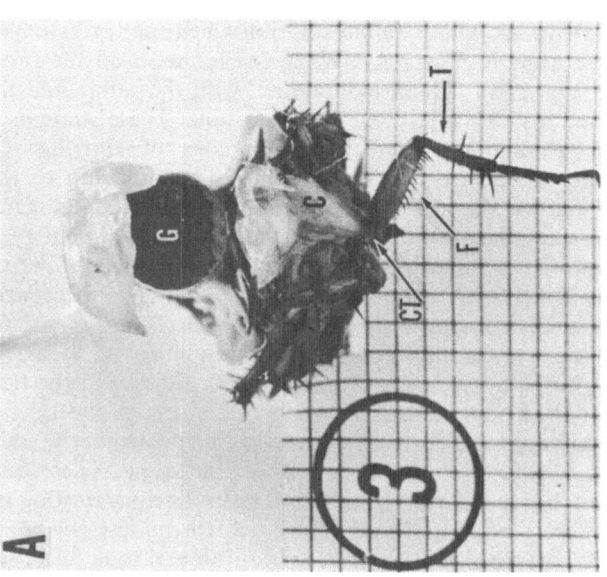

Fig. 2. Anterior view of the isolated ganglion preparation mounted for the learning experiments. The vaseline sealing the cut made by removing the head is seen just above the coxa. (A) Position of the left prothoracic leg before training. (B) Position of the same leg at the end of training. The leg has been lifted out of contact with the saline bath by flection at the coxal-trochanter joint. C, coxa; CT, coxal-trochanter joint; F, femur; T, tibia; G, glass rod on which preparation is mounted. Grid squares, 1.5 by 1.5 mm (from Eisenstein and Cohen, 1965).

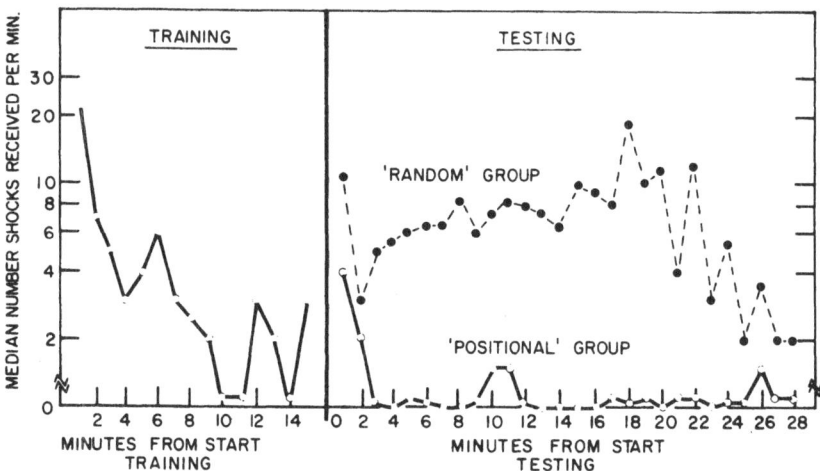

Fig. 3. The left curve shows the decrease in the median number of shocks taken per minute by a group of *P* animals during training. Each *P* animal was connected in series with an *R* animal that received the same shock pattern. The right curves show the difference in the median number of shocks taken by the *P* and *R* pair of animals during the testing procedure. The members of a *P* and *R* pair were disconnected from each other during testing so that each could now avoid shock by lifting its own leg free of the saline bath (from Eisenstein and Cohen, 1965).

taking fewer shocks and approached the performance of the *P* group. The different response curves shown by the *R* and *P* groups during the test period indicate that the drop in number of shocks taken was not simply due to a general excitatory phenomenon. If this had been so, both groups would have behaved similarly during the test period because each had the same number and pattern of shocks. We have interpreted this data from the isolated ganglion just as Horridge (1962) did for the intact ventral nerve cord by assuming that the *P* animals retained some information about the relationship between leg position and shock from their initial training session, i.e., they learned to avoid shock by lifting their leg.

RNA AND CELL MAPS

We next proceeded to examine the distribution of nerve-cell bodies in a thoracic ganglion so that some of the units involved in specific behavioral acts might be identified. We decided to look at cytoplasmic ribonucleic acid (RNA) changes in neurons whose axons had been injured as a means of relating central nerve-cell bodies to their peripheral terminations. This type of neuronal response to injury is well known in the vertebrate as chromatolysis (Nissl, 1892, and Brattgård, Edström, and Hydén, 1957). We hoped that similar changes in the neurons of the cockroach would allow us to identify which nerve-cell body sends its axon in a given peripheral nerve trunk and eventually which cell body innervates a particular leg muscle.

Although cockroach central neurons do not exhibit the classical chromatolysis of the vertebrate cells, they do show marked alterations in the distribution of cytoplasmic RNA when their axons are injured. The cell body of an injured cockroach neuron develops a concentration of cytoplasmic RNA around the nucleus within 12 hours after the axon is cut (Cohen and Jacklet, 1965; Cohen, 1967). The response is specific to the cell body whose axon is injured and reaches a peak at about 3 days after injury (Fig. 4). Using this response as a marker, we were able to construct cell maps from serial sections of the ganglion which identified the peripheral nerve trunk containing the axon of a given motor nerve-cell body (Cohen and Jacklet, in press). We could do this for approximately 50 cell bodies on each side of the ganglion. The large motor cells are bilaterally distributed within the ganglion such that one can identify matched members of a pair, as seen in Fig. 5. By comparing cell maps made from serial sections of five ganglia, we observed that these large motor nerve-cell bodies are consistently located in the ganglion. We have assigned numbers to these cells (Fig. 5) and can identify them from one animal to another. For some of the cells we have been able to identify the particular muscle that they innervate by removing the muscle and destroying the attached nerve terminals. This causes a similar perinuclear RNA ring in the cell body whose terminals have been destroyed.

A three-dimensional representation of the mapped motor nerve-cell bodies is shown in Fig. 6. The cell bodies are distributed in a peripheral rind around the ventral-lateral area of the ganglion. By comparing Fig. 6 with the numbered map of Fig. 5, the spatial orientation of certain cells can be determined, such as the large anteriorly located cell-3 pair and the posteriorly arranged cells of the "50" group. The localization of these cells within the ganglion should make it possible to apply micro recording and injection techniques to specific cells, which can then be marked and identified from preparation to preparation.

REGENERATION

We then used our information about the distribution of motor nerve cells and their peripheral termination to investigate some of the factors that determine connections between a neuron and the muscle it innervates. Bodenstein (1957) showed that cockroach leg muscles whose nerves had been cut first stop contracting when stimulated directly with surface electrodes and then after 30 to 40 days once again contract in response to electrical stimulation. He concluded from this that the nerve terminals attached to the distal stump degenerate and then regenerating axons from the proximal stump grow out to the affected muscle and re-establish functional neuromuscular junctions. We extended Bodenstein's initial observations by correlating histochemical, electrical, and behavioral events that

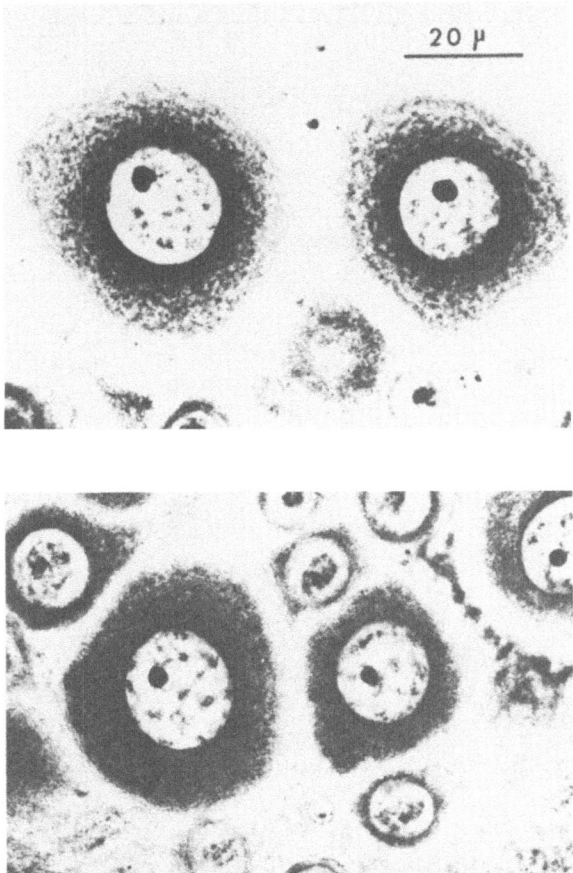

Fig. 4. *Above:* Metathoracic ganglion cells whose axons were cut 2 days previous to fixation. Stained with pyronine malachite-green to show RNA. Note the dense perinuclear ring of pyronine stain in these cells. *Below:* Bilaterally matched control cells from opposite side of the same ganglion. Axons of these cells were left intact. Note the relative absence of perinuclear pyronine ring. The scale is the same for both photographs.

occur during degeneration and subsequent regeneration of specific neuromuscular elements in *Periplaneta*.

Behavior

If a normal animal is suspended and allowed to grasp a cork ball with its legs, it will hold the ball securely with its pretarsal claws as seen in

Fig. 5. A cell map of the metathoracic ganglion of a young adult male cockroach. Reconstructed from 10-μ serial sections. Dorsal view with anterior connectives at the top and posterior connectives at the bottom. The peripheral nerve trunks are numbered II to VI. Note the bilateral distribution of cells indicated by the same numbers for cells on opposite sides of the ganglion (from Cohen and Jacklet, *in press*).

the upper photo of Fig. 7. When nerve 5 to the metathoracic leg is cut as shown in Fig. 8(D), the leg assumes a fixed elevated position as seen in the lower photo of Fig. 7 and cannot grasp the cork ball. Within 30 to 70 days after injury, the affected leg may show some abortive movement. In

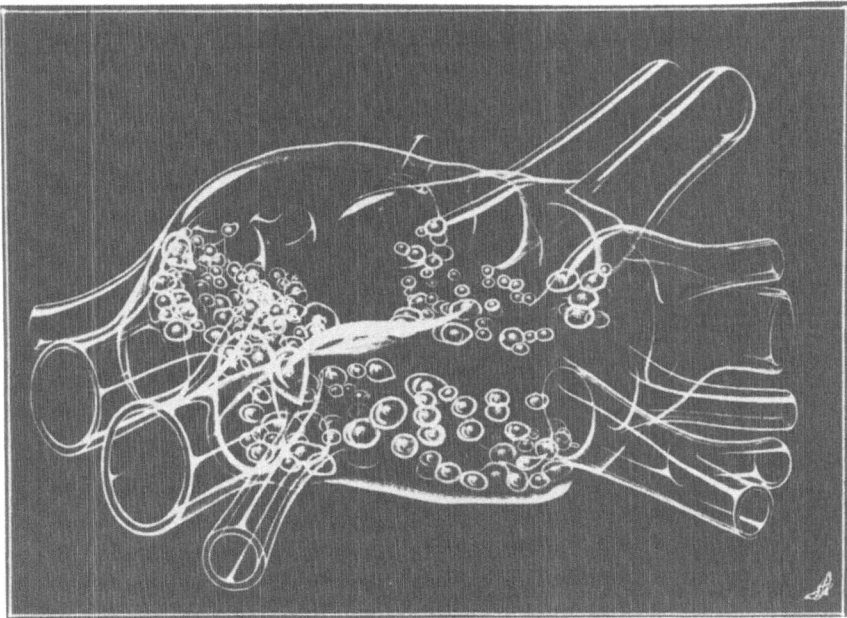

Fig. 6. A three-dimensional representation of the same cell map shown in Fig. 5 viewed from an anterior-lateral position. The large medial pair of anterior connectives are seen at the left flanked by the trunks of Nerve II. The posterior connectives are seen pointed toward the upper-right corner. Some of the numbered cells of Fig. 5 can be identified, such as the large cell 3 directly over Nerve II (from Cohen and Jacklet, *in press*).

many instances this progresses to full functional recovery when the operated limb can once again firmly grasp the cork ball (Jacklet and Cohen, 1967b).

Electrical and Histological Correlates

If one follows the electrical activity in the affected muscle from the time of cutting the motor nerve trunk, it is possible to correlate the behavioral changes described above with alterations in the electrical activity of the muscle. One can also monitor changes in the distribution of cytoplasmic RNA within the injured and regenerating motor neuron soma. The relationships of these various factors are summarized in Fig. 8 (Jacklet and Cohen, 1967b).

In the normal animal, intracellular records from a coxal muscle show a spontaneous background of miniature end-plate potentials (m.e.p.p.'s) as seen in the upper record of Fig. 8(B). Reflex stimulation results in a burst of fast electrical activity that causes contraction of the muscle [lower record of Fig. 8(B)]. In such a normally innervated muscle, the motor nerve-cell

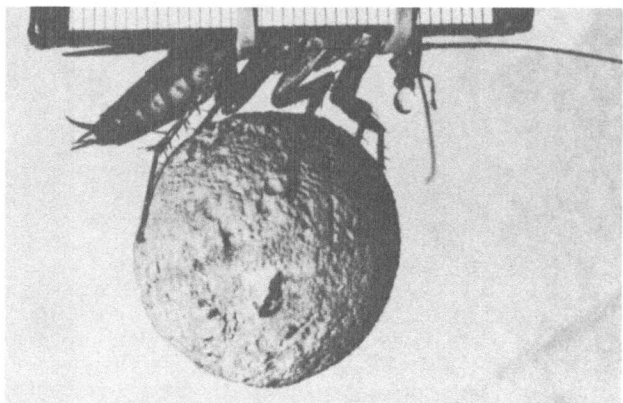

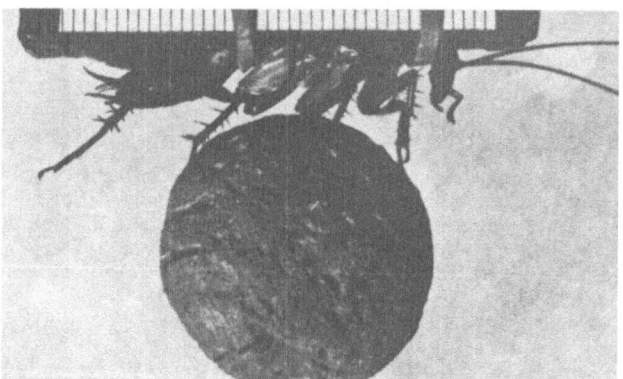

Fig. 7. *Above:* Normal suspended cockroach shown grasping a cork ball. Note that the last (metathoracic) leg is grasping the ball with the terminal pretarsal segment. *Below:* Suspended cockroach in which Nerve V to the right metathoracic leg was cut 2 days previous to taking the picture. Note that the distal segments of the leg are elevated and it does not grasp the cork ball. Within 40 to 70 days after such an injury, many of these animals can once again grasp the ball with the injured leg just as the animal does in the upper photograph. This is due to regeneration of the injured axons and their neuromuscular junctions. The scale is 1 mm.

bodies have the cytoplasmic RNA in a uniformly finely dispersed state, as seen in Fig. 8(C).

If a peripheral nerve trunk is cut as shown in Fig. 8(D), all reflex activity immediately disappears as would be expected. The m.e.p.p.'s are reduced in amplitude within 2 days after injury. Their frequency is down by 3 days, and, by 5 days after cutting the nerve, the m.e.p.p.'s have disappeared entirely [Fig. 8(E)]. Within 2 days after cutting the nerve, the electrical potentials evoked in the muscle by stimulating the attached distal nerve stump are reduced. By 5 days after nerve section, it is not possible to evoke any electrical activity in the muscle either by direct stimulation or by stimulating the distal nerve stump. This indicates a degeneration of the distal nerve fibers or the neuromuscular junctions or both. Between 12 hours and approximately a week after nerve section, during the time when degeneration is occurring, the cytoplasmic RNA of the injured motor nerve-cell bodies aggregates into a perinuclear ring, as shown in Fig. 8(F). No obvious evidence of nerve sprouting from the proximal stump is apparent at this time.

About 2 weeks after cutting the nerve trunk, sprouting axons can be seen emerging from the proximal stump, as also described by other workers (Bodenstein, 1957, and Guthrie, 1962). This may result in the axons cleanly spanning the gap between proximal and distal stump and forming a swollen bridge, as shown in Fig. 8(G). Return of spontaneous m.e.p.p.'s may occur as early as 30 days after nerve section in muscles which previously showed the degenerative changes [Fig. 8(H)]. During this later stage when regeneration of the axon and the neuromuscular junction is taking place, the perinuclear RNA ring disappears. The nucleus of the nerve-cell body shifts to an eccentric position in the cell, as shown in Fig. 8(I). This eccentric position seems to be associated with high-level synthetic activity needed to regenerate the axon and reestablish functional neuromuscular junctions.

GANGLION TRANSPLANTS

Bodenstein (1957) demonstrated in *Periplaneta* that a thoracic ganglion from a donor cockroach could be transplanted into the coxa of a host cockroach. After a suitable interval, the host coxal muscles surrounding the transplanted ganglion showed fibrillation even after the normal motor innervation to these muscles had been cut. Bodenstein interpreted these findings as indicating that axons from the transplanted ganglion formed connections with the coxal muscles of the host.

With the background provided by the cell maps and the regeneration studies, we examined the use of central nervous tissue transplants as a means for investigating the factors controlling the connections between cells in an excitable system (Jacklet and Cohen, 1967a). The experimental procedure is diagrammed in Fig. 9. The coxa of a metathoracic leg in the host animal was prepared to receive a ganglion implant by removing a portion of the *flexor trochanteris* muscles. This procedure also severed nerve 3B which innervates the *extensor tibia* muscle located in the femur. A donor metathoracic ganglion was obtained from a separate animal and placed in the prepared host coxa. Approximately 30 days after implanting the ganglion, the host leg showed spontaneous clonic twitching movements which were unrelated to locomotion. The movements consisted primarily of tibial extension and some elevation of the femur due to flection at the coxal-trochanter joint.

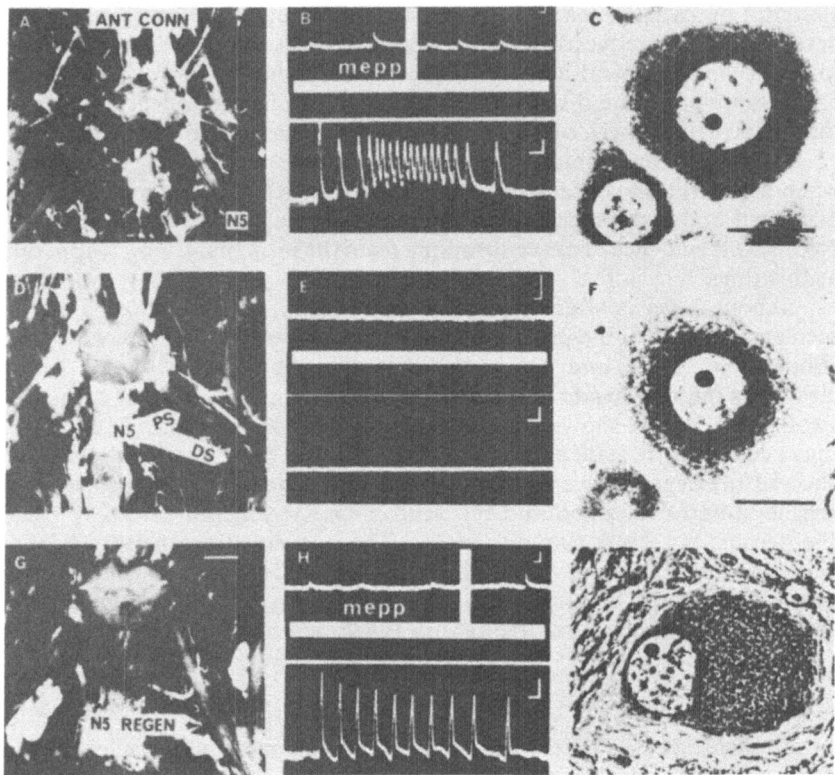

Fig. 8. (A) Ventral view of normal metathoracic ganglion of the cockroach. The left
nerve trunk 5 (N5) and the anterior connectives (ant. conn.) are labeled. The tracheole
trunks have been removed in all ganglion photographs to expose the peripheral nerves.
The scale on all ganglion photographs is 0.5 mm, as indicated in Fig. 1(G). (B) Upper
record, spontaneous miniature end-plate potentials (m.e.p.p.'s) from normal coxal
muscle. Lower record, intracellular recording of "fast" activity in normal coxal muscle
evoked by reflex stimulation. Scales in all electrical figures: upper record, 1 mv, 10 msec;
lower record, 10 mv, 10 msec. (C) Normal motor neuron soma from metathoracic
ganglion showing the uniform distribution of cytoplasmic RNA. Stain in all cell photographs is pyronine malachite-green; the scale is 20 μ. (D) Ventral view of experimental
ganglion 1 day after section of N5. The proximal stump (PS) and distal stump (DS)
have sprung apart. (E) Intracellular recording from muscles, as in Fig. 1(B), 5 days after
denervation. Note the lack of spontaneous m.e.p.p.'s in the upper record. The lower
record shows the lack of evoked muscle response when the distal stump is stimulated
electrically. (F) Motor neuron soma 5 days after cutting its axon. Note the densely
stained pyronine ring in the perinuclear cytoplasm, indicating a concentration of RNA
in this region. (G) Ventral view of metathoracic ganglion 45 days after cutting the left N5.
The gap between the nerve stumps has been bridged by regenerating axons (N5 regen.).
(H) Intracellular records from reinnervated coxal muscle at 140 days after nerve section.
Note the near normal m.e.p.p.'s in the upper record and the reflexly evoked fast activity
in the lower record. (I) Motor nerve-cell body 24 days after section of its axon when
regeneration is well along. Note that the perinuclear RNA ring has disappeared and
the nucleus has shifted to an eccentric position (from Jacklet and Cohen, 1967b).

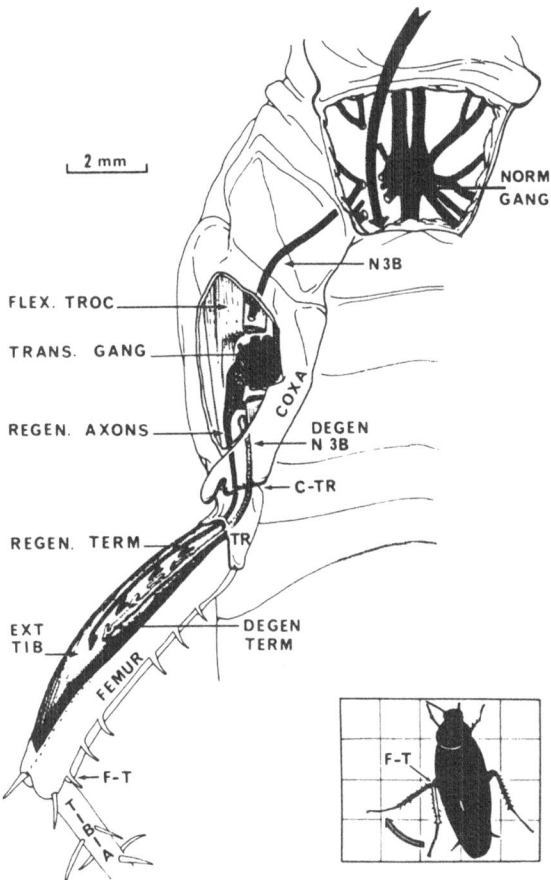

Fig. 9. Diagram of the ventral surface of the cockroach metathorax and the right metathoracic leg. The cuticle over the metathoracic ganglion (norm. gang.) and the coxa is removed. Portions of the *flexor trochanteris* muscle (*flex. troc.*) were removed and a metathoracic ganglion from a donor cockroach (trans. gang.) transplanted into the space provided. Regenerating axons (regen. axons) grow from the transplanted ganglion through the coxal-trochanter (c-tr) joint to form functional neuromuscular junctions (regen. term.) with the *extensor tibia* muscle. Within 30 days after ganglion implantation clonic extensions of the tibia are observed. At this time all the peripheral nerve trunks to the operated leg are cut close to the normal ganglion, as indicated by the sweeping black arrow. The clonic extensions of the tibia continue. *Inset:* Superimposed tracings from frames of a moving film showing the nature of the clonic extensions driven by the implanted ganglion. Dorsal view: Ganglion is implanted in the left metathoracic leg; right metathoracic leg is shown in a resting position. The hatched position of the left leg is the rest position when all normal innervation to this leg is cut. During the clonic extension movements the tibia moves forward at the femural-tibial (f-t) joint to the solid position and then returns to the resting position. Scale: 1-cm grid squares. The ganglia are slightly enlarged in the diagram for better illustration (from Jacklet and Cohen, 1967a).

There was the possibility that these twitching movements might be due to regeneration from the nerve trunks of the host that were cut during implantation of the ganglion. To rule out this possibility, all the peripheral nerve trunks normally innervating the twitching limb were cut close to the host's metathoracic ganglion 40 to 70 days after the donor ganglion was implanted. The spontaneous twitching movements persisted but all other movements of the limb stopped. These twitching movements were followed up to 16 days after cutting the normal innervation. We know from our previous studies (Jacklet and Cohen, 1967b) that the cut motor axons of the distal stump have degenerated by 5 days after nerve section. The persistence of these twitching movements for as long as 16 days after section of the host motor nerves clearly indicates that the movements were not caused by innervation from the normal metathoracic ganglion of the host.

By 40 days after ganglion implantation, nerve fibers can be seen streaming from the donor ganglion toward the distal end of the coxa, as shown in Fig. 9. This indicated that regenerating axons from the transplanted ganglion might have formed neuromuscular junctions with the *extensor tibia* muscle of the host and might be responsible for the twitchlike extensions of the host tibia. We therefore obtained intracellular electrical recordings from the *extensor tibia* muscle of the implanted leg while the clonic extensions of the tibia were taking place. The recordings were taken 40 to 60 days after the ganglion had been implanted. All normal innervation to the leg had been cut 5 to 15 days before recording to allow time for degeneration of the host motor fibers. Under these experimental conditions, depolarizing and hyperpolarizing junctional potentials can be recorded from the *extensor tibia* muscle responsible for the twitching movements of the host limb (Fig. 10). These junctional potentials may summate to produce high-amplitude depolarizations which can be visually correlated with the twitching of the limb. Since all normal motor innervation to this muscle had been eliminated, we concluded that the junctional potentials recorded under these conditions were evoked by regenerating axons from the transplanted ganglion, which have formed new synaptic junctions with the *extensor tibia* muscle of the host.

Our evidence indicates that only host muscles which are denervated at the time of ganglion implantation will form functional synaptic connections with regenerating axons from the implanted ganglion (Jacklet and Cohen, 1967a). We have observed some evidence of reflex activity from the transplanted ganglion, indicating that the transplant may have captured some regenerating sensory fibers of the host, as suggested by Guthrie (1966).

A SPECULATIVE SUMMARY

Cytoplasmic RNA and Functional State

Changes in the quantity and distribution of cytoplasmic RNA in the neurons of vertebrates have been correlated with many kinds of functional

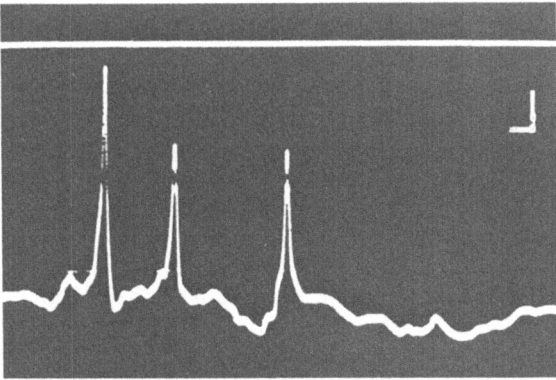

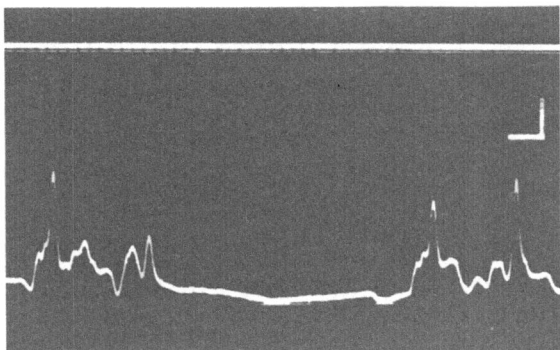

Fig. 10. Intracellular records of synaptic activity in the *extensor tibia* muscle driven by a donor ganglion implanted in the coxa. All normal innervation to this muscle has degenerated. The larger fast-rising depolarizations are correlated with a twitch extension of the tibia. The lower record shows periodic bursts of electrical activity associated with clonic extensions of the tibia. Summation of the small junctional potentials frequently occurs. The upper line in each record is zero potential level. Scales: 10 msec, 10 mv (from Jacklet and Cohen, 1967a).

alterations ranging from nerve regeneration through learning (Hydén, 1943 and 1960). Caspersson (1947) pointed out that adult vertebrate neurons are synthesizing proteins at a very high rate. He associates the Nissl body organization, consisting of ribosomes attached to stacked layers of endoplasmic reticulum (Palay and Palade, 1955), with a state of readiness for massive protein synthesis. During vertebrate chromatolysis, the shift in cytoplasmic RNA from the large Nissl aggregates to the finely dispersed state has been associated with a shift from an active to a superactive state of protein synthesis (Bråttgard, Edström, and Hydén, 1957, and Hydén, 1960).

These findings are particularly interesting when related to the cockroach neurons. The cytoplasmic RNA of the normal cockroach neuron exists in a finely dispersed state rather than in the large aggregates of the vertebrate Nissl bodies (Hess, 1958; Ashhurst, 1961; and Cohen and Jacklet, 1965). Before the cockroach neuron can engage in extraordinary protein synthesis, such as repairing an injured axon, it must first shift its cytoplasmic RNA into a Nissl-like organization which takes the form of an increase in rough endoplasmic reticulum (Cohen, 1967). This appears to be a preparatory stage preceding axon repair and is correlated with the perinuclear ring of RNA. The breakdown of the perinuclear RNA ring can be considered analagous to chromatolysis in the vertebrates where the ribosomes of the rough endoplasmic reticulum appear to pull off of the reticular cisternae and give rise to the finely dispersed cytoplasmic RNA of the chromatolyzed cell (Porter and Bowers, 1963). In cockroach and vertebrate neurons, a dispersion of ribosomes from their aggregation on endoplasmic reticulum seems necessary for very high level protein synthesis. The major difference between the cockroach and vertebrate neuron seems to be that the vertebrate neuron is normally primed for rapid protein production by having its cytoplasmic RNA organized into Nissl aggregates. The cockroach must first shift into the Nissl-like aggregation of ribosomes to achieve high-level protein synthesis. This implies that the ribosomes recently dispersed from the Nissl-like aggregation of the RNA ring differ from those normally dispersed in the uninjured cockroach neuron. Perhaps the orientation of ribosomes on a scaffolding of endoplasmic reticulum facilitates their combination with messenger RNA from the nucleus, which will control the production of protein needed for axon regeneration.

Regeneration and Synaptic Specificity

The questions posed at the start of this paper center around the factors controlling the establishment of synaptic connections in excitable systems. I feel that this problem is fundamental to our understanding the cellular basis of behavior and that a knowledge of how junctions are established initially during development will bear strongly on the question of how they may be altered through use and disuse.

The use of regenerating systems offers the possibility of studying selected junctions while they are undergoing the plastic changes of degeneration and subsequent reformation. The series of studies presented here on the cockroach is primarily in the form of a progress report. They are intended to provide some of the groundwork for developing a preparation that may answer certain of the critical questions about the control of connections between excitable units. The cockroach preparation has the advantage of being able to define some of the individual cells involved in discrete bits of behavior. The bilaterally symmetrical distribution of cells within the ganglion permits the use of one member of a pair as an experimental object and the other member as a control. This seems particularly advantageous because our results indicate that the metabolic properties of adjacent cells in the same ganglion may differ significantly, as shown by Bodian (1947) for cells of the vertebrate central nervous system. The cells comprising a bilaterally symmetrical pair, on the other hand, seem to have similar metabolic characteristics (Cohen and Jacklet, *in press*).

The question of the specificity of connections seems particularly open to attack in the cockroach. The increasing evidence for chemical differences between neurons in the same individual (Levi-Montalcini, 1966), supports the concept that a chemical compatibility may be involved in controlling the linkage between two excitable cells (Weiss, 1936). Does a given motor neuron that normally innervates a specific muscle always reconnect with that muscle during regeneration? If such a specificity is found in the regenerated neuromuscular connections of the cockroach, then the possibility of determining the nature of the specificity at the level of single cells seems feasible.

The transplanted thoracic ganglion provides an *in vivo* tissue culture situation where a bit of central nervous system and its peripheral connections can be examined in isolation from other neural influences. The transplant apparently has some intrinsic activity, as evidenced by the twitching movements of the host leg, which are driven by regenerated motor axons from the transplanted ganglion. The preparation offers the fascinating possibility of constructing an excitable system where the connections can be specified and related to discrete acts of behavior. The similarity of neural regeneration in the cockroach and vertebrate neurons at all levels of organization indicate that the information gained here may lead to some general conclusions about the factors which determine the connections between units of an excitable system.

REFERENCES

Ashhurst, D. E., The cytology and histochemistry of the neurones of *Periplaneta americana. Quart. J. Microscop. Sci.*, 1961, **102**, 399.

Bodenstein, D., Studies on nerve regeneration in *Periplaneta americana. J. Exp. Zool.*, 1957, **136**, 89.

Bodian, D., Nucleic acid in nerve cell regeneration. *Symp. Soc. Exp. Biol.*, 1947, **1**, 163.

Brattgård, S. O., J. E. Edström, and H. Hydén, The chemical changes in regenerating neurons. *J. Neurochem.*, 1957, **1**, 316.

Bullock, T. H., and G. A. Horridge, *Structure and Function in the Nervous System of Invertebrates*, Vol. 1, Chaps. 3 and 4. San Francisco: W. H. Freeman and Co., 1965.

Caspersson, T., The relations between nucleic acid and protein synthesis. *Nucleic acid. Symp. Soc. Exp. Biol.*, 1947, **1**, 127.

Cohen, M. J., Correlations between structure, function and RNA metabolism in central neurons of insects, *in* C. A. G. Wiersma, ed., *The Invertebrate Nervous System*. Chicago: University of Chicago Press, 1967, pp. 65–78.

Cohen, M. J., and J. W. Jacklet, Neurons of insects: RNA changes during injury and regeneration. *Science*, 1965, **148**, 1237.

Cohen, M. J., and J. W. Jacklet, The functional organization of motor neurons in an insect ganglion. *Phil. Trans. Roy. Soc. (London), Ser. B,* **252,** 571.

Eisenstein, E. M., and M. J. Cohen, Learning in an isolated prothoracic insect ganglion. *Anim. Behav.,* 1965, **13,** 104.

Ekholm, R., and H. Hydén, Polyribosomes in nerve cells. *J. Ultrastruct. Res.,* 1965, **12,** 239.

Guthrie, D. M., Regenerative growth in insect nerve axons. *J. Insect Physiol.,* 1962, **8**(1), 79.

Guthrie, D. M., Physiological competition between host and implanted ganglia in an insect. *Nature,* 1966, **210,** 312.

Hess, A., Experimental anatomical studies of pathways in the severed central nerve cord of the cockroach. *J. Morph.,* 1958, **103,** 479.

Horridge, G. A., Learning of leg position by the ventral nerve cord in headless insects. *Proc. Roy. Soc. (London), Ser. B,* 1962, **152,** 33.

Hydén, H., Protein metabolism in the nerve cell during growth and function. *Acta Physiol. Scand.,* 1943, **6,** *Suppl.* 17, 1.

Hydén, H., The neuron, *in* J. Brachet and A. E. Mirsky, eds., *The Cell,* Vol. 4. New York: Academic Press, 1960, pp. 215–323.

Jacklet, J. W., and M. J. Cohen, Neuronal regeneration: electrophysiological, histological and behavioral correlates. *Amer. Zool.,* 1965, **6**(4), 94.

Jacklet, J. W., and M. J. Cohen, Synaptic connections between a transplanted insect ganglion and muscles of the host. *Science,* 1967a, **156,** 1638.

Jacklet, J. W., and M. J. Cohen, Nerve regeneration: correlation of electrical, histological and behavioral events. *Science,* 1967b, **156,** 1640.

Levi-Montalcini, R., The nerve growth factor: its mode of acting on sensory and sympathetic nerve cells. *Harvey Lectures,* 1966, **60,** 217.

Nissl, F., Ueber die Veranderungen der Ganglienzellen am Facialiskern des Kaninchens nach Ausreissung der Nerven. *Allgemeine Z. für Psychiatrie,* 1892, **48,** 197.

Palay, S. L., and G. E. Palade, The fine structure of neurons. *J. Bioph. Biochem. Cytol.,* 1955, **1,** 69.

Porter, K. R., and M. B. Bowers, A study of chromatolysis in motor neurons of the frog *Rana pipiens. J. Cell Biol.,* 1963, **19,** 56A.

Pringle, J. W. S., The reflex mechanism of the insect leg. *J. Exp. Biol.,* 1940, **17,** 8.

Weiss, P., Selectivity controlling the central-peripheral relations in the nervous system. *Biol. Rev.,* 1936, **11,** 494.

Wilson, D. M., Proprioceptive leg reflexes in cockroaches. *J. Exp. Biol.,* 1965, **43,** 397.

Chapter 26

Interocular Transfer, Brain Lesions, and Maze Learning in the Wood Ant, *Formica rufa*

D. M. Vowles

Institute of Experimental Psychology
Oxford, England

One of the major evolutionary trends in the Insecta has resulted in the increase and size and complexity of the mushroom bodies (corpora pedunculata) in the brains of social hymenoptera. Since ants and bees are known to learn well, it seems reasonable to suppose that these lobes are involved. In a previous study (Vowles, 1964a) it was shown that, in wood ants which had been trained in an olfactory task, lesions within the mushroom bodies did not disrupt performance, but severing tracts between the antennal lobes and the alpha lobes (Vowles, 1955) of the mushroom bodies caused failure to recognize the olfactory cues. Unilateral lesions disrupted performance only if the contralateral antenna had been amputated.

The object of the present study was to investigate the role of the mushroom bodies in visual learning. Following training in a visual task, lesions were placed in the mushroom bodies themselves or in the tracts connecting the optic ganglia with the mushroom bodies or in the optic ganglia. During the early stages of the study it became clear that unilateral lesions produced no adverse effects on a simple learning task (to orient by means of light direction), and a task of greater complexity (involving pattern discrimination) was therefore introduced. All experiments were duplicated on groups of ants which had learned either the simple or the complex task. The effects of lesions were also compared with unilateral blinding, carried out by occluding one eye with paint.

A problem of general interest, which became urgent when interpreting the results of the lesion experiments, concerns the integration of information from two eyes which have largely independent visual fields (Werringloer, 1932, and *personal observation*). Some standard tests on interocular transfer were therefore performed with intact ants. As a further extension blinding and unblinding experiments were also carried out. The total experimental design is summarized in Table 1.

TABLE 1

Summary of Experiments Performed

Training conditions	Locus of lesions	Testing conditions
Binocular	None	Monocular
Monocular	None	Monocular (transfer)
Monocular	None	Binocular
Binocular	Within mushroom bodies; left or right side	Binocular
Binocular	Between mushroom bodies and optic ganglia; left or right side	Binocular
Monocular	Within mushroom bodies; blind or exposed side	Monocular
Monocular	Between mushroom bodies and optic ganglia; blind or exposed side	Monocular
Monocular (clear maze only)	Outer optic chiasma; blind side	Monocular
Monocular (clear maze only)	Inner and middle optic ganglia; blind side	Monocular

TRAINING METHODS

A simple technique for training ants has been described in an earlier paper (Vowles, 1965). Briefly, the ants are trained individually to find their way home to their nest through a small T maze. The motivation is to escape from peppermint odor at the entrance to the maze, and the reward, to return home. Each ant is given one trial per day, and the exit alley is alternately left and right on successive trials. There is no punishment, such as an electric shock, for entering a blind alley. Odor seems relatively unimportant in this situation, but is controlled by changing and cleaning the cardboard linings of the maze. Under these conditions ants will learn to use visual cues in order to return home; they do not learn to alternate. A pretraining period of 3 to 7 days exposure to a similar screened maze with white walls is given, after which full training is started using the visual cues to which the ants are to be trained.

Two experimental conditions were used for training and testing. In the first of these, the guiding cues were provided by vertical and horizontal stripes upon the walls of the maze, these stripes extending back from the arms and along the entrance alley. The maze was screened from its surroundings by a white cardboard tube and was illuminated by a bench lamp placed vertically above the junction of the T. In half the experiments vertical stripes were positive, and in half, horizontal.

In the second experimental condition a transparent perspex maze was used and the ant guided itself by landmarks in the room outside the maze.

The main cues were the direction of light from a window and from a bench lamp placed opposite the junction of the T, at an angle of approximately 30° to the horizontal. Since the direction in which the ant had to turn at the choice point to reach home was alternated on successive trials, this meant that, although visual conditions in the arms of the T were constant from trial to trial, the initial direction taken in the entrance alley was alternately toward or away from the light in successive trials.

Both conditions were used in all subsequent experiments. Ants were trained and tested both binocularly and monocularly (i.e., with both eyes uncovered and also with one eye blinded by a cap). The blinding was done by placing a thick blob of red cellulose paint over the cornea of the compound eye and adjacent chitin of the head, while the ant was anaesthetized with carbon dioxide. This did not seem to hurt the ant in any way, not did it produce any behavioral deficiencies such as circling; the ants merely appeared blind on the painted side.

The efficiency of blinding was tested in two ways. In all the ants, their reaction to an object moving in the visual field was observed before and after blinding. When an object is moved past them they normally react by turning toward the object in an aggressive posture. When they have been blinded on one side, the ants continue to react on the exposed side, but cease to show any response to an object moving on the blinded side. If they did show such a response, the eye cover was closely reexamined under a microscope and repainted. When training was finished, the eye cover was removed and usually came off intact, showing the imprint of the compound eye on its inner side. This was examined under a microscope and was usually found to contain no pores nor to transmit any detectable light; those few results obtained with a suspect eye cover were discarded. Immediately after removing the cover, the ants showed normal visual responses; this suggests that the visual system was not harmed by blinding. When the experiment involved successive blinding of the two eyes, the original cover was removed and a fresh one painted on the opposite side.

A second method of testing the efficiency of blinding was electrophysiological. The electroretinogram (ERG) of the ant, recorded with fine platinum-wire electrodes, consists of a sustained negative plateau of approximately 250 μv under the experimental conditions of illumination. The size of the ERG was recorded before blinding, after blinding, and after subsequent removal of the eye cover. After insertion of the electrode, the eye was dried, and a little beeswax was moulded around the electrode at its junction with the cornea. This was done in order to eliminate any subsequent anaesthetic effect of the paint solvent. The paint was then placed over the cornea, the wax, and the electrode. While the cover was in place, the ERG was undetectable, the noise level of the system being about 20 μv. After removal of the eye cover, the ERG returned to normal. It appears, therefore, that the method of blinding is both efficient and harmless.

The criteria adopted for learning and the reasons for these have been described in detail in the previous paper (Vowles, 1965). Briefly, it was shown that the maze presents at least three choices to the ant. It may enter the exit alley and return to the nest *or* enter the exit alley but later turn back *or* enter the blind alley and turn back. The probability of making a correct run by chance after becoming familiar with the maze was found empirically to be 0.34. Hence the probability of making five correct error-free choices is $(0.34)^5 \backsimeq 0.004$. The criterion for learning which was used was five consecutive error-free trials. An error is scored both following a complete entry (i.e., the ant is wholly within the alley) into the blind alley and for a complete entry into the exit alley when the ant turned and left again via the alley entrance.

Difficulties which arise in these experiments are the following:

1. The performance of ants under these conditions usually deteriorates after 25 to 30 trials, and the experiments were therefore abandoned if the ant had not completed its test within 30 trials.

2. If an ant is disturbed after a change in conditions and errors recur, it may relearn the new situation, which will cause a decrease in the number of errors. The crucial test period used for analysis must therefore be at least shorter than the number of trials for normal learning to occur, and, if possible, short enough to exclude "savings."

3. If test trials are alternated with trials retaining the original conditions and the test conditions produce a disturbance, this effect unfortunately spreads to the training trials and causes errors to recur even under the original conditions. Hence this type of control could not be used.

For statistical analysis it is necessary therefore to select those trials immediately after the experimental change in conditions which, while sufficiently numerous to permit valid testing, do not extend over so long a period as to bias the results. The procedure finally adopted was to select the six trials immediately following a change in experimental conditions and to apply the Mann-Whitney U test (Siegel, 1956) in order to compare performance on these trials with that of a control group of ants, which had achieved the learning criteria but which had experienced no change in conditions. Ants are extremely consistent after learning, showing a mean of only 1.6 trials with errors in 10 trials following learning. However the test is crude; although it will show whether or not a disturbance occurs, it is not sufficiently sensitive to test degrees of disturbance.

In Fig. 1 are shown some typical learning graphs for individual ants. In the graphs on the left ants were undisturbed by the change in experimental conditions, and on the right they were disorientated and showed a recurrence of errors on the first manipulation, but not on the second.

Special mention should be made of the use of control groups in this study. It was found that there was considerable variability in the speed of learning between different nests and at different times of the year as well as between individuals. There were also unpredictable periods when all the individuals in a particular nest tended to give a poor performance. It was therefore very important that both control and experimental groups

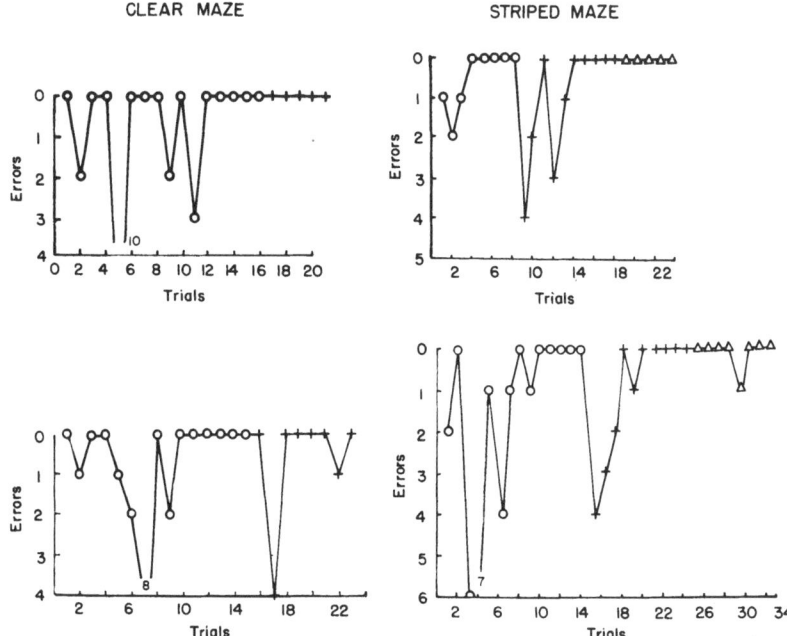

Fig. 1. Graphs showing individual examples of interocular transfer tests. The hollow circles show performance during training with one eye covered, and the crosses show performance with the opposite eye covered. The two graphs on the left refer to ants trained in a clear maze, and the two graphs on the right, to ants trained in a striped maze. In the latter two groups the ants finally had the eye cover removed and were tested binocularly (triangles).

should come from the same populations (i.e., from the same nests and at the same time). Similarly, when comparing the results of different experiments, one should be sure that differences were due to the experimental conditions and not to variability in the ant population. In any one nest, therefore, there were always at least two groups of ants being trained simultaneously, one being an experimental and the other a control group. More often three groups were present in each nest, two used for different experiments and one as controls; this enabled comparisons to be made between the results of different experiments, as well as between experimentally disturbed ants and their controls.

SURGICAL TECHNIQUES

The technique for making brain lesions has already been described (Vowles, 1955 and 1964). Briefly, it consisted of anaesthetizing the ant lightly with CO_2; embedding it temporarily in soft plasticine with its head exposed; folding back a thin strip of chitin over that part of the brain which was to be damaged; making the lesion, using a sliver of razor blade; replacing the flap of chitin; drying the wound with filter paper; and finally sealing the wound with a mixture of beeswax and resin.

The immediate survival rate from such operations has been about 75 %, of which about two-thirds have lived indefinitely, the remainder dying within 5 to 10 days. Only those ants which survived the total duration of the experiment were used in the quantitative treatment of results. Since some survivors became behaviorally abnormal (usually very aggressive) after operations, the final survival rate of experimental subjects has been very low.

The locus of the brain lesion was determined in two ways. Immediately after killing the ant, its fresh brain was examined under a dissecting microscope, with a squared grid in the eye piece. The lesion could usually be detected by the presence of brown deposits in the brain tissue around it. The locus of the lesion was then marked on a previously prepared map of the ant's brain constructed both from serial microscope sections and from transparent whole mounts. The process was assisted by the fact that, in the fresh brain, after desheathing, the outlines of the calyces and alpha lobes of the mushroom bodies can be clearly seen. Subsequently, the brain was fixed and serial sections made to confirm the position of the lesion and to check the depth of the damage.

Lesions do not appear to be irritant, except for a brief period of several minutes after the operation, when an ant will show abnormal behavior. After such abnormal behavior (e.g., circling) has subsided (usual with small lesions), most ants appear qualitatively normal, and in no way disturbed or lethargic. After a lesion the ants were always tested to see if they reacted aggressively to a moving stimulus on the operated side. If they did not (which was rare), they were discarded as they were considered blinded by the operation. It is stressed that the successful operates did not appear to be blind.

RESULTS

Difficulty of Problem and Speed of Learning

The speed of learning was studied using both normal ants and monocularly blinded individuals. As already described, two experimental conditions were used—a clear maze, in which light direction was the main cue, and a maze in which cues were provided by vertical and horizontal striped patterns upon the walls. Learning graphs for two individuals under each of these conditions were shown in Fig. 1.

Taking the number of trials to reach the first of five consecutive error-free trials as a measure of the speed of learning, a histogram was plotted of the number of ants learning at different speeds under the four conditions.

These histograms are shown in Fig. 2. The distribution of three of these populations appears skewed owing to rapid learning, while with monocular training conditions in a clear maze the population seems to be bimodal.

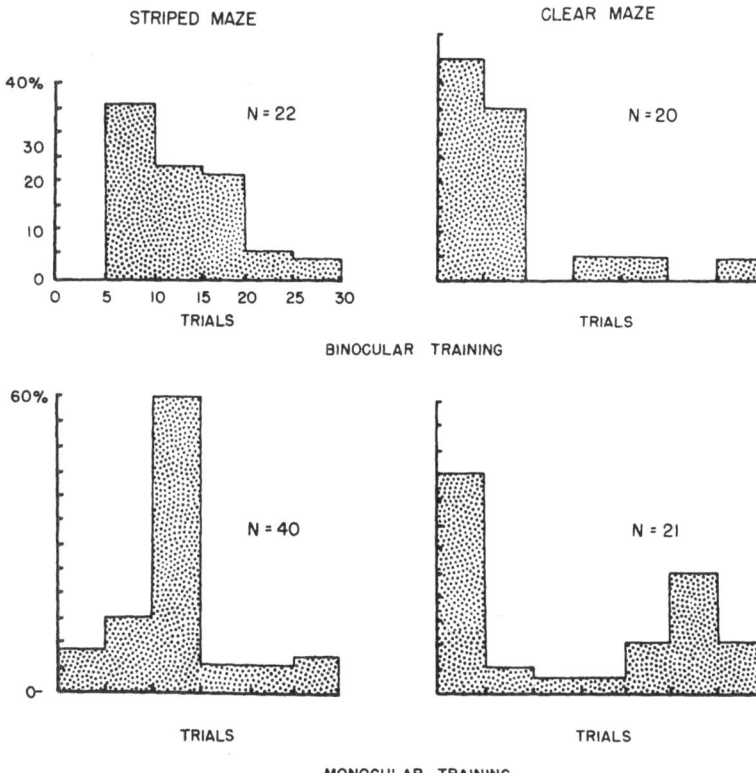

Fig. 2. Histograms showing speed of learning in a striped and a clear maze
under binocular and monocular training conditions.

Analysis of the behavior of this last group of ants showed that, before
reaching their criterion, the slow learning individuals gave a long sequence
of runs with either no errors or one error occurring on alternate trials.
On those runs where single errors occurred, it was found that the ants were
turning first into the alley on their blind side. This might be expected,
since, when returning to the nest, ants are basically negatively phototactic
(Jander, 1958). By turning toward their blinded side, they will show alternate errors and correct runs owing to the experimental alternation of the
direction in which they have to run. It would obviously not be sensible
to quote a mean figure for the speed of learning in such a bimodal population.

The other three distributions appear to be skewed, but the two groups
of ants trained in stripe discrimination did not differ significantly ($p > 0.05$)
from normality, although the ants trained binocularly in the clear maze
did. However, for the purpose of comparison this last group was treated

TABLE 2

Mean Learning Speeds (X) Under Different Conditions

Maze	Training	X^*	SD†
Striped	Binocular	14.5	7.3
	Monocular	13.0	5.8
Clear	Binocular	6.3	3.7

$^*X$ = Mean number of trials to criterion (the first of five consecutive error-free trials).
†SD = Standard deviation.

as if it had a normal distribution. According to Fisher (1949) this does not distort the results greatly. The speeds of learning with standard deviations for the three conditions are shown in Table 2.

The significance of the difference between groups was assessed by a t test. This showed that in the striped maze the speeds of learning under monocular and binocular conditions did not differ significantly ($p > 0.1$). There was, however, a significant difference between the speeds of learning under binocular conditions in the striped and clear mazes. Learning in the clear maze takes only about half the number of trials needed for the striped maze.

When ants were trained monocularly in the striped maze, they showed a type of behavior which might be interpreted as visual scanning. In the entrance alley the two patterns which they have to distinguish are on the two opposite side walls. Ants were frequently seen to walk a little way along one wall with their exposed eye toward the pattern and then to cross to the other wall, at the same time turning around and thus bringing their good eye against the new pattern. Similarly, at the T junction, where both patterns are presented next to each other, an ant would often walk along past the two patterns with its exposed eye toward them, later turning back if this was necessary to go into the positive alley. This suggests that ants were making a temporal comparison of the two patterns with the same eye. At a later stage in learning, this type of scanning movement was not shown, but the ants ran unhesitatingly through the maze.

Binocular Integration

The usual test for interocular transfer involves training an animal with one eye covered and testing it with the originally trained eye covered and the previously covered eye exposed. This type of test was used here, and in addition the effects of monocular blinding or unblinding on learned performance were also studied. Some graphs showing typical results are shown in Fig. 3, and the statistical summary is presented in Table 3. The control groups for both the clear and the striped maze did not differ among themselves for any of the three experimental conditions and have therefore been grouped to give two common control populations, one for each type of maze.

TABLE 3
Effect of Transfer Tests and Monocular Blinding and Unblinding

Conditions	Controls	Monocular training Monocular testing (transfer test)	Monocular training Binocular testing	Binocular training Monocular testing
Striped Maze	$N = 15$ $X = 0.99$	$N = 7$ $X = 4.4$ $P < 0.01$	$N = 8$ $X = 3.8$ $P < 0.01$	$N = 9$ $X = 5$ $P < 0.005$
Clear Maze	$N = 17$ $X = 0.67$	$N = 10$ $X = 2.1$ $P < 0.7$	$N = 11$ $X = 1.0$ $P < 0.7$	$N = 10$ $X = 1.1$ $P < 0.9$

N = Number of ants.
X = Mean number of trials with errors out of six trials following change in eye cover.
P = Probability of groups coming from the same population as controls on a chance basis.

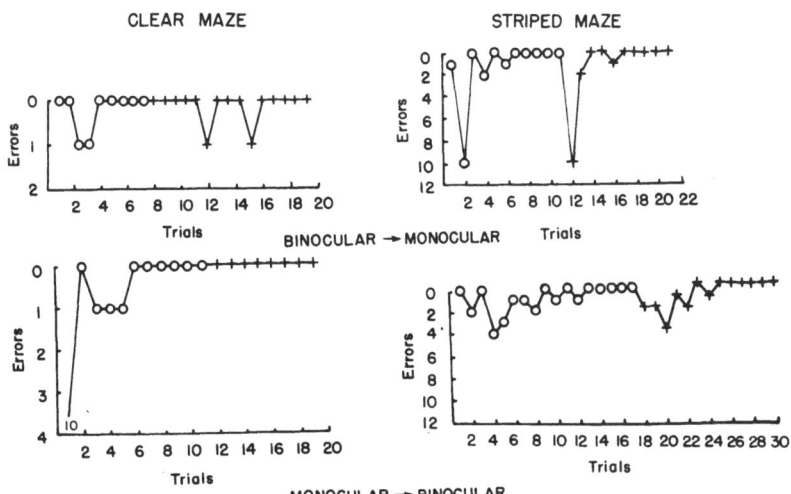

Fig. 3. Typical graphs showing types of response under various experimental conditions. The graphs on the left refer to ants trained in a clear maze, and those on the right, to ants trained in a striped maze. The sequences of blinding are shown on the graphs. Open circles indicate training trials, and crosses indicate test trials.

Examination of the results shows that, when using the clear maze, ants are not disturbed by any of the three experimental manipulations. Although the figures suggest a slight deterioration when testing for interocular transfer, this is not significant ($p > 0.7$). In the striped maze, however, the ants were disturbed in all three tests and had to relearn the task. There was no evidence of savings, although ants which had been trained successively with each one of the two eyes, subsequently performed correctly when tested binocularly.

These results are consistent with those obtained for other animals tested for the degree of interocular transfer. In octopuses (Muntz, 1961), cats (Meikle and Sechzer, 1960), and monkeys (Myers, in Mountcastle, 1962), the degree of interocular transfer also depends upon the difficulty of the discrimination involved.

The Effect of Brain Lesions

The mushroom bodies. Two types of lesion were inflicted in these experiments. In the first, the lesions were placed to sever at least some of the tracts between the optic ganglia and the mushroom bodies on one side of the brain, and, in the second, the lesions were placed within the mushroom bodies. Since placing a lesion was inaccurate, the actual locus not being known until after histological examination, the criterion for allocating an ant to either of these two groups was solely histological. Hence, at the time of testing it was not actually known into which group an ant would fall, particularly since the same nest contained individuals in which both types of operation had been intended. Some sham operations were also performed as controls; in these, all the surgical procedures for the experimentals were used with the exception that no lesions were inflicted on the brain.

Some attempts were made to produce bilaterial lesions. Unfortunately most of the ants died (probably from excessive loss of blood), and the results with the small remainder were inconclusive. Some midline transections were also made, but the operates showed abnormal postures and locomotor inhibition (see also Roeder, 1953 and 1963) and could not be tested in the maze.

Separate groups of subjects were trained in each type of maze and under binocular and monocular training. A factorial design was employed involving type of maze, positive and negative cues, binocular and monocular training, side of lesion, and side of blinding. Although the design was symmetrical, operations on the right side of the brain were more often successful than those on the left (presumably owing to the handedness of the experimenter), but there was no significant difference between the groups operated on (or blinded or both) on the left side and those so treated on the right side. No changes were made in the eye covers during an experiment; ants were both trained and tested under the same monocular and binocular conditions. The loci of the lesions are shown in Fig. 4.

In Fig. 5 some individual results are shown, and the statistical summary is given in Tables 4 and 5.

TABLE 4

Lack of Disturbance Produced by Control Operations

Experimental conditions	Number of ants	Mean number of trials with errors*	P
Striped maze			
Lesions within mush-room bodies	13	0.97	
Sham operations	16	0.64	< 0.4
Clear maze			
Lesions within mush-room bodies	9	1.2	
Sham operations	15	0.93	< 0.6

*The scores given are the mean number of trials on which errors occurred out of the six trials following the operation.

TABLE 5

Effect of Unilateral Section of Tracts Between Mushroom Bodies and Optic Ganglia

Training conditions	Binocular training	Monocular training, lesion on exposed side	Monocular training, lesion on blind side
Striped maze	$N = 18$ $X = 2.8$ $P < 0.001$	$N = 13$ $X = 4.4$ $P \leqslant 0.001$	$N = 10$ $X = 3.8$ $P \leqslant 0.001$
Clear maze	$N = 7$ $X = 0.71$ $P \leqslant 0.1$	$N = 5$ $X = 3.6$ $P \leqslant 0.001$	$N = 7$ $X = 3.7$ $P \leqslant 0.001$

N = Number of ants.
X = Mean number of trials with errors.
P = Probability of chance difference from control groups.

A comparison of the sham-operated controls with ants which merely continued normal maze running, following their first five consecutive error-free trials, showed no significant difference ($p < 0.6$). There was no significant difference between ants receiving lesions within the mushroom bodies under monocular training and those under binocular training, and the scores are grouped in Table 4. These do not differ significantly from the sham-operated control groups in either type of maze. In Fig. 4 the loci of five lesions outside the mushroom body complex and its tracts are shown; the results were considered too few to permit statistical analysis, but these subjects showed no obvious difference from the controls.

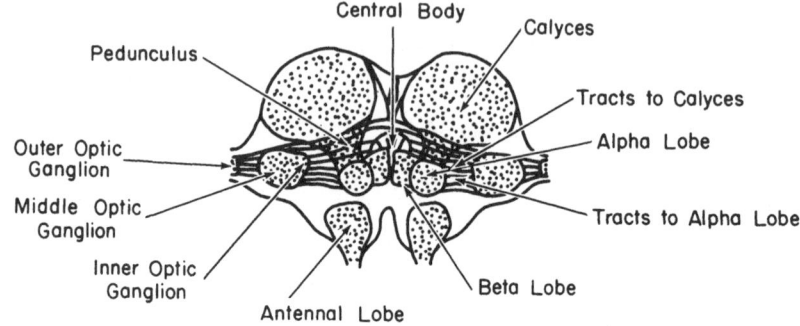

LOBES OF THE BRAIN AND OPTIC TRACTS

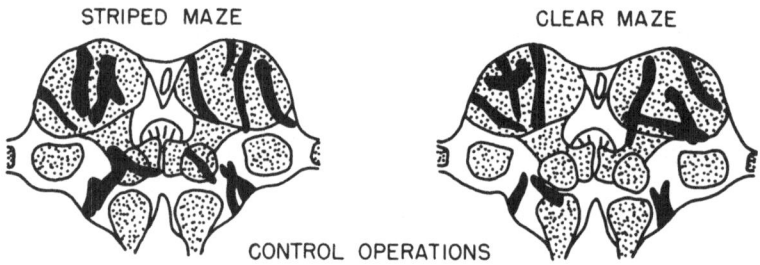

CONTROL OPERATIONS

Fig. 4(a). Maps of the ant brain as seen from the frontal surface, showing loci and extent of lesions which do not affect the tracts between the mushroom bodies and the optic ganglia. Each coarse line indicates an individual lesion.

In Table 5 the effect of severing (at least partially) the tracts between the mushroom bodies and the optic ganglia on one side of the brain are summarized. The scores are compared with the grouped results of lesions within the mushroom bodies for each type of maze. It is clear that under monocular training conditions a unilateral lesion disrupts performance whether the lesion is on the blinded or the exposed side. In the striped maze a unilateral lesion disrupts performance under binocular training, but it has no such effect in a clear maze.

Some individuals relearned the maze after an initial decrement in performance following a lesion. However, since the lesions did not completely sever all the tracts involved, it is not clear how much of this recovery is due to use of the remaining connections. It seems unlikely, however, that the disruptive effect of the lesion can be due merely to temporary irritation, since the lesions elsewhere would be expected to have a similar effect. Moreover it appears that degeneration is complete within 3 days (Vowles, 1955), and there was no evidence of a greater frequency of errors during the first 3 days following the lesion than during the next 3 days.

The optic ganglia. The experiments described in this section were designed for studying the effect on learning of lesions placed within the optic ganglia. The most detailed studies of the optic ganglia of the Hymenoptera have been made by Kenyon (1896) and by Cajal and Sanchez (1915). Three main ganglia on each side of the grain make up the optic system of ants and bees. The most peripheral ganglion (*retina intermediaria*), which receives connections directly from the retinula cells, lies adjacent to the basement membrane of the retina against which it is closely apposed. The retinula cells synapse in this outer ganglion with monopolar cells, whose axons pass to the middle ganglion (*retina profunda*). The collection of axons passing between the outer and the middle ganglia is called the *outer optic chiasma*. The outer optic chiasma is relatively lengthy in ants and can be damaged or cut without injury to either the outer or the middle ganglia.

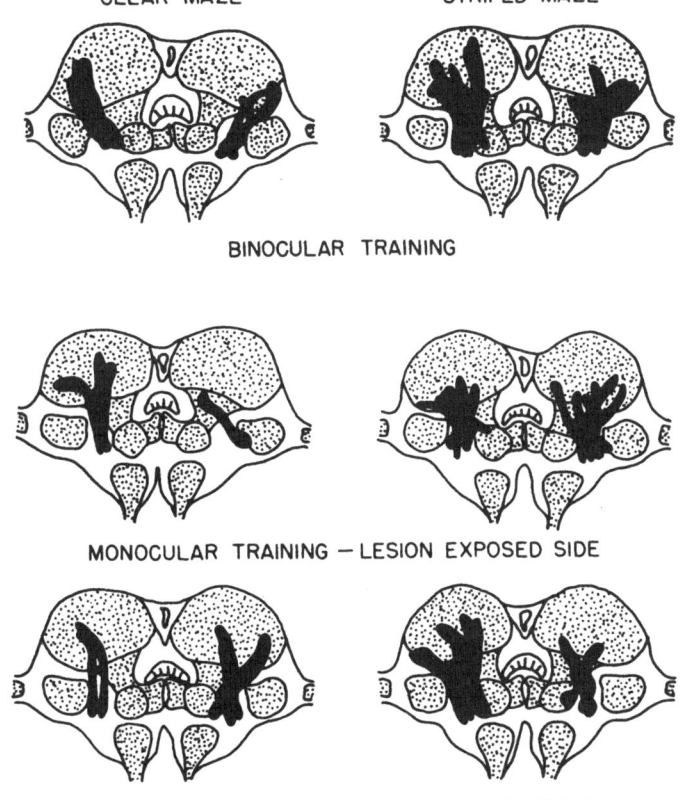

Fig. 4(b). Maps of the ant brain as seen from the frontal surface, showing loci and extent of lesions which severed tracts between the mushroom bodies and the optic ganglia.

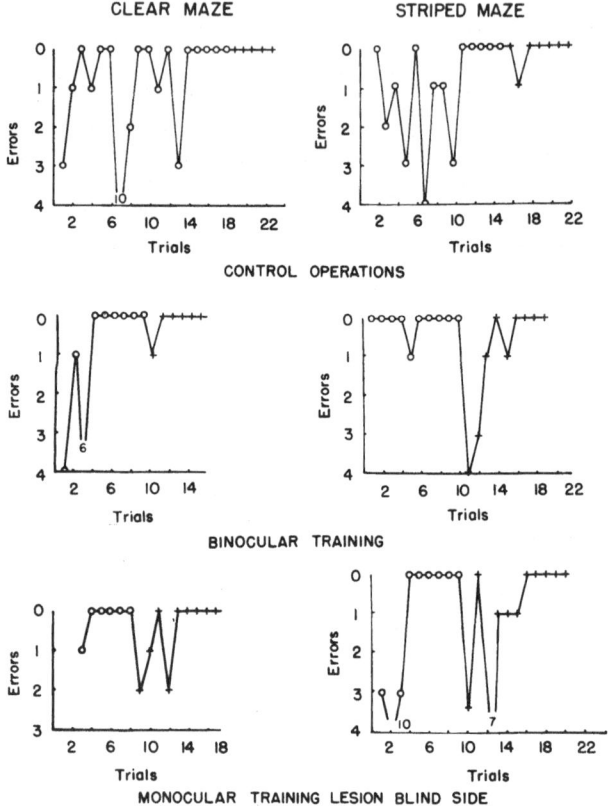

Fig. 5. Graphs showing examples of performance following
lesions. The graphs in the left column were obtained from the
clear maze, and those on the right, from the striped maze.
The conditions are shown on the graphs. Circles indicate
training; crosses show performance after operation.

Between the middle and the inner optic ganglia lies the inner optic
chiasma, but the ganglia are so closely apposed to each other that it is very
difficult to damage one without partially destroying the other. In the
operations subsequently described, therefore, no attempt was made to
damage the middle or the inner ganglia separately.

Both the middle and the inner optic ganglia give and receive tracts
to the mushroom bodies on both their own and the opposite side of the
brain. They also have a few connections to the optic ganglia of the opposite
side of the brain. In addition, they send fibers directly to other parts of the
nervous system, such as the suboesophageal ganglion.

Damage to the optic ganglia will obviously lead to a disturbance
of visual function on that side of the brain, and any disturbance in behavior

as a result of such damage would obviously be of little help in clarifying the contribution of the optic ganglia to learning. However, this difficulty can be partly overcome in the following way. It has already been shown that, in ants which were trained monocularly, cutting the tracts between the optic ganglia and the mushroom bodies on the blinded side produced a failure in maze performance. This suggests that even on the blinded side the optic ganglia are themselves actively contributing to the performance of the animal. An experiment was therefore done to study the effect of lesions in the optic ganglia on the blinded side of the monocularly trained ant.

Two types of lesion were attempted. In one of these the connections between the outer optic ganglion and the middle optic ganglion were interrupted by inserting a sharp scalpel at the back of the eye on the blinded side. The outer optic ganglion (*retina intermediaria*) lies adjacent to the back of the eye, and the connections (outer optic chiasma) between this ganglion and the middle optic ganglion (*retina profunda*) are accessible for surgery without damaging other tissue. The second type of lesion involved damage to the middle and inner optic ganglia. Damage to these two ganglia is more drastic in its pathological effects than lesions placed elsewhere. It was found that lesions in the middle and inner optic ganglia produced gross, widespread degeneration, which was obvious to the most superficial examination. The reasons for this are not understood.

In the experiments ants were trained monocularly in a clear perspex maze. Half the ants were blinded on the left and half on the right side, and within each group half had to keep the light on their left in order to home correctly and the other half had to keep the light on their right. After learning had occurred, the operations upon the optic ganglia of the blind side were performed. The subsequent performance of the ants in the maze was studied, and the effects of the lesions upon their behavior were analyzed as previously described. After the operations ants appeared behaviorally normal, except that, following damage to the middle and inner ganglia, there was at times a tendency to circle. No maze running was attempted when an ant showed circling behavior, and, since the bouts of circling only lasted for short periods of a few minutes, it was possible to perform test trials in the usual way.

TABLE 6

Effect of Lesions on the Visual System on the Blind Side Following Monocular Training*

Conditions	N	X	P
Controls	9	1.2	
Section of outer optic chiasma	8	1.5	< 0.1
Damage to middle and inner ganglia	7	4.0	< 0.02

*Notation: N, number of ants; X, mean number of trials with errors in six trials following the operation; P, probability of chance difference for control group.

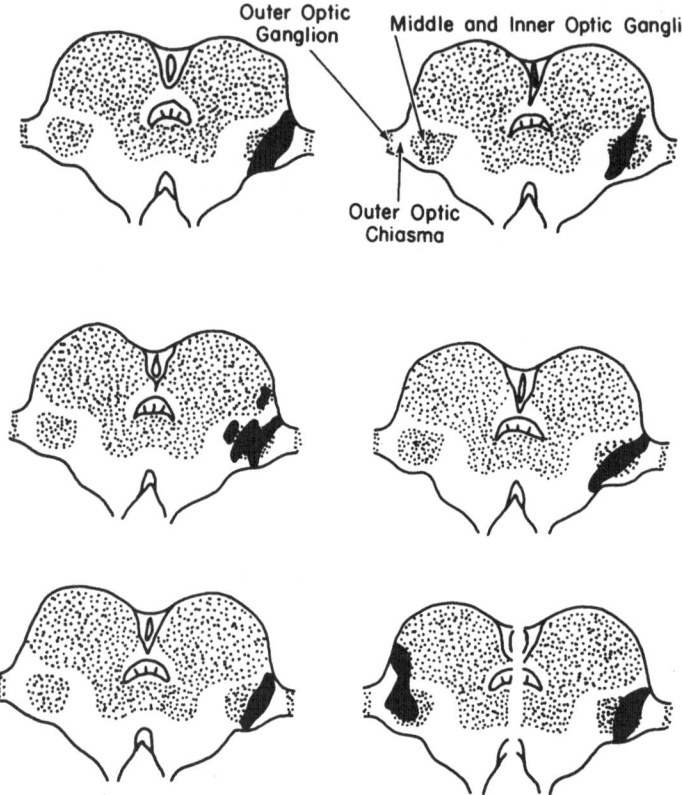

Fig. 6. Maps of the ant brain with loci of lesions damaging the middle and inner optic ganglia. The individual lesions of seven subjects are shown.

The loci of the lesions to the middle and inner ganglia are shown in Fig. 6, and the results of the experiment are summarized in Table 6. It will be seen that, when the optic chiasma was severed between the outer and middle optic ganglia, there was no effect upon the ant's performance in the maze. However, when there was damage to the middle and inner optic ganglia, there was a gross disruption of the ant's performance. This shows clearly, therefore, that the middle and inner optic ganglia have an active role to play in the control of the ant's learned performance, even on the blinded side of the brain.

DISCUSSION

In this section an attempt will be made to survey all the results described in this paper, and a possible explanation will be proposed. The salient results may be summarized as follows:

1. In experiments on interocular transfer, blinding, and unblinding, ants trained in a clear maze learned more rapidly and were undisturbed by the experimental manipulations; whereas ants trained in a striped maze learned more slowly and were disturbed by all three procedures.

2. Ants trained binocularly in a clear maze were undisturbed by a lesion placed unilaterally between the optic ganglia and the mushroom bodies, but ants trained monocularly in a clear maze or monocularly or binocularly in a striped maze were disturbed by cutting the tracts between the optic ganglia and the mushroom bodies on one side of the brain, whether on the trained or the untrained side. Perhaps one critical point to be noted here is that a unilateral lesion in an ant trained *binocularly* in a clear maze had no effect, whereas a similar lesion on the blinded side of an ant trained *monocularly* in a clear maze did have an effect.

3. When ants were trained monocularly in a clear maze, severing the outer optic chiasma on the blinded side did not disturb them, but damage to the middle and inner optic ganglia disturbed their learned performance.

4. Lesions in the mushroom bodies themselves did not disrupt learned performance.

All these results are summarized graphically in Fig. 7.

These results will first be discussed in terms of the possible neural mechanism underlying learning, and then with reference to a more specific hypothesis concerning the integration of activities on the two sides of the brain.

In a previous paper (Vowles, 1965) it was suggested that, when an ant learns a maze by using visual cues, one is not strictly studying visual discrimination but the ability of the ant to control the details of its orientation by the visual cues in the environment. Under natural conditions (MacGregor, 1948, and Carthy, 1951) ants show circling and backtracking behavior when a familiar landmark is removed or disturbed. The backtracking and circling can be used as measures of disorientation, the function of which may be to scan the environment. In the maze it leads to recurrent errors.

Learning to orient will presumably involve both the visual analyzers in the ant's brain (presumably the optic ganglia) and those parts of the brain concerned with control of the motor mechanisms involved in locomotion. In insects the mechanisms for the integration of stepping and flying lie in the thoracic ganglia, but ablation studies (Roeder, 1953 and 1963) and localized electrical stimulation of the brain (Oberholzer and Huber, 1957; Huber, 1959; Rowell, 1963; and Vowles, 1964b) show that the mushroom bodies control the orientation of locomotion by their effect on lower centers. Anatomical studies have shown that in ants and bees (Vowles, 1955) the mushroom bodies are connected to lower and motor centers via their β lobe, while the calyces and the α lobe have apparently duplicate connections with all the main sensory centers. An electrophysiological analysis (Vowles, 1964b) suggests that the main circuits in this system

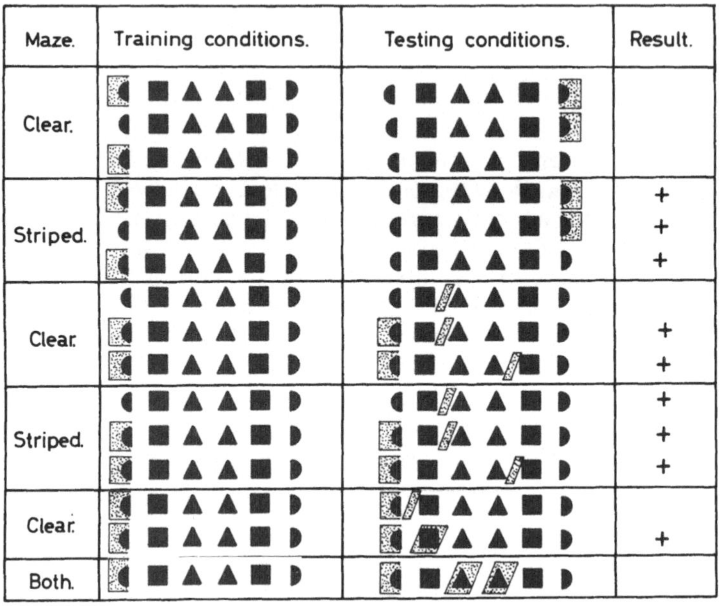

Maze.	Training conditions.	Testing conditions.	Result.
Clear.			
Striped.			+ + +
Clear.			+ +
Striped.			+ + +
Clear.			+
Both.			

Fig. 7. Diagram summarizing the experimental results.
◖ = eye and outer optic ganglion.
■ = middle and inner optic ganglia.
▲ = mushroom body.
╱ = locus of lesion.
▨ = eye covered.
✦ = performance disrupted by experimental change.

carry information from the sensory centers to the calyces, where it is integrated to select a given output dispatched via the β lobe, while at the same time the transformed information is fed back via the α lobe to the sensory centers. This pattern is indicated in Fig. 8. Thus a possible reverberatory circuit is provided between the mushroom bodies and the optic ganglia and it is suggested that the repeated use of such circuits leads to some structural changes of the neurons involved which lays down a permanent engram, as suggested by Young (1964 and 1965) in the octopus. The changes might occur in the optic ganglia, the mushroom bodies, or both. In any case it would be anticipated that severing the tracts between the optic ganglia and mushroom bodies would disrupt learned performance, since the circuits involved would be interrupted (see the discussion of olfactory learning in Vowles, 1964). Individuals in some pilot experiments with gross damage to the mushroom bodies, sufficient to produce circling, did not appear to show learning loss; the ants circled their way correctly through the maze. This would suggest either that the engram is not located within the mushroom bodies or that it is diffuse, giving the equivalent of Lashley's equipotentiality effects.

The theory can now be made a little more specific on the way in which information from the two sides of the brain is integrated. The hypothesis proposed can be summarized as follows:

1. When an ant learns its way through the maze, it is learning to make a series of orientations guided by visual cues on the two sides of the body.

2. The mechanism for steering the ant depends on the interaction between the optic ganglia and the mushroom bodies, both ipsilaterally and contralaterally.

3. An engram is set up primarily within the optic ganglia by the circular flow of information between the optic ganglia and the mushroom bodies, both contralaterally and ipsilaterally. The mushroom body on one

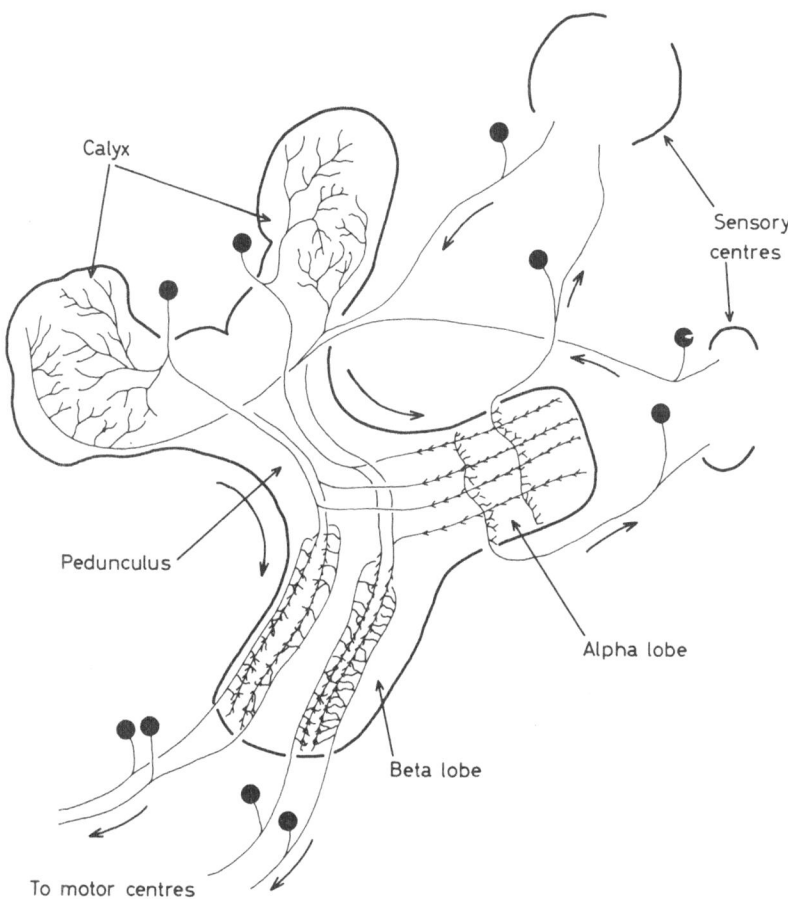

Fig. 8. Diagram showing circuits and connections of the mushroom bodies.

side is dominantly but not completely controlled by the ipsilateral optic ganglia.

4. The engram on one side of the brain is established under both ipsi- and contralateral influences, which may reinforce each other.

5. Information passing from one side of the brain to the other is degraded in information content. In the case of the clear maze, which is a simple problem, it is less degraded than in the case of the striped maze.

6. Severing the tracts between the optic ganglia and the mushroom bodies on one side will disrupt the control of orientation unless contralateral influences are still strong enough to ensure adequate performance.

Perhaps the crux of the hypothesis is the suggestion that information is degraded in passing from one side of the brain to the other (Muntz, 1961). This could be expected on two grounds: First, tracts crossing the midline, both between the optic ganglia on the two sides and between the mushroom bodies, are relatively poorly developed, compared, e.g., with those of the octopus or with the interhemispheric connectives in vertebrates (Kenyon, 1896; Cajal and Sanchez, 1915; Sanchez and Sanchez, 1941; and Vowles, 1955). Secondly, when recording from the calyces in response to visual stimulation in bees, it is more common to find units which are excited by ipsilateral stimulation than those excited by contralateral stimulation or by both (*personal observation*).

If the hypothesis is accepted, it will explain all the results of experiments involving lesions. For example, under assumption 6 one would predict that lesions cutting the tracts between the optic ganglia and the mushroom bodies on one side would disrupt performance in every case except one. This case is following binocular training in a clear maze, where the degradation of information passing to the opposite side of the brain will be least important and hence the reinforcement of traces from the two sides will be maximal. It is suggested that the reinforcement is then strong enough for the optic ganglia on one side to adequately control the mushroom bodies on both sides.

Similar explanations can be given for the results of the experiments on interocular transfer. Thus in the case of the clear maze an adequate engram is set up on both sides, and this can subsequently be used by the previously blinded eye, sending information to both sides. In the case of the striped maze however, information passing to the blind side is too impoverished to establish an adequate engram, and, when the blind eye is subsequently used, it has no adequate memory trace on its own side, and, further it cannot adequately activate the engram on the other. A similar explanation can be applied to the results obtained with binocular training and monocular testing, where again one would predict that in a clear maze there would be no disturbance but in a striped maze the exposed eye would be unable to activate the engram on the blind side sufficiently to control the mushroom bodies on both sides.

In the case of monocular training and binocular testing, there is a paradox. For the clear maze there is no difficulty, and one would predict no disturbance following unblinding; but in the case of the striped maze there seems no a priori reason why unblinding should disturb the ant, since the trained eye is still exposed and presumably could still theoretically control the behavior. However, this might be explained by the mechanism by which an ant "decides" which way to turn on approaching the choice point. In a clear maze, where light direction is the important cue, the decision is very simple, e.g., if the light in the entrance alley is ahead, turn to the left, and, if behind, to the right. This clearly does not require both eyes to be excited, and in any case uncovering one would not confuse the ant. Consider, however, the case of an ant trained in the striped maze with vertical stripes positive and the left eye covered. The choice will then be if vertical stripes are seen, turn right; if horizontal stripes are seen, turn left. However, after uncovering the naive eye it will be seeing the opposite stimulus to that seen by the trained eye, and by the decision process suggested above the ant will receive conflicting information from the two eyes and a decrement in performance would be expected.

The hypotheses proposed above must clearly be regarded as tentative. Further refinements would clearly necessitate the devising of tests which would allow the quantification of performance and its disturbance, rather than the mere identification of the presence of adverse effects. More precise methods of lesioning are also required. However, the proposed explanation has the merit of economy, and the results show that in experiments of this sort one cannot neglect factors such as the difficulty of the learning task or the role of apparently "untrained" parts of the brain.

SUMMARY

1. A method is described for training ants in simple and difficult visual discriminations.

2. The ants are tested for the effects of interocular transfer monocular blinding following binocular training, and binocular exposure following monocular training. In a simple problem no disturbance is caused by the experimental changes. In the difficult problem all three changes produced a disturbance.

3. The effect was studied of severing the tracts unilaterally between the optic ganglia and the mushroom bodies following binocular and monocular training in both a difficult and a simple problem. With one exception a disturbance in learned performance was caused by the operation, even if on the blinded side. The exception was following binocular training in the simple problem, control lesions did not produce any disturbance.

4. Lesions to the optic ganglia showed that these were involved in the engram.

5. A hypothesis based on neurophysiological evidence is proposed to explain the results. It is suggested that the mushroom bodies on both sides of the brain are controlled by the optic ganglia on both sides and that an engram is set up in the optic ganglia on both sides under both ipsilateral and contralateral influences. However, the degradation of information on its passage across the brain produces a less adequate engram on the contralateral side, and this affects the relative efficiency of information storage and retrieval for simple and complex tasks.

ACKNOWLEDGMENTS

The research reported in this paper has been sponsored in part by the Air Force Research Division of the Air Research and Development Command, United States Air Force, through its European Office, under Contract No. AF 61(052)-420, AF61(052)-114. I also wish to express my gratitude to the Royal Society and the Nuffield Foundation for their generous support and to Miss A. Gawadi, Mrs. J. Wallace, and Mrs. G. Glusker for technical assistance. J. Z. Young and R. W. Sperry kindly read and criticized the manuscript.

REFERENCES

Cajal, R. S., and D. Sanchez, Contribucion a conociminento de los centros nerviosis de los insectos. *Trab. Lab. Invest. Biol., Madrid*, 1915, **13**, 1–169.
Carthy, J. D., The orientation of two allied species of British ants, I. *Behaviour*, 1951, **3**, 275–303.
Fisher, R. A., *Statistical Methods of Research Workers*. London: Oliver and Boyd, 1949.
Huber, F., Auslosung von Bewegungsmustern durch elektrische Reizung des Oberschlundganglions bei Orthopteren. *Verhandl. Deut. Zool. Ges. Munster Zool. Anz. Suppl.*, 1959, **23**, 248–269.
Jander, R., Die Optische Orientierung der Roten Waldameise. *Z. Vergleich. Physiol.*, 1958, **40**, 162–238.
Kenyon, C. F., The anatomy of the corpora pedunculata in the Honey Bee. *J. Comp. Neurol.*, 1896, **6**, 133–196.
MacGregor, E. C., Odour as a basis for orientated movements in ants. *Behaviour*, 1948, **1**, 267–297.
Meikle, T. H., and J. A. Sechzer, Interocular transfer of brightness discrimination in "split-brain" cats. *Science*, 1960, **131**, 734–735.
Mountcastle, V. B., ed., *Interhemispheric Relations and Cerebral Dominance*. Baltimore, Md.: Johns Hopkins and Oxford University Press, 1962.
Muntz, W. R. A., Interocular transfer in octopus. *J. Comp. Physiol. Psychol.*, 1961, **54**, 192–195.
Oberholzer, R. J., and F. Huber, Methodike der elektrischen Reizung und Auschaltung in Gehirn nichtnerkotisierten Grillen. *Helv. Physiol. Pharmacol. Acta.*, 1957, **15**, 185–192.
Roeder, K. D., Chapters 17 and 18, *in* K. D. Roeder, ed., *Insect Physiology*. New York: John Wiley & Sons, Inc., 1953.
Roeder, K. D., *Nerve Cells and Insect Behavior*. Cambridge, Mass.: Harvard University Press, 1963.
Rowell, C. H. F., A technique for chronically implanting electrodes in the brains of Locusts. *J. Exp. Biol.*, 1963, **40**, 271–284.

Sanchez, D., and D. Sanchez, Contribution à la connaissance des centres nerveux des Insectes. *Trab. Inst. Cajal Invest. Biol.*, 1941, **33**, 165–213.

Siegel, S., *Non-parametric Statistics for the Behaviour Sciences.* New York: McGraw-Hill Book Company, 1956.

Vowles, D. M., The structure and connexions of the corpora pedunculata in bees and ants. *Quart. J. Microscop. Sci.*, 1955, **96**, 239–255.

Vowles, D. M., Olfactory learning and brain lesions in the wood ant (*F. rufa*). *J. Comp. Physiol. Psychol.*, 1964a, **58**, 105–111.

Vowles, D. M., Models and the Insect Brain, *in* R. F. Reiss, ed., *Neural Theory and Modelling.* Stanford, Calif.: Stanford University Press, 1964b.

Vowles, D. M., Maze learning and visual discrimination in the wood ant (*F. rufa*). *Brit. J. Psychol.*, 1965, **56**, 15–31.

Werringloer, A., Die Sehorgane und Sehzentren der Dorylinen nebst Untersuchungen über die Facettenaugen der Formiciden. *Z. wiss. Zool.*, 1932, **141**, 432–520.

Young, J. Z., *A Model of the Brain.* London: Oxford University Press, 1964.

Young, J. Z., The organization of a memory system. *Proc. Roy. Soc. London Ser. B*, **163**, 285–320.

Discussion

W. Corning: I would like to suggest another experiment which is possible with the preparation described by Dr. Cohen. Hoyle has recently outlined procedures in which he is able to use the output (EMG) of a single excitatory axon in the leg of the locust as a cue for shocking the leg. For example, when there is an increase in the discharges recorded from the muscle units, the leg is shocked and this eventually leads to a decrease in the spontaneous activity of the units. Now with Dr. Cohen's techniques for "marking" central cell bodies, it is possible to use electrophysiological techniques to define whether conditioning has occurred in a single fiber and then to perform histochemical and radioautographic analyses on relevant cell bodies.

D. Rahmann: We have worked on RNA labeling by using tritiated uridine and I am not sure that there is newly synthesized RNA in Dr. Cohen's experiment. Also, was there any movement of RNA along the axon? In the vertebrate, RNA exists only in the cell body and in the dendritic terminations.

M. Cohen: We ran a systematic series and found uptake of the uridine within an hour and a half, so I think we have newly synthesized RNA. We have not found any RNA along the axon; it stops at the axon hillock. We are also starting to investigate the protein changes in these cells and the outgrowth of new terminals. We can correlate a shift in the nucleus with the arrival of the regenerating axon at the distal stump. When that happens, the nucleus begins to shift and the miniature end-plate potentials return. This indicates that the regenerating terminals have made contact with the muscle and that the nucleus is induced into that area. I might mention that the eccentric position of the nucleus is the same thing that one sees in lymphocytes that are rapidly producing antibodies.

We have not made comparisons with other cockroaches, but we have done some work on the locust. The locust does not have as marked lability in RNA as the cockroach, but I might point out that the cockroach is a peculiar kind of insect in many ways, especially in its CNS morphology. For example, there are special kinds of collagen and polysaccharides that other insects don't have.

E. Bennett: In the scheme of how the control of RNA synthesis is being exercised in terms of injury or activity, I assume that Cohen is really expressing the extreme conditions and not implying that in the absence of these extremes these RNA changes did not occur.

M. Cohen: These are extremes. Obviously the cockroach neuron is producing protein in the course of its normal activities—it must make transmitters, etc.

M. Balaban: In Dr. Cohen's hypothesis about the dispersion of the substance prior to increased protein synthesis, could this be due to a damming-up of the normal process of protein production? The dispersion could be an artifact of this damming-up rather than the other way around.

M. Cohen: This is possible. Another interpretation is that the structuring of the ribosomes on the endoplasmic reticulum around the nuclear membrane makes them more available to the new messenger RNA and that the ribosomes are different after they leave the scaffolding of endoplasmic reticulum because they now have a new messenger on them. Now the signal for these changes might come from the terminal points; these signals could be antidromic firing due to the injury or of a chemical nature.

S. Ratner: Dr. Applewhite has demonstrated in his interesting animals a variety of learning situations. I was wondering, if he were to proceed now with his analyses and move toward the chemistry of learning, which of these situations would be selected for more intensive study?

P. Applewhite: First, we need to replicate all these experiments, so I'm not prepared at present to say which one I would like to pursue. Habituation is easy to obtain, but it is likely that the effects might be peripheral and we might not detect any changes in central cells.

W. Corning: Many have felt that with our present level of sensitivity we shall not observe any changes in the synaptic junctions between cells as a consequence of stimulation. The circuit diagrams of a system might help, but with respect to the chemistry of learning I wonder what sort of expectations Dr. Applewhite has for his electron microscopy studies?

P. Applewhite: I can't say that I am going to find something, but I do maintain that analyses should be carried out for a total system. There have been studies on single cells and small groups of cells but never on an entire system. If morphological changes are found, the next step might be to fractionate the organisms and try to extract these structures for chemical analyses.

H. Brown: Dr. Cohen has mentioned that extended electrical stimulation of the nerves resulted in essentially the same changes that are observed when the axon is cut. Is this a normal physiological situation?

M. Cohen: Our stimulation is not physiological with respect to the duration, which is at 50 per second for 3 to 6 hours. It is physiological in the sense that we are sending information through an afferent pathway. Antidromic stimulation is relatively ineffective in evoking these changes; there has to be orthodromic stimulation plus synaptic transmission. I should also point out that the effects we observe are not all-or-none, i.e., one cell has a greater density than the other. But, to repeat, I believe that orthodromic stimulation does cause the kinds of changes in RNA that I've described.

J. McConnell: Evans (1966a and b) has recently published two papers on *Nereis*, in which he appears to be convinced that he was not getting classical conditioning in these animals but, rather, a form of pseudo-learning. I'm not sure I like the stimulus parameters and the experimental setup, but this might be the beginning of a controversy in the annelid world. I wonder if Dr. Ratner had any comments on these studies.

S. Ratner: I have not yet completely grasped the test situations which were used. One of the reasons that the controversy is deemphasized is that, when you move from one species to another, the animal is so entirely different that you have to expect different things. In planarians, the differences between species may not be as great as in annelids.

A. Jacobson: I'd like to ask Dr. Ratner another question concerning earth-worms. There are several papers in the literature on maze learning. One of the better papers is one of Datta's (1962). It is possible that the interpretation Dr. Ratner has suggested is the real one, but I wonder if there is any evidence that supports this. Unless we are fairly sure that the results were contaminated, I am reluctant to dismiss maze learning so quickly.

S. Ratner: In the paper referred to by Jacobson, there are several points at which I begin to feel uncomfortable about fine-grain analysis. The behavioral changes within a session were great, with the response increasing—there was considerable variability. This indicates that other things are happening and that it is a noisy experimental setting.

A. Jacobson: The performance of the subjects at the beginning of each session was not very good, but after training the animals went from a low level to a very high rate of correct performance during a session. As far as fine-grain analysis is concerned, I don't think this should be terribly disturbing. If the animal shows something like the warm-up effect that we see in rats, I don't see how this should prevent Ratner from accepting the data.

S. Ratner: In the summary of the Datta paper you find statements like: "The probability of correct choice might reflect the operation of learning processes"; "measures of the probability of initial correct choice gave evidence of two changes in behavior which are at least independent. One of these changes is the development of a preference for the correct arm." This doesn't sound like learning. Would Dr. Jacobson think that there would be any reason that an investigator would want to stay with a T maze? (See pp. 394–397.)

A. Jacobson: No, but I think that, if we are considering the general question of demonstrations of learning in annelids, we ought not to accept the interpretation you offer without a careful scrutiny to see whether these factors are operative.

S. Ratner: I would agree. As vehicles for the study of chemical or biological changes associated with learning, I would set them aside as weak preparations.

REFERENCES

Datta, Lois-Ellin, Learning in the earthworm *Lumbricus terrestris. Amer. J. Psychol.,* 1962, **75**, 531.
Evans, S. M., Non-associative avoidance learning in nereid polychaets. *Anim. Behav.,* 1966a, **14**(1), 102.
Evans, S. M., Non-associative behavioral modifications in the polychaet *Nereis diversicolor. Anim. Behav.,* 1966b, **14**(1), 107.

Author Index

Subject Index